CISM COURSES AND LECTURES

The series presents lecture notes, monographs, edited works and proceedings in the field of Mechanics, Engineering, Computer Science and Applied Mathematics.
Purpose of the series is to make known in the international scientific and technical community results obtained in some of the activities organized by CISM, the International Centre for Mechanical Sciences.

INTERNATIONAL CENTRE FOR MECHANICAL SCIENCES

COURSES AND LECTURES - No. 432

INELASTIC BEHAVIOUR OF STRUCTURES UNDER VARIABLE REPEATED LOADS

DIRECT ANALYSIS METHODS

EDITED BY

DIETER WEICHERT
RHEINISCH-WESTFÄLISCHE TECHNISCHE HOCHSCHULE AACHEN

GIULIO MAIER
POLITECNICO DI MILANO

Springer-Verlag Wien GmbH

This volume contains 166 illustrations

Originally published by Springer-Verlag Wien New York in 2002

SPIN 10877360

In order to make this volume available as economically and as rapidly as possible the authors' typescripts have been reproduced in their original forms. This method unfortunately has its typographical limitations but it is hoped that they in no way distract the reader.

ISBN 978-3-211-83687-3 ISBN 978-3-7091-2558-8 (eBook)
DOI 10.1007/978-3-7091-2558-8

PREFACE

The integrity assessment of structural elements susceptible to operate beyond the elastic limits under variable repeated thermo-mechanical loads is an important task in many situations of structural engineering. Examples of such situations can be found in both mechanical and civil engineering (transportation technologies, pressure vessels, general machinery, constructions such as dams, pavements, offshore platforms, pipelines and buildings in seismic zones).

The so-called "direct" or "simplified" methods based on the shakedown theorems and their specialisation to limit theorems are receiving increasing attention for the prediction of structural failure in the inelastic range, although they basically originated decades ago within classical plasticity. Reasons for such revival are, first, that these methods provide the information which are essential in practice (e.g. safety factor and collapse mechanisms) by means of procedures potentially more economical than step-by-step inelastic analysis; second, they only need a minimum of information on the evolution of loads as functions of time (many technical problems do not permit to assume the precise knowledge of the loading history for time-marching simulations).

These methods are the subject of the present book. It reflects the lectures held in the framework of the CISM-Course "Inelastic Behaviour of Structures under Variable Repeated Loads" which took place at Udine in October 2000. The volume includes the presentation of the foundations and recent developments of shakedown theory and of limit analysis as particular case. Both the so-called static and the kinematic approach and related bounding methods are discussed, as well as new developments based on advanced material models. The addressed audience are engineers and scientists active in research and development in the following fields: structural engineering (mechanical and civil engineering), inelastic behaviour of structures, safety and reliability of structures and structural elements, computational and materials mechanics.

The book is composed of seven main chapters, according to the number of lecturers. Each lecturer is the responsible author of his individual part, eventually thematically subdivided and co-authored.

Basically, the structure of the book follows that of the course: The first chapter is mainly devoted to the mathematical foundations and the variational formulation of shakedown theory, followed by a second chapter on limit- and shakedown analysis by using the kinematic approach and on new extensions to poroplasticity. The third chapter deals with shakedown analysis for dynamic systems and materials with temperature-dependent yield limits, including damage. Also upper bound techniques on history dependent quantities are presented in this part. Material modelling, the influence of geometrical effects as well as finite-element formulations for shakedown and limit analysis are subject of the fourth chapter. The fifth chapter focuses on the numerically particularly efficient linear matching method for shakedown analysis with examples of application to technical problems. The combination of shakedown- and reliability analysis as well as structural optimisation under constraints are the subject of the sixth chapter. In the seventh chapter, the relation between high cycle fatigue modelling and shakedown theory is discussed, including the problem of multiaxial polycyclic fatigue criteria.

The editors express their gratitude to all the authors who made possible by their efforts this state-of-the-art presentation of the theory of direct methods and their applications to engineering problems. Although this volume is not intended to be comprehensive, we believe that it gives to the reader a fruitful introduction into the fast moving field of research in direct methods.

Special thanks go to Dr. A. Hachemi for his valuable help in putting the book into final shape.

Dieter Weichert
Giulio Maier

CONTENTS

Basic Definitions and Results

Géry de Saxcé

University of Lille, Lille, France

Abstract. The experimental tests show that many engineering structures subjected to variable repeated loads exhibit plastic strains. Above a critical value α^a of the load factor, they collapse by ratchet or alternating plasticity. On the contrary, below α^a, the plastic deformations are stabilized and the dissipation is bounded in time. We say that the structure shakes down. Firstly, the standard plasticity model is briefly recalled. Next, we define the basic tools of the fictitious elastic and residual fields. The statical approach due to Melan allows to characterize the shakedown. The kinematical approach due to Halphen gives a description of the collapses. The end of the lecture is devoted to useful concepts of mathematical programming and non smooth mechanics.

1 Compatibility and Equilibrium

Before presenting the main ideas of the theory, some classical hypothesis concerning the structure, the loading and the material are recalled. We consider a body (or structure) Ω occupying an open domain of the space $\boldsymbol{R}^3$, limited by the surface Γ. The part Γ_0 corresponds to the supports, while the remaining part is $\Gamma_1 = \Gamma - \Gamma_0$. The mechanical fields are generally depending on the position vector $x \in \Omega$ and the time variable $t > 0$, namely the displacement field $(x,t) \mapsto u(x,t)$, the strain field $(x,t) \mapsto \varepsilon(x,t)$ and the stress field $(x,t) \mapsto \sigma(x,t)$. The structure is subjected to time dependent actions :

1. body forces : $(x,t) \mapsto \bar{f}(x,t)$ prescribed in $\Omega \times [0,+\infty[$,
2. imposed surface forces : $(x,t) \mapsto \bar{p}(x,t)$ prescribed in $\Gamma_1 \times [0,+\infty[$,
3. imposed displacements : $(x,t) \mapsto \bar{u}(x,t)$ prescribed in $\Gamma_0 \times [0,+\infty[$.

The strain field is related to the displacement one by the internal compatibility equations :

$$\varepsilon_{ij} = \frac{1}{2}\left(\frac{\partial u_i}{\partial x_j} + \frac{\partial u_j}{\partial x_i}\right) \quad in \ \Omega .$$

Using intrinsic notation, we write equivalently :

$$\varepsilon(u) = grad_s u \quad in \ \Omega . \tag{1.1}$$

The displacements are subjected to boundary kinematical conditions :

$$u=\bar{u} \ on \ \Gamma_0. \tag{1.2}$$

A displacement field u is said to be *kinematically admissible* (K.A.) if it satisfies (1.2) and if there exists a strain field ε associated to u by (1.1).

On the other hand, the stress field has to fulfill the internal equilibrium equations :

$$\frac{\partial \sigma_{ij}}{\partial x_j}+\bar{f}_i=0 \ in \ \Omega,$$

or, in brief :

$$div\,\sigma+\bar{f}=0 \ in \ \Omega. \tag{1.3}$$

Let n be the unit normal vector to Γ. The stress field is also subjected to boundary equilibrium equations :

$$p_i=\sigma_{ij}n_j=\bar{p}_i \ on \ \Gamma_1.$$

In a more compact way, it holds :

$$p(\sigma)=\sigma.n=\bar{p} \ on \ \Gamma_1. \tag{1.4}$$

A stress field σ is said to be *statically admissible* (S.A.) if it satisfies (1.3) and (1.4).

Obviously, ε and σ are both symmetric tensors of order two :

$$\varepsilon_{ij}=\varepsilon_{ji}, \qquad \sigma_{ij}=\sigma_{ji}.$$

The 6-dimensional vector space of the strains (resp. of the stresses) is denoted E (resp. S). These two spaces are putted into duality by means of the following bilinear form (double contracted tensor product) :

$$\mathrm{S}\times\mathrm{E}\rightarrow \boldsymbol{R}:(\sigma,\varepsilon)\mapsto\sigma:\varepsilon=\sigma_{ij}\varepsilon_{ij},$$

representing the work by unit volume. Now, it is easy to prove that for any K.A. displacement field u and any S.A. stress field σ, Green's formula holds :

$$\int_\Omega \sigma:\varepsilon(u)\,d\Omega=\int_\Omega \bar{f}.u\,d\Omega+\int_{\Gamma_1}\bar{p}.u\,d\Gamma+\int_{\Gamma_0}p(\sigma).\bar{u}\,d\Gamma, \tag{1.5}$$

where the dot symbol denotes the ordinary scalar product in $\boldsymbol{R}^3$.

2 Elastic Perfectly Plastic Material

Convex Sets and Functions. Let E be a real vector space. A subset K of E is convex if $(1-\lambda)x_1+\lambda x_2 \in K$ when $x_1, x_2 \in K$ and $0 \le \lambda \le 1$. A real valued function $x \mapsto f(x)$ on E is convex if for all $x_1, x_2 \in K$ and all $0 \le \lambda \le 1$, one has :

$$f((1-\lambda)x_1+\lambda x_2) \le (1-\lambda)f(x_1)+\lambda f(x_2). \tag{2.1}$$

Normality Plastic Yielding Rule. The first usual assumption is concerned by the existence of a non empty, convex and closed subset K of S, called elastic domain, defined at any point of the structure Ω (Figure 1). It is supposed that the stress field must be plastically admissible (P.A.) :

$$\sigma \in K \ in \ \Omega.$$

The second basic assumption is that the plastic yielding rule is a *normality law*. In other words, the plastic strain rate field $\dot{\varepsilon}^p$ associated to σ by the plastic yielding rule satisfies *Hill's maximum power principle* (Figure 1):

$$\forall \sigma' \in K, \ (\sigma'-\sigma):\dot{\varepsilon}^p \le 0. \tag{2.2}$$

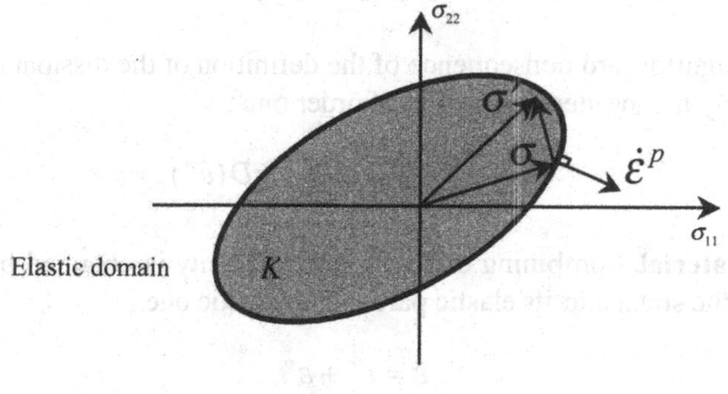

Figure 1. Elastic domain and normality law.

Thus, the dissipation power function is defined by :

$$D(\dot{\varepsilon}^p) = \sup\{\sigma':\dot{\varepsilon}^p; \sigma' \in K\}. \tag{2.3}$$

Three fundamental properties immediately result from the definition. Firstly, one can state the so-called *maximum dissipation principle* :

$$\forall \sigma' \in K, \ D(\dot{\varepsilon}^p) \ge \sigma':\dot{\varepsilon}^p. \tag{2.4}$$

For thermodynamical requirements, it is additionally supposed that the dissipation is supplied by the volume element to the environment. Because of the usual convention of the Mechanics of continua, it is non negative[1] :

$$\forall \dot{\varepsilon}^p, \; D(\dot{\varepsilon}^p) \geq 0. \tag{2.5}$$

Comparing inequalities (2.2) and (2.4), if $\dot{\varepsilon}^p$ is associated to σ, one has :

$$D(\dot{\varepsilon}^p) = \sigma : \dot{\varepsilon}^p \tag{2.6}$$

Secondly, D is a convex function as superior envelop of linear functions of $\dot{\varepsilon}^p$. From inequation (2.4) and equation (2.6), it is easy to deduce the following *convexity inequality* :

$$\forall \dot{\varepsilon}^{p'}, \; D(\dot{\varepsilon}^{p'}) - D(\dot{\varepsilon}^p) \geq \sigma : (\dot{\varepsilon}^{p'} - \dot{\varepsilon}^p). \tag{2.7}$$

By swap of $\dot{\varepsilon}^{p'}$ for $\dot{\varepsilon}^p$, we obtain also :

$$D(\dot{\varepsilon}^p) - D(\dot{\varepsilon}^{p'}) \geq \sigma' : (\dot{\varepsilon}^p - \dot{\varepsilon}^{p'}).$$

From the two previous inequalities, we deduce the *monotonicity condition* :

$$(\sigma - \sigma') : (\dot{\varepsilon}^p - \dot{\varepsilon}^{p'}) \geq 0. \tag{2.8}$$

Thirdly, a straightforward consequence of the definition of the dissipation power function is that D is a positively homogeneous function of order one :

$$\forall \lambda \geq 0, \; D(\lambda \dot{\varepsilon}^p) = \lambda D(\dot{\varepsilon}^p). \tag{2.9}$$

Elastic Plastic Material. Combining elasticity and plasticity is obtained by the usual additive decomposition of the strain into its elastic part and its plastic one :

$$\varepsilon = \varepsilon^e + \varepsilon^p.$$

The elastic strains are related to the stresses through Hooke's law :

$$\varepsilon^e_{ij} = S_{ijkl}\sigma_{kl},$$

with $S_{ijkl} = S_{jikl} = S_{ijlk} = S_{klij}$. In a more compact way, we write:

$$\varepsilon^e = S\sigma.$$

[1] This is equivalent to impose that the elastic domain K contains zero. Additionally, we claim that zero belongs to the interior of K.

Additionally, we suppose that the (complementary) elastic energy is positive :

$$\forall \sigma \neq 0, \; W(\sigma) = \tfrac{1}{2}\sigma : S\sigma > 0.$$

Thus, S is equipped with the elastic norm : $|\sigma|_e = \sqrt{2W(\sigma)}$.

Druckerian Materials. In Drucker's sense, an inelastic material is stable if under any stress increment $\Delta\sigma$, it absorbs a positive work $\Delta\sigma : \Delta\varepsilon$. Such Druckerian materials exhibit monotonic increasing loadings (no softening). The elastic plastic materials previously defined in the present section (i.e. satisfying the convexity of the elastic domain and the normality law) are Druckerian (see for instance Save, Massonnet, de Saxcé, 1997).

3 Shakedown

Many engineering structures or structural elements are exposed to simultaneous actions of loads and temperature fields. Moreover, the loads as well the temperature fields vary, usually in a more or less cyclic way, within wide limits.

Obviously, the design of such structures or structural elements should take account of inelastic material response, in particular of plastic effects. The oldest and most developed approach is the theory of limit analysis which allows to predict the ultimate load α^l, called the *limit load* (Save, Massonnet and de Saxcé, 1997). This theory, developed intuitively in the 1930's, theoretically and experimentally based in the 1950's, is widely used and has become recommended by design Codes on pressure vessels (ASME) and on reinforced concrete slabs (European Committee for Concrete).

However, in the case of variable repeated loads, the magnitude of the ultimate load is not the only factor characterizing the structural safety, as depicted in Figure 2. Let α be a positive real number governing the intensity of the loading, said the load factor. The experimental tests show that, for certain structures, above a particular value α^a of the load factor, we can observe an accumulation of plastic deformations leading to excessive structural deflections. This kind of failure is called *incremental collapse*, *ratchet* or *ratchetting*. On the contrary, below the value α^a, we observe a stabilization of the plastic deformations. For such an event, the response of the structure to the cyclic loading becomes purely elastic. This behavior is called (*elastic*) *shakedown* or *adaptation*. We say that the structure *shakes down*.

For other structures subjected to variable loading, the deflection remains small but it can be observed that, beyond a characteristic value α^a, alternating plastic strain increments occur, leading to fracture after a number of cycles N_F which depends on the local plastic strain amplitude $\Delta\varepsilon_p$ according to laws of which the most simple has the following form :

$$\Delta\varepsilon_p = \varepsilon_F \, (N_F)^{-\gamma},$$

where the exponent γ and the fatigue-ductility coefficient ε_F are material dependent positive constants. This law of low cycle fatigue (or plastic fatigue) is due to Manson (1954) and Coffin (1954). The behavior observed above the load α^a is called *alternating plasticity, accomodation* (or *plastic shakedown*, because it does not result in continuously increasing permanent deflections, though it is likely to lead to fracture by exhaustion or ductility).

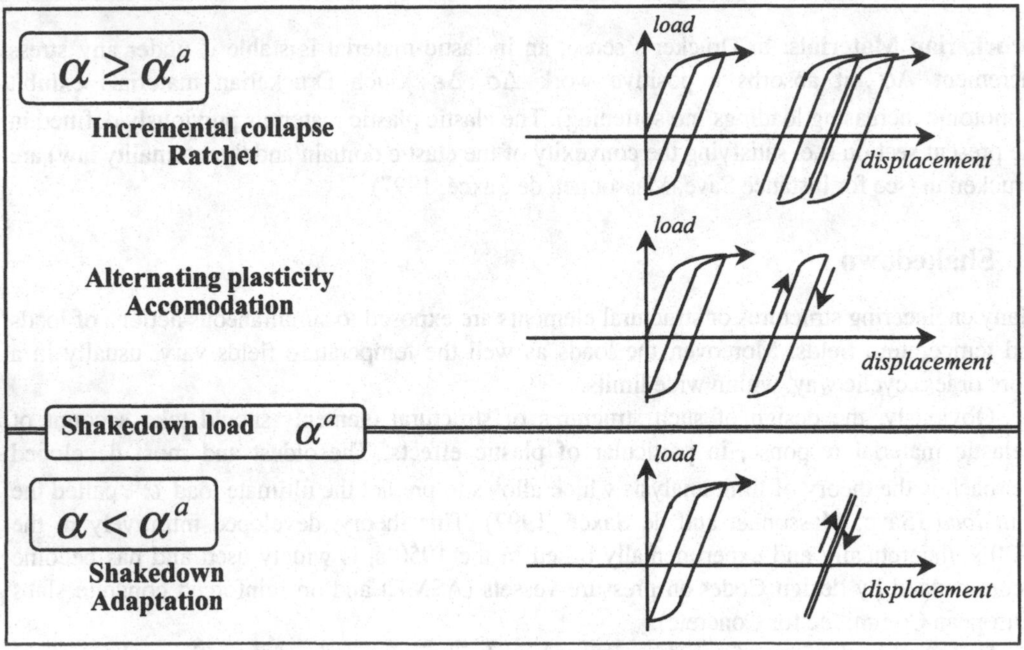

Figure 2. Incremental collapse, alternating plasticity and shakedown.

Therefore the structural safety requires that the power dissipated by the plastic deformation due to repeated load or temperature changes should eventually cease. The corresponding safety load α^a is called *shakedown load.* The experimental observations show that α^a is always situated below the limit load α^l, sometimes with a significant difference. Above this value, the structure fails by ratchet or alternating plasticity. Below α^a, the structure shakes down, that is: there exists a real constant $C > 0$ such that :

$$\forall t \geq 0, \quad \int_0^t \int_\Omega \sigma : \dot{\varepsilon}^p \, d\Omega dt \leq C \, .$$

In other words, the total dissipation is *bounded in time.*

4 Fictitious Elastic Fields and Residual Fields

For convenience, only mechanical loads are considered, avoiding thermal effects. We recall that α is a positive real number called *load factor*. The reference load $(\bar{f}^0, \bar{p}^0)$ is supposed to belong to a convex reference load domain P^0 which, for instance, can be described by an arbitrary combination of elementary loads varying between prescribed limits (Figure 3) :

$$\bar{f}^0(x,t) = \sum_{\alpha=1}^{n} \mu_\alpha(t)\, \bar{f}_\alpha^0(x), \quad \bar{p}^0(x,t) = \sum_{\alpha=1}^{n} \mu_\alpha(t)\, \bar{p}_\alpha^0(x), \quad (\underline{\mu}_\alpha \le \mu_\alpha(t) \le \bar{\mu}_\alpha).$$

Next, the load $(\bar{f}, \bar{p})$ belongs to an homothetic domain $P = \alpha P^0$ (Figure. 3).

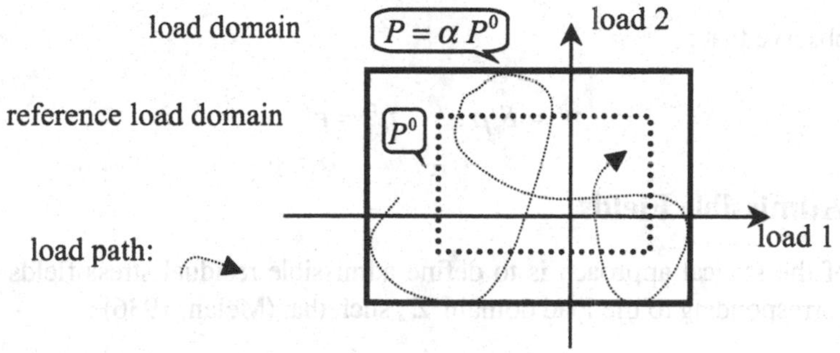

Figure 3. Load domain.

For the previous actions, the elastic responses in displacement u^E, in strain ε^E and in stress σ^E, in a corresponding fictitious perfectly elastic structure, satisfy Hooke's law. Because of the linearity of the elastic response, the stress field σ^E belongs to a convex polyhedral load domain $\Sigma = \alpha \Sigma^0$, homothetic to the reference load domain Σ^0 corresponding to the load domain P^0. Then, the field defined by :

$$\rho = \sigma - \sigma^E,$$

belongs to the set of residual stress fields :

$$N = \left\{ \rho \,\middle|\, div\, \rho = 0 \;\; in \;\; \Omega \;\; and \;\; p(\rho) = \rho.n = 0 \;\; on \;\; \Gamma_1 \right\}.$$

Moreover, the field defined by :

$$\eta = \varepsilon - \varepsilon^E,$$

belongs to the set of residual strain fields :

$$N^* = \left\{ \eta \,\middle|\, \exists v \;\; such \;\; that \;\; \eta = grad_s v \;\; in \;\; \Omega \;\; and \;\; v = 0 \;\; on \;\; \Gamma_0 \right\}.$$

of course, applying Green's formula (1.5), one obtain the *virtual power principle* :

$$\forall \rho \in N, \; \forall \eta \in N^*, \; \int_\Omega \rho : \eta \, d\Omega = 0 . \tag{4.1}$$

Besides, it can be remarked that :

$$\eta = (S\sigma + \varepsilon^p) - S\sigma^E = S(\sigma - \sigma^E) + \varepsilon^p .$$

Thus, we observe that :

$$\eta^e = S\rho, \qquad \eta^p = \varepsilon^p .$$

5 Statical Admissible Fields

The key idea of the statical approach is to define admissible residual stress fields (in Melan's sense) $\overline{\rho}(x)$, corresponding to the load domain Σ, such that (Melan, 1936) :

1. $\overline{\rho}$ is time-independent,
2. $\overline{\rho}$ is a residual stress field : $\overline{\rho} \in N$,
3. $\overline{\rho}$ is P.A. in the sense that : $\forall \sigma^E \in \Sigma, \; \overline{\sigma} = \sigma^E + \overline{\rho} \in K \;\; in \;\; \Omega, \;\; at \;\; any \;\; time$.

Moreover, if $\overline{\sigma}$ belongs to the interior of the closed set K in Ω at any time, $\overline{\rho}(x)$ is said to be a strictly admissible residual stress field . Hence, the following proposition was proved by Melan (1936) :

Theorem 1 (*Melan's theorem*) : if a strictly admissible residual stress field $\overline{\rho}$ can be found, the structure shakes down.

Proof. Let us suppose that an admissible residual stress field $\overline{\rho}$ exists. We consider the fictitious residual elastic energy of the stress difference $(\rho - \overline{\rho})$:

$$\mathrm{R} = \tfrac{1}{2}\int_\Omega (\rho - \overline{\rho}) : S(\rho - \overline{\rho}) \, d\Omega \geq 0 .$$

The stress difference $(\rho-\overline{\rho})$ is associated to the strain difference $(\eta^e-\overline{\eta}^e)$ by Hooke's law, and, because $\overline{\eta}^e$ is time independent, the time derivative of the energy is equal to :

$$\dot{\mathrm{R}}=\int_\Omega(\rho-\overline{\rho}):\dot{\eta}^e d\Omega .$$

Owing to the strain decomposition, one has :

$$\dot{\mathrm{R}}=\int_\Omega(\rho-\overline{\rho}):\dot{\eta}\,d\Omega-\int_\Omega(\rho-\overline{\rho}):\dot{\varepsilon}^p d\Omega .$$

As $\overline{\rho}\in N$ and $\dot{\eta}\in N^*$, because of virtual power principle (4.1), the first term of the right hand side vanishes. According to the definition of the residual stress fields:

$$\rho-\overline{\rho}=(\sigma-\sigma^E)-(\overline{\sigma}-\sigma^E)=\sigma-\overline{\sigma},$$

one has :

$$\dot{\mathrm{R}}=-\int_\Omega(\sigma-\overline{\sigma}):\dot{\varepsilon}^p d\Omega .$$

Equation (2.6) gives :

$$\dot{\mathrm{R}}=-\int_\Omega(D(\dot{\varepsilon}^p)-\overline{\sigma}:\dot{\varepsilon}^p)\,d\Omega .$$

Because of $\overline{\sigma}$ being P.A., maximum dissipation principle (2.4) leads to :

$$\dot{\mathrm{R}}\le 0 .$$

Next, we prove that the total dissipation is bounded[2]. It can be remarked that, as 0 and $\overline{\sigma}$ belongs to the interior of the elastic domain *K*, there exists a scalar $m>1$ such that $m\overline{\sigma}$ is P.A. Because of Hill's maximum power principle (2.2), it holds :

$$m(\overline{\sigma}-\sigma):\dot{\varepsilon}^p+(m-1)\sigma:\dot{\varepsilon}^p=(m\overline{\sigma}-\sigma):\dot{\varepsilon}^p\le 0 .$$

Consequently, one has :

$$\int_0^t\int_\Omega\sigma:\dot{\varepsilon}^p d\Omega\,dt\le\frac{m}{m-1}\int_0^t\int_\Omega(\sigma-\overline{\sigma}):\dot{\varepsilon}^p d\Omega\,dt=-\frac{m}{m-1}\int_0^t\dot{\mathrm{R}}dt=\frac{m}{m-1}(\mathrm{R}(0)-\mathrm{R}(t)) .$$

[2] This part of the proof is due to Koiter.

Because of its definition, $R(t)$ is non negative and one has :

$$R(0) - R(t) \leq R(0),$$

that leads to :

$$\forall t \geq 0, \quad \int_0^t \int_\Omega \sigma : \dot{\varepsilon}^p \, d\Omega \, dt \leq \frac{m}{m-1} R(0).$$

The right hand member is time independent and has finite value, that achieves the proof.

6 Kinematical Admissible Fields

The key idea of the dual approach is to characterize the collapse mode of the structure over a load cycle for a periodic action of period T. Then, the admissible plastic strain rates (in Koiter's sense) $\dot{\varepsilon}^p(x,t)$, corresponding to the load domain Σ, are defined by (Koiter, 1960) :

1. for the plastic strain increment, one has : $\Delta\varepsilon^p = \oint \dot{\varepsilon}^p \, dt \in N^*$,
2. $\dot{\varepsilon}^p$ is P.A. in the sense that : $\int_\Omega \oint \sigma^E : \dot{\varepsilon}^p \, dt \, d\Omega > 0$.

Indeed, the following dual theorem of Melan's one can be demonstrated (Halphen, 1978) :

Theorem 2 (*Halphen's theorem*) : if the structure does not shake down, the plastic strain rate field tends to an admissible field.

Proof. Let us assume a periodic loading of period T and let $\rho'(x,t)$ and $\rho''(x,t)$ be two evolutions of the residual stress field corresponding to different initial conditions, respectively $\rho'(x,t_0)$ and $\rho''(x,t_0)$. We characterize the distance between the two fields by the corresponding residual elastic energy :

$$R = \tfrac{1}{2}\int_\Omega (\rho' - \rho'') : S(\rho' - \rho'') \, d\Omega.$$

By a reasoning similar to the one developed in Melan's theorem, it can be easily proved that :

$$\dot{R} = \int_\Omega (\rho' - \rho'') : (\dot{\eta}' - \dot{\eta}'') d\Omega - \int_\Omega (\rho' - \rho'') : (\dot{\varepsilon}^{p'} - \dot{\varepsilon}^{p''}) \, d\Omega.$$

According to the virtual power principle (4.1), it holds :

$$\dot{R} = -\int_{\Omega} (\sigma' - \sigma'') : (\dot{\varepsilon}^{p'} - \dot{\varepsilon}^{p''})\, d\Omega .$$

Because of the monotonicity condition (2.8), we obtain :

$$\dot{R} \leq 0 .$$

Let us remark that the vector space N is equipped with the elastic norm :

$$\|\rho\|_e = \left| \int_{\Omega} |\rho|_e^2 d\Omega \right|^{1/2} .$$

As the structure does not shake down, the energy does not cease to be dissipated. The mapping $\rho(x,t_0) \mapsto \rho(x,t)$ is a contraction from the set of P.A. residual stress field into itself in the sense that for all t_1, there exists $t_2 > t_1$ such that :

$$\|\rho'(t_2) - \rho''(t_2)\|_e < \|\rho'(t_1) - \rho''(t_1)\|_e .$$

If this set is bounded, Brouwer's theorem states that the mapping has a fixed point and, thus, there exists a periodic solution (Halphen, 1978) :

$$\rho(x, t_0 + T) = \rho(x, t_0) .$$

The residual strains are associated to the periodic solution by Hooke's law :

$$\dot{\eta} - \dot{\varepsilon}^p = S\dot{\rho} .$$

Integrating over the loading cycle and taking account of the fact that the residual stress increment vanishes, we prove that the plastic strain rate is a residual strain field :

$$\Delta\varepsilon^p = \oint \dot{\varepsilon}^p\, dt = \oint \dot{\eta}\, dt \in N^* .$$

7 Decomposition in Spherical and Deviatoric Parts

The behavior of isotropic materials has the same mathematical form in all the orthogonal frames of the space. It is convenient to consider, in the Euclidean vector space E (for the scalar product : $(\varepsilon|\varepsilon') = \varepsilon_{ij}\varepsilon'_{ij}$), the one dimensional subspace of hydrostatic strains :

$$E_m = \{ \varepsilon_m \in S \quad such \quad that \quad \varepsilon_m = \tfrac{1}{3} e_m I, \quad e_m \in \boldsymbol{R} \},$$

and the 5-dimensional subspace of deviatoric strains :

$$\mathrm{E}_d = \{ e \in \mathrm{E} \quad such \quad that \quad Tr(e) = e_{ii} = 0 \}.$$

Thus, E is decomposed in the direct sum of the two previous orthogonal vector subspace E_m and E_d. In other words, one has the following unique decomposition :

$$\varepsilon = \varepsilon_m + e, \quad \varepsilon_m = proj(\varepsilon, \mathrm{E}_m) = \tfrac{1}{3}Tr(\varepsilon)I, \quad e = proj(\varepsilon, \mathrm{E}_d) = \varepsilon - \tfrac{1}{3}Tr(\varepsilon)I.$$

On the other hand, we introduce, in the Euclidean vector space S (for the scalar product : $(\sigma|\sigma') = \sigma_{ij}\sigma'_{ij}$), the one dimensional subspace of hydrostatic stresses :

$$\mathrm{S}_m = \{ \sigma_m \in \mathrm{S} \quad such \quad that \quad \sigma_m = s_m I, \quad s_m \in \boldsymbol{R} \},$$

and the 5-dimensional subspace of deviatoric stresses :

$$\mathrm{S}_d = \{ s \in \mathrm{S} \quad such \quad that \quad Tr(s) = s_{ii} = 0 \quad \}.$$

Thus, S is decomposed in the direct sum of the two previous orthogonal vector subspace S_m and S_d. In other words, one has the following unique decomposition :

$$\sigma = \sigma_m + s, \quad \sigma_m = proj(\sigma, \mathrm{S}_m) = \tfrac{1}{3}Tr(\sigma)I, \qquad s = proj(\sigma, \mathrm{S}_d) = \sigma - \tfrac{1}{3}Tr(\sigma)I.$$

The slightly different definition of the variables e_m and s_m is motivated by the fact that they are putted in duality, according to :

$$\varepsilon : \sigma = e_m s_m + e : s.$$

For instance, Hooke's law for isotropic materials can be written :

$$e_m = s_m / K, \quad e = s / 2\mu,$$

where the bulk modulus K and the shear modulus μ are material dependent positive constants. This law admits a potential (complementary elastic energy) :

$$W(s_m s) = \frac{s_m^2}{2K} + \frac{s_{ij}\, s_{ij}}{4\mu},$$

such that :

$$e_m = \frac{\partial W}{\partial s_m}, \qquad e = \frac{\partial W}{\partial s}.$$

8 Formulation of Plasticity Using Differentiable Optimization

Differentiable Convex Functions. The characterization (2.1) of the convexity is not very convenient. If f is moreover differentiable, it can be replaced by the following one. E is a topological vector space if the mappings $E \times E \to E:(x_1,x_2) \mapsto x_1 + x_2$ and $\boldsymbol{R} \times E:(\lambda,x) \mapsto \lambda x$ are continuous. The vector space of the continuous linear forms on E is called the (topological) dual of E and is denoted E^*, the scalar product being defined by: $E^* \times E \to \boldsymbol{R}:(y,x) \mapsto y.x$. Let $x \mapsto f(x)$ be a real valued function on E. If there exists, the gradient of f at x is the continuous linear form :

$$y = \frac{\partial f}{\partial x}(x) \in E^*,$$

such that :

$$y.\delta x = \lim_{\lambda \to 0_+} \frac{f(x+\lambda\delta x) - f(x)}{\lambda}.$$

Thus, if f is convex, it holds :

$$f((1-\lambda)x + \lambda x') - f(x) \le (1-\lambda)f(x) + \lambda f(x') - f(x) = \lambda(f(x') - f(x)).$$

λ being positive, one has :

$$\frac{f(x+\lambda(x'-x)) - f(x)}{\lambda} \le f(x') - f(x).$$

Considering the limit as $\lambda \to 0_+$, we obtain another characterization of the convexity for differentiable functions :

$$\forall x' \in E,\ f(x') - f(x) \ge \frac{\partial f}{\partial x}(x).(x'-x). \tag{8.1}$$

Moreover , it can be proved that this condition is sufficient for convexity of differentiable functions f.

Prager's Formulation of Plasticity. The usual definition of the elastic domain is based on a convex real valued function $\sigma \mapsto f(\sigma)$ said the *loading function* :

$$K = \{\sigma \in \mathrm{S} \quad \textit{such} \quad \textit{that} \quad f(\sigma) \le 0\}.$$

Obviously, K is convex and closed. The boundary ∂K of K, defined by the equation : $f(\sigma)=0$, is called the *plastic yielding surface*. We suppose that f is continuously differentiable (C^1) on ∂K. If σ belongs to the plastic yielding surface, according to inequation (8.1), one has :

$$\forall \sigma' \in K, \quad (\sigma'-\sigma) : \frac{\partial f}{\partial \sigma}(\sigma) \le f(\sigma') - f(\sigma) = f(\sigma') \le 0 .$$

Comparing to Hill's maximum power principle (2.2) leads to :

$$\exists \lambda \ge 0 \ \ such\ \ that\ \ \dot{\varepsilon}^p = \lambda \frac{\partial f}{\partial \sigma}(\sigma) .$$

λ is called the plastic multiplier. If σ belongs to the interior of K, there exists a basis $(\delta \sigma_\alpha)$ of S such that, for all α, $\sigma' = \sigma \pm \delta \sigma_\alpha$ belongs to K. Applying Hill's maximum power principle (2.2) gives :

$$\delta \sigma_\alpha : \dot{\varepsilon}^p \le 0, \quad \text{and} \quad \delta \sigma_\alpha : \dot{\varepsilon}^p \ge 0 .$$

Consequently, we have : $\delta \sigma_\alpha : \dot{\varepsilon}^p = 0$, for any α. Hence, $\dot{\varepsilon}^p$ vanishes. Finally, we can state Prager's formulation of the plastic yielding rule :

if $f(\sigma) < 0$ ***then*** ! *plasticity criterion*
$\dot{\varepsilon}^p = 0$! *elastic loading (or unloading)*
Else if $f(\sigma) = 0, \ \exists \lambda \ge 0 \ \ such\ \ that\ \ \dot{\varepsilon}^p = \lambda \frac{\partial f}{\partial \sigma}(\sigma)$! *plastic yielding*

A classical example is Von Mises model. For isotropic behaviors, we put :

$$\dot{e}_m^p = \lambda \frac{\partial f}{\partial s_m}, \qquad \dot{e}^p = \lambda \frac{\partial f}{\partial s} .$$

For the metallic materials, the experimental testing shows that : $\dot{e}_m^p = 0$. The load function depends only on the deviatoric stress. It is usual to introduce the equivalent stress :

$$\sigma_{eq}(s) = \left| \tfrac{3}{2} s_{ij}\, s_{ij} \right|^{1/2} ,$$

which is a norm in S_d. It allows to define the loading function : $f(\sigma) = \sigma_{eq}(s) - \sigma_Y$ where the yield stress σ_Y is a material dependent positive constant. For convenience, we introduce the equivalent strain rate :

$$\varepsilon_{eq}(\dot{e}^p) = \left|\tfrac{2}{3}\dot{e}^p_{ij}\,\dot{e}^p_{ij}\right|^{1/2},$$

which is a norm in E_{ij}. Owing to Cauchy-Schwartz inequality, one obtains :

$$\forall \sigma \in K,\ \sigma : \dot{\varepsilon}^p = s : \dot{e}^p \le \sigma_{eq}(s)\,\varepsilon_{eq}(\dot{e}^p) \le \sigma_Y\,\varepsilon_{eq}(\dot{e}^p).$$

Using definition (2.3), the dissipation power function corresponding to Von Mises model is:

$$D(\dot{\varepsilon}^p) = \sigma_Y\,\varepsilon_{eq}(\dot{e}^p). \tag{8.2}$$

Comment. As the value of the plastic multiplier λ is undetermined, for any $\sigma \in \partial K$, the associated $\dot{\varepsilon}^p$ can takes various values. The correspondence between stress and plastic strain rate is not one-to-one. We say that it is a *multivalued constitutive law*.

It is worthwhile quoting that D is convex but is not differentiable anywhere. The singular value is: $\dot{e}^p = 0$, for which one the gradient is not defined. Physically, it corresponds to the elastic behavior and will be discussed again in the next. In order to overcome numerical troubles in applying differentiable mathematical programming techniques, suitable regularizations of the dissipation function are performed in the vicinity of this singular point (see Jospin, Nguyen Dang and de Saxcé, 1991, and Maier's lecture notes in the same book).

Lagrange Multiplier Technique. We hope to solve a constrained minimization problem with equality constraints :

$$inf\left\{F(x); x \in E,\ g_\alpha(x) = 0,\ \alpha = 1,\ldots,n\right\},$$

where E is a finite dimensional topological vector space, F and g_α are C^1 real valued functions. The following saddle point problem is equivalent to the previous problem :

$$sup\left\{inf\left\{L(x,\lambda_\alpha) = F(x) + \sum_\alpha \lambda_\alpha\, g_\alpha(x)\ ; x \in E\right\}; \lambda_\alpha \in \boldsymbol{R}\right\}.$$

Indeed, the corresponding stationarity conditions are :

$$\frac{\partial L}{\partial x}=\frac{\partial F}{\partial x}+\sum_{\alpha}\lambda_{\alpha}\frac{\partial g_{\alpha}}{\partial x}=0, \qquad \frac{\partial L}{\partial \lambda_{\alpha}}=g_{\alpha}=0.$$

λ_{α} are said to be Lagrange multipliers. A constrained minimization problem with inequality constraints :

$$inf\left\{F(x);x\in E,\ g_{\alpha}(x)\leq 0,\ \alpha=1,\ldots,n\right\},$$

can be presented as a minimization problem with equality constraints by introducing slack variables ω_{α} :

$$g_{\alpha}(x)\leq 0 \Leftrightarrow g_{\alpha}(x)+\omega_{\alpha}^{2}=0.$$

Thus, one has :

$$sup\left\{inf\left\{L(x,\omega_{\alpha},\lambda_{\alpha})=F(x)+\sum_{\alpha}\lambda_{\alpha}(g_{\alpha}(x)+\omega_{\alpha}^{2})\,;x\in E,\omega_{\alpha}\in \boldsymbol{R}\right\};\lambda_{\alpha}\in \boldsymbol{R}\right\}.$$

The stationarity conditions are :

$$\frac{\partial L}{\partial x}=\frac{\partial F}{\partial x}+\sum_{\alpha}\lambda_{\alpha}\frac{\partial g_{\alpha}}{\partial x}=0, \qquad \frac{\partial L}{\partial \omega_{\alpha}}=2\,\lambda_{\alpha}\,\omega_{\alpha}=0, \quad \frac{\partial L}{\partial \lambda_{\alpha}}=g_{\alpha}+\omega_{\alpha}^{2}=0,$$

from which ones it results that if $\lambda_{\alpha}\neq 0$, then $g_{\alpha}(x)=0$, and if $g_{\alpha}(x)<0$, then $\lambda_{\alpha}=0$. Moreover, it can be proved Lagrange multipliers are non negative. Hence, we obtain the so-called *complementary relations* :

$$g_{\alpha}\leq 0, \qquad \lambda_{\alpha}\geq 0, \qquad \lambda_{\alpha}\,g_{\alpha}=0.$$

Finally, we state :

$$sup\left\{inf\left\{L(x,\lambda_{\alpha})=F(x)+\sum_{\alpha}\lambda_{\alpha}\,g_{\alpha}(x)\,;x\in E\right\};\lambda_{\alpha}\geq 0\right\}.$$

Accounting for the rule : *sup* F = - *inf* (- F), we can also show that the constrained maximization problem :

$$sup\left\{ F(x); x \in E,\ g_\alpha(x) \le 0,\ \alpha = 1,\dots,n \right\},$$

becomes :

$$inf\left\{ sup\left\{ L(x,\lambda_\alpha) = F(x) - \sum_\alpha \lambda_\alpha\, g_\alpha(x)\ ; x \in E \right\}; \lambda_\alpha \ge 0 \right\}.$$

Examples. As application, we can calculate the dissipation power function by solving :

$$inf\left\{ sup\left\{ L(\sigma,\lambda) = \sigma : \dot{\varepsilon}^p - \lambda f(\sigma); \sigma \in \mathrm{S} \right\}; \lambda \ge 0 \right\}.$$

The stationarity condition provides the normal plastic yielding rule :

$$\dot{\varepsilon}^p = \lambda \frac{\partial f}{\partial \sigma}(\sigma).$$

The plastic multiplier appears as Lagrange multiplier associated to the plastic yielding condition.

The reason to prefer Hill's formulation of section 2 to Prager's formulation is that the former one is much more general as the latter one. For instance, let us consider an elastic domain defined by a family of convex C^1 loading functions $(\sigma \mapsto f_\alpha(\sigma), \alpha = 1,\dots,n)$:

$$K = \left\{ \sigma \in \mathrm{S}\quad such\quad that\quad f_\alpha(\sigma) \le 0,\quad \alpha = 1,\dots,n \right\}.$$

This suggests to state the saddle-point problem :

$$inf\left\{ sup\left\{ L(\sigma,\lambda) = \sigma : \dot{\varepsilon}^p - \sum_\alpha \lambda_\alpha f_\alpha(\sigma); \sigma \in \mathrm{S} \right\}; \lambda_\alpha \ge 0 \right\}.$$

Hence, we obtain Koiter's plastic yielding rule :

$$\dot{\varepsilon}^p = \sum_\alpha \lambda_\alpha \frac{\partial f_\alpha}{\partial \sigma}(\sigma).$$

As example, we quote Tresca condition defined with respect to principal stresses $\sigma_I, \sigma_{II}, \sigma_{III}$:

$$K = \left\{ (\sigma_I, \sigma_{II}, \sigma_{III}) \in \boldsymbol{R}^3\ \ such\ that\ \ -\sigma_Y \le \sigma_\alpha - \sigma_\beta \le \sigma_Y,\ \ \alpha \ne \beta,\ \alpha,\beta = I, II, III \right\}.$$

Counterexample. Despite of the usefulness of Lagrange multiplier technique, it does not allows to treat models involving loading functions which are not differentiable on the plastic yielding surface. For instance, let us consider Drucker-Prager model for the plasticity of soils and rocks :

$$K = \left\{ \sigma = (s_m, s) \in \mathrm{S} \quad such \quad that \quad f(s_m, s) = \sigma_{eq}(s) - r(c - s_m \tan\phi) \leq 0 \right\},$$

where the cohesion stress $c > 0$, and the friction angle $\phi > 0$ are material dependent constants, and :

$$r = 3/\sqrt{3 + 4\tan^2\phi}\,.$$

The corresponding plastic yielding surface is a cone. The relevant point to quote is that the vertex $\sigma = (c/\tan\phi, 0)$ is a singular point of the plastic yielding surface with no definite gradient.

9 Non Smooth Mechanics

Discussion. In the previous section, we met some mathematical difficulties arising in the mathematical modeling of the plasticity using tools of differentiable optimization :

- the plastic yielding rule is a multivalued constitutive law,
- the dissipation power function is generally non differentiable,
- the plastic yielding surface can contain singular points where the gradient is undefined.

An elegant alternative is provided by the so-called non smooth Mechanics using tools of convex analysis. Basic results are summarized in the sequel.

Subgradient. Let V be a topological vector space of velocities $\dot{\kappa}$, and F be its dual space collecting the associated variables π, the scalar product being: $(\pi, \dot{\kappa}) \mapsto \pi.\dot{\kappa}$. Let $\overline{\boldsymbol{R}} = \boldsymbol{R} \cup \{+\infty\}$ be the set of extended real numbers equipped with the natural operations and ordering it inherits from the set $\boldsymbol{R}$ with particular rules concerning $+\infty$ (Taylor, 1965, and Moreau, 1966). Extending the concept of dissipation power function, we postulate the existence of a lower semi-continuous[3] and convex (but not necessarily differentiable) function $\varphi: V \to \overline{\boldsymbol{R}}$. We call effective domain of φ the set $dom\varphi$ of $\dot{\kappa}$ for which ones φ takes a finite value. Generalizing (8.1), we say that $\pi \in F$ is a *subgradient* of φ at $\dot{\kappa} \in dom\varphi$ if :

$$\forall \dot{\kappa}' \in V,\ \varphi(\dot{\kappa}') - \varphi(\dot{\kappa}) \geq \pi.(\dot{\kappa}' - \dot{\kappa}). \tag{9.1}$$

[3] The function φ is lower semi-continuous if for all $\alpha \in \boldsymbol{R}$, the set of $\dot{\kappa} \in V$ such that $\varphi(\dot{\kappa}) \leq \alpha$ is closed.

Hence, the subgradient is a generalized gradient which can be not unique. The set of all the subgradients at $\dot{\kappa}$ is called the *subdifferential* of φ at $\dot{\kappa}$ and is denoted $\partial\varphi(\dot{\kappa})$. Saying that π is a subgradient can be symbolized by the *differential inclusion* :

$$\pi \in \partial\varphi(\dot{\kappa}) . \tag{9.2}$$

If φ is differentiable at $\dot{\kappa}$, the subdifferential is reduced to the usual gradient :

$$\partial\varphi(\dot{\kappa}) = \left\{ \frac{\partial\varphi}{\partial\dot{\kappa}}(\dot{\kappa}) \right\}.$$

This mathematical frame allows to define multivalued constitutive laws by the mapping :

$$\dot{\kappa} \mapsto \partial\varphi(\dot{\kappa}) \subset F .$$

The concept of subgradient is due to Moreau (1963).

Polar Function. In order to write the inverse constitutive law, one introduces the polar (or conjugate) function χ of φ defined on F by Legendre-Fenchel transform :

$$\chi(\pi) = sup\left\{ \pi.\dot{\kappa} - \varphi(\dot{\kappa}); \dot{\kappa} \in V \right\}.$$

As superior envelop of affine (then convex) functions : $\dot{\kappa} \mapsto \pi.\dot{\kappa} - \varphi(\dot{\kappa})$, χ is a convex function too. A straightforward consequence of the definition is the relation :

$$\forall \dot{\kappa}' \in V,\ \forall \pi' \in F,\ \varphi(\dot{\kappa}') + \chi(\pi') \geq \pi'.\dot{\kappa}', \tag{9.3}$$

said *Fenchel's inequality* (Fenchel, 1949).

Next, a fruitful viewpoint consists in defining couples $(\dot{\kappa}, \pi)$ as *extremal* in the sense that the equality is exactly reached in the previous relation :

$$\varphi(\dot{\kappa}) + \chi(\pi) = \pi.\dot{\kappa} . \tag{9.4}$$

Hence, taking the value $\pi' = \pi$ in inequation (9.3) and subtracting member by member (9.4) from (9.3), we obtain (9.1). In other words, π is a subgradient of φ at $\dot{\kappa}$. A similar reasoning allows to prove that $\dot{\kappa}$ is a subgradient of χ at π. Conversely, if (9.2) is true, one has :

$$\forall \dot{\kappa}' \in V,\ \varphi(\dot{\kappa}) - \pi.\dot{\kappa} \leq \varphi(\dot{\kappa}') - \pi.\dot{\kappa}' .$$

Using the definition of the polar function, it holds :

$$\varphi(\dot\kappa)-\pi.\dot\kappa = inf\left\{\varphi(\dot\kappa')-\pi.\dot\kappa';\dot\kappa'\in V\right\} = -sup\left\{\pi.\dot\kappa'-\varphi(\dot\kappa');\dot\kappa'\in V\right\} = -\chi(\pi).$$

Hence, the couple $(\dot\kappa,\pi)$ is extremal. Finally, we can conclude to the double equivalence :

$$\varphi(\dot\kappa)+\chi(\pi)=\pi.\dot\kappa \;\Leftrightarrow\; \pi\in\partial\varphi(\dot\kappa) \;\Leftrightarrow\; \dot\kappa\in\partial\chi(\pi). \tag{9.5}$$

The differential inclusion $(9.5)_2$ can be interpreted as the inverse constitutive law of $(9.5)_3$. Functions such as φ and χ generalize the classical notion of potential (as the potential of elasticity, section 7), and are said to be superpotentials. The dissipative materials admitting a superpotential are often qualified of standard materials (Halphen and Nguyen Quoc Son, 1975). The standard materials are Druckerian. For more information concerning convex analysis and non smooth mechanics, the reader can see for instance Ekeland and Temam (1975) and Panagiatopoulos (1985).

Application to Plasticity. We call indicator function of $K\subset \mathrm{S}$, the function $\coprod_K : \mathrm{S}\mapsto \overline{\mathbf{R}}$ equal to :

$$\coprod\nolimits_K(\sigma)=0,$$

when $\sigma\in K$, and equal to $+\infty$ otherwise. Physically, the indicator functions can be interpreted as well (or pit) potentials. The value of the potential is zero at the bottom and infinite for the non admissible states. The idea is that an infinite power should be supplied to reach such states. If K is convex (resp. closed) , the indicator function is convex (resp. lower semi-continuous). Let us take $F=\mathrm{S}$ and $\chi=\coprod_K$ where K is the elastic domain and let us consider a subgradient $\dot\varepsilon^p$ of χ at $\sigma\in K$:

$$\dot\varepsilon^p\in\partial\coprod\nolimits_K(\sigma). \tag{9.6}$$

As σ belongs to K, $\chi(\sigma)=0$, and owing to (9.1), one has :

$$\forall\sigma'\in\mathrm{S},\quad \coprod\nolimits_K(\sigma')\geq(\sigma'-\sigma):\dot\varepsilon^p .$$

If $\sigma'\notin K$, the value of the left hand member is infinite while the one of the other member is finite, whence the previous inequation is trivially satisfied. Thus, we find Hill's maximum power principle (2.2) :

$$\forall\sigma'\in K,\ (\sigma'-\sigma):\dot\varepsilon^p\leq 0.$$

In other word, saying that $\dot{\varepsilon}^p$ is related to σ by the plastic yielding rule is equivalent to differential inclusion (9.6). Next, let us calculate the polar function of χ :

$$\varphi(\dot{\varepsilon}^p) = sup\left\{\sigma' : \dot{\varepsilon}^p - \coprod_K(\sigma'); \sigma' \in S\right\}$$

When $\sigma' \notin K$, the expression to maximize has the value $-\infty$. Otherwise, it has a finite value, strictly greater than $-\infty$. Thus, the superior value is reached for $\sigma' \in K$, and we find that the polar function :

$$\varphi(\dot{\varepsilon}^p) = sup\left\{\sigma' : \dot{\varepsilon}^p ; \sigma' \in K\right\} = D(\dot{\varepsilon}^p),$$

is exactly the dissipation power function. According to (9.5), the inverse plastic yielding rule is the differential inclusion :

$$\sigma \in \partial D(\dot{\varepsilon}^p), \tag{9.7}$$

equivalent to inequation (2.7). If $\dot{\varepsilon}^p$ is related to σ by the plastic yielding rule, the couple $(\dot{\varepsilon}^p, \sigma)$ is extremal :

$$D(\dot{\varepsilon}^p) + \coprod_K(\sigma) = \sigma : \dot{\varepsilon}^p .$$

When $\sigma \in K$, this reduces to definition (2.3) of the dissipation power function. For any couple $(\dot{\varepsilon}^p, \sigma')$, Fenchel's inequality holds :

$$D(\dot{\varepsilon}^p) + \coprod_K(\sigma') \geq \sigma' : \dot{\varepsilon}^p ,$$

which is equivalent to maximum dissipation principle (2.4).

As example, let us consider Drucker-Prager model for the plasticity of soils and rocks. At the vertex ($s_m = c / tan\,\phi, s = 0$) of the plastic yielding surface, the plastic strain rate being defined by the differential inclusion (9.6), Hill's maximum power principle (2.2) gives :

$$\forall \sigma' = (s'_m, s') \in K,\ (s'_m - c / tan\,\phi)\,\dot{e}^p_m + s' : \dot{e}^p \leq 0. \tag{9.8}$$

Let us consider the cone defined by :

$$K^* = \left\{\dot{\varepsilon}^p_m = (\dot{e}^p_m, \dot{e}^p) \in \mathrm{E} \quad such \quad that \quad \dot{e}^p_m \geq r tan\phi . \varepsilon_{eq}(\dot{e}^p)\right\},$$

called the dual (or polar) cone of K. Thus, for any $\sigma' \in K$ and $\dot{\varepsilon}^p \in K^*$, one has :

$$(s'_m - c/\tan\phi)\dot{e}^p_m + s':\dot{e}^p \leq (s'_m - c/\tan\phi)\dot{e}^p_m + \sigma_{eq}(s')\varepsilon_{eq}(\dot{e}^p) \leq f(\sigma')\,\dot{e}^p_m / r\,\tan\phi$$

Because $f(\sigma') \leq 0$ and $\dot{e}^p_m \geq 0$, we conclude that (9.8) is satisfied, that proves the differential inclusion :

$$\dot{\varepsilon}^p \in \partial \coprod_K (c/\tan\phi, 0) = K^*,$$

generalizes the normality law at the vertex. Moreover, we can remark that :

$$\partial \coprod_{K^*} (0,0) = K .$$

Plasticity With Kinematical Hardening. The most simple model is Prager linear kinematical hardening. The state of the material is described by the plastic strain $\dot{\varepsilon}^p$ and additional internal variables $\alpha \in \mathrm{E}$, said the kinematical hardening variables. This suggests to introduce the set $V = \mathrm{E} \times \mathrm{E}$ of the velocities $\kappa = (\dot{\varepsilon}^p, -\dot{\alpha})$ and the corresponding dual space $F = \mathrm{S} \times \mathrm{S}$ of the associated variables $\pi = (\sigma, X)$. The elastic domain is :

$$K_0 = \left\{ \pi = (\sigma, X) \in F \;\; such \;\; that \;\; \sigma - X \in K \right\},$$

where K is a non empty convex closed subset of S. The X variable is called the *back-stress*. Physically, it can be interpreted as internal residual stress at a micro-scale smaller than the one of the reference elementary volume. The corresponding dissipation power function is :

$$\begin{aligned}\varphi(\dot{\varepsilon}^p, -\dot{\alpha}) &= \sup\left\{ \sigma : \dot{\varepsilon}^p - X : \dot{\alpha} \;\; ; \;\; (\sigma, X) \in K_0 \right\} \\ &= \sup\left\{ (\sigma - X) : \dot{\varepsilon}^p \;\; ; \;\; \sigma - X \in K \right\} + \sup\left\{ X : (\dot{\varepsilon}^p - \dot{\alpha}) \;\; ; \;\; X \in \mathrm{S} \right\}\end{aligned}$$

The first term of the right hand member is the dissipation power function (2.3). The second term vanishes if $\dot{\varepsilon}^p = \dot{\alpha}$ and is equal to $+\infty$ if $\dot{\varepsilon}^p \neq \dot{\alpha}$. Finally, φ is equal to :

$$\varphi(\dot{\varepsilon}^p, -\dot{\alpha}) = D(\dot{\varepsilon}^p),$$

if $\dot{\varepsilon}^p = \dot{\alpha}$, and is equal to $+\infty$ otherwise. Using an indicator function, this expression can be recast as :

$$\varphi(\dot{\varepsilon}^p, -\dot{\alpha}) = D(\dot{\varepsilon}^p) + \coprod_{\{0\}} (\dot{\varepsilon}^p - \dot{\alpha}).$$

The linear kinematical hardening rule is characterized *by Prager's hardening rule* :

$$\dot{\varepsilon}^p = \dot{\alpha} . \tag{9.9}$$

Moreover, we assume the kinematical variable is related to the back-stress by a linear mapping :

$$\alpha = B\,X , \tag{9.10}$$

such that for all $X \neq 0$, one has :

$$X : B\,X > 0 . \tag{9.11}$$

10 Exercises

Exercise 1.1. Let K defined by the convex loading function *f*, supposed to be C^1 on ∂K. For the corresponding model with linear kinematical hardening, using Lagrange multiplier technique, find Prager's hardening rule (9.9).

Exercise 1.2. (*extension of Melan's theorem to models with linear kinematical hardening*). We say that a field $\overline{\pi}(x) = (\overline{\rho}(x), \overline{X}(x)) \in F$ is admissible (resp. strictly admissible) if it is time independent, $\overline{\rho} \in N$ and $\overline{\pi}$ is P.A. in the sense that for all $\sigma^E \in \Sigma$, $(\overline{\sigma}, \overline{X}) = (\sigma^E + \overline{\rho}, \overline{X})$ belongs to K_0 (resp. the interior of K_0) anywhere in Ω and at any time. Prove that if a strictly admissible field $\overline{\pi}$ can be found, the structure shakes down.
Hint : introduce the residual elastic energy :

$$\mathrm{R} = \tfrac{1}{2}\int_\Omega \left[(\rho - \overline{\rho}) : S(\rho - \overline{\rho}) + (X - \overline{X}) : B(X - \overline{X})\right] d\Omega \geq 0 .$$

Exercise 1.3. (*extension of Halphen's theorem to the models with linear kinematical hardening*). We say that a field $\dot{\varepsilon}^p(x,t)$ is admissible for this model if it is P.A. in the sense defined in section 6 and if the alternating plasticity occurs : $\Delta \varepsilon^p = 0$ (obviously, it is admissible in Koiter's sense). Prove that if the structure does not shakedown, the plastic strain rate field tends to an admissible field as previously defined.

Exercise 1.4. (*limited linear kinematical hardening model*[4]) Let us consider the following elastic domain defined by a two-surface yield condition :

$$K_1 = \left\{ \pi = (\sigma, X) \in F \ \ such\ \ that \ \ f(\sigma - X) \leq 0 \ \ and \ \ f_\infty(\sigma) \leq 0 \right\},$$

where f_∞ is an additional loading function defining the limit surface by : $f_\infty(\sigma) = 0$. Under suitable hypothesis of differentiability and convexity, and using Lagrange multiplier techniques, find the plastic yielding law and show that Prager's hardening rule (9.9) occurs when $f_\infty(\sigma) = 0$.

Exercise 1.5. (*extension of Melan's theorem to the limited linear kinematical hardening model).* We say that a field $\overline{\pi}(x) = (\overline{\rho}(x), \overline{X}(x)) \in F$ is admissible (resp. strictly admissible) if it is time independent, $\overline{\rho} \in N$ and $\overline{\pi}$ is P.A. in the sense that for all $\sigma^E \in \Sigma$, $(\overline{\sigma}, \overline{X}) = (\sigma^E + \overline{\rho}, \overline{X})$ belongs to K_1 (resp. the interior of K_1) anywhere in Ω and at any time. Under assumptions (9.10) and (9.11), prove that if a strictly admissible field $\overline{\pi}$ can be found, the structure shakes down.

Exercise 1.6. (*extension of Halphen's theorem to the limited linear kinematical hardening model).* Prove that if the structure does not shakedown, the plastic strain rate field tends to an admissible field in Koiter's sense.

Exercise 1.7. (*isotropic hardening*). This model is characterized by a scalar additional internal variable p. Let us introduce the set $V = \mathrm{E} \times \boldsymbol{R}$ of velocities $\dot{\kappa} = (\dot{\varepsilon}^p, -\dot{p})$ and the dual space $F = \mathrm{S} \times \boldsymbol{R}$ of associated variables $\pi = (\sigma, R)$. The elastic domain is define by :

$$K_2 = \left\{ \pi = (\sigma, R) \in F \ \ such\ \ that \ \ f(\sigma, R) = \sigma_{eq}(s) - R \leq 0 \right\}.$$

The associated variable R is interpreted as a variable yield stress. Using Lagrange multiplier technique, prove that p increases according to :

$$p(t) = p(0) + \int_0^t \varepsilon_{eq}(\dot{e}^p)\, dt\,.$$

Exercise 1.8. For the isotropic hardening model of the previous exercise, prove that the corresponding superpotential is equal to : $\varphi(\dot{\varepsilon}^p, -\dot{p}) = 0$, if $\varepsilon_{eq}^{\cdot}(\dot{e}^p) \leq \dot{p}$, and equal to $+\infty$ otherwise. Prove that for the extremal couples : $\sigma : \dot{\varepsilon}^p = R\,\dot{p}$.

---

[4] This model was extensively used by Weichert and his collaborators for numerical applications in shakedown theory (see for instance Weichert and Gross-Weege, 1988, and Weichert's lecture notes in the same book).

References

Coffin, L. F. (1954). *Transaction ASME*, 76:923.

Ekeland, I., and Temam, R. (1975). *Convex analysis and variational problems*, NY: North Holland.

Fenchel, W. (1949). On conjugate convex functions. *Canadian Journal of Mathematics* 1:73–77.

Halphen, B. and Nguyen Quoc Son (1975). Sur les matériaux standard généralisés, *Journal de Mécanique* 14:39–63.

Halphen, B. (1978). Periodic solutions in plasticity and viscoplasticity, in Nemat-Nasser ed., *Proceedings of IUTAM Symposium on Variational Methods in Mechanics of Solids*, Oxford :Pergamon Press.

Jospin, R., Nguyen Dang, H. and de Saxcé, G. (1991). Direct limit analysis of elbows by finite element and mathematical programing. *Proceedings of the Int. Symposium "Plasticity 91"*, Grenoble (France).

Koiter, W.T. (1960). General theorems for elasto-plastic solids. In *Progress in Solid Mechanics*, Volume 1.

Manson, S.S. (1954). *Nat. Advise. Comm. Aero. Tech.*, Note 2.933.

Melan, E. (1936). Theory statisch unbestimmter Systeme aus ideal plastischen baustoff. *Sitz Ber. Akad. Wiss. Wien IIa* 145:195.

Moreau, J.J. (1963). Fonctionnelles sous-différentiables. *Comptes-rendus de l'Académie des Sciences de Paris*, Série A 257:4117–4119.

Moreau, J.J. (1966). *Séminaire sur les Equations aux Dérivées Partielles, Fonctionnelles Convexes*, Paris: Collège de France.

Panagiatopoulos, P.D. (1975). *Inequality Problems in Mechanics and Applications*, Boston: Birkhauser.

Save, M., Massonnet, C. and de Saxcé, G. (1997). *Plastic Limit Analysis of Plates, Shells and Disks*, NY: North Holland.

Taylor, A. E. (1965). *General theory of functions and integration*, Blaisdell: Waltham.

Weichert, D., Gross-Weege, J. (1988). Assessment of elasto-plastic sheets under variable mechanical and thermal loads using a simplified two-surface yield condition. *International Journal of Mechanical Science* 30:757–767.

Variational Formulation

Géry de Saxcé

University of Lille, Lille, France

Abstract. We present the dual variational principles of the rigid perfectly plastic material due to Markov and Hill, from which ones the bound theorems of limit analysis can be deduced. Following the same kind of method, the bound theorems can be extended to variable repeated loads. They provide two dual ways to calculate the shakedown load α^a by solving constrained optimization problems. Finally, the duality between the variational principles of shakedown is discussed from the viewpoint of the non smooth mechanics.

1 Mandel's Approach for the Limit and Shakedown Analysis

For the rigid-perfectly plastic material, the boundary value problem (B.V.P.) corresponding to given external actions $(\bar{f}, \bar{p}, \dot{\bar{u}})$, as defined in section 2 of lecture 1, which are time independent and characterize the collapse, is :

(P₁) Find some couple $(\dot{u}, \sigma)$ such that :

1. $\dot{u}$ is admissible, that is K.A. and P.A. in the sense that : $\int_\Omega \bar{f}^0 . \dot{u} \, d\Omega + \int_{\Gamma_1} \bar{p}^0 . \dot{u} \, d\Gamma > 0$,
2. σ is admissible (statically admissible and P.A.).
3. $\dot{\varepsilon}(\dot{u})$ and σ are associated by the normality law in Ω.

The corresponding variational formulation consists in stating the two following propositions (see Markov, 1947; and Hill, 1948) :

Theorem 3 (*Markov's principle*) : if $\dot{u}$ is a solution of the B.V.P. ***(P₁)***, it is solution of the following problem :

Find some $\dot{u}$ such that the functional

$$\Phi(\dot{u}) = \int_\Omega D(\dot{\varepsilon}(\dot{u})) \, d\Omega - \int_\Omega \bar{f} . \dot{u} \, d\Omega - \int_{\Gamma_1} \bar{p} . \dot{u} \, d\Gamma,$$

is minimum among the admissible velocity fields.

Theorem 4 (*Hill's principle*) : if σ is a solution of the B.V.P. ***(P$_l$)***, it is a solution of the following problem :

Find some σ such that the functional

$$\Pi(\sigma) = -\int_{\Gamma_0} p(\sigma).\dot{\bar{u}}\, d\Gamma,$$

is minimum among the admissible stress fields.

Proof. It can be remarked that the two variational principles are dual one of each other in the sense that for any couple $(\dot{u}^k, \sigma^s)$ of admissible fields, it holds :

$$\Phi(\dot{u}^k) + \Pi(\sigma^s) \geq 0, \tag{1.1}$$

according to inequation (2.4) and Green's formula (1.5) of lecture 1. In particular, for a solution $(\dot{u}, \sigma)$ of the B.V.P., the equality is achieved in equation (1.1) because of equation (2.6) of lecture 1 :

$$\Phi(\dot{u}) + \Pi(\sigma) = 0.$$

Moreover, as particular case of inequation (1.1), one has :

$$\Phi(\dot{u}^k) \geq -\Pi(\sigma), \qquad \Phi(\dot{u}) \geq -\Pi(\sigma^s)$$

Combining the two previous results leads to :

$$\Phi(\dot{u}^k) \geq \Phi(\dot{u}) = -\Pi(\sigma) \geq -\Pi(\sigma^s),$$

that proves theorem 3 and 4.

Now, let us consider the usual assumptions of the limit analysis theory (Save, Massonnet and de Saxcé, 1997) :

1. proportional loading : $\bar{f} = \alpha \bar{f}^0, \qquad \bar{p} = \alpha \bar{p}^0,$
2. fixed supports : $\dot{\bar{u}} = 0$ *on* Γ_0.

It can be remarked that because of hypothesis 2, the problem of theorem 4 degenerates into :

Find some σ among the admissible stress fields,
where no variational functional has to be minimized. Only admissibility conditions have to be fulfilled.

The aim of the limit analysis is to define the conditions to be satisfied by the external loading in order that the B.V.P. *(P$_l$)* has a solution. Following an approach initially proposed by Mandel (1966), the so-called lower and upper bound problems can be deduced from Markov's and Hill's principles.

Let us come back again to the shakedown analysis. The considerations developed in the previous lecture suggest to characterize the collapse corresponding to given external actions $\bar{p} \in P$ by means of the following B.V.P. :

(P$_s$) Find some couple $(\dot{\varepsilon}^p, \bar{\rho})$ such that :

1. $\dot{\varepsilon}^p$ is an admissible plastic strain rate field (lecture 1, section 6),
2. $\bar{\rho}$ is an admissible residual stress field (lecture 1, section 5),
3. $\dot{\varepsilon}^p$ and $(\sigma^E + \bar{\rho})$ are associated by the normality law in Ω at any time.

In fact, the shakedown analysis is a powerful generalization of the classical limit analysis. Indeed, the last theory can be interpreted as the particular event of the reference load domains P^0 and Σ^0 being singletons. This remark suggests to follow for the shakedown analysis the same approach as the one due to Mandel for the limit analysis and, first of all, to state and prove new variational principles corresponding to Markov's and Hill's ones (G. de Saxcé, 1986, 1995).

2 Markov's Principle Over a Cycle

In order to construct the ad hoc functional, Markov's principle is modified in the following way :

1. the history of the structure over the whole collapse cycle is considered,
2. the admissible field is now $\dot{\varepsilon}^p$,
3. the external actions are replaced by the equivalent stress field σ^E in the corresponding fictitious elastic body.

Then, the following proposition may be stated (see de Saxcé, 1986,1995) :

Theorem 5 (Markov's principle over a cycle) : if $\dot{\varepsilon}^p$ is solution of the B.V.P. ***(P$_s$)***, it is a solution of the following problem :

Find some $\dot{\varepsilon}^p$ such that the functional

$$\Phi_s(\dot{\varepsilon}^p) = \int_\Omega \oint D(\dot{\varepsilon}^p)\, dt\, d\Omega - \int_\Omega \oint \sigma^E : \dot{\varepsilon}^p\, dt\, d\Omega \tag{2.1}$$

is minimum among the admissible plastic strain rate fields.

Proof. Taking into account the definition of the residual stress fields and convexity inequality (2.7) of lecture 1, one has, for any admissible field $\varepsilon^{p'}$:

$$\Phi_s(\dot{\varepsilon}^{p'}) - \Phi_s(\dot{\varepsilon}^{p}) \geq \int_\Omega \oint \bar{\rho} : (\dot{\varepsilon}^{p'} - \dot{\varepsilon}^{p})\, dt\, d\Omega .$$

$\bar{\rho}$ is admissible, therefore time independent, whence, one has :

$$\Phi_s(\dot{\varepsilon}^{p'}) - \Phi_s(\dot{\varepsilon}^{p}) \geq \int_\Omega \bar{\rho} : (\Delta\varepsilon^{p'} - \Delta\varepsilon^{p})\, d\Omega .$$

As $\Delta\varepsilon^{p'}, \Delta\varepsilon^{p} \in N^*$ and $\bar{\rho} \in N$, the proof is achieved by applying the virtual power principle (4.1) of lecture 1:

$$\Phi_s(\dot{\varepsilon}^{p'}) \geq \Phi_s(\dot{\varepsilon}^{p}) .$$

"Markov's principle over a cycle" means that it is an extension to shakedown problems of the well-known principle originally proposed by Markov (1947) for the rigid perfectly plastic material. It would be also justified to say "Ponter's principle" since this result was independently derived by Ponter, although it appeared only recently in the literature (Ponter and Chen, 2000, Chen and Ponter, 2000). The reader is also referred to Ponter's lecture notes in the same book.

3 Hill's Principle Over a Cycle

Similar considerations can be developed for the statical formulation, leading to the following proposition (see de Saxcé, 1986, 1995) :

Theorem 6 (*Hill's principle over a cycle*) : if $\bar{\rho}$ is a solution of the B.V.P. (P_s), it is a solution of the following problem :

Find some $\bar{\rho}$ among the admissible residual stress fields.

The proposition is trivial. Let us remark that, as for Hill's principle restricted to the limit analysis, the statical problem is degenerated in the sense that no variational functional has to be minimized. Conventionally, one may assume that the statical problem is a variational principle with admissible constraints and a vanishing functional :

$$\Pi_s(\bar{\rho}) = 0 . \tag{3.1}$$

As for the limit analysis, it can be proved that the two variational principles are dual one of each other in the sense that for any couple $(\dot{\varepsilon}^{pk}, \bar{\rho}^{s})$ of admissible fields, it holds :

$$\Phi_s(\dot{\varepsilon}^{pk}) + \Pi_s(\bar{\rho}^{s}) \geq 0 . \tag{3.2}$$

Indeed, taking account of maximum dissipation principle (2.4) of lecture 1, the definition of residual stress fields, equations (2.1) and (3.1), and that $\overline{\rho}^s$ is time independent, it occurs :

$$\Phi_s(\dot{\varepsilon}^{pk}) + \Pi_s(\overline{\rho}^s) \geq \int_\Omega \overline{\rho} : \Delta\varepsilon^{pk} d\Omega .$$

Owing to virtual power principle (4.1) of lecture 1, it is concluded that inequation (3.2) holds.

In particular, for a limit state $(\dot{\varepsilon}^p, \overline{\rho})$, solution of the B.V.P. ***(P$_s$)***, equation (2.4) replaces inequation (2.6) of lecture 1 and the equality is achieved in inequation (3.2) :

$$\Phi_s(\dot{\varepsilon}^p) + \Pi_s(\overline{\rho}) = 0 . \tag{3.3}$$

4 Lower Bound Theorem

Taking into account equations (3.1),(3.3), it can be quoted that for the solution $\dot{\varepsilon}^p$ of the B.V.P. ***(P$_s$)*** corresponding to a load domain $\Sigma = \alpha^a \Sigma^0$, one has :

$$\Phi_s^a(\dot{\varepsilon}^p) = \int_\Omega \oint D(\dot{\varepsilon}^p) dt d\Omega - \alpha^a \int_\Omega \oint \sigma^{E0} : \dot{\varepsilon}^p dt d\Omega = 0 . \tag{4.1}$$

In a similar way, following Martin (1975), the kinematical load factor α^k associated to an admissible plastic strain rate fields $\dot{\varepsilon}^{pk}$ is defined by the equality :

$$\int_\Omega \oint D(\dot{\varepsilon}^{pk}) dt d\Omega = \alpha^k \int_\Omega \oint \sigma^{E0} : \dot{\varepsilon}^{pk} dt d\Omega . \tag{4.2}$$

The relation has a sense because of inequation (2.5) of lecture 1 and $\dot{\varepsilon}^{pk}$ is P.A. (section 6 of lecture 1). Thus, the following proposition holds :

Theorem 7 (*Koiter's lower bound theorem*) : if a solution $(\dot{\varepsilon}^p, \overline{\rho})$ of the B.V.P. ***(P$_s$)*** exists for the load domain $\Sigma = \alpha^a \Sigma^0$, α^a is the lower bound of the kinematical factors α^k :

$$\alpha^a \leq \alpha^k . \tag{4.3}$$

Moreover, Markov's principle over a cycle has no solution for $\alpha^a < \alpha^k$.

Proof. Firstly, let us demonstrate that : $\alpha^a \leq \alpha^k$. Because of theorem 5 and equation (4.1), for any admissible plastic strain rate field $\dot{\varepsilon}^{pk}$ and the load domain $\Sigma = \alpha^a \Sigma^0$, one has :

$$\Phi_s^a(\dot{\varepsilon}^{pk}) = \int_\Omega \oint D(\dot{\varepsilon}^{pk}) dt d\Omega - \alpha^a \int_\Omega \oint \sigma^{E0} : \dot{\varepsilon}^{pk} dt d\Omega \geq \Phi_s^a(\dot{\varepsilon}^p) = 0 .$$

Taking account of definition (4.2) of the kinematical factors, one has :

$$(\alpha^k - \alpha^a)\int_\Omega \oint \sigma^{E0} : \dot{\varepsilon}^{pk}\, dt\, d\Omega \geq 0 .$$

Because of the second admissibility condition for $\dot{\varepsilon}^{pk}$ (lecture 1, section 6), bound property (4.3) is proved.

Finally, let us demonstrate that Markov's principle over a cycle has no solution for $\alpha^a < \alpha^k$. It is sufficient to prove that the functional (2.1) is not coercive (see Ekeland and Temam, 1975). Indeed, one has for the load domain $\Sigma = \alpha^k \Sigma^0$:

$$\Phi_s^k(\dot{\varepsilon}^p) = \int_\Omega \oint D(\dot{\varepsilon}^p)\, dt\, d\Omega - \alpha^k \int_\Omega \oint \sigma^{E0} : \dot{\varepsilon}^p\, dt\, d\Omega .$$

Taking account of equality (4.1) and the second admissibility condition, it holds :

$$\Phi_s^k(\dot{\varepsilon}^p) = (\alpha^a - \alpha^k)\int_\Omega \oint \sigma^{E0} : \dot{\varepsilon}^p\, dt\, d\Omega < 0 .$$

Thus, by choosing the minimizing sequence $(\dot{\varepsilon}_n^p)$ defined by : $\dot{\varepsilon}_n^p = n\,\dot{\varepsilon}^p$ $(n = 1,2,\dots)$, it is easy to prove that $\Phi_s^k(\dot{\varepsilon}_n^p) < 0$ tends to $-\infty$ when $\|\dot{\varepsilon}_n^p\|$ tends to $+\infty$, the space of the plastic strain rate fields being equipped with a suitable norm[1].

A more pervasive study by Ponter and Chen (2000) shows that, even when $\alpha^a < \alpha^k$, the functional Φ_s^k is coercive and has a vanishing minimum provided that the plastic strain field is subjected to other admissibility conditions in addition to the one of theorem 5.

5 Upper Bound Theorem

The following proposition has to be proved :

Theorem 8 (*Koiter's upper bound theorem*) : if a solution $(\dot{\varepsilon}^p, \bar{\rho})$ of the B.V.P. (P_s) exists for the load domain $\Sigma = \alpha^a \Sigma^0$, Hill's principle over a cycle has a solution $\bar{\rho}^s$ for $\Sigma = \alpha^s \Sigma^0$ if and only if

$$\alpha^s \leq \alpha^a . \tag{5.1}$$

Moreover, if $\alpha^s < \alpha^a$, the structure shakes down for the load factor α^s.

Proof. *Necessary condition*. Introducing :

$$\sigma = \alpha^a \sigma^{E0} + \bar{\rho} \in K, \qquad \sigma^s = \alpha^s \sigma^{E0} + \bar{\rho}^s \in K, \tag{5.2}$$

[1] The nature of the space and the definition of the norm will be discussed in section 2 of lecture 3.

and taking account that $\bar{\rho}$ and $\bar{\rho}^s$ are time independent, it holds :

$$\int_\Omega \oint (\sigma^s - \sigma):\dot{\varepsilon}^p \, dt \, d\Omega = \int_\Omega (\bar{\rho}^s - \bar{\rho}):\Delta \, \varepsilon^p \, d\Omega + (\alpha^s - \alpha^a) \int_\Omega \oint \sigma^{E0}:\dot{\varepsilon}^p \, dt \, d\Omega . \quad (5.3)$$

The first term of the right hand side member vanishes because of the second admissibility condition for residual stress fields (lecture 1, section 5) and virtual power principle (4.1) of lecture 1. Owing to Hill's maximum power principle (2.2) of lecture 1, one has :

$$(\alpha^s - \alpha^a) \int_\Omega \oint \sigma^{E0}:\dot{\varepsilon}^p \, dt \, d\Omega \le 0 .$$

Thus, it is concluded that inequation (5.1) holds because of the admissibility conditions for the plastic strain rate field.

Sufficient condition. According to $0 \in K$, taking account of inequality (5.1) and the convexity of the elastic domain *K*, one has :

$$\forall \lambda \in [0,1], \; \lambda \, (\bar{\rho} + \alpha^a \sigma^{E0}) \in K .$$

Because of inequation (5.1), the following particular value : $\lambda = \alpha^s / \alpha^a$ can be chosen. Hence, the following residual stress field : $\bar{\rho}^s = \lambda \bar{\rho}$ is plastically admissible ($\bar{\rho}^s + \alpha^s \sigma^{E0} \in K$) and consequently is admissible in Melan's sense. Thus, $\bar{\rho}^s$ is a solution of Hill's principle over a cycle.

Finally, one has to prove that the structure shakes down for the load factor $\alpha^s < \alpha^a$. Let us assume that there exists a plastic strain rate $\dot{\varepsilon}^{ps}$ satisfying the first admissibility condition (lecture 1, section 5) and associated to $(\bar{\rho}^s + \alpha^s \sigma^{E0})$ by the normality law in Ω at any time. Thus, it is allowed to replace $\dot{\varepsilon}^p$ by $\dot{\varepsilon}^{ps}$ and σ by σ^s in equation (5.3) and, by similar arguments, it is concluded that :

$$(\alpha^a - \alpha^s) \int_\Omega \oint \sigma^{E0}:\dot{\varepsilon}^{ps} \, dt \, d\Omega \le 0 . \quad (5.4)$$

Moreover, because $\dot{\varepsilon}^{ps}$ is associated to $\bar{\rho}^s$ and equation (2.6) of lecture 1, one has :

$$D(\dot{\varepsilon}^{ps}) = \sigma^s : \dot{\varepsilon}^{ps} .$$

According to virtual power principle (4.1), it holds :

$$\int_\Omega \oint D(\dot{\varepsilon}^{ps}) \, dt \, d\Omega = \alpha^s \int_\Omega \oint \sigma^{E0}:\dot{\varepsilon}^{ps} \, dt \, d\Omega . \quad (5.5)$$

From inequation (5.4) and equation (5.5), it is deduced that :

$$\left(\frac{\alpha^a}{\alpha^s}-1\right)\int_\Omega \oint D(\dot{\varepsilon}^{ps})\,dt\,d\Omega \leq 0 .$$

The first factor of the left hand member is positive. Hence :

$$\int_\Omega \oint D(\dot{\varepsilon}^{ps})\,dt\,d\Omega = 0 .$$

Because of inequation (2.5) of lecture 1, $\dot{\varepsilon}^{ps}$ vanishes and the structure shakes down for the load factor α^s.

Comments. As it was shown by de Saxcé (1986), it is not very necessary to consider a time periodic loading. In this context, the load cycle integrals are replaced by average values :

$$\lim_{\tau\to+\infty}\frac{1}{\tau}\int_\Omega\int_0^\tau D(\dot{\varepsilon}^p)\,dt\,d\Omega, \qquad \lim_{\tau\to+\infty}\frac{1}{\tau}\int_\Omega\int_0^\tau \sigma:\dot{\varepsilon}^p\,dt\,d\Omega$$

More pervasive developments about the numerical methods based on the variational approach and the finite element method can be found in Save, Massonnet et de Saxcé (1997), and Save, de Saxcé and Borkowski (1991).

6 Kinematical and Statical Bound Problems

According to equation (4.2), the load factor is equal to the total dissipation over a cycle. Hence Koiter's lower bound theorem leads to solve the so-called *kinematical bound problem* :

$$inf\left\{\alpha^k = \int_\Omega \oint D(\dot{\varepsilon}^{pk})\,dt\,d\Omega \Big/ \int_\Omega \oint \sigma^{E0}:\dot{\varepsilon}^{kp}\,dt\,d\Omega \;;\quad \dot{\varepsilon}^{pk}\in \mathrm{E}_\Omega,\quad \Delta u^k\in \mathrm{U}_\Omega, \right.$$
$$\left. such\ \ that\quad \oint \dot{\varepsilon}^{kp}dt = grad_s\Delta u^k \quad in\quad \Omega,\quad \Delta u^k = 0\quad on\quad \Gamma_0,\quad \sigma^{E0}\in\Sigma^0\right\} \tag{6.1}$$

E_Ω and U_Ω are suitable topological vector spaces of which the nature will be discussed in lecture 3.

Next, let us remark that if $\dot{\varepsilon}^p$ is admissible in Koiter's sense and $\lambda>0$, $\lambda\dot{\varepsilon}^p$ is admissible too. Hence, if $(\dot{\varepsilon}^p,\bar{\rho})$ is a solution of the B.V.P. (P_s), $\sigma=\bar{\rho}+\sigma^E$ is associated by the plastic yielding normality law to $\dot{\varepsilon}^p$ and, consequently, to $\lambda\dot{\varepsilon}^p$, because Hill's maximum power principle (2.2) of lecture 1, applied to $\dot{\varepsilon}^p$, implies :

$$\forall\sigma'\in K,\ (\sigma'-\sigma):(\lambda\dot{\varepsilon}^p)\leq 0 .$$

Thus, $(\lambda \dot{\varepsilon}^p, \bar{\rho})$ is a solution of the B.V.P. (P_s) too.

Solving (P_s), the value of the previous factor λ is not relevant and it is allowed to fix it by enforcing an additional *normalization condition*, for instance :

$$\int_\Omega \oint \sigma^E : \dot{\varepsilon}^p \, dt \, d\Omega = 1. \tag{6.2}$$

Finally, the *kinematical bound problem* can be recast as :

$$inf \Big\{ \int_\Omega \oint D(\dot{\varepsilon}^{pk}) \, dt \, d\Omega \quad ; \quad \dot{\varepsilon}^{pk} \in \mathrm{E}_\Omega, \quad \Delta u^k \in \mathrm{U}_\Omega \quad such \quad that$$
$$\oint \dot{\varepsilon}^{kp} dt = grad_s \Delta u^k \quad in \quad \Omega, \quad \Delta u^k = 0 \quad on \quad \Gamma_0, \quad \int_\Omega \oint \sigma^{E0} : \dot{\varepsilon}^{kp} dt \, d\Omega = 1, \quad \sigma^{E0} \in \Sigma^0 \Big\}$$

In a similar way, Koiter's upper bound theorem leads to solve the *statical bound theorem* :

$$sup \big\{ \alpha^s \quad ; \quad \alpha^s \in \boldsymbol{R}, \quad \bar{\rho}^s \in \mathrm{S}_\Omega, \quad such \quad that \quad div \bar{\rho}^s = 0 \quad in \quad \Omega,$$
$$p(\bar{\rho}^s) = 0 \quad on \quad \Gamma_1, \quad \alpha^s \sigma^{E0} + \bar{\rho}^s \in K \quad in \quad \Omega, \quad at \quad any \quad time, \quad \sigma^{E0} \in \Sigma^0 \big\} \tag{6.3}$$

7 Non Smooth Mechanics and Variational Theory of Shakedown

Shakedown Constitutive Law. In the B.V.P. (P_s), $\dot{\varepsilon}^p$ and $\sigma = \bar{\rho} + \sigma^E$ are associated by the plastic yielding normality law. According to results of non smooth Mechanics (lecture 1, section 9), it was seen that it is equivalent to state :

$$\dot{\varepsilon}^p \in \partial \textstyle\coprod_K (\bar{\rho} + \sigma^E).$$

This suggests to introduce the superpotential :

$$\chi_s(\bar{\rho}) = \textstyle\coprod_K (\bar{\rho} + \sigma^E).$$

In short, we obtain the differential inclusion :

$$\dot{\varepsilon}^p \in \partial \chi_s(\bar{\rho}), \; in \;\; \Omega, \; at \;\; any \;\; time, \tag{7.1}$$

which can be interpreted as a multivalued constitutive law involving the dual variables $\dot{\varepsilon}^p$ and $\bar{\rho}$, characterizing the shakedown problem. The polar function of χ_s is defined as :

$$\varphi_s(\dot{\varepsilon}^p) = sup \left\{ \bar{\rho} : \dot{\varepsilon}^p - \chi_s(\bar{\rho}) ; \bar{\rho} \in \mathrm{S} \right\} = sup \left\{ \bar{\rho} : \dot{\varepsilon}^p ; \bar{\rho} + \sigma^E \in K \right\}.$$

Accounting to the changing of variable : $\bar{\rho} \mapsto \sigma = \bar{\rho} + \sigma^E$, one has :

$$\varphi_s(\dot{\varepsilon}^p) = sup\left\{ \sigma : \dot{\varepsilon}^p \; ; \sigma \in K \right\} - \sigma^E : \dot{\varepsilon}^p .$$

Using the definition of the dissipation power function (lecture 1, equation (2.3)) leads to :

$$\varphi_s(\dot{\varepsilon}^p) = D(\dot{\varepsilon}^p) - \sigma^E : \dot{\varepsilon}^p .$$

This allows to inverse the constitutive law :

$$\bar{\rho} \in \partial \varphi_s(\dot{\varepsilon}^p), \; in \; \Omega, \; at \; any \; time . \tag{7.2}$$

Besides, φ_s and χ_s verify Fenchel's inequality :

$$\forall \bar{\rho}' \in \mathrm{S}, \quad \forall \dot{\varepsilon}^{p'} \in \mathrm{E}, \quad \varphi_s(\dot{\varepsilon}^{p'}) + \chi_s(\bar{\rho}') \geq \bar{\rho}' : \dot{\varepsilon}^{p'} . \tag{7.3}$$

For the couples related by the shakedown constitutive law, it holds :

$$\varphi_s(\dot{\varepsilon}^p) + \chi_s(\bar{\rho}) = \bar{\rho} : \dot{\varepsilon}^p \;\Leftrightarrow\; \dot{\varepsilon}^p \in \partial \chi_s(\bar{\rho}) \;\Leftrightarrow\; \bar{\rho} \in \partial \varphi_s(\dot{\varepsilon}^p) . \tag{7.4}$$

Calculus of Variations. If $(\dot{\varepsilon}^p, \bar{\rho})$ is a solution of the B.V.P. (***P****s*), $\dot{\varepsilon}^p$ must be admissible in Koiter's sense and satisfy (7.2). This suggests to introduce the functional :

$$\overline{\Phi}_s(\dot{\varepsilon}^p) = \int_\Omega \oint \varphi_s(\dot{\varepsilon}^p) \, dt \, d\Omega ,$$

which is obviously identical to (2.1). Thus, we state the corresponding variational principle :

$$inf \left\{ \overline{\Phi}_s(\dot{\varepsilon}^p) \quad ; \quad \dot{\varepsilon}^p \in \mathrm{E}_\Omega, \quad \dot{\varepsilon}^p \quad admissible \right\},$$

which is identical to Markov's principle over a cycle (theorem 5). On the other hand, if $(\dot{\varepsilon}^p, \bar{\rho})$ is a solution of the B.V.P. (***P****s*), $\bar{\rho}$ must be admissible in Melan's sense and satisfy the differential inclusion (7.2). Hence, we introduce the corresponding functional :

$$\overline{\Pi}_s(\bar{\rho}) = \int_\Omega \oint \chi_s(\bar{\rho}) \, dt \, d\Omega . \tag{7.5}$$

Next, we state the corresponding variational principle :

$$inf \left\{ \overline{\Pi}_s(\bar{\rho}) \quad ; \quad \bar{\rho} \in \mathrm{S}_\Omega, \quad \bar{\rho} \in N \right\}. \tag{7.6}$$

Let us remark that if $\overline{\rho}+\sigma^E \notin K$, χ_s takes infinite values and $\overline{\Pi}_s$ is infinite. Thus, the minimum is realized only for $\overline{\rho}$ such that $\overline{\rho}+\sigma^E \in K$ and $\overline{\rho} \in N$. In other words, the problem (7.5) is equivalent to Hill's principle over a cycle (theorem 6).

Finally, we can prove once again the duality of Markov's and Hill's principles over a cycle, but in a very short and synthetic way. If $\dot{\varepsilon}^{pk}$ is admissible and $\overline{\rho}^s \in N$, according to Fenchel's inequality (7.3) and to the virtual power principle, one has :

$$\overline{\Phi}_s(\dot{\varepsilon}^{pk})+\overline{\Pi}_s(\overline{\rho}^s)\geq 0. \tag{7.7}$$

If $(\dot{\varepsilon}^p,\overline{\rho})$ additionally satisfies the B.V.P. (P_s), it holds :

$$\overline{\Phi}_s(\dot{\varepsilon}^p)+\overline{\Pi}_s(\overline{\rho})=0.$$

As particular case of inequality (7.7), we have :

$$\overline{\Phi}_s(\dot{\varepsilon}^{pk})\geq -\overline{\Pi}_s(\overline{\rho}), \qquad \overline{\Phi}_s(\dot{\varepsilon}^p)\geq -\overline{\Pi}_s(\overline{\rho}^s).$$

Combining the two previous results leads to :

$$\overline{\Phi}_s(\dot{\varepsilon}^{pk})\geq \overline{\Phi}_s(\dot{\varepsilon}^p)=-\overline{\Pi}_s(\overline{\rho})\geq -\overline{\Pi}_s(\overline{\rho}^s)$$

that proves theorem 5 and 6.

8 Exercises

Exercise 2.1. If the elastic domain K is defined by a convex loading function *f*, which is C^1 on ∂K, using Lagrangean multiplier technique, prove that Euler-Lagrange variational equations of the statical bound problem (6.3) are normalization condition (6.2) and the admissibility conditions in Koiter's sense.

Hint : introduce Lagrange function :

$$L(\alpha,\overline{\rho},\lambda,\Delta u,\Delta v)=\alpha-\int_\Omega \oint \lambda f(\overline{\rho}+\alpha\sigma^{E0})\,dt\,d\Omega-\int_\Omega div\,\overline{\rho}.\Delta u\,d\Omega+\int_{\Gamma_1} p(\overline{\rho}).\Delta v\,d\Gamma,$$

where $x\mapsto \Delta u(x)$ is a vector field defined on Ω, $x\mapsto \Delta v(x)$ is a vector field defined on Γ_1, and $(x,t)\mapsto \lambda(x,t)$ is a scalar field defined on $\Omega\times[0,+\infty[$.

Exercise 2.2. For the models with linear kinematical hardening rule (lecture 1, section 9), consider the following B.V.P. :

(P$_{s0}$) Find some couple $(\dot{\kappa},\overline{\pi})$ such that :

1. $\dot{\kappa}=(\dot{\varepsilon}^{p},-\dot{\alpha})$ is such that $\dot{\varepsilon}^{p}$ is an admissible plastic strain rate field (lecture 1, exercise 1.3),
2. $\overline{\pi}=(\overline{\rho},\overline{X})$ is admissible (lecture 1, exercise 1.2),
3. the couple $(\dot{\kappa},\overline{\pi})$ is extremal for the law of plasticity with linear kinematical hardening in Ω at any time.

Prove Markov's principle over a cycle for this model : if $\dot{\varepsilon}^{p}$ is such that $(\dot{\kappa},\overline{\pi})$ is solution of the B.V.P. ***(P$_{s0}$)***, it is a solution of the following problem :
find some $\dot{\varepsilon}^{p}$ such that functional (2.1) of lecture 1 is minimum among the admissible fields.

Exercise 2.3. For the models with linear kinematical hardening rule, Hill's principle over a cycle can be stated as : if $\overline{\pi}=(\overline{\rho},\overline{X})$ is such that $(\dot{\kappa},\overline{\pi})$ is solution of the B.V.P. ***(P$_{s0}$)***,it is a solution of the following problem :

find some $\overline{\pi}$ among the admissible fields.

State and prove the corresponding statical bound theorem. If the elastic domain K is defined by a convex loading function f, which is C^1 on ∂K, using Lagrangean multiplier technique, prove that Euler-Lagrange variational equations of this problem are normalization condition (6.2) and the admissibility conditions of exercise 1.3.

References

Chen, H. and Ponter, A.R.S. (2000). A method for the evaluation of a ratchet limit and the amplitude of plastic strain for bodies subjected to cyclic loading. *European Journal of Mechanics A/Solids*, accepted for publication.

de Saxcé, G. (1986). *Sur Quelques Problèmes de Mécanique des Solides Considérés Comme Matériaux à Potentiels Convexes*, Doctor thesis, Université de Liège, Collection des Publications de la Faculté des Sciences Appliquées, 118.

de Saxcé, G. (1995). A variational deduction of upper and lower bound shakedown theorems by Markov's and Hill's principles over a cycle. In Mröz, Z., et al., eds., *Inelastic Behavior of Structures Under Variable Loads*, Dordrecht :Kluwer Academic Publishers.

Ekeland, I. and Temam, R. (1975). *Convex analysis and variational problems*, NY :North Holland.

Hill, R. (1948). A variational principle of maximum plastic work in classical plasticity. *Quarterly Journal of Mechanics and Applied Mathematics* 1:18.

Koiter, W.T. (1960). General theorems for elasto-plastic solids. In *Progress in Solid Mechanics*, Volume 1.

Mandel, J. (1966). *Cours de Mécanique des Milieux Continus, Tome 2 : Mécanique des Solides.* Paris :Gauthier-Villars.

Markov, A.A. (1947). On variational principles in theory of plasticity. *Prek. Math. Mech.* 11:339.

Martin, J.B. (1975). *Plasticity, fundamentals and general results*. MA :MIT Press.

Ponter, A.R.S. and Chen, H. (2000). A minimum theorem for cyclic load in excess of shakedown with application to the evaluation of a ratchet limit. *European Journal of Mechanics A/Solids*, accepted for publication.

Save, M., de Saxcé, G. and Borkowski, A. (1991). Computation of shake-down loads, feasibility study, Commission of the European Communities, Nuclear Science and Technology, report EUR 13618.

Save, M., Massonnet, C. and de Saxcé, G. (1997). *Plastic Limit Analysis of Plates, Shells and Disks*, NY: North Holland.

Nature of the Solutions

Géry de Saxcé

University of Lille, Lille, France

Abstract. This lecture is devoted to the determination of the regularity of the solutions of the shakedown variational problems and to the underlying topology of the solution spaces. It is showed that the kinematical solutions are not functions but bounded measures. A particular attention is paid to the problem of the thick wall tube subjected to an internal variable repeated pressure. It is shown how, when the load factor tends to α^a by upper values, the plastic strain rate in the tube converges to Dirac's measure.

1 Infinite Plate with Circular Hole

Let us consider the shakedown analysis of an infinite plate of unit thickness in plane stress state, perforated by a circular hole of radius a, and subjected far from the hole to an uniform hydrostatic stress : $\sigma_r = \sigma_\theta = q(t) = \alpha\, q^0(t)$, varying in the range :

$$-\underline{q}^0 \leq q^0(t) \leq 1 .$$

The hole is unloaded. The elastic behavior is isotropic. The elastic solution is well-known and given by :

$$\sigma_\theta^{E0}(r,t) = q^0(t)\big(1+\varphi(r)\big), \qquad \sigma_r^{E0}(r,t) = q^0(t)\big(1-\varphi(r)\big), \tag{1.1}$$

where we put for convenience : $\varphi(r) = (a/r)^2$. The plasticity is governed by Tresca model. Expressions (1.1) suggest to reduce the plastic yielding condition to :

$$-\sigma_Y \leq \sigma_\theta \leq \sigma_Y .$$

The corresponding dissipation power function is equal to :

$$D(\dot{\varepsilon}^p) = \sigma_Y \left|\dot{\varepsilon}_\theta^p\right| .$$

The normality rule entails that : $\dot{\varepsilon}_r^p = 0$. If $r \mapsto \Delta u_r(r)$ denotes the field of radial displacement increment, the compatibility equations are :

$$\Delta\varepsilon_r^p = \frac{d}{dr}(\Delta u_r), \qquad \Delta\varepsilon_\theta^p = \frac{\Delta u_r}{r} .$$

For this application, solving the statical bound problem (6.3) of lecture 2 gives the following value of the shakedown load factor :

$$\alpha^a = \frac{\sigma_Y}{1+\underline{q}^0}\,.$$

On the other hand, the kinematical bound problem (6.1) of lecture 2 is :

$$inf\left\{\alpha = \int_\Omega \oint \sigma_Y |\dot\varepsilon_\theta^p|\, dt\, d\Omega \,/ \int_\Omega (1+\varphi(r)) \oint q^0 \dot\varepsilon_\theta^p\, dt\, d\Omega \quad ; \quad (\dot\varepsilon_r^p, \dot\varepsilon_\theta^p) \in \mathrm{E}_\Omega, \quad \Delta u_r \in \mathrm{U}_\Omega, \right.$$
$$\left. \forall r > a, \quad \frac{d}{dr}(\Delta u_r) = 0, \quad \frac{\Delta u_r}{r} = \oint \dot\varepsilon_\theta^p\, dt, \quad -\underline{q}^0 \le q^0(t) \le 1 \right\}$$

Thus, Δu_r is uniform. If the loading would be proportional, the limit analysis shows that the displacement rate is uniform : $\dot u_r(r) = \dot u_0$, and we obtain : $\dot\varepsilon_\theta(r) = \dot u_0 / r$. The plastic yielding spreads within the whole plate. On the other hand, if the loading is cyclic, the compatibility conditions are not needed to be fulfilled at any time by the plastic strain rate $\dot\varepsilon^p$, but only by the increment $\Delta\varepsilon^p$ over the cycle. As we shall see, the plastic yielding can be concentrated in a very localized region.

Let us assume the collapse occurs by alternating plasticity :

$$\Delta\varepsilon_\theta^p = 0\,. \tag{1.2}$$

Thus, Δu_r vanishes. Let us consider a solution of the following form :

$$\dot\varepsilon_\theta^p(r,t) = \dot\varepsilon_0^p(t) f(r), f(r) \ge 0\,.$$

The kinematical bound problem becomes :

$$inf\left\{\sigma_Y \oint |\dot\varepsilon_0^p|\, dt \int_\Omega f(r)\, d\Omega / \oint q^0\, \dot\varepsilon_0^p\, dt \int_\Omega (1+\varphi(r)) f(r)\, d\Omega ; \right. \tag{1.3}$$
$$\left. f \ge 0, \dot\varepsilon_0^p, \ -\underline{q}^0 \le q^0(t) \le 1 \right\}$$

Using the decomposition in positive and negative part gives :

$$\dot\varepsilon_0^p = \dot\varepsilon_{0+}^p - \dot\varepsilon_{0-}^p, \qquad |\dot\varepsilon_0^p| = \dot\varepsilon_{0+}^p + \dot\varepsilon_{0-}^p, \qquad \dot\varepsilon_{0+}^p \ge 0, \qquad \dot\varepsilon_{0-}^p \ge 0\,.$$

In order to satisfy (1.2), one has :

$$\oint \dot\varepsilon_{0+}^p\, dt = \oint \dot\varepsilon_{0-}^p\, dt\,. \tag{1.4}$$

Besides, it holds :

$$\oint q^0\,\dot{\varepsilon}^p_{0+}\,dt \le \oint \dot{\varepsilon}^p_{0+}\,dt\,, \qquad -\oint q^0\,\dot{\varepsilon}^p_{0-}\,dt \le \underline{q}^0 \oint \dot{\varepsilon}^p_{0-}\,dt\,.$$

Hence taking account of (1.4) leads to the relations :

$$(1+\underline{q}^0)\oint \dot{\varepsilon}^p_{0+}\,dt = \oint (\dot{\varepsilon}^p_{0+} + q^0\,\dot{\varepsilon}^p_{0-})\,dt \ge \oint q^0\,\dot{\varepsilon}^p_0\,dt\,, \qquad \oint \left|\dot{\varepsilon}^p_0\right| dt = 2\oint \dot{\varepsilon}^p_{0+}\,dt\,,$$

from which ones we deduce :

$$\oint \left|\dot{\varepsilon}^p_0\right| dt \ge \frac{2}{1+\underline{q}^0}\oint q^0\,\dot{\varepsilon}^p_0\,dt\,. \tag{1.5}$$

The minimum is reached in (1.3) when the equality is reached in (1.5). For an optimal solution, the plastic yielding occurs

1. with a tensile hoop strain when $q^0(t)=1$,
2. and a compressive one when $q^0(t) = -\underline{q}^0$,
3. the behavior remains elastic when $-\underline{q}^0 < q^0(t) < 1$.

Next, our aim is to know how the plastic yielding is distributed by determining *f*. The problem is to minimize the functional representing the kinematical load factor :

$$\alpha(f) = \frac{2\sigma_Y}{1+\underline{q}^0}\,\frac{\int_\Omega f(r)\,d\Omega}{\int_\Omega (1+\varphi(r))\,f(r)\,d\Omega}\,. \tag{1.6}$$

According to the mean value theorem, one has :

$$\int_\Omega f(r)\varphi(r)\,d\Omega = \left|\int_\Omega f(r)\varphi(r)\,d\Omega\right| \le sup\left\{\left|\varphi(r)\right|; r>a\right\}\left|\int_\Omega f(r)\,d\Omega\right| = \int_\Omega f(r)\,d\Omega\,. \tag{1.7}$$

Finally, one has :

$$\alpha(f) \ge \frac{\sigma_Y}{1+\underline{q}^0}\,. \tag{1.8}$$

Let (f_i) be a convergent minimizing sequence of α in a suitable topological vector space. The limit realizes the minimum :

$$\alpha^a = \inf\{\alpha(f_i); i = 1,2,\ldots\} = \frac{\sigma_Y}{1+\underline{q}^0}$$

representing the shakedown load factor. We have now to determine the nature of the solution f and the underlying topology of the solution space. As it will be seen in the next, the solutions are not functions but measures.

2 Bounded Measures

Banach Spaces. Let E be a real vector space. A norm is a mapping $\varphi \mapsto \|\varphi\| \in \boldsymbol{R}$ defined in E, such that :

1. $\varphi \neq 0 \Rightarrow \|\varphi\| > 0$ and $\|0\| = 0$,
2. $\forall \lambda \in \boldsymbol{R},\ \|\lambda \varphi\| = |\lambda| \|\varphi\|$,
3. $\|\varphi + \varphi'\| \leq \|\varphi\| + \|\varphi'\|$.

Any normed space is naturally equipped with a distance : $d(\varphi, \varphi') = \|\varphi - \varphi'\|$, and the associated topology of metric space. A useful tool of topology is the concept of complete metric space, for which one any Cauchy sequence is convergent. A complete normed space is called a Banach space.

Example 1. Let Ω be an open subset of $\boldsymbol{R}^n$ and K be a compact subset of Ω. The vector space $D_K^0(\Omega)$ of the real valued continuous functions on Ω, vanishing in the exterior of K, is equipped with the norm :

$$\|\varphi\| = sup\{ |\varphi(x)|; x \in \Omega \}. \tag{2.1}$$

Hence a sequence (φ_i) converges to 0 in $D_K^0(\Omega)$ if it uniformly converges to 0. We say that $D_K^0(\Omega)$ is equipped with the uniform convergence topology. Classical theorems about the uniform convergence show that $D_K^0(\Omega)$ is a Banach space.

Example 2. $\boldsymbol{R}$ is naturally a normed vector space. The (topological) dual space E^* of a normed space E is equipped with the (dual) norm defined for any $f \in E^*$:

$$\|f\|_* = sup\{|f.\varphi| ; \varphi \in E, \ \|\varphi\| \le 1\}. \tag{2.2}$$

such that :

$$|f.\varphi| \le \|f\|_* \ \|\varphi\|.$$

As $\boldsymbol{R}$ is a Banach space, it can be proved that E^* is a Banach space too (even if E is not a Banach space).

Measures. Let $\mathrm{D}^0(\Omega)$ be the space of the continuous functions on Ω, vanishing in the exterior of a compact subset of Ω. $\mathrm{D}^0(\Omega)$ is the union of all $\mathrm{D}_K^0(\Omega)$ when K varies. $\mathrm{D}^0(\Omega)$ is a normed space but not a Banach space. As the suitable topology on $\mathrm{D}^0(\Omega)$ offers some complications[1], we just say that a measure is a linear form $\varphi \mapsto d\mu.\varphi \in \boldsymbol{R}$, defined in $\mathrm{D}^0(\Omega)$, which is continuous for the uniform convergence topology. In other words, for any compact $K \subset \Omega$, there exists $a_K \ge 0$ (depending on K) such that :

$$\forall \varphi \in \mathrm{D}_K^0(\Omega), \quad |d\mu.\varphi| \le a_K \|\varphi\|. \tag{2.3}$$

The value $d\mu.\varphi$ of $d\mu$ for φ is called the integral of φ and is often denoted $\int \varphi \, d\mu$.

If a sequence $(d\mu_i)$ of measures is such that for any $\varphi \in \mathrm{D}^0(\Omega)$, the sequence $(d\mu_i.\varphi)$ has a limit $d\mu.\varphi$, it can be proved that $d\mu$ is a measure. This property is generally easy to verify and allows to define new measures from known ones.

Regular Measures. It is possible to extend the notion of integral to a wider class of functions, namely to the set of lower semi-continuous functions $x \mapsto f(x) \in \overline{\boldsymbol{R}}$ defined in Ω, minorized by a function of $\mathrm{D}^0(\Omega)$, by defining :

$$\int f \, d\mu = sup\left\{\int \varphi \, d\mu \quad ; \quad \varphi \in \mathrm{D}^0(\Omega), \quad \varphi \le f \right\} \in \overline{\boldsymbol{R}}.$$

[1] The topology of $\mathrm{D}^0(\Omega)$ is defined by inductive limit of the topologies of the $\mathrm{D}_K^0(\Omega)$. Generally, $\mathrm{D}^0(\Omega)$ is not metrizable.

If $\int f\, d\mu = -\int (-f)\, d\mu$ and is finite, we say that f is integrable (with respect to $d\mu$). Next, let f be a locally integrable function (i.e. integrable on all compact subset of Ω). Hence, if $\varphi \in D^0(\Omega)$ vanishes outside of a compact K, one has :

$$\left| \int f \varphi\, d\mu \right| \leq \int_K |f|\, d\mu . \|\varphi\|,$$

that, according to (2.2), proves the mapping $\varphi \mapsto \int f \varphi\, d\mu$ is continuous for the uniform convergence topology and, consequently, is a measure. This provides a correspondence between locally integrable functions and measures. This kind of measure is said to be regular. In the sequel, a regular measure will be identified to the corresponding function.

Singular Measures. There exist measures which are not regular. If x_0 belongs to Ω, let δ_{x_0} be a linear form on $D^0(\Omega)$, defined by :

$$\delta_{x_0} . \varphi = \varphi(x_0) .$$

It is continuous because :

$$\left| \delta_{x_0} . \varphi \right| = \left| \varphi(x_0) \right| \leq \|\varphi\| \tag{2.4}$$

Thus, it is a measure called Dirac's measure. It is the simplest example of non regular measure (see Exercise 3.2). With abusive notations, it is sometimes written :

$$\delta_{x_0} . \varphi = \int \delta_{x_0}(x) \varphi(x)\, d\Omega .$$

It can be remarked that measures generalize functions or, more precisely, locally integrable functions. The measures are particular cases of the so-called distributions of which the rigorous theory is due to Schwartz (1967).

Bounded Measures. According to (2.2), for any measure, we define :

$$\|d\mu\|_* = sup \left\{ |d\mu . \varphi| \quad ; \quad \varphi \in D^0(\Omega), \quad \|\varphi\| \leq 1 \right\}. \tag{2.5}$$

Unfortunately, this expression is not a norm on the vector space $M(\Omega)$ of the measures because its value can eventually be infinite. This suggests to consider the subset $M^1(\Omega)$ of the so-called bounded measures $d\mu$ such that $\|d\mu\|_*$ is finite. It is a vector space equipped with the norm (2.5) such that :

$$|d\mu.\varphi| \le \|d\mu\|_* \|\varphi\|. \tag{2.6}$$

As dual of the normed space $D^0(\Omega)$, the space $M^1(\Omega)$ is a Banach space.
It is worthwhile to say that, according to (2.4) and (2.5), Dirac's measure is bounded :

$$\|\delta_{x_0}\|_* = 1. \tag{2.7}$$

For our purpose, the relevant point to quote is that if the absolute value of a given function φ reaches its maximum value at x_0, one has :

$$|\delta_{x_0}.\varphi| = |\varphi(x_0)| = sup\{|\varphi(x)|; x \in \Omega\} = \|\varphi\|.$$

Owing to (2.7), Dirac's measure is the bounded measure that allows to reach the equality in (2.6) :

$$|\varphi(x_0)| = sup\{|\varphi(x)|; x \in \Omega\} \quad \Rightarrow \quad |\delta_{x_0}.\varphi| = \|\delta_{x_0}\|_* \|\varphi\|. \tag{2.8}$$

Besides, the space $M^1(\Omega)$ can be also considered as the dual of the space of the continuous function φ which converge to zero at the infinity in the sense that for all $\varepsilon > 0$, there exists a compact $K \subset \Omega$ such that :

$$\forall x \in \Omega - K, \ |\varphi(x)| \le \varepsilon.$$

Application to Shakedown Theory. The previous developments shed a light on inequality (1.7) used in solving the problem of the infinite plate with circular hole. Let us remark that $\varphi(r) = (a/r)^2$ is a continuous function which converges to zero at the infinity and that the norm (2.1) appears in (1.7). This suggests to consider that f is not a function but a bounded measure. Thus, inequality (1.7) is recast as :

$$|f.\varphi| \le \|f\|_* \|\varphi\|. \tag{2.9}$$

In order to minimize the kinematical load factor (1.6), we must reach the equality in (2.9). As $\varphi(r) = (a/r)^2$ reaches its maximum value at $r = a$, we choose :

$$f = \delta_a .$$

More generally, we claim that the plastic strain rate fields $\dot{\varepsilon}^p$ belong to a Banach space of tensor valued bounded measures[2]. For more details about the use of bounded measures in plasticity, the reader is referred to Moreau (1976), Suquet (1978), Temam and Strang (1980). Concerning the functional analysis aspects of the shakedown problems, the pioneering works are due to Débordes and Nayroles (1976), and Débordes (1977).

3 The Thick Wall Tube : Statical Solutions

Statical Bound Problem. Let us consider the shakedown analysis of a thick wall tube with internal radius a and external one b, which is in plane strain state under the influence of an internal pressure $q(t) = \alpha q^0(t)$, the reference pressure varying within the range : $0 \leq q(t) \leq 1$. The elastic solution is well-known and given by :

$$\sigma_r^{E0}(r,t) = q^0(t)\frac{a^2}{b^2-a^2}\left(1-\left(\frac{b}{r}\right)^2\right), \qquad \sigma_\theta^{E0}(r,t) = q^0(t)\frac{a^2}{b^2-a^2}\left(1+\left(\frac{b}{r}\right)^2\right).$$

Eliminating the radius r between the two previous relations, it holds :

$$\sigma_r^{E0} + \sigma_\theta^{E0} = 2\frac{b^2-a^2}{a^2}q^0 .$$

With Tresca's model, this suggests to consider only the plastic yielding condition :

$$-\sigma_Y \leq \sigma_\theta - \sigma_r \leq \sigma_Y . \tag{3.1}$$

As remarked by Koiter (1953), this suppose in plane strain state that the axial stress is the intermediate principal stress. This condition is satisfied for usual materials and applications. For a detailed discussion of this point, the reader is referred to de Saxcé (1986). Under the previous assumptions, the statical bound problem (6.3) of lecture 2 is :

$$\begin{aligned} sup\{\alpha \ ; \quad & \alpha \in \boldsymbol{R}, \quad \bar{\rho} \in S_\Omega, \quad such \quad that \quad \bar{\rho}_r(a) = \bar{\rho}_r(b) = 0, \quad \forall r \in \,]a,b[, \\ r\frac{d\bar{\rho}_r}{dr} &= \bar{\rho}_\theta - \bar{\rho}_r, \quad -\sigma_Y \leq 2\alpha q_0(t) a^2 b^2 / r^2 (b^2-a^2) + \bar{\rho}_\theta - \bar{\rho}_r \leq \sigma_Y, \quad \underline{q_0} \leq q(t) \leq 1\} \end{aligned}$$

[2] Incidentally, this choice of the functional space determines the nature of the norm used in the proof of Koiter's lower bound theorem (lecture 2, section 4).

Combining the internal equilibrium equation and the plastic yielding condition gives :

$$-\sigma_Y \le 2\alpha q^0(t)\frac{a^2b^2}{(b^2-a^2)r^2}+r\frac{d\bar{\rho}_r}{dr}\le \sigma_Y.$$

For the given load domain, the previous inequalities are fulfilled, provided that :

$$-\sigma_Y \le 2\alpha\frac{a^2b^2}{(b^2-a^2)r^2}+r\frac{d\bar{\rho}_r}{dr}\le \sigma_Y, \qquad -\sigma_Y \le r\frac{d\bar{\rho}_r}{dr}\le \sigma_Y. \tag{3.2}$$

Considering the superior and inferior envelopes of the elastic solution, we retain only the stronger inequalities :

$$-\sigma_Y \le r\frac{d\bar{\rho}_r}{dr}\le \sigma_Y - 2\alpha\frac{a^2b^2}{(b^2-a^2)r^2}. \tag{3.3}$$

Limit Analysis Solution. When the loading is proportional, the collapse occurs when the pressure reach its maximum value, that is equivalent to assume that the equality is realized with the superior envelop :

$$r\frac{d\bar{\rho}_r}{dr}= \sigma_Y - 2\alpha\frac{a^2b^2}{(b^2-a^2)r^2}. \tag{3.4}$$

Taking account of the first boundary condition, the integration gives :

$$\bar{\rho}_r(r)=\sigma_Y\, ln(r/a)+\frac{\alpha b^2}{b^2-a^2}\left(\frac{a^2}{r^2}-1\right).$$

The plastic yielding occupying the whole tube, the last boundary condition gives :

$$\alpha=\alpha^l=\sigma_Y\, ln(b/a). \tag{3.5}$$

Next, we deduce the residual stress field :

$$\bar{\rho}_r(r)=\sigma_Y\left(ln(r/a)-\frac{b^2 ln(b/a)}{b^2-a^2}\left(1-\frac{a^2}{r^2}\right)\right),$$

$$\bar{\rho}_\theta(r)=\sigma_Y\left(1+ln(r/a)-\frac{b^2 ln(b/a)}{b^2-a^2}\left(1+\frac{a^2}{r^2}\right)\right).$$

This solution exists provided the inferior inequality is satisfied in inequation (3.3). Taking account of equations (3.4) and (3.5), this is true if we have :

$$\frac{b^2}{r^2} ln(b/a) \leq \frac{b^2 - a^2}{a^2} .$$

This condition is fulfilled for all value of r between a and b if one has :

$$ln(b/a) \leq \frac{b^2 - a^2}{b^2} . \tag{3.6}$$

Otherwise, one has to find another solution.

Alternating Plasticity Solution (Hodge, 1954). Owing to inequation (3.3), the statical bound problem has a solution only if one has :

$$-\sigma_Y \leq \sigma_Y - \frac{2\alpha b^2}{(b^2 - a^2)}$$

Thus, the solution is obtained by considering the equality in the previous condition :

$$\alpha = \alpha^f = \sigma_Y \frac{b^2 - a^2}{b^2} \tag{3.7}$$

Next, we have to show the existence of a residual stress field associated to this value of the load factor. For this purpose, we choose it such that, before the collapse, the plastic flow occurred only in the internal ring $a \leq r \leq c$, the external part $c \leq r \leq b$ of the tube remaining elastic. Combining $(3.2)_1$ and (3.7), we find by integration :

$$a \leq r \leq c, \qquad \bar{\rho}_r(r) = \sigma_Y \left(ln(r/a) + \frac{a^2}{r^2} - 1 \right), \qquad \bar{\rho}_\theta(r) = \sigma_Y \left(ln(r/a) - \frac{a^2}{r^2} \right).$$

In the external part, we use the elastic solution :

$$c \leq r \leq b, \qquad \bar{\rho}_r(r) = \rho_0 \left(1 - \frac{b^2}{r^2} \right), \qquad \bar{\rho}_\theta(r) = \rho_0 \left(1 + \frac{b^2}{r^2} \right).$$

The continuity of $\bar{\rho}_r$ at $r = c$ is ensured provided that :

$$\rho_0 = \sigma_Y F(\gamma)/(1 - \beta^2/\gamma^2),$$

introducing the ratios : $\beta = b/a$, $\gamma = c/a \le \beta$, and the function :

$$F(\xi) = ln\xi + 1/\xi^2 - 1.$$

On the other hand, taking account of relations (3.3) and (3.7), the solution is subjected to the condition :

$$-\sigma_Y \le \bar{\rho}_\theta - \bar{\rho}_r \le \sigma_Y\left(1 - 2\frac{a^2}{r^2}\right), \tag{3.8}$$

which is obviously satisfied in the internal plastic ring $a \le r \le c$. Next, it can be shown that the alternating plasticity solution exists and satisfies inequation (3.8) only if one has :

$$ln(b/a) > \frac{b^2 - a^2}{b^2}. \tag{3.9}$$

This discussion is rather technical and is proposed in exercise 3.3. . Finally, it results from inequalities (3.6) and (3.9) that the limit analysis and alternating plasticity solutions are exclusive one of the other and we observe that :

$$\alpha^a = \inf\{\alpha^l, \alpha^f\}.$$

4 The Thick Wall Tube : Kinematical Solutions

Kinematical Bound Problem. The dissipation function corresponding to plastic yielding condition (3.1) is :

$$D(\dot{\varepsilon}^p) = \sigma_Y |\dot{\varepsilon}^p_\theta|.$$

The normality rule entails that : $\dot{\varepsilon}^p_r = -\dot{\varepsilon}^p_\theta$. The compatibility conditions are the same as in section 1. The kinematical bound problem (6.1) of lecture 2 is :

$$inf\left\{\alpha = \frac{b^2-a^2}{2a^2b^2}\frac{\int_a^b \oint \sigma_Y|\dot{\varepsilon}^p_\theta|\,dt\,r\,dr}{\int_a^b \oint q^0\,\dot{\varepsilon}^p_\theta\,dt\frac{dr}{r}};\dot{\varepsilon}^p_\theta,\ a<r<b,\ -\frac{d}{dr}(\Delta u_r) = \frac{\Delta u_r}{r} = \oint \dot{\varepsilon}^p_\theta\,dt,\ 0 \le q^0(t) \le 1\right\}$$

Integrating the internal compatibility equation gives :

$$\oint \dot{\varepsilon}_\theta^p \, dt = \Delta \varepsilon_0^p \, b^2 / r^2 .$$

Limit Analysis Solution. It is assumed that the collapse results from an incipient unrestrained flow mechanism :

$$\dot{\varepsilon}_\theta^p (r,t) = \dot{\varepsilon}_0^p (t) \, b^2 / r^2 ,$$

such that :

$$\oint \dot{\varepsilon}_0^p \, dt = \Delta \varepsilon_0^p .$$

Thus, the load factor becomes :

$$\alpha = \sigma_Y \ln(b/a) \oint \left| \dot{\varepsilon}_0^p \right| dt / \oint q^0 \dot{\varepsilon}_0^p \, dt .$$

On the other hand, for the considered load domain, we obtain the inequality

$$\oint q^0 \dot{\varepsilon}_0^p \, dt = \oint q^0 \dot{\varepsilon}_{0+}^p \, dt - \oint q^0 \dot{\varepsilon}_{0-}^p \, dt \leq \oint \dot{\varepsilon}_{0+}^p \, dt \leq \oint \left| \dot{\varepsilon}_0^p \right| dt , \tag{4.1}$$

from which one we deduce :

$$\alpha \geq \sigma_Y \ln(b/a) .$$

The minimum is reached when the equality is satisfied in condition (4.1). For an optimal solution, the plastic yielding occurs with a tensile hoop stress when $q^0(t) = 1$, and otherwise the behavior is elastic. Finally, the load factor is :

$$\alpha = \alpha^l = \sigma_Y \ln(b/a) . \tag{4.2}$$

Alternating Plasticity Solution. Following the considerations of section 1 and 2, we choose the following solution :

$$\dot{\varepsilon}_\theta^p (r,t) = \dot{\varepsilon}_0^p (t) \, \delta_c (r) ,$$

where δ_c denotes Dirac's measure at point $r = c$, and $\dot{\varepsilon}_0^p$ is such that :

$$\oint \dot{\varepsilon}_0^p \, dt = 0 ,$$

that entails :

$$\oint \dot{\varepsilon}_{0-}^{p}\, dt = \oint \dot{\varepsilon}_{0+}^{p}\, dt$$

For the considering load domain, it holds :

$$2\oint q^0 \dot{\varepsilon}_0^p\, dt = 2\oint q^0 \dot{\varepsilon}_{0+}^p\, dt - 2\oint q^0 \dot{\varepsilon}_{0-}^p\, dt \le 2\oint \dot{\varepsilon}_{0+}^p\, dt = \oint \left|\dot{\varepsilon}_0^p\right| dt . \tag{4.3}$$

Finally, one has :

$$\alpha = \sigma_Y \frac{b^2 - a^2}{2a^2 b^2} \frac{\int_a^b \delta_c(r) r\, dr}{\int_a^b \delta_c(r) \frac{dr}{r}} \frac{\oint \left|\dot{\varepsilon}_\theta^p\right| dt}{\oint q^0 \dot{\varepsilon}_\theta^p\, dt} \ge \sigma_Y \frac{b^2 - a^2}{b^2} \frac{c^2}{a^2} .$$

The minimum value :

$$\alpha = \alpha^f = \sigma_Y \frac{b^2 - a^2}{b^2}, \tag{4.4}$$

is obtained when $c = a$, and when the equality is reached in condition (4.3), that is the plastic yielding occurs with a tensile hoop stress when $q^0(t) = 1$, and a compressive one when $q^0(t) = 0$, the behavior remaining elastic when $0 < q^0(t) < 1$. Taking account of (4.2), (4.4), and Koiter's lower bound theorem (lecture 2, section 4), we obtain :

$$\alpha^a = \inf\left\{ \alpha^l, \alpha^f \right\}.$$

It is worthwhile to remark that both statical and kinematical bound methods lead to the same value of the shakedown load factor.

5 The Thick Wall Tube : Collapse by Alternating Plasticity

When condition (3.9) is satisfied, the collapse occurs by alternating plasticity. A careful analysis of the statical approach for the evolution of the thick wall tube shows that the tube shakes down even if α is exactly equal to α^a. As remarked by Martin (1975), the collapse actually occurs by alternating plasticity for $\alpha > \alpha^a$. In the sequel, we shall prove that when α tends to α^a by upper values, the plastic strain rate field in the tube converges to Dirac's measure.

For this purpose, we analyze the actual collapse by alternating plasticity. For α such that $\alpha^a = \alpha^f < \alpha < \alpha^l$, the residual stress field $\rho(r,t)$ cannot be stabilized. When the pressure is large enough, the plastic yielding occurs in a small internal ring $a < r \le c$ while, in the external region $c < r < b$, the behavior remains elastic. For an isotropic material of elastic modulus E and Poisson's ration ν, the solving of the corresponding rate problem gives :

$$\dot{\varepsilon}_\theta(r,t) = \frac{2}{E}\dot{q}(t)\left(\frac{b^2}{b^2-c^2} - \nu^2\right)\frac{c^2}{r^2}, \tag{5.1}$$

within the ring $a < r \le c$ (Exercise 3.4). As when establishing (3.3) and owing to (3.7), one has to satisfy :

$$-\sigma_Y \le \sigma_Y\left(1 - 2\frac{\alpha}{\alpha^a}q^0\frac{a^2}{r^2}\right).$$

The equality being realized at $r = c$, one has :

$$c(t) = a\sqrt{\frac{\alpha}{\alpha^a}q^0(t)}. \tag{5.2}$$

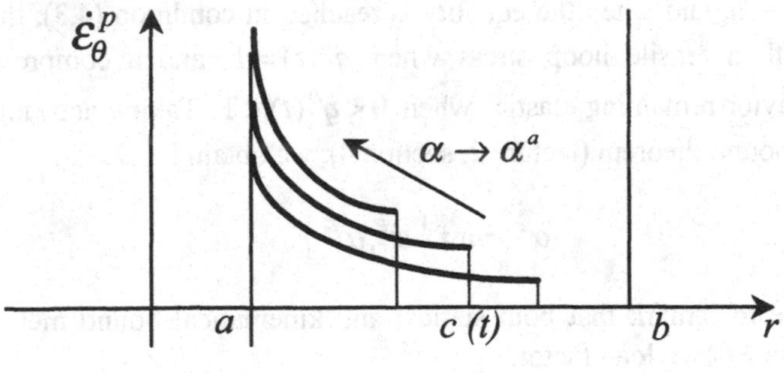

Figure 1. The thick wall tube: collapse by alternating plasticity.

The plastic yielding occurs if $c(t) > a$, hence if $q^0(t) > \alpha^a / \alpha$. Finally, the plastic strain rate is equal to :

$$\dot{\varepsilon}_\theta(r,t) = \frac{2}{E}\dot{q}(t)\left(\frac{b^2}{b^2 - a^2 q^0(t)\alpha/\alpha^a} - \nu^2\right)\frac{\alpha}{\alpha^a}q^0(t)\frac{a^2}{r^2},$$

if $r < a\sqrt{q^0(t)\alpha/\alpha^a}$, and $\alpha^a/\alpha < q^0(t) \le 1$, and equal to zero otherwise (Figure 1). Thus, we define :

$$\dot{\eta}^p(r,t) = \dot{\varepsilon}_\theta^p(r,t)/I(t),$$

where :

$$I(t) = \int_\Omega \dot{\varepsilon}_\theta^p(r,t)\,d\Omega = \frac{4\pi}{E}\dot{q}\,c^2\left(\frac{b^2}{b^2-c^2} - v^2\right)ln(c/a).$$

Thus, $r \mapsto \dot{\eta}^p(r,t)$ is a piecewise continuous function, integrable on Ω, vanishing outside the plastic ring $a < r \le c(t)$, and such that :

$$\int_\Omega \dot{\eta}^p(r,t)\,d\Omega = 1.$$

Let us consider a time for which the plastic yielding occurs when $\dot{q} > 0$. As $E > 0$, $v < 1/2$, and $b > c$, $\dot{\varepsilon}_\theta^p$ and $\dot{\eta}^p$ are non negative. By the mean value theorem, we see that for any $\varphi \in D^0(\Omega)$, one has :

$$inf\{\varphi(x)\,;a \le r \le c\} \le \int_\Omega \dot{\eta}^p.\varphi\,d\Omega \le sup\{\varphi(x)\,;a \le r \le c\}.$$

Thus, let us consider a sequence (α_i) of load factors, which converge to α^a by upper values, and the corresponding sequence of functions $(\dot{\eta}_i^p)$. Accounting for (5.2), the external radius *c* of the plastic ring converges to *a*. The function φ being *continuous*, it holds :

$$lim_{i\to+\infty} \int_\Omega \dot{\eta}_i^p.\varphi\,d\Omega = \varphi(a).$$

Hence the measures associated to the integrable functions $\dot{\eta}_i^p$ converge to Dirac's measure at $r = a$. Generally speaking, singular measures are typically associated to alternating plasticity collapses. On the other hand, the limit analysis and incremental collapses are represented by regular measures. The reason is that, as the plastic strain increment does not vanish, the solution must be regular enough to ensure that the displacement increment is K.A. . Such a circumstance does not occur for the event of alternating plasticity and, consequently, the solution space is wider than for the other collapses.

6 Exercises

Exercise 3.1. For the problem of the infinite plate with circular hole presented in section 1, find the solution of the statical bound theorem :

$$sup\{\alpha \quad ; \quad \alpha \in \boldsymbol{R}, \quad \overline{\rho} \in \mathrm{S}_{\Omega}, \quad such \quad that \quad \overline{\rho}_r(a) = 0, \quad \forall r > a, \quad r\frac{d\overline{\rho}_r}{dr} = \overline{\rho}_\theta - \overline{\rho}_r,$$
$$-\sigma_Y \leq \overline{\rho}_\theta + \alpha q_0(1+\varphi(r)) \leq \sigma_Y, \quad \underline{q_0} \leq q(t) \leq 1\}$$

Exercise 3.2. Show that Dirac's measure cannot be represented by a locally integrable function.

Hint : consider the functions $\varphi_\varepsilon \in \mathrm{D}^0(\boldsymbol{R})$ equal to :

$$\varphi_\varepsilon(x) = \exp(-\varepsilon^2/(\varepsilon^2 - x^2)),$$

when $|x| \leq \varepsilon$, and equal to zero otherwise. Show that for any locally integrable function *f*, the value $f.\varphi_\varepsilon$ of the associated measure tends to zero when ε tends to zero, while $\delta_0 .\varphi_\varepsilon = 1/e$.

Exercise 3.3. For the thick wall tube problem, prove that the alternating plasticity statical solution exists only if the condition (3.9) is satisfied.

Hint : Under condition (3.9), prove that ρ_0 is non negative and that the plastic yielding condition (3.8) is verified anywhere in the tube.

Exercise 3.4. Prove (5.1) by solving the rate boundary problem of a thick wall tube subjected to an internal pressure rate $\dot{q}$.

Hint : using the plane strain condition, the plastic yielding condition, the internal compatibility conditions and Hooke's law, prove that, in the internal ring, prove that :

$$2\dot{\varepsilon}_\theta^p = \frac{\dot{u}_r}{r} - \frac{d\dot{u}_r}{dr}, \qquad \frac{(1+\nu)(1-2\nu)}{E}(\dot{\sigma}_r + \dot{\sigma}_\theta) = \frac{\dot{u}_r}{r} + \frac{d\dot{u}_r}{dr}.$$

Besides, using the equilibrium equations and integrating, prove that, for $r \in]a, c[$, one has :

$$\dot{\varepsilon}_\theta^p = \left(\frac{\dot{u}_r(c)}{c} + \frac{(1+\nu)(1-2\nu)}{E}\dot{q}\right)\frac{c^2}{r^2}.$$

Considering the elastic solution in the external region and using the continuity conditions at $r = c$, prove (5.1).

References

Débordes, O. and Nayroles, B. (1976). Sur la théorie et le calcul à l'adaptation plastique des structures élasto-plastiques. *Journal de Mécanique*, 15:1–53.

Débordes, O. (1977). *Contribution à la Théorie et au Calcul de l'Elasto-Plasticité Asymptotique*. Doctor thesis, Université de Provence, Aix-Marseille.

de Saxcé, G. (1986). *Sur Quelques Problèmes de Mécanique des Solides Considérés Comme Matériaux à Potentiels Convexes*, Doctor thesis, Université de Liège, Collection des Publications de la Faculté des Sciences Appliquées, 118.

de Saxcé, G. (1995). A variational deduction of upper and lower bound shakedown theorems by Markov's and Hill's principles over a cycle. In Mröz, Z., et al., eds., *Inelastic Behavior of Structures Under Variable Loads*, Dordrecht: Kluwer Academic Publishers, 153–167.

Hodge, P.G. (1954). Shake-down of elasto-plastic structures. In *Residual Stresses in Metals and Metal Structures*, NY :Reinhold Pub. .

Koiter, W.T. (1953). On partially plastic thick-wall tubes. In *Anniversary Volume on Applied Mechanics Dedicated to Biezeno*, Haarlem :N.V. De Technische Uitgeverij H. Stam.

Martin, J.B. (1975). *Plasticity, fundamentals and general results*. MA :MIT Press.

Moreau, J.J. (1976). Application of convex analysis to the treatment of elasto-plastic systems. In Germain, P., et al., eds., *Lecture Notes in Mathematics*, 503, Berlin :Springer-Verlag.

Schwartz, L. (1967). *Théorie des Distributions*, Paris :Hermann.

Suquet, P.M. (1978). Existence et régularité des solutions des équations de la plasticité. *Comptes-rendus de l'Académie des Sciences de Paris*, série A 286:1129-1132 and 286:1201–1204.

Suquet, P.M. (1978). *Existence et Régularité des Solutions des Equations de la Plasticité*, Doctor thesis, Université de Paris.

Temam, R. and Strang, G. (1980). Functions of bounded deformations. *Archive for Rational Mechanics and Analysis*, 75:7–21.

Implicit Standard Materials

Géry de Saxcé [1] and Lahbib Bousshine [2]

[1] University of Lille, Lille, France
[2] University Hassan II, Casablanca, Morocco

Abstract. We present a formulation for the non associated constitutive laws based on a generalization of Fenchel's inequality and the concept of bipotential. Then, the normality rule appears as an implicit relation between dual variables. Examples of applications to the plasticity of soils and rocks, the cyclic plasticity of metals, the frictional contact and the plasticity with damage are presented.

1 Non Associated Laws

Non Associated Drucker-Prager Model. For realistic soils and rocks materials, the plastic yielding flow rule is generally non associated. For instance, materials obeying to Drucker-Prager model (Lecture 1, section 8 and 9) with an elastic domain defined by :

$$K_\sigma = \{\sigma = (s_m, s) \in \mathrm{S} \quad such \quad that \quad f(s_m, s) = \sigma_{eq}(s) - r(c - s_m \tan\phi) \leq 0 \},$$

are characterized by the plastic dilatancy angle θ, a material dependent constant such that $0 \leq \theta \leq \phi$. For any regular stress point of the plastic yielding surface, one has :

$$\dot{e}^p_m = r \tan\theta \,.\, \varepsilon_{eq}(\dot{e}^p)\,. \tag{1.1}$$

At the vertex ($s_m = c / \tan\phi, s = 0$), the plastic strain rate belongs to a convex cone :

$$K_\varepsilon = \{\dot{\varepsilon}^p_m = (\dot{e}^p_m, \dot{e}^p) \in \mathrm{E} \quad such \quad that \quad \dot{e}^p_m \geq r \tan\theta \,.\, \varepsilon_{eq}(\dot{e}^p) \}.$$

When $\theta \neq \phi$, the law is said to be non associated.

Melan's Plastic Potential. More generally, let us assume that the elastic domain is defined by a convex loading function :

$$K_\sigma = \{\sigma \in \mathrm{S} \quad such \quad that \quad f(\sigma) \leq 0 \}.$$

It is currently admitted that the non associated law can be described by introducing a real valued function $\sigma \mapsto g(\sigma)$, defined on S, supposed to be C^1 on ∂K, called the plastic potential. The corresponding constitutive law is :

if $f(\sigma) < 0$ *then* ! *plasticity criterion*

$\dot{\varepsilon}^p = 0$! *elastic loading (or unloading)*

Else if $f(\sigma) = 0, \ \exists \lambda \geq 0$ *such that* $\dot{\varepsilon}^p = \lambda \dfrac{\partial g}{\partial \sigma}(\sigma)$! *plastic yielding*

When $g \neq f$, the constitutive law is non associated. It is worthwhile quoting that, conversely to the loading function f, the plastic potential g can by arbitrary modified by an additive constant. As example, it is easy to verify that the non associated Drucker-Prager model admits the following plastic potential family :

$$g(s_m, s) = \sigma_{eq}(s) + r\, s_m \, tan\theta + C^{te} . \tag{1.2}$$

2 Bipotential and Implicit Standard Materials

Bipotential. We showed in section 9 of Lecture 1 that Fenchel's inequality (9.5) generalizes Hill's maximum power principle (2.2). A more general theory providing tools to model non associated laws is the hemivariational inequation approach proposed by Panagiatopoulos (1975). In the sequel, we present an alternative formulation based on a generalization of Fenchel's inequality and the concept of bipotential (de Saxcé and Feng, 1991, de Saxcé, 1992). Let V be a topological vector space of velocities $\dot{\kappa}$, and F be its dual space collecting the associated variables π, the scalar product being: $(\pi, \dot{\kappa}) \mapsto \pi . \dot{\kappa}$. A bipotential is a function $b: V \times F \to \overline{\boldsymbol{R}}$, separately convex and satisfying the fundamental inequality :

$$\forall (\dot{\kappa}', \pi') \in V \times F, \quad b(\dot{\kappa}', \pi') \geq \pi' . \dot{\kappa}' . \tag{2.1}$$

Next, the couples $(\dot{\kappa}, \pi)$ for which ones the variable are related by the dissipative constitutive law are qualified as extremal in the sense that the equality is reached in the previous relation :

$$b(\dot{\kappa}, \pi) = \pi . \dot{\kappa} . \tag{2.2}$$

Hence, taking the value $\dot{\kappa}' = \dot{\kappa}$ in inequation (2.1) and subtracting member by member (2.2) from (2.1), we obtain :

$$\forall \pi' \in F, \quad b(\dot{\kappa}, \pi') - b(\dot{\kappa}, \pi) \geq (\pi' - \pi) . \dot{\kappa}' . \tag{2.3}$$

Similarly, we have :

$$\forall \dot{\kappa}' \in V, \quad b(\dot{\kappa}', \pi) - b(\dot{\kappa}, \pi) \geq \pi . (\dot{\kappa}' - \dot{\kappa}) . \tag{2.4}$$

Briefly, the extremal couples are characterized by the differential inclusions :

$$\dot{\kappa} \in \partial_{\pi} b(\dot{\kappa},\pi), \qquad \pi \in \partial_{\dot{\kappa}} b(\dot{\kappa},\pi). \tag{2.5}$$

This kind of relation can be qualified as implicit normality laws in the sense that (by reference to the implicit function theorems) the unknown $\dot{\kappa}$ (resp. π) belongs to both left and right hand members of the relation. For this reason, the dissipative materials admitting a bipotential are qualified of implicit standard materials. Generally speaking, this kind of material is non-Druckerian because the constitutive law is non associated. This behavior is unstable and softening may occur (Bousshine et al., 2001). In the particular case of the standard materials (as defined in section 9 of Lecture 1), the bipotential is separated in two parts : the superpotential of dissipation φ and its polar function χ :

$$b(\dot{\kappa},\pi) = \varphi(\dot{\kappa}) + \chi(\pi). \tag{2.6}$$

Hence fundamental inequality (2.1) degenerates into Fenchel's one.

Application to Non Associated Drucker-Prager Model (de Saxcé and Berga, 1994). Let us prove that the plastic yielding rule of this model is equivalent to the differential inclusion :

$$(\dot{e}^p_m + r(tan\phi - tan\theta)\,\varepsilon_{eq}(\dot{e}^p), \dot{e}^p) \in \partial \coprod_{K_\sigma} (s_m, s). \tag{2.7}$$

To be brief, let us introduce :

$$\dot{\tilde{\varepsilon}}^p = (\dot{e}^p_m + r(tan\phi - tan\theta)\,\varepsilon_{eq}(\dot{e}^p), \dot{e}^p).$$

Hence, (2.7) becomes :

$$\dot{\tilde{\varepsilon}}^p \in \partial \coprod_{K_\sigma} (\sigma),$$

which is equivalent to the inequality :

$$\forall \sigma' \in K_\sigma, \; (\sigma' - \sigma) : \dot{\tilde{\varepsilon}}^p \leq 0. \tag{2.8}$$

Next, three following events have to be examined.

Firstly, σ belongs to the interior of K_σ. By similar reasoning to the ones used section 8 of Lecture 1, we conclude that $\dot{\tilde{\varepsilon}}^p$ vanishes, and because of the uniqueness of the decomposition into spherical and deviatoric parts, it holds :

$$\dot{e}^p = 0, \qquad \dot{e}^p_m + r(tan\phi - tan\theta)\,\varepsilon_{eq}(\dot{e}^p) = 0.$$

Hence, $\dot{\varepsilon}^p$ vanishes. We are in case of elastic loading.

Secondly, σ is a regular point of the plastic yielding surface ∂K_σ. Inequation (2.8) entails that there exists $\lambda \geq 0$ such that :

$$\dot{e}^p = \dot{\tilde{e}}^p = \lambda \frac{\partial f}{\partial s}(s_m, s) = \lambda \frac{\partial \sigma_{eq}}{\partial s}(s),$$

$$\dot{e}^p_m + r(\tan\phi - \tan\theta)\,\varepsilon_{eq}(\dot{e}^p) = \tilde{e}^p_m = \lambda \frac{\partial f}{\partial s_m}(s_m, s) = \lambda\, r \tan\phi\,.$$

Because of : $\lambda = \varepsilon_{eq}(\dot{e}^p)$, it holds :

$$\dot{\varepsilon}^p = (\dot{e}^p_m, \dot{e}^p) = \lambda\,(r \tan\theta, \frac{\partial \sigma_{eq}}{\partial s}(s)) = \lambda \frac{\partial g}{\partial \sigma}(\sigma)\,.$$

Thirdly, $\sigma = (c/\tan\phi, 0)$ is the vertex of the cone K_σ. Hence $\dot{\tilde{\varepsilon}}^p$ belongs to the cone :

$$K^*_\sigma = \{ \dot{\varepsilon}^p_m = (\dot{e}^p_m, \dot{e}^p) \in \mathrm{E} \quad such \quad that \quad \dot{e}^p_m \geq r \tan\phi\,.\varepsilon_{eq}(\dot{e}^p) \}.$$

Thus, $\dot{\varepsilon}^p$ belongs to the cone K_ε.

Now, we get to the heart of the matter by introducing the following function defined on $\mathrm{E} \times \mathrm{S}$, and equal to :

$$b(\dot{\varepsilon}^p, \sigma) = \frac{c\,\dot{e}^p_m}{\tan\phi} + r(\tan\theta - \tan\phi)\left(\mathrm{s_m} - \frac{c}{\tan\phi}\right)\varepsilon_{eq}(\dot{e}^p), \tag{2.9}$$

if $\sigma \in K_\sigma$ and $\dot{\varepsilon}^p \in K_\varepsilon$, equal to $+\infty$ otherwise.

Let us prove it is a bipotential. One has to verify condition (2.1), that is :

$$\forall \dot{\varepsilon}^p \in K_\varepsilon,\ \forall \sigma \in K_\sigma,\ \frac{c\,\dot{e}^p_m}{\tan\phi} + r(\tan\theta - \tan\phi)\left(\mathrm{s_m} - \frac{c}{\tan\phi}\right)\varepsilon_{eq}(\dot{e}^p) \geq s_m\,\dot{e}^p_m + s{:}\dot{e}^p\,. \tag{2.10}$$

Firstly, for any $\sigma \in K_\sigma$, taking account Cauchy-Schwartz inequality, one has :

$$-\mathrm{r}\tan\phi\left(\mathrm{s_m} - \frac{c}{\tan\phi}\right)\varepsilon_{eq}(\dot{e}^p) \geq \sigma_{eq}(s)\,\varepsilon_{eq}(\dot{e}^p) \geq s{:}\dot{e}^p\,. \tag{2.11}$$

On the other hand, for any $\dot{\varepsilon}^p \in K_\varepsilon$ and any $\sigma \in K_\sigma$, one has :

$$\mathrm{s_m} \leq \frac{c}{tan\,\phi}, \quad \text{and:} \quad \dot{e}^p_m \geq r\,tan\theta.\varepsilon_{eq}(\dot{e}^p).$$

Hence, it holds :

$$r\,tan\theta\left(\mathrm{s_m} - \frac{c}{tan\,\phi}\right)\varepsilon_{eq}(\dot{e}^p) \geq \left(\mathrm{s_m} - \frac{c}{tan\,\phi}\right)\dot{e}^p_m. \qquad (2.12)$$

Hence condition (2.10) results from inequalities (2.11) and (2.12), and this proves that (2.9) is a bipotential.

Next, let us prove that the extremal couples for the previously defined bipotential satisfy the flow rule (2.7) and conversely. Indeed, applying $(2.5)_1$ to the considered bipotential, it can be deduced that the extremal couple satisfy the flow rule (2.7). For the inverse proposition, a couple $(\dot{\varepsilon}^p, \sigma)$ satisfying the flow rule (2.7) is considered and the satisfaction of (2.2) has to be demonstrated. Of course, in order to fulfill (2.2), the value of the bipotential has to be finite, hence $\dot{\varepsilon}^p \in K_\varepsilon$, $\sigma \in K_\sigma$, and the bipotential is given by expression (2.9). After some algebraic manipulations, it holds :

$$b(\dot{\varepsilon}^p,\sigma) = \frac{c}{tan\,\phi}(\dot{e}^p_m - r\,tan\theta\,\varepsilon_{eq}(\dot{e}^p)) + r\,tan\theta\,\varepsilon_{eq}(\dot{e}^p)\,s_m + r\,(c - tan\,\phi\,s_m)\,\varepsilon_{eq}(\dot{e}^p).$$

When the plastic strain rate vanishes, (2.2) is trivially fulfilled. Otherwise, the stress point is on the plastic yielding surface. If σ is a regular point of the plastic yielding surface, $f(\sigma) = 0$ and (1.1) hold, hence one has :

$$b(\dot{\varepsilon}^p,\sigma) = s_m\,\dot{e}^p_m + \sigma_{eq}(s)\,\varepsilon_{eq}(\dot{e}^p).$$

Next, the flow rule (2.7) implies that s and $\dot{e}^p$ are collinear and (2.2) is satisfied. Finally, for the vertex, $s_m = c / tan\,\phi$, $s = 0$ and (2.2) is again fulfilled.

In conclusion, the non associated Drucker-Prager model can be described by an unique function of both dual variables, the bipotential equal to (2.9) if $\sigma \in K_\sigma$ and $\dot{\varepsilon}^p \in K_\varepsilon$, equal to $+\infty$ otherwise. Using indicator functions, the bipotential can be recast as :

$$b(\dot{\varepsilon}^p,\sigma) = \frac{c\,\dot{e}^p_m}{tan\,\phi} + r\,(tan\theta - tan\,\phi)\left(\mathrm{s_m} - \frac{c}{tan\,\phi}\right)\varepsilon_{eq}(\dot{e}^p) + \coprod\nolimits_{K_\varepsilon}(\dot{\varepsilon}^p) + \coprod\nolimits_{K_\sigma}(\sigma). \qquad (2.13)$$

Because of the non associated law, the material is non-Druckerian. The possibility of softening phase in the loading curves is discussed in de Saxcé and Berga (1994).

3 Non Linear Kinematical Hardening Rule for Cyclic Plasticity of Metals

As seen in Lecture 1, the elastic perfectly plastic model allows to model both incremental and alternating plasticity collapses. The modern trend to use harden metal exhibiting higher failure strength needs to take into account the hardening effects. From the viewpoint of the shakedown analysis, it requires a careful choice of the suitable model.

Prager kinematical hardening (Lecture 1, section 9) is the simplest one because of its linearity. Unfortunately, under variable repeated loading, it does not allows to represent the incremental collapse (Exercise 1.3). Thus, its use is restricted to particular engineering applications for which ones the Codes only require to verify the structure safety against alternating plasticity collapse.

On the other hand, the limited linear kinematical hardening model using a two-surface yield condition is able to represent the incremental collapse too and provides a powerful tool for shakedown analysis (Exercises 1.4 and 1.6). Nevertheless, the previous hardening being linear under the limit surface, leads to a rather rough idealization of the plasticity limit cycles observed in experimental testing.

A more realistic representation of the cyclic plasticity of metals is given by the so-called non linear kinematical hardening rules. A simple and efficient one was proposed by Armstrong and Frederick (1966) and was popularized by Lemaitre and Chaboche (1990). The hardening variables are the kinematical ones $\alpha \in \mathbb{E}_d$ and the isotropic one $p \in \boldsymbol{R}$. The corresponding associated variable are respectively the back-stresses $X \in \mathbb{S}_d$ and the yield stress $R \in \boldsymbol{R}$. Thus, we consider the velocities $\dot{\kappa} = (\dot{\varepsilon}^p, -\dot{\alpha}, -\dot{p}) \in \mathbb{E} \times \mathbb{E}_d \times \boldsymbol{R}$ and the corresponding associated variables $\pi = (\sigma, X, R) \in \mathbb{S} \times \mathbb{S}_d \times \boldsymbol{R}$. First, let us introduce :

$$K_2 = \left\{ (\sigma, R) \in \mathbb{S} \times \boldsymbol{R} \quad such \quad that \quad \sigma_{eq}(s) - R \leq 0 \right\}.$$

Its polar cone is :

$$K_2^* = \left\{ (\dot{\varepsilon}^p, -\dot{p}) \in \mathbb{E} \times \boldsymbol{R} \quad such \quad that \quad \dot{p} \geq \varepsilon_{eq}(\dot{e}^p) \right\}.$$

The elastic domain is defined as the convex cone :

$$K = \left\{ \pi = (\sigma, X, R) \in F \ such \ that \ (\sigma - X, R) \in K_2 \right\}.$$

Both plastic yielding hardening law and isotropic hardening rule are given by the differential inclusion :

$$(\dot{\varepsilon}^p, -\dot{p}) \in \partial \coprod\nolimits_{K_2} (\sigma - X, R). \tag{3.1}$$

Equivalently, we can write the inverse law in the form:

$$(\sigma - X, R) \in \partial \coprod_{K_2^*} (\dot{\varepsilon}^p, -\dot{p}). \tag{3.2}$$

The non linear kinematical hardening rule is given by :

$$\dot{\alpha} = \dot{\varepsilon}^p - \frac{3}{2}\frac{X}{X_\infty}\dot{p}, \tag{3.3}$$

where $X_\infty > 0$ is a material dependent constant. Moreover, we assume that the kinematical variable is related to the back-stress by :

$$X = \frac{2}{3} C\alpha$$

where $C > 0$ is a material dependent constant. In order to represent only the stabilized cycles, the plastic threshold R is supposed to be equal to a constant denoted σ_Y. In uniaxial loading, the only vanishing stress is σ_{11}, denoted σ. For convenience, the corresponding plastic strain rate $\dot{\varepsilon}^p_{11}$ is denoted $\dot{\varepsilon}_p$, and $\dot{\alpha}_{11}$ is denoted $\dot{\alpha}$, while we put : $X_{11} = 2X/3$. Tensorial relation (3.3) degenerates into the following scalar one :

$$\dot{\alpha} = \dot{\varepsilon}_p - \frac{X}{X_\infty}\dot{p} = \dot{\varepsilon}_p - \frac{X}{X_\infty}\left|\dot{\varepsilon}_p\right|.$$

For an uniaxial tensile loading, the flow rule gives : $\dot{\varepsilon}_p > 0$ and :

$$dX = C\,d\alpha = C\left(1 - \frac{X}{X_\infty}\right)d\varepsilon_p.$$

Integrating this differential equation leads to the solution :

$$X(\varepsilon_p) = X_\infty - (X_\infty - X(\varepsilon_{p0}))\,exp\left[-\frac{C}{X_\infty}(\varepsilon_p - \varepsilon_{p0})\right], \qquad \sigma(\varepsilon_p) = X(\varepsilon_p) + \sigma_Y.$$

The back-stress monotonically increases during the loading up to the saturation value X_∞. In case of uniaxial compressive loading, $\dot{\varepsilon}_p < 0$, and we obtain :

$$X(\varepsilon_p) = -X_\infty + (X_\infty + X(\varepsilon_{p0}))\,exp\left[\frac{C}{X_\infty}(\varepsilon_p - \varepsilon_{p0})\right], \qquad \sigma(\varepsilon_p) = X(\varepsilon_p) - \sigma_Y.$$

The back-stress monotonically decreases during the loading and is bounded by the asymptotic value $(-X_\infty)$. This provides a realistic representation of the smooth hysteresis observed in alternating plasticity of metals (Figure 1).

In uniaxial loading, the stress is bounded within the limit domain : $|\sigma_{11}| < \sigma_\infty = \sigma_Y + X_\infty$, and, more generally, within the limit domain : $\sigma_{eq}(s) \le \sigma_\infty$. It can be shown this constitutive law is equivalent to Mróz multilayer model (see Lemaitre and Chaboche, 1990). The main drawback of this model is its non-associated nature. Nevertheless, it admits a bipotential equal to (de Saxcé, 1992, de Saxcé and Hjiaj, 1997) :

$$b(\dot{\kappa},\pi) = \frac{(\sigma_{eq}(X))^2}{X_\infty}\dot{p}\,, \tag{3.4}$$

when $(\sigma - X, R) \in K_2$, $(\dot{\varepsilon}^p, -\dot{\alpha}) \in K_2^*$, and (3.3) is satisfied, and equal to $+\infty$ otherwise. Using indicator functions leads to the more compact definition :

$$b(\dot{\kappa},\pi) = \frac{(\sigma_{eq}(s))^2}{X_\infty}\dot{p} + \coprod\nolimits_{K_2}(\sigma - X, R) + \coprod\nolimits_{K_2^*}(\dot{\varepsilon}^p, -\dot{\alpha}) + \coprod\left(\dot{\alpha} - \dot{\varepsilon}^p + \frac{3}{2}\frac{X}{X_\infty}\dot{p}\right). \tag{3.5}$$

Indeed, let us prove that the previous function is a bipotential. For this, we must show that for any set of generalized velocities and stresses such that $(\sigma - X, R) \in K_2$, $(\dot{\varepsilon}^p, -\dot{\alpha}) \in K_2^*$, and (3.3) is satisfied, we have :

$$\frac{(\sigma_{eq}(X))^2}{X_\infty}\dot{p} \ge \sigma : \dot{\varepsilon}^p - X : \dot{\alpha} - R\dot{p}\,.$$

According to $(\sigma - X, R) \in K_2$, $(\dot{\varepsilon}^p, -\dot{\alpha}) \in K_2^*$, and remembering Cauchy-Schwartz inequality, we can write :

$$R\dot{p} \ge R\varepsilon_{eq}(\dot{\varepsilon}^p) \ge \sigma_{eq}(\sigma - X)\varepsilon_{eq}(\dot{\varepsilon}^p) \ge (\sigma - X) : \dot{\varepsilon}^p\,.$$

but :

$$(\sigma - X) : \dot{\varepsilon}^p = \sigma : \dot{\varepsilon}^p - X : \left(\dot{\alpha} + \frac{3}{2}\frac{X}{X_\infty}\dot{p}\right).$$

By accounting for the flow rule again, we obtain :

$$\frac{(\sigma_{eq}(X))^2}{X_\infty}\dot{p} \ge \sigma : \dot{\varepsilon}^p - X : \dot{\alpha} - R\,\dot{p}\,.$$

This last inequality proves that the function considered above is a bipotential.

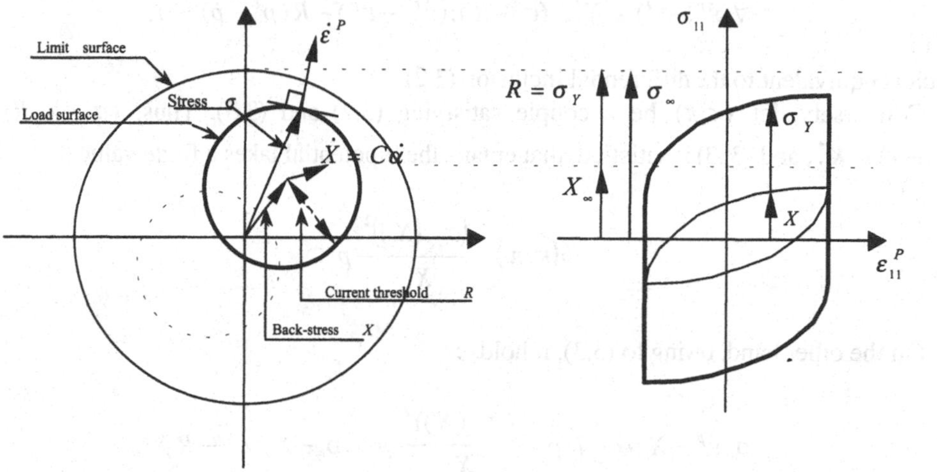

Figure 1. Non linear kinematical hardening rule.

Finally, it is possible to show that any extremal couple for the bipotential fulfills the constitutive law (3.2), (3.3) and, conversely, that any couple satisfying the constitutive law is extremal for the bipotential function. Indeed, let $(\dot{\kappa},\pi)$ be an extremal couple. Thus, $(\sigma - X, R) \in K_2$, $(\dot{\varepsilon}^p, -\dot{\alpha}) \in K_2^*$, and (3.3) is satisfied because the bipotential must take finite values for the extremal couples. Let us consider any $\dot{\kappa}' = (\dot{\varepsilon}^{p'}, -\dot{\alpha}', -\dot{p}')$ such that $(\dot{\varepsilon}^p, -\dot{\alpha}') \in K_2^*$ is verified and :

$$\dot{\alpha}' = \dot{\varepsilon}^p - \frac{3}{2}\frac{X}{X_\infty}\dot{p}' . \tag{3.6}$$

Taking account of inequation (2.4) applied to the bipotential (3.5), one has :

$$\frac{(\sigma_{eq}(X))^2}{X_\infty}(\dot{p}' - \dot{p}) \geq \sigma:(\dot{\varepsilon}^{p'} - \dot{\varepsilon}^p) - X:(\dot{\alpha}' - \dot{\alpha}) - R(\dot{p}' - \dot{p}) .$$

On the other hand, from (3.3) and (3.6), it results :

$$-X:(\dot{\alpha}' - \dot{\alpha}) = -X:(\dot{\varepsilon}^{p'} - \dot{\varepsilon}^p) + \frac{(\sigma_{eq}(X))^2}{X_\infty}(\dot{p}' - \dot{p}) .$$

Finally, one obtains :

$$\forall(\dot{\varepsilon}^{p},-\dot{\alpha}')\in K_2^*,\ (\sigma-X):(\dot{\varepsilon}^{p'}-\dot{\varepsilon}^{p})-R(\dot{p}'-\dot{p})\leq 0,$$

which is equivalent to the differential inclusion (3.2).

Conversely, let $(\dot{\kappa},\pi)$ be a couple satisfying (3.2) and (3.3). Thus, $(\sigma-X,R)\in K_2$, $(\dot{\varepsilon}^{p},-\dot{\alpha})\in K_2^*$, and (3 .3) is satisfied, that entails the bipotential takes a finite value :

$$b(\dot{\kappa},\pi)=\frac{(\sigma_{eq}(X))^2}{X_\infty}\dot{p}.$$

On the other hand, owing to (3.3), it holds :

$$\sigma:\dot{\varepsilon}^{p}-X:\dot{\alpha}-R\,\dot{p}=\frac{(\sigma_{eq}(X))^2}{X_\infty}\dot{p}+(\sigma-X):\dot{\varepsilon}^{p}-R\,\dot{p}.$$

Accounting for (3.3), the couple $((\dot{\varepsilon}^{p},-\dot{\alpha}),(\sigma-X,R))$ is extremal for the indicator functions of K_2 and K_2^*, polar one of the other. The sum of the two last terms of the right hand member vanishes, that achieves the proof.

Finally, let us remark that the extension of the non linear kinematical hardening to the viscoplastic behavior is straightforward. For this one, a bipotential was proposed by Hjiaj, Bodovillé and de Saxcé (2000).

4 Other Models

Cauchy-Schwartz Bipotential. As previously, the dot symbol denotes the ordinary scalar product in the Euclidean vector space $V=\boldsymbol{R}^n$, identified to its dual F, and $\|.\|$ is the corresponding norm. For the velocity $\dot{\kappa}\in V$ and the associated variable $\pi\in F$, the constitutive law defined by: "$\dot{\kappa}$ and π are colinear", admits the following bipotential:

$$b(\dot{\kappa},\pi)=\|\dot{\kappa}\|\|\pi\|,$$

because of Cauchy-Schwartz inequality:

$$\forall\dot{\kappa}'\in V,\quad \forall\pi'\in F,\quad \dot{\kappa}'.\pi'\leq\|\dot{\kappa}'\|\|\pi'\|.$$

Hill's Bipotential. Let V be the set of the $n \times n$ symmetric real matrices. As Euclidean vector space for the scalar product: $(\varepsilon|\varepsilon') = \varepsilon_{ij}\,\varepsilon'_{ij}$, it can be identified to its dual F. Any symmetric matrix ε has n real eigenvalues denoted: $\lambda_1(\varepsilon) \geq \ldots \geq \lambda_n(\varepsilon)$. We say that $\varepsilon \in V$ and $\sigma \in F$ are coaxial if they have the same eigenvectors. In other words, there exists an orthogonal $n \times n$ matrix Q such that:

$$Q^T \varepsilon\, Q = diag\,(\lambda_1(\varepsilon), \ldots, \lambda_n(\varepsilon)), \quad Q^T \sigma\, Q = diag\,(\lambda_1(\sigma), \ldots, \lambda_n(\sigma)).$$

Then, it can be proved that the constitutive law defined by "ε and σ are coaxial" admits the following bipotential:

$$b(\varepsilon, \sigma) = \sum_{i=1}^{n} \lambda_i(\varepsilon)\,\lambda_i(\sigma). \tag{4.1}$$

The convexity results from Rayleigh-Courant theory and the corner stone inequality of the bipotential from an inequality due to Hill (Hiriart-Urruty, 1998).

Linear Law (Vallée et al., 1997). Let us consider a linear elastic law: $\sigma_{ij} = D_{ijkl}\,\varepsilon_{kl}$ (in short: $\sigma = D\varepsilon$). If $D_{ijkl} = D_{klij}$, the law admits a potential:

$$U(\varepsilon) = \tfrac{1}{2}\varepsilon : D\varepsilon .$$

Hence, the law is elastic in the sense that the behavior is reversible. If $D_{ijkl} \neq D_{klij}$, the law is not elastic and does not admit a potential. Let S (resp. C) be a mapping from S into E (resp. from E into S). the two mappings are *positive* in the sense that:

$$\forall \sigma \neq 0, \quad \sigma : S\sigma > 0, \qquad \forall \varepsilon \neq 0, \quad \varepsilon : C\varepsilon > 0.$$

Then, S and C are regular but generally: $S \neq C^{-1}$. Next, let F be a mapping from E into E and F^T be its transposed mapping. Thus, the continuously differentiable function:

$$b(\varepsilon, \sigma) = \tfrac{1}{2}\sigma : S\sigma + \tfrac{1}{2}\varepsilon : C\varepsilon + \sigma : F\varepsilon + \sigma : \varepsilon , \tag{4.2}$$

is a bipotential if and only if:

$$S = F C^{-1} F^T .$$

The corresponding extremal couples (ε,σ) satisfy:

$$\sigma = D\varepsilon ,$$

with: $D = -S^{-1}F$. The bipotential is separable if and only if: $F = -I$, and: $S = C^{-1}$.

Coaxial Linear Law. We say that a linear law: $\sigma = D\varepsilon$ is coaxial if ε and σ are coaxial. Then, the constitutive law can be written as follows:

$$\sigma = (\Lambda : \varepsilon) I + 2\mu\varepsilon , \tag{4.3}$$

where: $\Lambda \in \mathrm{S}$, I is the identity, and $\mu \in \boldsymbol{R}$. Of course, if moreover the material is isotropic, we find Hooke's law with: $\Lambda = \lambda I$, λ and μ being Lamé's coefficients. More generally, the coaxial linear laws which are not necessarily isotropic admit the following bipotential (Vallée et al., 1997):

$$b(\varepsilon,\sigma) = \tfrac{1}{2}(\sigma - (\Lambda : \varepsilon) I - 2\mu\varepsilon) : S(\sigma - (\Lambda : \varepsilon) I - 2\mu\varepsilon) + \sigma : \varepsilon , \tag{4.4}$$

provided that S is symmetric ($S = S^{T}$) and positive, C and F being given by:

$$C\varepsilon = 2\mu(\Lambda : \varepsilon) SI + 4\mu^2 S\varepsilon + 2\mu(I : S\varepsilon)\Lambda + (\Lambda : \varepsilon)(I : SI)\Lambda , \tag{4.5}$$

$$F\varepsilon = -(\Lambda : \varepsilon) SI - 2\mu S\varepsilon . \tag{4.6}$$

Unilateral Contact with Coulomb's Dry Friction. Let Ω_1 and Ω_2 be two bodies in contact at a point *M* for some value of the time. The instantaneous velocity of the particles of Ω_1 and Ω_2 passing at point *M* being respectively $\dot{u}_1$ and $\dot{u}_2$, the relative velocity is : $\dot{u} = \dot{u}_1 - \dot{u}_2$. Let r be the contact reaction at *M* from Ω_2 onto Ω_1. Then Ω_2 is subjected to the reaction $-r$, acting from Ω_1. Let *n* denote the normal unit vector at point *M* to the bodies, directed towards Ω_1 and let *T* denote the orthogonal plane to *n* in $\boldsymbol{R}^3$. Any vector v can be uniquely decomposed in the form :

$$v = v_t + v_n\, n,\ \ v_t \in T,\ \ v_n \in \boldsymbol{R} .$$

Let K_μ denote Coulomb's cone with friction coefficient μ :

$$K_\mu = \left\{ r \in \boldsymbol{R}^3 \ \ such\ \ that\ \ \| r_t \| \le \mu r_n \right\}$$

The complete contact law is a non smooth dissipative law including three statuses : no contact, contact with sticking and contact with sliding. It can be analytically represented by two overlapped « **if…then…else** » statements :

if $r = 0$ ***then*** $\dot{u}_n \geq 0$! *no contact*

Else if $\|r_t\| < \mu r_n$ ***then*** $\dot{u} = 0$! *sticking*

Else $(r_n > 0$ and $\|r_t\| = \mu r_n)$! *sliding*

$$\dot{u}_n = 0, \; \exists \lambda \geq 0 \; such \; that \; \dot{u}_t = -\lambda \frac{r_t}{\|r_t\|}$$

A modeling of this well-known law in the frame of the implicit standard material approach was developed by de Saxcé and Feng (1991), de Saxcé (1992), de Saxcé and Feng (1998). Indeed, this constitutive law can be written in a more compact way as the following differential inclusion :

$$-(\dot{u} + \mu \|\dot{u}_t\| n) \in \partial \coprod_{K_\mu} (r) . \tag{4.7}$$

This law admits a bipotential equal to :

$$b_c(-\dot{u}, r) = \mu r_n \|\dot{u}_t\| ,$$

if $\dot{u}_n \geq 0$ and $r \in K_\mu$, and equal to $+\infty$ otherwise. Using indicator functions, one has :

$$b_c(-\dot{u}, r) = \mu r_n \|\dot{u}_t\| + \coprod_{]-\infty,0]} (-\dot{u}_n) + \coprod_{K_\mu} (r) . \tag{4.8}$$

A piecewise linearization of this bipotential was proposed by Klärbring (1992).

Modified Cam-Clay Model (Burland and Roscoe, 1968). This is typically a non Druckerian material: softening may occur as performing, for instance, drained triaxial tests on clay specimens. This can be predicted by the Modified Cam-Clay model (or MCC model), with an elliptical plastic yielding curve controlled by an internal scalar variable α, linked to the plastic bulk strain. For heavily overconsolidated specimens, the plastic dilatancy entails the contraction of the yield curve, that explains the *softening* phase before reaching the critical state. In this model, the dual variables are $\dot{\kappa} = (\dot{\varepsilon}^p, -\dot{\alpha}) \in \mathrm{E} \times \boldsymbol{R}$ and $\pi = (\sigma, p_{cr}) \in \mathrm{S} \times \boldsymbol{R}$. The spaces E and S are respectively equipped with the dual norms :

$$\|\dot{\varepsilon}^p\| = \left| \frac{1}{M^2} (\dot{e}^p_m)^2 + (\varepsilon_{eq}(\dot{e}^p))^2 \right|^{1/2} , \qquad \|\sigma\|_* = \left| \; M^2 \; s_m^2 \; + \; (\sigma_{eq}(s))^2 \; \right|^{1/2},$$

where $M > 0$ is a material dependent constant. Introducing the back-stress vector $X = (-p_{cr}, 0) \in S$, the elastic domain is defined by :

$$K = \{\pi = (\sigma, p_{cr}) \in F \text{ such that } f(\sigma, p_{cr}) = \|\sigma - X\|_* - M\, p_{cr} \leq 0\}.$$

The MCC yielding rule claims that if $\sigma \in K$,

$$\exists\, \lambda \geq 0 \text{ such that } \dot{\varepsilon}^p = \lambda \frac{\partial f}{\partial \sigma} . \tag{4.9}$$

The hardening rule is :

$$\dot{e}^p_m + \dot{\alpha} = 0 . \tag{4.10}$$

The non associativity of the dissipative rule results from the fact that :

$$- \dot{\alpha} \neq \lambda \frac{\partial f}{\partial p_{cr}} .$$

It can be verified that the corresponding family of plastic potentials is :

$$g(\sigma, p_{cr}) = \|\sigma - X\|_* + C^{te} ,$$

restating the constitutive law as :

$$\exists \lambda \geq 0 \;\; \text{such} \;\; \text{that} \;\; \dot{\kappa} = \lambda \frac{\partial g}{\partial \pi}(\pi) .$$

This constitutive law admits a bipotential (de Saxcé, 1995). Indeed, it can be proved that the set of the yielding rule (4.9) and the hardening rule (4.10) can be transcript into the following more compact statement :

$$(\dot{\varepsilon}^p, -(\dot{\alpha} + M\|\dot{\varepsilon}^p\|)) \in \partial \coprod\nolimits_K (\sigma, p_{cr}) \tag{4.11}$$

The next point of interest is that the previous constitutive law can be considered as an implicit standard material law. For this purpose, the following function is proposed :

$$b(\dot{\kappa}, \pi) = M\, p_{cr} \, \|\dot{\varepsilon}^p\|$$

if $\dot{\alpha} = -\dot{e}^p_m$ and $\pi \in K$, equal to $+\infty$ otherwise. In a more compact way, it can be recast as :

$$b(\dot{\kappa}, \pi) = M\, p_{cr} \left\|\dot{\varepsilon}^p\right\| + \coprod_{\{0\}} (\dot{e}^p_m + \dot{\alpha}) + \coprod_K (\sigma, p_{cr}) \,. \qquad (4.12)$$

Lemaitre's Isotropic Damage Law (Lemaitre, 1987). Various damage behaviors of the metallic materials can be predicted by introducing a scalar damage variable ω varying within the range from 0 to ω_c, the value $\omega = 0$ corresponding to the undamaged material, and $\omega = \omega_c$ to the failure for a material dependent critical value $\omega_c \leq 1$. The modified complementary elastic energy potential :

$$\tilde{W}(\sigma,\omega) = \frac{W(\sigma)}{1-\omega} = \frac{1}{1-\omega}\left(\frac{s_m^2}{2K} + \frac{s_{ij}\, s_{ij}}{4\mu}\right),$$

allows to deduce the expression of the elastic strain :

$$e^e_m = \frac{\partial \tilde{W}}{\partial s_m} = \frac{1}{K}\frac{s_m}{1-\omega}, \qquad e^e = \frac{\partial \tilde{W}}{\partial s} = \frac{1}{2\mu}\frac{s}{1-\omega},$$

that is Hooke's law where the actual stress σ is replaced by the effective stress : $\tilde{\sigma} = \sigma/(1-\omega)$. Next, we define the strain energy density release rate Y by :

$$-Y = \frac{\partial \tilde{W}}{\partial \omega} = \frac{W(\sigma)}{(1-\omega)^2},$$

which must be obviously non positive. The scalar isotropic hardening variable is denoted r. For the dissipative part of the model, the dual variables are $\dot{\kappa} = (\dot{\varepsilon}^p, -\dot{r}, -\dot{\omega}) \in \mathrm{E} \times \boldsymbol{R} \times \boldsymbol{R}$ and $\pi = (\sigma, R, Y) \in \mathrm{S} \times \boldsymbol{R} \times \boldsymbol{R}$. The elastic domain is defined as the convex cone:

$$K_\omega = \left\{ (\sigma, R) \in \mathrm{S} \times \boldsymbol{R} \quad such \quad that \quad f_\omega(\sigma, R) = \sigma_{eq}(s) - (1-\omega)R \leq 0 \right\},$$

where ω occurs as parameter. Its polar cone is :

$$K^*_\omega = \left\{ (\dot{\varepsilon}^p, -\dot{r}) \in \mathrm{E} \times \boldsymbol{R} \quad such \quad that \quad \dot{r} \geq (1-\omega)\varepsilon_{eq}(\dot{e}^p) \right\}.$$

The plastic yielding law is described by the following differential inclusion :

$$(\dot{\varepsilon}^p, -\dot{r}) \in \partial \coprod_{K_\omega} (\sigma, R) \,. \qquad (4.13)$$

The evolution of the damage variable (from $\omega = 0$ to $\omega = \omega_c$) is monotonic increasing and governed by the rule :

$$\dot{\omega} = \frac{(-Y)^{\beta}}{S} \frac{\dot{r}}{(1-\omega)^{\alpha+1}}, \tag{4.14}$$

if $r \geq r_\omega$, and $\dot{\omega} = 0$ otherwise. The parameters α, β, S and r_ω are positive material dependent constants. In the sequel, we suppose that the damage threshold is overcame : $r \geq r_\omega$. As shown by Bodovillé (1999), the previous plastic yielding law (4.13) coupled with the damage evolution rule (4.14) admits a bipotential equal to :

$$b(\dot{\kappa}, \pi) = \frac{(-Y)^{\beta+1}}{S} \frac{\dot{r}}{(1-\omega)^{\alpha+1}},$$

if $(\dot{\varepsilon}^p, -\dot{r}) \in K_\omega^*$, $(\sigma, R) \in K_\omega$, and (4.14) is satisfied, and equal to $+\infty$ otherwise. Using indicator functions, it can be written as :

$$b(\dot{\kappa}, \pi) = \frac{(-Y)^{\beta+1}}{S} \frac{\dot{r}}{(1-\omega)^{\alpha+1}} + \coprod_{K_\omega^*} (\dot{\varepsilon}^p, -\dot{r}) + \coprod_{K_\omega} (\sigma, R) + \coprod_{\{0\}} \left(\dot{\omega} - \frac{(-Y)^{\beta}}{S} \frac{\dot{r}}{(1-\omega)^{\alpha+1}} \right). \tag{4.15}$$

5 Exercises

Exercise 4.1 (*Non linear kinematical hardening rule with a threshold*) The hardening rule (3.3) can be replaced by (Chaboche et al., 1991, Chaboche, 1994) :

$$\dot{\alpha} = \dot{\varepsilon}^p - \left(\frac{(\sigma_{eq}(X) - X_l)_+}{X_\infty} \right)^m \frac{3}{2} \frac{X}{\sigma_{eq}(X)} \dot{p}, \tag{5.1}$$

where m and X_l are positive material dependent constants, all the other constitutive relations being unchanged. Prove this model admits a bipotential equal to :

$$b(\dot{\kappa}, \pi) = \left(\frac{(\sigma_{eq}(X) - X_l)_+}{X_\infty} \right)^m \sigma_{eq}(X) \dot{p},$$

if $(\sigma - X, R) \in K_2$, $(\dot{\varepsilon}^p, -\dot{\alpha}) \in K_2^*$, and (5.1) is satisfied, and equal to $+\infty$ otherwise (Bodovillé and de Saxcé, 2001).

Exercise 4.2. Prove that the function (4.1) is separately convex and satisfy Hill's inequality:

$$\forall \varepsilon' \in V, \quad \forall \sigma' \in F, \quad \varepsilon' : \sigma' = \sum_{i=1}^{n} \lambda_i (\varepsilon' \sigma') \leq \sum_{i=1}^{n} \lambda_i (\varepsilon') \lambda_i (\sigma') .$$

Prove that the equality is reached if and only if ε' and σ' are coaxial.

Exercise 4.3 (*linear law*). Prove that (4.2) is a bipotential.

Exercise 4.4 (Vallée et al., 1997). Prove that any coaxial linear law $\sigma = D\varepsilon$ can be represented by expression (4.3).

Hint: necessarily, it holds: $DI = kI$, with: $k \in \boldsymbol{R}$. Next, use the spectral decomposition:

$$I = \sum_{i=1}^{n} v_i v_i^T ,$$

in an orthonormal basis (v_1, v_2, v_3) of $\boldsymbol{R}^3$. Prove that:

$$\forall j = 1,2,3, \quad D(v_j v_j^T) = \beta I + \alpha v_j v_j^T ,$$

with β and α depending on v_j. Finally, show that: $\alpha = 2\mu$ is constant, $\beta(v_j)$ being the eigenvalues of Λ.

Exercise 4.5 (Vallée et al., 1997). Prove that the coaxial linear law (4.3) admits the bipotential (4.4) under conditions (4.5) and (4.6).

Exercise 4.6. Prove that the unilateral contact law with Coulomb's dry friction can be represented by the differential inclusion (4.7), that the function (4.8) is a bipotential and that a couple is extremal for this bipotential if and only if it satisfies (4.7).

Exercise 4.7. Prove that the Modified Cam-Clay law can be represented by the differential inclusion (4.11), that the function (4.12) is a bipotential and that a couple is extremal for this bipotential if and only if it satisfies (4.11).

Exercise 4.8. Prove that Lemaitre's isotropic damage admits the function (4.15) as a bipotential and that a couple is extremal for this bipotential if and only if it satisfies (4.13) and (4.14).

References

Armstrong, P.J. and Frederick, C.O. (1966). A Mathematical Representation of the Multiaxial Bausshinger Effects. *C.E.G.B. Report* RD/B/N 731.

Bodovillé, G. (1999). Sur l'endommagement et les matériaux standard implicites. *Comptes-rendus de l'Académie des Sciences de Paris*, série II 327:715–720.

Bodovillé, G. and de Saxcé, G. (2001). Plasticity with non-linear kinematic hardening: modelling and shakedown analysis by the bipotential approach. *European Journal of Mechanics A/Solids* 20:99–112.

Bousshine, L., Chaaba, A. and de Saxcé, G. (2001). Softening in stress-strain curve for Drucker-Prager non-associated plasticity. *International Journal of Plasticity* 17:21–46.

Burland, I.B. and Roscoe, K.H. (1968). On the generalized stress strain behavior of wet clay. In Heyman et al. eds., *Engineering Plasticity*, Cambridge Press.

Chaboche, J.L., Nouailhas, D., Pacou, D. and Paulmier, P. (1991). Modeling of the cyclic response and ratchetting effects on inconel-718 alloy. *European Journal of Mechanics A/Solids* 10:101–121.

Chaboche, J.L. (1994) Modeling of ratchetting : evaluation of various approaches. *European Journal of Mechanics A/Solids* 10:101–121.

de Saxcé, G. and Feng, Z.Q. (1991). New Inequation and Functional for Contact with Friction : the Implicit Standard Material Approach. *International Journal Mechanics of Structures and Machines* 19:301–325.

de Saxcé, G. (1992). Une généralisation de l'inégalité de Fenchel et ses applications aux lois constitutives. *Comptes-rendus de l'Académie des Sciences de Paris*, série II 314:125–129.

de Saxcé, G. and Berga, A. (1994). Elastoplastic finite element analysis of soil problems with implicit standard material constitutive law. *Revue Européenne des Eléments Finis* 3:411–456.

de Saxcé, G. (1995). The Bipotential Method, a New Variational and Numerical Treatment of the Dissipative Laws of Materials. *Proceedings of the 10th International Conference on Mathematical and Computer Modeling and Scientific Computing*, Boston.

de Saxcé, G. and Hjiaj, M. (1997). Sur l'intégration numérique de la loi d'écrouissage cinématique non-linéaire. *Actes du 3ème Colloque National en Calcul des Structures, Giens (Var)*, 773–778.

de Saxcé, G. and Feng, Z.Q. (1998). The bipotential method : a constructive approach to design the complete contact law with friction and improved numerical algorithms. *International Journal of Mathematical and Computer Modelling* 28:225–245.

Hiriart-Urruty, J.B. (1998). *Optimisation et analyse convexe*. Collection Mathématiques, Paris: PUF.

Hjiaj, M., Bodovillé, G. and de Saxcé, G. (2000). Matériaux viscoplastiques et lois de normalité implicites. *Comptes-rendus de l'Académie des Sciences de Paris*, série IIb 328:519–524.

Klärbring, A. (1992). Mathematical programming and augmented Lagrangean methods for frictional contact problems. In Curnier ed., *Proceedings Contact Mechanics International Symposium*, PPUR, 369–390.

Lemaitre, J. (1987). Formulation and identification od damage kinetic constitutive equation. In Krajcinovic et al., eds., *Continuum Damage Mechanics, Theory and Application*, CISM Courses and Lectures, 295, NY: Springer-Verlag.

Lemaitre, J. and Chaboche, J.L. (1990). *Mechanics of Solid Materials*, Cambridge University Press.

Panagiatopoulos, P. D. (1975). *Inequality Problems in Mechanics and Applications*, Boston :Birkhauser.

Vallée, C., Fortuné, D., Jessin, K. and Lainé, E. (1997). Lois de comportement coaxiales linéaires et matériaux standards implicites. *Actes du 13ème Congrès Français de Mécanique*, Poitiers, 541–544.

Shakedown with Non Associated Flow Rule

Géry de Saxcé and Jean-Bernard Tritsch
University of Lille, Lille, France

Abstract. First, we present the concept of bifunctional which allows to extend the calculus of variation in case of a material admitting a bipotential. Next, the bound theorems of the shakedown analysis are generalized for this class of plastic materials. The key of the proof is that the normality rule is conserved but in an implicit form. The theory is illustrated by the problem of a thin walled tube under constant tension and alternating cyclic torsion. We recover the value of the shakedown factor given by Lemaitre and Chaboche and we prove that it is the exact one.

1 Bifunctional for Implicit Standard Materials

The topic considered in the present Lecture was originally proposed by de Saxcé and Tritsch (2000). For generality, we define the generalized velocities as $\dot{\kappa} = (\dot{\varepsilon}^p, \dot{\kappa}') \in V$, including the velocities $\dot{\kappa}'$ of additional internal variables (hardening, damage,...), and the corresponding associated variables $\pi = (\sigma, \pi') \in F$. Let Ω be a solid body with elastoplastic materials admitting bipotentials, subjected to variable periodic external actions varying between given limits controlled by a load factor λ. The following question arises : under which conditions the body shakes down ? By numerical step-by-step analysis on simple examples, with a non associated Drucker-Prager material (Lecture 4, sections 1 and 2), the existence of time-independent residual stress fields was observed under a critical value λ^a of the load factor by Bousshine et al. (2001). General theorems concerning the non standard materials were proposed by Pycko and Maier (1995). Unfortunately, no generalization of Melan's and Halphen's theorems to material admitting a bipotential has been rigorously proved up to now.

In spite of this, we admit the existence of admissible stress fields $(\bar{\rho}, \bar{\pi}')$ in the sense that :

1. $\bar{\rho}$ is a residual stress field,
2. $\bar{\rho}$ and $\bar{\pi}'$ are time-independent and plastically admissible when adding to $\bar{\rho}$ the stress response $\sigma^E = \lambda \sigma^{Eo}$ in the fictitious elastic body :

$$\forall x \in \Omega, \forall t, \quad \left(\sigma^E(x,t) + \bar{\rho}(x), \bar{\pi}'(x)\right) = \left(\lambda \sigma^{Eo}(x,t) + \bar{\rho}(x), \bar{\pi}'(x)\right) \in K .$$

On the other hand, we define admissible generalized velocity fields $(\dot{\varepsilon}^P, \dot{\kappa}')$ in the sense that :

1. for the plastic strain increment : $\Delta\varepsilon^p = \oint \dot{\varepsilon}^p \, dt \in N^*$,
2. $\dot{\varepsilon}^P$ is plastically admissible : $\int_\Omega \oint \sigma^E : \dot{\varepsilon}^P dt \, d\Omega > 0$.

As usual, the admissible velocity fields are normalized :

$$\int_\Omega \oint \sigma^{Eo} : \dot{\varepsilon}^P dt \, d\Omega = 1. \tag{1.1}$$

Let us consider the following B.V.P. :

(P$_s$) Find some couple $(\dot{\kappa}, \overline{\pi})$ such that :

1. $\dot{\kappa} = (\dot{\varepsilon}^p, \dot{\kappa}')$ is an admissible generalized velocity field,
2. $\overline{\pi} = (\overline{\rho}, \overline{\pi}')$ is an admissible stress field,
3. the couple $(\dot{\kappa}, \overline{\pi})$ is extremal for the bipotential in Ω at any time.

A possible variational formulation of shakedown problems results from introducing the so-called bifunctional :

$$\beta_S(\dot{\varepsilon}^p, \dot{\kappa}', \overline{\rho}, \overline{\pi}', \lambda) = \int_\Omega \oint \{ b[(\dot{\varepsilon}^p, \dot{\kappa}'), (\overline{\rho} + \lambda\sigma^{Eo}, \overline{\pi}')] - \lambda\sigma^{Eo} : \dot{\varepsilon}^P - \overline{\pi}' . \dot{\kappa}' \} \, dt \, d\Omega . \tag{1.2}$$

By virtue of the principle of virtual work, one has, for admissible fields :

$$\int_\Omega \oint \overline{\rho} : \dot{\varepsilon}^P dt \, d\Omega = \int_\Omega \overline{\rho} : \Delta\varepsilon^P d\Omega = 0 . \tag{1.3}$$

A straightforward consequence of (2.1) of Lecture 4, (1.2) and (1.3) is that for any admissible fields :

$$\beta_S\left(\dot{\varepsilon}^{P*}, \dot{\kappa}'^*, \overline{\rho}^*, \overline{\pi}'^*, \lambda\right) \geq 0 . \tag{1.4}$$

In particular, for the exact solution of the B.V.P. ***(P$_s$)***, the constitutive law is exactly satisfied anywhere in Ω and at any time. As consequence of (2.2) of Lecture 4, it holds :

$$\beta_S(\dot{\varepsilon}^P, \dot{\kappa}', \overline{\rho}, \overline{\pi}', \lambda^a) = 0 . \tag{1.5}$$

As we shall seen, the previous observation is crucial in the sequel. Special events are usual standard materials with no hardening variable κ' and separable bipotentials :

$$b(\dot{\varepsilon}^P,\sigma) = D(\dot{\varepsilon}^P) + \coprod_K(\sigma). \tag{1.6}$$

The bifunctional splits up into two terms :

$$\beta_s(\dot{\varepsilon}^P,\overline{\rho},\lambda) = \overline{\Phi}_s(\dot{\varepsilon}^P,\lambda) + \overline{\Pi}_s(\overline{\rho},\lambda), \tag{1.7}$$

where :

$$\overline{\Phi}_s(\dot{\varepsilon}^P,\lambda) = \int_\Omega \oint D(\dot{\varepsilon}^P)\,dt\,d\Omega - \lambda\int_\Omega \oint \sigma^{0E}:\dot{\varepsilon}^P\,dt\,d\Omega, \tag{1.8}$$

is the functional of Markov's principle over a cycle as defined in theorem 5 of Lecture 2 and :

$$\overline{\Pi}_s(\overline{\rho},\lambda) = \int_\Omega \oint \coprod_K(\overline{\rho} + \lambda\sigma^{E0})\,dt\,d\Omega, \tag{1.9}$$

is the one of Hill's principle over a cycle, as given by (7.5) of Lecture 2.

First of all, let us remark that, for a proportional loading, the classical theorems of the limit analysis can be extended to the implicit standard materials (de Saxcé and Bousshine, 1998). On this ground, our goal is to extend to materials admitting bipotentials the method of Lecture 2 for usual materials in order to establish bound theorems similar to Koiter's ones. For the exact solution $\left((\dot{\varepsilon}^P,\dot{\kappa}'),(\overline{\rho},\overline{\pi}')\right)$ of the B.V.P. ($\boldsymbol{P_s}$), condition (1.5) combined with the normalization condition (1.1) allows to calculate the value of the shakedown load factor :

$$\lambda^a = \int_\Omega \oint \left\{ b\left[(\dot{\varepsilon}^P,\dot{\kappa}'),(\lambda^a\sigma^{Eo} + \overline{\rho},\overline{\pi}')\right] - \dot{\kappa}'.\overline{\pi}' \right\} dt\,d\Omega. \tag{1.10}$$

Comments. Because of conditions (1.4) and (1.5), the bifunctional can be interpreted as a variational estimator of the error. On this ground, it is possible to construct a posteriori estimators of the error in constitutive law (Hjiaj, 1999) generalizing to the implicit standard materials the method proposed by Ladevèze (1996) for the standard materials. On the other hand, a mixed variational method based on the bifunctional was proposed by Pontes et al. (2000) for the numerical computation of the limit load in non associated soils, using mathematical programming algorithms.

2 Kinematical Bound Theorem

By analogy with (1.10), for any admissible velocity field $\left(\dot{\varepsilon}^{P^*}, \dot{\kappa}'^*\right)$, the corresponding kinematical load factor is defined by :

$$\lambda^k = \int_\Omega \oint \left\{ b\left[\left(\dot{\varepsilon}^{P*}, \dot{\kappa}'^*\right), \left(\lambda^a \sigma^{Eo} + \bar{\rho}, \bar{\pi}'\right)\right] - \dot{\kappa}'^* . \bar{\pi}' \right\} dt\, d\Omega . \qquad (2.1)$$

As $\left(\dot{\varepsilon}^{P*}, \dot{\kappa}'^*\right)$ is an admissible velocity field and $\left(\bar{\rho}, \bar{\pi}'\right)$ is the admissible stress field corresponding to the shakedown load λ^a, one has :

$$\beta_S\left(\dot{\varepsilon}^{P^*}, \dot{\kappa}'^*, \bar{\rho}, \bar{\pi}', \lambda^a\right) \geq 0 .$$

Taking into account that $\dot{\varepsilon}^P$ is normalized by (1.1) and definition (2.1), it holds :

$$\lambda^k \geq \lambda^a .$$

That can be considered as the extension of the usual kinematical bounding theorem to materials admitting a bipotential.

3 Statical Bound Theorem

Let $\left(\bar{\rho}^*, \bar{\pi}'^*\right)$ be any admissible stress field corresponding to the statical load factor λ^s :

$$\forall x \in \Omega, \forall t \quad \left(\lambda^s \sigma^{Eo}(x,t) + \bar{\rho}^*(x), \bar{\pi}'^*(x)\right) \in K .$$

Let $\left(\dot{\varepsilon}^P, \dot{\kappa}'\right)$ be the exact admissible velocity field. Then, (1.4) gives :

$$\beta_S\left(\dot{\varepsilon}^P, \dot{\kappa}', \bar{\rho}^*, \bar{\pi}'^*, \lambda^s\right) \geq 0 .$$

More explicitly, one has :

$$\lambda^s \leq \int_\Omega \oint \left\{ b\left[\left(\dot{\varepsilon}^P, \dot{\kappa}'\right), \left(\lambda^s \sigma^{Eo} + \bar{\rho}^*, \bar{\pi}'^*\right)\right] - \dot{\kappa}' . \bar{\pi}'^* \right\} dt\, d\Omega . \qquad (3.1)$$

From (1.10) and (3.1), we deduce the inequality :

$$\begin{aligned} &\lambda^a - \int_\Omega \oint \left\{ b\left[\left(\dot{\varepsilon}^P, \dot{\kappa}'\right), \left(\lambda^a \sigma^{Eo} + \bar{\rho}, \bar{\pi}'\right)\right] - \dot{\kappa}' . \bar{\pi}' \right\} dt\, d\Omega \geq \\ &\lambda^s - \int_\Omega \oint \left\{ b\left[\left(\dot{\varepsilon}^P, \dot{\kappa}'\right), \left(\lambda^s \sigma^{Eo} + \bar{\rho}^*, \bar{\pi}'^*\right)\right] - \dot{\kappa}' . \bar{\pi}'^* \right\} dt\, d\Omega \end{aligned} \qquad (3.2)$$

For the special event of usual standard materials with separable bipotential (1.6) and no hardening variables, the previous relation degenerates, taking account of (1.7), (1.8) and (1.9) :

$$\lambda^a - \int_\Omega \oint \chi\left(\lambda^a \sigma^{Eo} + \bar{\rho}\right) dt\, d\Omega \geq \lambda^s - \int_\Omega \oint \chi\left(\lambda^s \sigma^{Eo} + \bar{\rho}^*\right) dt\, d\Omega . \tag{3.3}$$

For admissible stress fields, the values of the complementary dissipation superpotential vanish :

$$\lambda^a \geq \lambda^s .$$

Then, the inequality (3.2) represents an extension of the usual statical bounds property for standard materials to material admitting bipotentials.

4 Thin Walled Tube Under Constant Tension and Alternating Cyclic Torsion

The analytical example concerns a thin walled tube subjected to constant tension σ_{11} and alternating torsion generating an alternating cyclic shear stress state σ_{12} (Figure 1) The material behavior is governed by the non linear kinematical hardening rule of Lecture 4, section 3. Only considering the stabilized cycle, the plastic threshold R is supposed to be equal to the constant value σ_Y :

$$R = \sigma_Y .$$

On the other hand, the back stress is linearly dependent on the kinematical variables through :

$$X = \frac{2}{3} C \alpha .$$

where C is a constant kinematical hardening modulus. In this simple problem, there is no residual stress. Nevertheless, according to the remark of Lecture 1, section 9, the back stress X is an internal residual stress and plays the role of $\bar{\rho}$. Because of the plane stress state, the shifted stress tensor is as follows :

$$\sigma - X = \begin{bmatrix} \sigma_{11} - \frac{2}{3}X_{11} & \sigma_{12} - X_{12} & 0 \\ \sigma_{12} - X_{12} & \frac{1}{3}X_{11} & 0 \\ 0 & 0 & \frac{1}{3}X_{11} \end{bmatrix}$$

So, the deviatoric part is :

$$s - X = \begin{bmatrix} \frac{2}{3}(\sigma_{11} - X_{11}) & \sigma_{12} - X_{12} & 0 \\ \sigma_{12} - X_{12} & -\frac{1}{3}(\sigma_{11} - X_{11}) & 0 \\ 0 & 0 & -\frac{1}{3}(\sigma_{11} - X_{11}) \end{bmatrix}$$

Accounting for the Von-Mises criterion, the yield function is of the form :

$$f(\sigma_{11}, \sigma_{12}, X_{11}, X_{12}) = (\sigma_{eq}(\sigma - X))^2 - \sigma_Y{}^2 = (\sigma_{11} - X_{11})^2 + 3(\sigma_{12} - X_{12})^2 - \sigma_Y{}^2,$$

such that the yield criterion gives :

$$(\sigma_{11} - X_{11})^2 + 3(\sigma_{12} - X_{12})^2 \leq \sigma_Y{}^2.$$

Let us take the following transformed variables :

$$\sigma_{11} = \sigma \,,\; X_{11} = X \,,\; \sqrt{3}\sigma_{12} = \tau \;\; et \;\; \sqrt{3}X_{12} = Y. \tag{4.1}$$

The yield function then becomes :

$$f(\sigma, \tau, X, Y) = (\sigma - X)^2 + (\tau - Y)^2 - \sigma_Y{}^2.$$

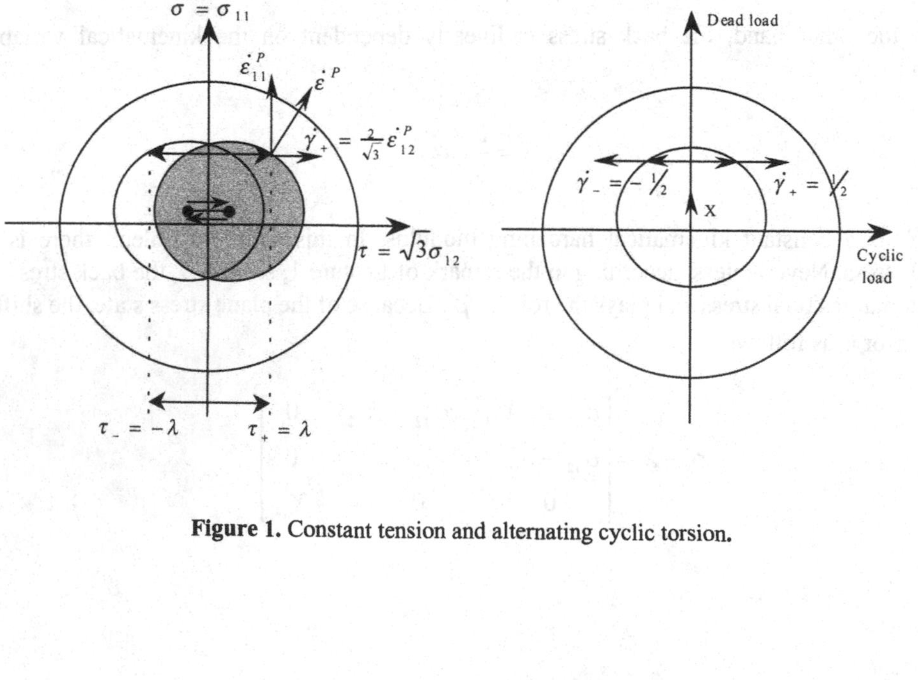

Figure 1. Constant tension and alternating cyclic torsion.

Assuming incompressibility of plastic strains, the tensor of the plastic strain rates is :

$$\dot{\varepsilon}^P = \begin{bmatrix} \dot{\varepsilon}^P_{11} & \dot{\varepsilon}^P_{12} & 0 \\ \dot{\varepsilon}^P_{12} & -\frac{1}{2}\dot{\varepsilon}^P_{11} & 0 \\ 0 & 0 & -\frac{1}{2}\dot{\varepsilon}^P_{11} \end{bmatrix}.$$

The tensor of the kinematical internal variable rates has the same form :

$$\dot{\alpha} = \begin{bmatrix} \dot{\alpha}_{11} & \dot{\alpha}_{12} & 0 \\ \dot{\alpha}_{12} & -\frac{1}{2}\dot{\alpha}_{11} & 0 \\ 0 & 0 & -\frac{1}{2}\dot{\alpha}_{11} \end{bmatrix}.$$

In the same spirit as previously for stresses, we take now :

$$\dot{\varepsilon}^P_{11} = \dot{\varepsilon} \quad and \quad \frac{2}{\sqrt{3}}\dot{\varepsilon}^P_{12} = \dot{\gamma}. \tag{4.2}$$

so, the cumulate plastic deformation becomes :

$$\dot{p} = \varepsilon_{eq}\left(\dot{\varepsilon}^P\right) = \sqrt{\frac{2}{3}\left(\frac{3}{2}\left(\dot{\varepsilon}^P_{11}\right)^2 + 2\left(\dot{\varepsilon}^P_{12}\right)^2\right)} = \sqrt{\dot{\varepsilon}^2 + \dot{\gamma}^2} \quad .$$

For the non linear kinematical hardening rule, we can write :

$$\dot{\alpha}_{11} = \dot{\varepsilon}^P_{11} - \frac{X_{11}}{X_\infty}\dot{p},\ \dot{\alpha}_{12} = \dot{\varepsilon}^P_{12} - \frac{X_{12}}{X_\infty}\dot{p}.$$

Putting :

$$\dot{\alpha} = \alpha_{11},\ \dot{\beta} = \frac{2}{\sqrt{3}}\dot{\alpha}_{12},$$

and owing to (4.1), (4.2), one gets the following condensed form :

$$\dot{\alpha} = \dot{\varepsilon} - \frac{X}{X_\infty}\dot{p}, \quad \dot{\beta} = \dot{\gamma} - \frac{Y}{X_\infty}\dot{p}.$$

For the sets of dual variables such that $(\sigma - X, R) \in K_2$, $(\dot{\varepsilon}^p, -\dot{\alpha}) \in K_2^*$, and the non-linear hardening rule (3 .3) of Lecture 4 is satisfied, the bipotential reduces to (3.4) of Lecture 4 :

$$b = \frac{(\sigma_{eq}(X))^2}{X_\infty}\dot{p} = \frac{X^2 + Y^2}{X_\infty}\sqrt{\dot{\varepsilon}^2 + \dot{\gamma}^2}\,. \tag{4.3}$$

The unit value is taken as reference shear stress, that allows to identify the load factor to the maximum shear stress. Therefore, considering cyclic loading, the state will alternate between two shear stress extremum such that, for the maximum of the cycle :

$$\sigma = \sigma, \quad \tau = \lambda \ (\tau^0 = 1), \quad \dot{\varepsilon} = \dot{\varepsilon}_+, \quad \dot{\gamma} = \dot{\gamma}_+, \quad \dot{\alpha} = \dot{\alpha}_+, \quad \dot{\beta} = \dot{\beta}_+. \tag{4.4}$$

For the minimum of the cycle, we will get :

$$\sigma = \sigma\ , \quad \tau = -\lambda \ (\tau^0 = -1)\ , \quad \dot{\varepsilon} = \dot{\varepsilon}_-\ , \quad \dot{\gamma} = \dot{\gamma}_-\ , \quad \dot{\alpha} = \dot{\alpha}_-\ , \quad \dot{\beta} = \dot{\beta}_-. \tag{4.5}$$

For sake of simplicity, a unit volume sample Ω is now considered, in order to avoid the volume integrals.

5 Calculation of the Shakedown Factor

Step 1. It is assumed that the collapse occurs by incremental collapse only in traction :

$$\oint \dot{\gamma}\, dt = 0.$$

Let us remark that non vanishing contributions to the time integral are related to the extrema of the collapse cycle. At each extremum, we consider that the velocities are constant during a unit time interval, that leads to :

$$\dot{\gamma}_+ + \dot{\gamma}_- = 0.$$

On the other hand, because of the normalization condition and (4.4), (4.5), one has :

$$\oint \tau\dot{\gamma}\,dt = \lambda\oint \tau^{o}\dot{\gamma}\,dt = \lambda(\dot{\gamma}_{+} - \dot{\gamma}_{-}) = \lambda. \tag{5.1}$$

Consequently, we get :

$$\dot{\gamma}_{+} = -\dot{\gamma}_{-} = \frac{1}{2}\text{ , and } \dot{p} = \sqrt{\dot{\varepsilon}^{2} + \frac{1}{4}}. \tag{5.2}$$

Step 2. We suppose the maxima of the cycle are located on the load surface

$$(\sigma - X)^2 + (\lambda - Y)^2 = \sigma_Y^2,$$

$$(\sigma - X)^2 + (-\lambda - Y)^2 = \sigma_Y^2.$$

The difference between the two equations gives

$$(\lambda - Y)^2 - (-\lambda - Y)^2 = -4\lambda Y = 0.$$

Because λ is non negative, $Y=0$ and the yield criteria becomes :

$$(\sigma - X)^2 + \lambda^2 = \sigma_Y^2, \tag{5.3}$$

with the following positive solution :

$$\lambda = \sigma_Y\sqrt{1 - \left(\frac{\sigma - X}{\sigma_Y}\right)^2}. \tag{5.4}$$

Step 3. Our goal now is to determine the value of X at collapse accounting for the plastic flow and hardening rules. Therefore, we calculate $\dot{\varepsilon}^P$ and $\dot{p}$, in order to get an explicit expression of X through the hardening rule. The plastic yielding rule gives :

$$\dot{\varepsilon} = \dot{\xi}\frac{\partial f}{\partial \sigma} = 2\dot{\xi}(\sigma - X), \tag{5.5}$$

$$\dot{\gamma} = \dot{\xi}\frac{\partial f}{\partial \tau} = 2\dot{\xi}(\tau - Y) = 2\dot{\xi}\tau.$$

In particular, at the extrema of the cycle, it holds :

$$\dot{\gamma}_{\pm} = \pm 2 \dot{\xi}_{\pm} \lambda .$$

Combining with relation (5.2) of step 1, the plastic multiplier is equal to :

$$\dot{\xi}_{\pm} = \frac{1}{4\lambda} .$$

then, taking account of the yield criterion (5.3), one has :

$$\dot{\varepsilon}_{\pm} = \frac{(\sigma - X)}{2\lambda} \text{, and } \dot{p}_{\pm} = \sqrt{\dot{\varepsilon}_{\pm}^{\,2} + \frac{1}{4}} = \sqrt{\frac{1}{4\lambda^2}\left((\sigma - X)^2 + \lambda^2\right)} = \frac{\sigma_Y}{2\lambda} . \tag{5.6}$$

On the other hand, the non linear hardening rule allows now to write :

$$\dot{\alpha}_{\pm} = \frac{1}{2\lambda}(\sigma - X) - \frac{X}{X_\infty}\frac{\sigma_Y}{2\lambda} = \frac{1}{2\lambda}\left(\sigma - \frac{\sigma_\infty}{X_\infty} X\right) . \tag{5.7}$$

A straightforward consequence of the previous developments is :

$$\dot{\alpha}_{+} = \dot{\alpha}_{-} . \tag{5.8}$$

Step 4. As shown by Martin (1975), the actual collapse by rachetting occurs only for a load factor greater than the shakedown one. After a transient phase, the back-stress field tends to a time periodic solution. In other words, the back-stress increment over the collapse cycle vanishes :

$$\Delta X = \oint \dot{X}\, dt = C \oint \dot{\alpha}\, dt = C(\dot{\alpha}_{+} + \dot{\alpha}_{-}) = 0 .$$

Therefore, one has :

$$\dot{\alpha}_{+} + \dot{\alpha}_{-} = 0 .$$

Combining with (5.8) gives :

$$\dot{\alpha}_{+} = \dot{\alpha}_{-} = 0 .$$

Consequently, from the expression (5.5) of $\alpha_{\pm}$, we deduce the value of the back-stress :

$$X = X_{\infty} \frac{\sigma}{\sigma_{\infty}}.$$

Then, putting it into (5.4) gives :

$$\lambda = \sigma_Y \sqrt{1 - \left(\frac{\sigma(\sigma_{\infty} - X_{\infty})}{\sigma_{\infty} \sigma_Y} \right)^2},$$

which leads to the following expression :

$$\lambda = \sigma_Y \sqrt{1 - \left(\frac{\sigma}{\sigma_{\infty}} \right)^2}. \qquad (5.9)$$

This solution is the same as the one given by Lemaitre and Chaboche (1990). In the present work, the solution is deduced from shakedown theory.

6 The Previous Solution is the Exact one

Step 1. The key idea is to consider the corresponding bifunctional :

$$\beta_S = \oint \{ b[(\sigma, X, R), (\dot{\varepsilon}^p, -\dot{\alpha}, -\dot{p})] - \sigma \dot{\varepsilon} - \tau \dot{\gamma} + X \dot{\alpha} + Y \dot{\beta} + R \dot{p} \} dt,$$

and to prove its value is zero. Accounting for the expression (4.3) of the bipotential, and the normalization condition (5.1), we simplify :

$$\beta_S = \oint \left[\frac{X^2 + Y^2}{X_{\infty}} \dot{p} - \sigma \dot{\varepsilon} + X \dot{\alpha} + Y \dot{\beta} + R \dot{p} \right] dt - \lambda.$$

Step 2. Moreover, for a stabilized cycle, $R = \sigma_Y$, and $Y = 0$ as demonstrated in the previous calculations of section 5, step 2. Then :

$$\beta_S = \oint \left[\left(\frac{X^2}{X_{\infty}} + \sigma_Y \right) \dot{p} - \sigma \dot{\varepsilon} + X \dot{\alpha} \right] dt - \lambda.$$

Because the tension stress σ acts as a dead load and the back stress X is time indépendant :

$$\beta_S = \oint \left(\frac{X^2}{X_\infty} + \sigma_Y \right) \dot{p}\, dt - \sigma \oint \dot{\varepsilon}\, dt + X \oint \dot{\alpha}\, dt - \lambda .$$

Taking into account the remark of section 5, step 1 concerning the time integrals, one has :

$$\beta_S = \left(\frac{X^2}{X_\infty} + \sigma_Y \right) (\dot{p}_+ + \dot{p}_-) - \sigma (\dot{\varepsilon}_+ + \dot{\varepsilon}_-) + X (\dot{\alpha}_+ + \dot{\alpha}_-) - \lambda .$$

Step 3. Accounting for the explicit expressions (5.6) and (5.7) of $\dot{\varepsilon}$, $\dot{p}_+$, $\dot{\alpha}_+$ previously found, the bifunctional reduces to :

$$\beta_S = 2 \left[\left(\frac{X^2}{X_\infty} + \sigma_Y \right) \frac{\sigma_Y}{2\lambda} - \sigma \frac{(\sigma - X)}{2\lambda} + \frac{X}{2\lambda} \left(\sigma - \frac{\sigma_\infty}{X_\infty} X \right) \right] - \lambda$$

$$\beta_S = \frac{1}{\lambda} \left[-(\sigma - X)^2 - \lambda^2 + \sigma_Y^2 \right]$$

Finally, taking into account that the yield criterion (5.3) at the extrema of the collapse cycle is satisfied, we prove that the bifunctional vanishes :

$$\beta_S = 0 .$$

The theoretical considerations show that the previous analytical solution (5.9) is the exact one. For the same problem, statical and kinematical approximations of the shakedown factor were previously given by Pycko and Maier (1995).

7 Exercises

Exercise 5.1. The elastic behavior of the material is isotropic with Poisson's coefficient ν. The cyclic plasticity is governed by the non linear kinematical hardening rule. For the thin walled tube under constant tension and alternating cyclic torsion in *plane strain*, prove that the shakedown factor is :

$$\lambda = \sigma_Y \sqrt{1 - ((1 - \nu + \nu^2)(4\sigma_\infty^2 + X_\infty^2) - 2(1 - 4\nu + \nu^2) X_\infty \sigma_\infty) \left(\frac{2\sigma}{4\sigma_\infty^2 - X_\infty^2} \right)^2} .$$

Exercise 5.2. (Bodovillé and de Saxcé, 2001) The cyclic plasticity is governed by the non linear kinematical hardening rule with a threshold (see Exercise 4.8). For the thin walled tube under constant tension and alternating cyclic torsion in plane stress, prove that the shakedown factor is given by (5.4), the value X of the back-stress satisfying the condition :

$$\left(\frac{(X - X_l)_+}{X_\infty}\right)^m \sigma_Y + X - \sigma = 0 .$$

If $X > X_l$, prove that this equation has one and only one solution within the interval $]X_l, \sigma[$.

References

Bodovillé, G. and de Saxcé, G. (2001). Plasticity with non-linear kinematic hardening: modelling and shakedown analysis by the bipotential approach. *European Journal of Mechanics A/Solids* 20:99–112.

Bousshine, L., Chaaba, A. and de Saxcé, G. (2001). Softening in stress-strain curve for Drucker-Prager non-associated plasticity. *International Journal of Plasticity* 17:21–46.

de Saxcé, G., Bousshine, L. (1998). The Limit Analysis Theorems for the Implicit Standard Materials: Application to the Unilateral Contact with Dry Friction and the Non Associated Flow Rules in Soils and Rocks. *International Journal of Mechanical Science*, 40:387–398.

de Saxcé, G. and Tritsch, J. B. (2000). Shakedown of elastic-plastic structures with non linear kinematical hardening by the bipotential approach. In Weichert, D., and Maier, G., eds., *Inelastic Analysis of Structures under Variable Loads : Theory and Engineering Applications*, Solid Mechanics and its Applications, 83. Dordrecht :Kluwer Academic Publishers, 167–182.

Hjiaj, M. (1999). *Algorithmes adaptés à l'analyse de structures constituées de matériaux non standards et à l'estimation a posteriori de l'erreur.* Thèse de Doctorat de la Faculté Polytechnique de Mons, Belgique.

Ladevèze, P. (1996). *Mécanique non linéaire des structures: nouvelle approche et méthodes de calcul non incrémentales.* Paris: Hermès.

Lemaitre, J. and Chaboche, J.L. (1990). *Mechanics of Solid Materials*, Cambridge University Press.

Martin, J.B. (1975). *Plasticity, fundamentals and general results.* MA: MIT Press.

Pontes, I.D.S., Borges, L.A., Zouain, N. and Andrade, I.J.P. (2000). A variational formulation and algorithm for collapse in softening materials. *Proceedings of the European Congress on Computational Methods in Applied Sciences and Engineering ECCOMAS 2000*, Barcelona.

Pycko, S. and Maier, G. (1995). Shakedown Theorems for some Classes of Nonassociative Hardening Elastic-plastic Material Models. *International Journal of Plasticity* 11:367–395.

Fundamentals of Direct Methods in Poroplasticity *

Giulio Maier and Giuseppe Cocchetti

Technical University (Politecnico) of Milan, Milan, Italy

Abstract. A nonlinear initial-boundary-value coupled problem, central to poroplasticity, is formulated under the hypotheses of small deformations, quasi-static regime, full saturation, linear Darcy diffusion law and piecewise-linearized stable and hardening poroplastic material model. After a preliminary nonconventional multifield (mixed) finite element modelling, shakedown and upper bound theorems are presented and discussed, numerically tested and applied to dam engineering situations using commercial linear and quadratic programming solvers. Limitations of the presented methodology and future prospects are discussed in the conclusions.

1 Introduction

In classical shakedown (SD) theory (see e.g. Kaliszky, 1989; König, 1987) and in several relatively recent generalizations of it (see e.g. Lloyd Smith, 1990; Maier et al., 2000; Weichert and Maier, 2000), the physical time has no role because of two basic hypotheses. In fact, (*a*) the external actions are supposed to vary slowly, so that inertia forces and viscous damping can be ignored; on the other hand, (*b*) the material constitutive model is assumed time-independent (or "inviscid" in Prager's terminology), in the sense that the material behaviour depends on the sequence of the input events (prescribed strain or stress history) but not on their rates, i.e. not on the time scale. Thus, time plays the role of ordering variable only in step-by-step analyses; it plays no role when both the above hypotheses (*a*) and (*b*) are adopted to assess structural "strength" by direct methods (then only the assigned "loading domain" matters, not the loading history within it, as shown in preceding Chapters of this book).

However, in many engineering situations, either one or the other of the above restrictive assumptions are unrealistic and must be relaxed.

The generalization of the classical SD theory to dynamics, i.e. the removal of hypothesis (*a*), started about three decades ago: Ceradini (1969) extended to dynamics Melan's lower bound "static" theorem; the dynamical counterpart of Koiter's "second" or upper bound "kinematic" theorem was established in Corradi and Maier (1973 and 1974); various Authors proposed in dynamics techniques apt to generate upper bounds on post-shakedown history-dependent quantities, such as residual displacements and plastic strains, see e.g. Capurso (1979); Corigliano et al. (1995); Genna (1991); Maier (1973); Maier and Vitiello (1974); Ponter (1975). Since the early Seventies, several important research results have been provided to dynamic SD theory and relevant analysis methods: they are dealt with elsewhere in this book.

---

* Dedicated to the memory of Professor Andrzej Gawecki.

Parallel developments (but not analogous, mainly leading to bounding procedures) occurred with reference to viscous behaviour of materials (primarily creep of metals at high temperatures), thus relaxing the above hypothesis (b), see e.g. Ponter (1972); Ponter and Williams (1973).

A third extension of SD theory and analysis to time-dependent phenomena occurring in two-phase poroplastic systems, was tackled in the last few years and is the subject of the present Chapter.

Coupled problems concerning deformable porous solids with permeating fluids slowly moving inside them, represent nowadays an important and fast-growing area of engineering mechanics. Pioneering works in this area are due to Terzaghi, Fillunger and others in the Twenties and were motivated by geotechnical and dam engineering problems. However, the classical and lasting foundation kernel for the mechanics of porous media is basically represented by Biot's linear poroelasticity theory shaped mainly in the Forties. Incentives for further developments in this area, in terms both of problem formulations and theorizations and of computational solution techniques, arise from a number of engineering situations: soil and rock mechanics implications of foundation and tunnel design; consolidation and subsidence control in environmental engineering; biomechanics of bones and soft tissues; exploitation of oil and gas deposits; design and rehabilitation of new and existing earth, masonry and concrete dams. Two books, Coussy (1995), Lewis and Schrefler (1998), contain comprehensive treatments of the accumulated knowledge concerning fluid-saturated (fully or partially), linear elastic and nonlinear inelastic porous media: in the former book, emphasis is on constitutive modelling and continuum theory; in the latter on discretizations by finite elements in space and finite differences in time and on time-marching algorithms for the solution of poroelastic and poroplastic initial-boundary-value problems.

"Direct" methods of poroplastic analysis in the present sense (namely as methods apt to assess the safety margin of a system without computing its inelastic evolution under variable loads) have so far been developed, to the writers' knowledge, only recently and under rather restrictive hypotheses, see e.g. Cocchetti and Maier (1998, 2000a and 2000b). Poroplastic SD analysis, still in its infancy, is presented in this Chapter, as for its main, simplest and application-oriented theoretical aspects and numerical techniques.

2 Problem Formulation

The following basic hypotheses are assumed herein: (*a*) linear kinematics, namely deformations are regarded as infinitesimal; (*b*) full saturation of solid skeleton by a single viscous liquid; (*c*) quasi-static regime, i.e. slowly variable external actions (but possibly fast with respect to the diffusion process) and, hence, no inertia forces; (*d*) permeability constant in time; (*e*) no fluid generation by injection within the system.

Besides equilibrium and geometric compatibility between strains and displacements of the solid, additional requirements to satisfy throughout the considered solid are conservation of the fluid mass and Darcy's linear filtration law, see e.g. Coussy (1995), Lewis and Schrefler (1998). Specifically, the field equations read:

$$\varepsilon_{ij} = \tfrac{1}{2}\left(u_{i,j} + u_{j,i}\right) \qquad \text{in } \Omega \qquad (1)$$

$$\sigma_{ij,i} + \hat{b}_j\left(x_h, \tau\right) = 0 \qquad \text{in } \Omega \qquad (2)$$

$$\dot{\zeta} = -q_{i,i} \qquad \text{in } \Omega \tag{3}$$

$$q_i = -k_{ij}\, p_{,j} + k_{ij}\, \hat{f}_j(x_h, \tau) \qquad \text{in } \Omega \tag{4}$$

The initial and boundary conditions to be associated to equations (1)-(4) are, generally, as follows:

$$p = p^0(x_h) \qquad \text{in } \Omega \text{ at } \tau = 0 \tag{5}$$

$$u_i = \hat{u}_i(x_h, \tau) \quad \text{on } \Gamma_u, \qquad \sigma_{ij} n_i = \hat{t}_j(x_h, \tau) \quad \text{on } \Gamma_\sigma \tag{6}$$

$$p = \hat{p}(x_h, \tau) \quad \text{on } \Gamma_p \qquad q_i n_i = \hat{q}(x_h, \tau) \quad \text{on } \Gamma_q \tag{7}$$

The above adopted symbols have the following meanings. The (open) domain occupied by the system is denoted by Ω, its boundary by Γ. Symbols Γ_u and Γ_σ represent the (complementary, disjoint) portions of Γ where displacements $\hat{u}_i$ of the solid skeleton and tractions $\hat{t}_i$, respectively, are assigned as functions of time τ ($\Gamma_u \cup \Gamma_\sigma = \Gamma$, $\Gamma_u \cap \Gamma_\sigma = \emptyset$). Similarly, Γ_p and Γ_q are the parts of Γ on which fluid pressure $\hat{p}$ and flux $\hat{q} = n_i q_i$, respectively, are assigned ($\Gamma_p \cup \Gamma_q = \Gamma$, $\Gamma_p \cap \Gamma_q = \emptyset$), n_i being the outward unit normal to the boundary Γ, assumed as smooth.

Tensor notation is adopted below, together with the usual index summation rule, in a Cartesian reference (x_i, i=1,2,3). Commas mean space-coordinate derivatives, dots time derivatives. Strains of the solid phase are denoted by ε_{ij}; total stresses by σ_{ij} ("total" in the sense that they concern the two-phase, "bulk" solid). The fluid content ζ and flux q_i are defined as liquid volume per unit bulk volume and, respectively, per unit time and unit crossed area orthogonal to axis x_i; $k_{ij} = k_{ji}$ will denote the permeability tensor of the porous material.

To engineering purposes, the external actions include, besides the aforementioned boundary data ($\hat{u}_i$, $\hat{t}_i$, $\hat{p}$ and $\hat{q}$): bulk body forces $\hat{b}_i$ constant in time; fluid specific weight $\hat{f}_i$ (per unit liquid volume); initial (at $\tau = 0$) field p^0 over Ω.

The nonlinear inelastic behaviour of the two-phase medium considered has to be described at the macroscopic, average level (i.e. after a hypothetical homogenization of the local properties of fluid and skeleton at the microscale of the pores). In other terms, the constitutive law to be combined with the preceding field equations must relate static quantities σ_{ij} and p to the kinematic ones ε_{ij} and ζ, work-conjugate to the former, each one of the latter variables being the sum of a reversible, poroelastic addend and an irreversible, poroplastic addend (marked by superscript e and p, respectively), according to an additivity assumption like in (monophase) elastoplasticity:

$$\varepsilon_{ij} = \varepsilon_{ij}^e + \varepsilon_{ij}^p, \qquad \zeta = \zeta^e + \zeta^p \tag{8}$$

The former poroelastic addends are linearly related to the static variables according to Biot's poroelastic model, which is governed, for isotropic materials, by four material parameters to identify by means of laboratory experiments, see e.g. Coussy (1995), Lewis and Schrefler (1998). Biot's model reads:

$$2G\varepsilon_{ij}^{e} = \sigma_{ij} - \frac{\nu}{1+\nu}\sigma_{hh}\delta_{ij} + \alpha\frac{1-2\nu}{1+\nu}p\,\delta_{ij} \tag{9a}$$

$$\zeta^{e} = \frac{\alpha}{2G}\frac{1-2\nu}{1+\nu}\sigma_{hh}^{e} + m^{-1}p \tag{9b}$$

The material constant m can be shown to depend on the other four parameters (shear modulus G of the solid; drained Poisson ratio ν; Biot modulus M; Biot coefficient of effective stress α with $0 \le \alpha \le 1$ (for $\alpha = 1$ the l.h.s. of Eq. (9a) becomes Terzaghi's "effective stress"):

$$m = \left(\frac{1}{M} + \frac{3\alpha^2}{2G}\frac{1-2\nu}{1+\nu}\right)^{-1} \tag{10}$$

The latter, poroplastic addends in equation (8) reflect re-arrangements at the material microscale, which capture nonrecoverable fluid content ζ^p and cause residual, "permanent" strains in the solid skeleton. Their evolution is modelled by means of the following relations, similar to those recurrent in conventional (monophase) plasticity:

$$\dot{\varepsilon}_{ij}^{p} = \frac{\partial\tilde{\varphi}_\alpha}{\partial\sigma_{ij}}\dot{\lambda}_\alpha, \qquad \dot{\zeta}^{p} = \frac{\partial\tilde{\varphi}_\alpha}{\partial p}\dot{\lambda}_\alpha \tag{11}$$

$$\tilde{\varphi}_\alpha = \tilde{\varphi}_\alpha\left(\sigma_{ij}, p, \chi_r\right) \qquad \alpha = 1\ldots n_y \tag{12}$$

$$\varphi_\alpha = \varphi_\alpha\left(\sigma_{ij}, p, \chi_r\right) \le 0, \qquad \dot{\lambda}_\alpha \ge 0, \qquad \varphi_\alpha\dot{\lambda}_\alpha = 0 \tag{13}$$

$$\chi_r = \frac{\partial U}{\partial\eta_r}(\eta_s), \qquad \dot{\eta}_r = -\frac{\partial\tilde{\varphi}_\alpha}{\partial\chi_r}\dot{\lambda}_\alpha \qquad r,s = 1\ldots n_\nu \tag{14}$$

The basic features of nonassociative hardening (or softening) plasticity are preserved: for each (α-*th*) of n_y "yield modes" ($\alpha = 1\ldots n_y$), a plastic potential $\tilde{\varphi}_\alpha$ and a yield function φ_α are envisaged, Eqs. (12) and (13). The latter functions φ_α define, through inequality (13a), the current "poroelastic domain", convex by hypothesis, in the space of the static variables σ_{ij}, p; the former $\tilde{\varphi}_\alpha$ through Eqs. (11) define, in the superposed space of conjugate "kinematic" variables $\left\{\varepsilon_{ij}, \zeta\right\}$, the "direction" of the rates $\left\{\dot{\varepsilon}_{ij}^{p}, \dot{\zeta}^{p}\right\}$ and relate them to the plastic multipliers λ_α, which represent a measure of the contribution of each (α-*th*) yield mode to the irreversible processes. The complementarity relations (13) represent Prager's "consistency" rule of loading-unloading. The n_ν static internal variables χ_r ($r = 1\ldots n_\nu$), in the argument of $\tilde{\varphi}_\alpha$ and φ_α, govern the hardening (or softening) behaviour. They are related to their conjugate kinematic counterparts η_r by Eq. (14a) through the potential $U(\eta_s)$ of "free energy" (not dissipated into heat, but stored in the material texture because of possible irreversible rearrangements at the microscale). The kinematic internal variables η_r are controlled through Eq. (14b) by the plastic multipliers.

In order to simplify later theoretical developments and numerical solutions for SD analyses, like often in structural elastoplasticity, see e.g. Lloyd Smith (1990), Maier (1969, 1970 and 1976), Sloan (1988), Tin-Loi (1990), the poroplastic relations (11)-(14) are given a "piecewise linear" (PWL) approximation through the specializations which follow

(a) for each mode α, both the yield function φ_α and the plastic potential $\tilde{\varphi}_\alpha$ are assumed linear in their arguments (σ_{ij}, p, χ_r) and the gradients of the latter with respect to (σ_{ij}, p) will be normalized (by setting their moduli to one), namely:

$$\frac{\partial \varphi_\alpha}{\partial \sigma_{ij}} = N^\sigma_{ij\alpha}, \qquad \frac{\partial \varphi_\alpha}{\partial p} = N^p_\alpha; \qquad \frac{\partial \tilde{\varphi}_\alpha}{\partial \sigma_{ij}} = \tilde{N}^\sigma_{ij\alpha}, \qquad \frac{\partial \tilde{\varphi}_\alpha}{\partial p} = \tilde{N}^p_\alpha \tag{15}$$

(b) the gradients with respect to the internal variables χ_r are further specialized as follows (δ being the Kronecker symbol):

$$\frac{\partial \varphi_\alpha}{\partial \chi_r} = \frac{\partial \tilde{\varphi}_\alpha}{\partial \chi_r} = -\delta_{\alpha r}, \qquad \text{with} \quad \alpha = 1...n_y, \quad r = 1...n_\nu \tag{16}$$

which implies, through Eq. (14b), the coincidence of plastic multipliers and kinematic internal variables, namely: $\dot{\eta}_\alpha = \delta_{\alpha r}\,\dot{\lambda}_r$, whence $\eta_\alpha = \lambda_\alpha$ and $n_y = n_\nu$.

(c) the internal variable potential U is a quadratic function and, therefore, Eq. (14a) becomes:

$$\chi_\alpha = \frac{\partial U}{\partial \eta_\alpha} = H_{\alpha\beta}\,\eta_\beta, \qquad \alpha, \beta = 1...n_y \tag{17}$$

By virtue of specializations (a)-(c), equations (11)-(14) reduce to the following set of linear inequalities and equations:

$$\varphi_\alpha = N^\sigma_{ij\alpha}\,\sigma_{ij} + N^p_\alpha\,p - Y_\alpha \le 0, \qquad Y_\alpha = Y^0_\alpha + H_{\alpha\beta}\,\lambda_\beta \tag{18}$$

$$\dot{\varepsilon}^p_{ij} = \tilde{N}^\sigma_{ij\alpha}\,\dot{\lambda}_\alpha, \qquad \dot{\zeta}^p = \tilde{N}^p_\alpha\,\dot{\lambda}_\alpha \tag{19}$$

$$\dot{\lambda}_\alpha \ge 0, \qquad \varphi_\alpha \dot{\lambda}_\alpha = 0, \qquad \alpha = 1...n_y \tag{20}$$

where the quantities denoted by N, $\tilde{N}$ and H are constants and Y^0_α represents the (positive) initial "yield limit" of the α-*th* mode.

Interpreting Eqs. (18)-(20) geometrically in the $\{\sigma_{ij}, p\}$ and $\{\varepsilon_{ij}^{p}, \zeta^{p}\}$ spaces superposed, the poroelastic domain is the convex (hyper) polyhedron consisting of the intersection of the n_y half-spaces $\varphi_\alpha \le 0$. Each yield plane $\varphi_\alpha = 0$ ($\alpha = 1...n_y$) is defined by its unit normal $\{N_{ij}^{\sigma}\ N^{p}\}_{\alpha}$ and by its distance from the origin (and current yield limit) Y_α, Eqs. (18b) and (17); generally, it translates at yielding of mode $\beta = 1...n_y$ (forwards or backwards, depending on the sign of $H_{\alpha\beta}$); its possible contribution to plastic "deformations" $\{\varepsilon_{ij}^{p}, \zeta^{p}\}$ is directed as the (fixed) gradient $\{\tilde{N}_{ij}^{\sigma}, \tilde{N}^{p}\}_{\alpha}$ of the relevant plastic potential $\tilde{\varphi}_\alpha$.

In equation (17), the Hessian of the potential U is the "hardening matrix" $H_{\alpha\beta}$ which governs the translations of the yield planes and the interactions among them through Eqs. (18): its diagonal entries $H_{\alpha\alpha}$ are "direct" hardening moduli; the "indirect" ones $H_{\alpha\beta}$ with $\alpha \neq \beta$ describe the effect on mode α due to the "activation" of mode β. The symmetry $H_{\alpha\beta} = H_{\beta\alpha}$ ("reciprocal hardening") is entailed by the adoption of the potential U, but a nonsymmetric hardening matrix ("nonreciprocal" hardening) might be directly introduced in the definition of φ_α, Eq. (18a). Peculiar features of matrix $H_{\alpha\beta}$ apt to represent diverse types of hardening (kinematic, isotropic and mixed) have been pointed out in Maier (1970 and 1976). In associative cases ($\varphi_\alpha = \tilde{\varphi}_\alpha$), positive semi-definiteness of $H_{\alpha\beta}$ is sufficient for material stability in Drucker sense of classical plasticity.

It is worth noting that, in view of Eqs. (12) and (18), the plastic potentials read:

$$\tilde{\varphi}_\alpha = \tilde{N}_{ij\alpha}^{\sigma}\,\sigma_{ij} + \tilde{N}_{\alpha}^{p}\,p - H_{\alpha\beta}\,\lambda_\beta - Y_\alpha^{0} \le 0, \qquad \alpha,\beta = 1,\ldots,n_y \tag{21}$$

Denoting by $\dot{\mathcal{D}}$ the local (per unit volume) dissipation rate due to irreversible deformation processes, "adaptation" or SD will be characterized by the inequality:

$$D_\infty = \lim_{\tau\to\infty} \int_0^\tau \int_\Omega \dot{\mathcal{D}}(x_i,\tau')\,d\Omega\,d\tau' = \lim_{\tau\to\infty} \int_0^\tau \int_\Omega \left(\sigma_{ij}\dot{\varepsilon}_{ij}^{p} + p\dot{\zeta}^{p} - \chi_r\dot{\eta}_r\right) d\Omega\,d\tau' < \infty \tag{22}$$

Like in plasticity (Kaliszky, 1989; König, 1987) lack of SD, i.e. unbounded cumulative dissipation, manifests itself either as incremental collapse (unlimited growth of displacements u_i) or as alternating yielding (or low-cycle fatigue).

In the present PWL context, the energy characterization of shakedown by inequality (22) becomes:

$$\lim_{\tau\to\infty} = \int_0^\tau \int_\Omega \left[\sigma_{ij}\left(\tilde{N}_{ij\alpha}^{\sigma} - N_{ij\alpha}^{\sigma}\right)\dot{\lambda}_\alpha + p\left(\tilde{N}_{\alpha}^{p} - N_{\alpha}^{p}\right)\dot{\lambda}_\alpha + Y_\alpha^{0}\,\dot{\lambda}_\alpha\right] d\Omega\,d\tau' < \infty \tag{23}$$

In associative poroplasticity ($\varphi_\alpha = \tilde{\varphi}_\alpha$) with symmetric hardening ($H_{\alpha\beta} = H_{\beta\alpha}$), the SD condition reduces to:

$$\lim_{\tau\to\infty} \int_0^\tau \int_\Omega Y_\alpha \, \lambda_\alpha(\tau') \; d\Omega \, d\tau' = \\ = \lim_{\tau\to\infty} \left[Y_\alpha^0 \int_\Omega \lambda_\alpha(\tau) \; d\Omega + \tfrac{1}{2} \int_\Omega \lambda_\alpha(\tau) H_{\alpha\beta} \lambda_\beta(\tau) \; d\Omega \right] < \infty \tag{24}$$

Consider the boundedness in time of the plastic multipliers:

$$\lim_{\tau\to\infty} \lambda_\alpha(x_i, \tau) < \infty, \qquad \forall x_i \in \Omega, \;\; \forall \alpha \;\; \left(\alpha = 1 \ldots n_y\right) \tag{25}$$

Clearly, Eq. (25) implies (24) if Y_α^0 and Ω are bounded, a practically weak restriction which is assumed herein.

The purposes of what follows are: (a) to establish sufficient and necessary conditions for SD in the sense of equation (25), in order to determine the safety factor with respect to lack of SD by a direct (non time-marching) procedure; (b) to derive upper bounds on post-shakedown, history-dependent quantities by direct techniques.

Since both purposes will be pursued primarily as bases for practical cost-effective analysis tools, finite element modelling is preliminarily performed in the next Section.

3 Space Discretization

The continuum formulation presented in Section 2 will be discretized here by the non-conventional mixed finite element (FE) approximation proposed in Maier and Comi (1997) on a variational basis and further developed by a Galerkin weighted residuals approach and numerically tested in Cocchetti and Maier (2000a). Cumbersome formulae, available in Cocchetti and Maier (2000a), Maier and Comi (1997), are skipped here and replaced by the following conceptual outline of the particular multifield space-modelling procedure adopted.

(a) All fields and constitutive equations are enforced "in a weak sense", namely by setting to zero their integrals "weighted" by means of the relevant work-conjugate fields, for any choice of these. For instance, as for the compatibility equation (1):

$$\int_\Omega \sigma_{ij} \left[\varepsilon_{ij} - \tfrac{1}{2}\left(u_{i,j} + u_{j,i}\right)\right] d\Omega = 0 \qquad \forall \sigma_{ij} \tag{26}$$

(b) In each pair of conjugate fields (namely, in the pairs: $\varepsilon_{ij}, \sigma_{ij}$; ζ, p; η_r, χ_r; q_i, π_i; $\lambda_\alpha, \varphi_\alpha$) the shape functions (denoted by Ψ) are chosen, indipendently for each pair, by satisfying not only suitable continuity conditions at element interfaces, but also the condition that the dot-product over the domain Ω of conjugate modelled fields equals the one between the generalized variable vectors governing them, namely, e.g.:

$$\int_{\Omega} \sigma_{ij}\, \varepsilon_{ij}\, d\Omega = \bar{\boldsymbol{\sigma}}^T \bar{\boldsymbol{\varepsilon}} \qquad \forall\, \bar{\boldsymbol{\sigma}}, \bar{\boldsymbol{\varepsilon}} \tag{27}$$

(c) After the above finite element modelling, the weak enforcement (a) for any weight-governing vectors leads to a space-discrete approximation, in the Galerkin weighted residuals sense, of the relationships formulated in Section 2.

(d) The boundary data will be entered now in terms of constraints imposed on generalized variables.

The following formalism of matrix algebra is adopted henceforth: bold-face symbols denote matrices and, in particular, vectors; $\boldsymbol{\varepsilon}$ gathers the independent strains in "engineering sense"; superscript T means transpose; $\mathbf{B}$ represents the (linear, differential) operator of geometric compatibility (so that $\boldsymbol{\varepsilon} = \mathbf{B}\,\mathbf{u}$); $\mathbf{B}^T$ is the equilibrium operator; inequalities apply componentwise etc… Vectors of generalized variables governing the modelled fields, or generated by integrations and matrix operators concerning them, will be denoted by bars over the relevant symbols used for the same quantities in the continuum description. Symbols for vectors of data will be capped. The shape (interpolation) functions are understood to be defined over the whole finite element (FE) aggregate covering the domain Ω (not only over the "supports" of the relevant nodes), so that the assemblage is implicitly entailed and the vectors of generalized variables concern the entire space-discrete model.

With the above notation, the set of approximate governing relations, space-discretized according to the guidelines (a)-(c), and with piecewise linear (PWL) material models introduced earlier, reads:

$$\bar{\boldsymbol{\varepsilon}} = \bar{\mathbf{B}}\,\bar{\mathbf{u}}, \qquad \bar{\mathbf{B}}^T \bar{\boldsymbol{\sigma}} = \hat{\bar{\mathbf{b}}} + \bar{\mathbf{t}} \tag{28}$$

$$\dot{\bar{\boldsymbol{\zeta}}} = \bar{\mathbf{G}}^T \bar{\mathbf{q}} - \bar{\mathbf{q}}_{\Gamma}, \qquad \bar{\mathbf{q}} = \bar{\mathbf{k}}\,\bar{\boldsymbol{\pi}} = -\bar{\mathbf{k}}\,\bar{\mathbf{G}}\,\bar{\mathbf{p}} + \bar{\mathbf{k}}\,\hat{\bar{\mathbf{f}}} \tag{29}$$

$$\bar{\mathbf{p}} = \bar{\mathbf{p}}_0 \qquad \text{at } \tau = 0 \tag{30}$$

$$\bar{\boldsymbol{\varepsilon}} = \bar{\boldsymbol{\varepsilon}}^e + \bar{\boldsymbol{\varepsilon}}^p, \qquad \bar{\boldsymbol{\zeta}} = \bar{\boldsymbol{\zeta}}^e + \bar{\boldsymbol{\zeta}}^p \tag{31}$$

$$\bar{\boldsymbol{\varepsilon}}^e = \bar{\mathbf{E}}^{-1} \bar{\boldsymbol{\sigma}} + \bar{\mathbf{P}}\,\bar{\mathbf{p}}, \qquad \bar{\boldsymbol{\varepsilon}}^e = \bar{\mathbf{P}}^T \bar{\boldsymbol{\sigma}} + \bar{\mathbf{m}}^{-1}\bar{\mathbf{p}} \tag{32}$$

$$\bar{\boldsymbol{\varphi}} = \bar{\mathbf{N}} \begin{Bmatrix} \bar{\boldsymbol{\sigma}} \\ \bar{\mathbf{p}} \end{Bmatrix} - \bar{\mathbf{Y}} \le \mathbf{0}, \qquad \bar{\mathbf{Y}} = \bar{\mathbf{Y}}^0 + \bar{\mathbf{H}}\,\bar{\boldsymbol{\lambda}} \tag{33}$$

$$\begin{Bmatrix} \dot{\bar{\boldsymbol{\varepsilon}}}^p \\ \dot{\bar{\boldsymbol{\zeta}}}^p \end{Bmatrix} = \tilde{\bar{\mathbf{N}}}^T \dot{\bar{\boldsymbol{\lambda}}}, \qquad \dot{\bar{\boldsymbol{\lambda}}} \ge \mathbf{0}, \qquad \bar{\boldsymbol{\varphi}}^T \dot{\bar{\boldsymbol{\lambda}}} = 0 \tag{34}$$

having set:

$$\bar{\mathbf{N}} = \left[\bar{\mathbf{N}}^{\sigma} \;\; \bar{\mathbf{N}}^{p} \right], \qquad \tilde{\bar{\mathbf{N}}} = \left[\tilde{\bar{\mathbf{N}}}^{\sigma} \;\; \tilde{\bar{\mathbf{N}}}^{p} \right] \tag{35}$$

The coefficient matrices and vectors which show up in the above formulation arise quite naturally from the modelling process. For instance, if ∇ denotes the gradient operator, $\mathbf{n}$ the outward normal of the boundary Γ and the interpolation functions are gathered in matrices denoted by $\boldsymbol{\Psi}$ with appropriate subscript, we have:

$$\overline{\mathbf{B}} = \int_{\Omega} \boldsymbol{\Psi}_{\sigma}^{T} \mathbf{B} \boldsymbol{\Psi}_{\varepsilon} \, d\Omega; \qquad \hat{\overline{\mathbf{b}}} = \int_{\Omega} \boldsymbol{\Psi}_{u}^{T} \mathbf{b} \, d\Omega; \qquad \overline{\mathbf{t}} = \int_{\Gamma} \boldsymbol{\Psi}_{u}^{T} \mathbf{t} \, d\Gamma \tag{36}$$

$$\overline{\mathbf{G}} = \int_{\Omega} \boldsymbol{\Psi}_{q}^{T} \nabla \boldsymbol{\Psi}_{p} \, d\Omega; \qquad \overline{\mathbf{q}}_{\Gamma} = \int_{\Gamma} \boldsymbol{\Psi}_{p}^{T} \mathbf{n}^{T} \mathbf{q} \, d\Gamma; \qquad \hat{\overline{\mathbf{f}}} = \int_{\Omega} \boldsymbol{\Psi}_{q}^{T} \hat{\mathbf{f}} \, d\Omega \tag{37}$$

By substituting variables $\overline{\mathbf{q}}$, $\overline{\boldsymbol{\varepsilon}}$, $\overline{\boldsymbol{\zeta}}$, the overall (for the whole element aggregate) linear poroelastic constitution (32) and the discrete field equations (28)-(30) read, respectively:

$$\begin{Bmatrix} \overline{\boldsymbol{\varepsilon}}^{e} \\ \overline{\boldsymbol{\zeta}}^{e} \end{Bmatrix} = \overline{\mathbf{C}} \begin{Bmatrix} \overline{\boldsymbol{\sigma}} \\ \overline{\mathbf{p}} \end{Bmatrix}, \qquad \text{with } \overline{\mathbf{C}} = \begin{bmatrix} \overline{\mathbf{E}}^{-1} & \overline{\mathbf{P}} \\ \overline{\mathbf{P}}^{T} & \overline{\mathbf{m}}^{-1} \end{bmatrix} \tag{38}$$

$$\overline{\mathbf{B}}\,\overline{\mathbf{u}} = \overline{\boldsymbol{\varepsilon}}^{e} + \overline{\boldsymbol{\varepsilon}}^{p}, \qquad \overline{\mathbf{B}}^{T} \overline{\boldsymbol{\sigma}} = \hat{\overline{\mathbf{b}}} + \overline{\mathbf{t}} \tag{39}$$

$$\dot{\overline{\boldsymbol{\zeta}}}^{e} + \dot{\overline{\boldsymbol{\zeta}}}^{p} = -\overline{\mathbf{V}}\,\overline{\mathbf{p}} + \overline{\mathbf{F}}, \quad \text{where: } \overline{\mathbf{V}} \equiv \overline{\mathbf{G}}^{\mathrm{T}} \overline{\mathbf{k}}\, \overline{\mathbf{G}}, \quad \overline{\mathbf{F}} = \overline{\mathbf{G}}^{T} \overline{\mathbf{k}}\, \hat{\overline{\mathbf{f}}} - \overline{\mathbf{q}}_{\Gamma} \tag{40}$$

$$\overline{\mathbf{p}}(\tau = 0) = \overline{\mathbf{p}}_{0} \tag{41}$$

By further substituting vectors $\overline{\boldsymbol{\varepsilon}}^{e}$, $\overline{\boldsymbol{\zeta}}^{e}$ and $\overline{\boldsymbol{\sigma}}$, Eqs. (38)-(40) are re-cast below into the following more compact form (where the meaning of the new symbols trivially emerge from the performed substitutions):

$$\mathbf{K}\overline{\mathbf{u}} - \mathbf{L}\overline{\mathbf{p}} = \hat{\overline{\mathbf{b}}} + \overline{\mathbf{t}} + \mathbf{R}\overline{\boldsymbol{\varepsilon}}^{p} \tag{42}$$

$$\mathbf{L}^{\mathrm{T}} \dot{\overline{\mathbf{u}}} + \mathbf{S}\dot{\overline{\mathbf{p}}} + \overline{\mathbf{V}}\,\overline{\mathbf{p}} = \mathbf{Q}\dot{\overline{\boldsymbol{\varepsilon}}}^{p} - \dot{\overline{\boldsymbol{\zeta}}}^{p} + \overline{\mathbf{G}}^{\mathrm{T}} \overline{\mathbf{k}}\, \hat{\overline{\mathbf{f}}} - \overline{\mathbf{q}}_{\Gamma} \tag{43}$$

The external actions over the domain Ω are body forces of bulk $\hat{\overline{\mathbf{b}}}$ and fluid $\hat{\overline{\mathbf{f}}}$, regarded as "dead loads"(imposed strains such as thermal effects being ignored for brevity). External actions acting on the boundary Γ are: imposed displacements on Γ_u and fluxes on Γ_q, both assumed herein, in view of the examples of Section 7, as zero at any time (impervious Γ_q); tractions on Γ_σ and pressure on Γ_p, both regarded as combinations of "live loads" affected by a multiplier μ (load factor) and "dead loads" constant in time (not affected by μ).

Account taken of the above assumptions on the external actions, the discretized field equations (42) and (43) will be re-written below, Eqs. (44)-(45), with the following re-interpretation of the symbols: $\bar{\mathbf{u}}$ collects the unconstrained, unknown nodal displacements alone and, accordingly, all equations corresponding to constraints ($\hat{\bar{\mathbf{u}}} = \mathbf{0}$) will be dropped from Eq. (42); similarly, $\bar{\mathbf{p}}$ will henceforth gather only unknown nodal pressures and, accordingly, all equations corresponding to assigned boundary pressures ($\hat{\bar{\mathbf{p}}}$) will be eliminated from Eq. (43). After these provisions: vector $\bar{\mathbf{t}}$ contains only data (namely, given tractions and zeros) and, hence, will be indicated by $\hat{\bar{\mathbf{t}}}$; similarly, vector $\bar{\mathbf{q}}$ contains only data (given fluxes and zeros) and, hence, will be indicated by $\hat{\bar{\mathbf{q}}}$ and, in view of the above assumptions, will be set to zero ($\hat{\bar{\mathbf{q}}} = \mathbf{0}$); vector $\dot{\bar{\boldsymbol{\zeta}}}^p$ gathers only unknown fluid contents in nodes on Ω and on Γ_q.

As a consequence of the preceding definition of external actions and re-interpretation of symbols, Eqs. (42) and (43) become:

$$\bar{\mathbf{K}}\bar{\mathbf{u}} - \bar{\mathbf{L}}\bar{\mathbf{p}} = \bar{\mathbf{c}} + \bar{\mathbf{R}}\bar{\boldsymbol{\varepsilon}}^p \tag{44}$$

$$\bar{\mathbf{L}}^{\mathrm{T}}\dot{\bar{\mathbf{u}}} + \bar{\mathbf{S}}\dot{\bar{\mathbf{p}}} + \bar{\mathbf{V}}\bar{\mathbf{p}} = \bar{\mathbf{d}} + \bar{\mathbf{P}}\dot{\bar{\boldsymbol{\varepsilon}}}^p - \dot{\bar{\boldsymbol{\zeta}}}^p + \bar{\mathbf{G}}^{\mathrm{T}}\bar{\mathbf{k}}\hat{\bar{\mathbf{f}}} \tag{45}$$

having set:

$$\bar{\mathbf{c}} = \hat{\bar{\mathbf{b}}} + \hat{\bar{\mathbf{t}}} + \bar{\bar{\mathbf{L}}}\hat{\bar{\mathbf{p}}} = \mu\bar{\mathbf{c}}' + \bar{\mathbf{c}}'' \tag{46}$$

$$\bar{\mathbf{d}} = \bar{\bar{\mathbf{S}}}\dot{\hat{\bar{\mathbf{p}}}} + \bar{\bar{\mathbf{V}}}\hat{\bar{\mathbf{p}}} = \mu\bar{\mathbf{d}}' + \bar{\mathbf{d}}'' \tag{47}$$

The above outlined discretization procedure, called discretization in "Prager's generalized variables", was presented in detail in Cocchetti and Maier (2000a), Maier and Comi (1997) (and anticipated in Maier, 1970 for the special case of plasticity). It exhibits the following peculiar property: the energy meaning of dot products and the essential features of key operators (symmetry, sign (semi) definiteness) are preserved in passing from the continuum to the discrete formulation of the problem. As it is easily seen, the fulfilment of crucial requirements like (27) can be achieved simply by deriving, in each pair of conjugate fields, the interpolation matrix (say $\boldsymbol{\Psi}_\sigma$) for one field from that for the other (say $\boldsymbol{\Psi}_\varepsilon$), through a relation of the following kind (referring e.g. to stresses and strains):

$$\boldsymbol{\Psi}_\sigma = \boldsymbol{\Psi}_\varepsilon \left[\int_\Omega \boldsymbol{\Psi}_\varepsilon^T(\mathbf{x})\, \boldsymbol{\Psi}_\varepsilon(\mathbf{x})\, d\Omega \right]^{-1} \tag{48}$$

The afore mentioned preservation of essential features is especially advantageous here, because it legitimates theoretical developments based on the space-discretized formulation as those carried out in the subsequent Sections. With respect to the continuum formulation of Section 2, the discrete formulation in Prager's generalized variables turns out to be formally and mathematically simpler, more concise, more open to physical insight and closer to computational techniques and engineering applications.

4 Static Shakedown Theorems

The developments presented in this Section are based on the piecewise linear (PWL) approximation of the two-phase material model of Section 2, specialized to associative perfect plasticity, and on the discrete FE model formulated in Section 3. On this restricted basis, two statements (I and II) will be proved below. They represent a generalization to poroplasticity of the static (safe, lower-bound, Melan's) SD theorems of classical plasticity, which in turn can be regarded as generalizations of the static theorems of limit analysis.

(I) Sufficient Condition. Shakedown does occur if a steady-state poroelastic response to fictitious imposed constant strains $\bar{\boldsymbol{\varepsilon}}^{p}$ (specifically, consequent constant self-stresses $\bar{\boldsymbol{\sigma}}^{S}$) exists such that the superposition of it on the fictitious poroelastic response (marked by superscript *E*) to the external actions, after a finite time τ^{*}, strictly satisfies the constitutive yield inequalities, Eq. (33).

(II) Necessary Condition. If the poroplastic system shakes down in the sense that dissipative inelastic phenomena cease after a time τ^{*}, then there are time-independent self-stresses such that the superposition of them on the poroelastic response (superscript *E*) after a time τ^{*} satisfies strictly the yield inequalities with arbitrarily slight increases of the yield limits.

Proof of Statement I. Preliminarily, consider the solutions to the following fictitious linear-poroelastic problems.

(a) If we assume $\bar{\boldsymbol{\varepsilon}}^{p}=\mathbf{0}$ and $\bar{\boldsymbol{\zeta}}^{p}=\mathbf{0}$ identically, Eqs. (44)-(45) uniquely define, as functions of time τ, the poroelastic response to loads $\bar{\mathbf{u}}^{E}, \bar{\boldsymbol{\sigma}}^{E}, \bar{\mathbf{p}}^{E}, \bar{\boldsymbol{\varepsilon}}^{E}, \bar{\boldsymbol{\zeta}}^{E}$, referred to henceforth as "fictitious poroelastic" process (*E*). This can be computed preliminarily on the basis of the time history of the external actions.

(b) Let $\bar{\boldsymbol{\varepsilon}}^{p}(\tau)$ and $\bar{\boldsymbol{\zeta}}^{p}(\tau)$ be identified in Eqs. (44)-(45) as the actual irreversibility manifestations of the system in its poroplastic evolution and let them represent the only imposed external actions with homogeneous boundary and initial conditions ($\hat{\mathbf{t}}=\mathbf{0}$, $\hat{\mathbf{p}}=\mathbf{0}$ and $\hat{\mathbf{p}}_{0}=\mathbf{0}$). Clearly, the consequent poroelastic solution, or "complementary process" (*C*), provides the terms (marked by *C*) to be added by superposition to the solution (a) (process *E*) in order to obtain the actual poroplastic response. For stresses and strains, e.g.:

$$\bar{\boldsymbol{\sigma}}=\bar{\boldsymbol{\sigma}}^{E}+\bar{\mathbf{E}}_{\varepsilon}\left(\bar{\boldsymbol{\varepsilon}}^{C}-\bar{\boldsymbol{\varepsilon}}^{p}\right)+\bar{\mathbf{E}}_{\varepsilon\zeta}\left(\bar{\boldsymbol{\zeta}}^{C}-\bar{\boldsymbol{\zeta}}^{p}\right), \qquad \bar{\boldsymbol{\varepsilon}}=\bar{\boldsymbol{\varepsilon}}^{E}+\bar{\boldsymbol{\varepsilon}}^{C} \tag{49}$$

(c) Set $\hat{\mathbf{t}}=\mathbf{0}$, $\hat{\mathbf{p}}=\mathbf{0}$, $\hat{\mathbf{p}}_{0}=\mathbf{0}$ and assume as fictitious external actions $\bar{\boldsymbol{\varepsilon}}^{p*}$ and $\bar{\boldsymbol{\zeta}}^{p*}$ constant in time. The consequent fictitious poroelastic stationary state (superscript *S*) to which the system tends for $t \to \infty$ is readily seen, from Eqs. (44)-(45), to be independent from constant $\bar{\boldsymbol{\zeta}}^{p*}$ and to exhibit the following features:

$$\bar{\mathbf{u}}^S = \bar{\mathbf{K}}^{-1}\bar{\mathbf{R}}\,\bar{\boldsymbol{\varepsilon}}^p, \qquad \bar{\mathbf{p}}^S = \mathbf{0}, \qquad \bar{\mathbf{q}}^S = \mathbf{0} \qquad \bar{\boldsymbol{\varepsilon}}^{eS} = \bar{\mathbf{C}}\,\bar{\mathbf{u}}^S - \bar{\boldsymbol{\varepsilon}}^{p*} = \bar{\mathbf{E}}_\varepsilon^{-1}\bar{\boldsymbol{\sigma}}^S \tag{50}$$

Account taken of Eqs. (50), consider now the following energy norm of the difference between the complementary process (*C*), see solution (b), and the fictitious poroelastic steady state (*S*), see solutions (C):

$$\chi(\tau) \equiv \frac{1}{2}\left(\begin{bmatrix}\bar{\boldsymbol{\sigma}}^C \\ \bar{\mathbf{p}}^C\end{bmatrix} - \begin{bmatrix}\bar{\boldsymbol{\sigma}}^S \\ \mathbf{0}\end{bmatrix}\right)^T \bar{\mathbf{E}}^{-1}\left(\begin{bmatrix}\bar{\boldsymbol{\sigma}}^C \\ \bar{\mathbf{p}}^C\end{bmatrix} - \begin{bmatrix}\bar{\boldsymbol{\sigma}}^S \\ \mathbf{0}\end{bmatrix}\right) + \int_0^\tau \left(\bar{\mathbf{q}}^C\right)^T \bar{\mathbf{k}}^{-1}\bar{\mathbf{q}}^C\, d\tau' \geq 0 \tag{51}$$

The inequality ($\chi \geq 0$) is implied by the positive definiteness of the poroelastic matrix $\bar{\mathbf{E}}$ and of the permeability matrix $\bar{\mathbf{k}}$.

Making use of the poroelasticity constitutive equations and of Darcy law, the time derivative of function χ can be given the expression:

$$\dot{\chi}(\tau) = \begin{bmatrix}\dot{\bar{\boldsymbol{\varepsilon}}}^{eC} \\ \dot{\bar{\boldsymbol{\zeta}}}^{eC}\end{bmatrix}^T \left(\begin{bmatrix}\bar{\boldsymbol{\sigma}}^C \\ \bar{\mathbf{p}}^C\end{bmatrix} - \begin{bmatrix}\bar{\boldsymbol{\sigma}}^S \\ \mathbf{0}\end{bmatrix}\right) - \left(\bar{\mathbf{q}}^C\right)^T \bar{\mathbf{G}}\,\bar{\mathbf{p}}^C \tag{52}$$

As a generalization of the virtual work principle to poroelasticity (see e.g. Coussy, 1995), the following equation can be proven when equilibrated fields of static variables (marked by *a*) are associated to kinematic fields (marked by *b*) which satisfy geometric compatibility and mass conservation (again, setting to zero fluid sources and body forces):

$$\int_\Omega \left(\sigma_{ij}^a\,\dot{\varepsilon}_{ij}^b + p^a\dot{\zeta}^b - \frac{\partial p^a}{\partial x_i}q_i^b\right) d\Omega = \int_\Gamma \left(t_i^a\dot{u}_i^b - p^a q_n^b\right) d\Gamma \tag{53}$$

In view of the properties of the present mixed FE discretization (Section 3), Eq. (53) can directly be translated in terms of Prager's generalized variables, namely (a tilde marks vectors of pressure and flux on the boundary):

$$\bar{\boldsymbol{\sigma}}_a^T\dot{\bar{\boldsymbol{\varepsilon}}}_b + \bar{\mathbf{p}}_a^T\dot{\bar{\boldsymbol{\zeta}}}_b - \bar{\mathbf{p}}_a^T\bar{\mathbf{G}}^T\bar{\mathbf{q}}_b = \bar{\mathbf{t}}^T\dot{\bar{\mathbf{u}}}_b - \tilde{\bar{\mathbf{p}}}_a^T\tilde{\bar{\mathbf{q}}}_b \tag{54}$$

The static modelled fields (*a*) in Eq. (54) is now identified with the (self-equilibrated) difference between the complementary process (*C*) and the stationary state (*S*); the kinematic

The initial value $\chi(0)$ of the energy norm χ, Eq. (51), and each component of $\bar{\boldsymbol{\lambda}}(0)$, are certainly finite. As a consequence of this fact and of the nonnegativeness of function χ at any time, inequality (60b) entails that $\mathbf{V}^T\bar{\boldsymbol{\lambda}}(\tau)$ and, hence, that each one of the plastic multipliers in vector $\bar{\boldsymbol{\lambda}}$ is bounded in time. Since $\bar{\mathbf{Y}} > \mathbf{0}$, it can be concluded that SD, according to the criterion (25), is guaranteed.

(q.e.d.)

Proof of Statement II. By hypothesis the plastic strains $\bar{\boldsymbol{\varepsilon}}^{p*}$ and the irreversible variations of fluid content $\bar{\boldsymbol{\zeta}}^{p*}$ remain constant in time after an instant τ^*, i.e. SD occurs and all components of the generalized variable vector $\bar{\boldsymbol{\lambda}}$ are bounded in time. In other terms, the actual behaviour of the system after τ^* is purely poroelastic, with the modelled fields of stresses and pressure governed, say, by $\bar{\boldsymbol{\sigma}}^{E*}$ and $\bar{\mathbf{p}}^{E*}$. These fields, in view of the linearity of their governing equations, can be split into three addends pertaining to: the process (*E*) earlier defined in (a), i.e. $\bar{\boldsymbol{\sigma}}^E(\tau)$, $\bar{\mathbf{p}}^E(\tau)$; the stationary state $\bar{\boldsymbol{\sigma}}^S$, $\bar{\mathbf{p}}^S = \mathbf{0}$ (state of type *S*), which is the solution to the boundary value problem of time-independent poroelasticity with homogeneous boundary conditions and with $\bar{\boldsymbol{\varepsilon}}^{p*}$ as input fictitious external action; the transient response (*T*), say $\bar{\boldsymbol{\sigma}}^{**}(\tau)$, $\bar{\mathbf{p}}^{**}(\tau)$, to only the external action represented by the difference $\bar{\mathbf{u}}-\bar{\mathbf{u}}^E$, $\bar{\mathbf{p}}-\bar{\mathbf{p}}^E$ at instant τ^*, taken as initial conditions.

The yield inequalities for the actual evolution after time τ^*, in view of the above remarks, can be written in the form:

$$\bar{\mathbf{N}}^{\sigma^T}\left[\bar{\boldsymbol{\sigma}}^E(\tau)+\bar{\boldsymbol{\sigma}}^S+\bar{\boldsymbol{\sigma}}^{**}(\tau)\right]+\bar{\mathbf{N}}^{p^T}\left[\bar{\mathbf{p}}^E(\tau)+\bar{\mathbf{p}}^{**}(\tau)\right] \leq \bar{\mathbf{Y}}, \qquad \forall\, \tau \geq \tau^* \tag{61}$$

In Eq. (61) the last (transient) addend in both square brackets dies off asymptotically in time and, therefore, any perturbation which increases the yield limits $\bar{\mathbf{Y}}$ by an arbitrarily small (but finite) amount $\delta\bar{\mathbf{Y}}$ would imply the existence of some instant, say τ^{**}, such that:

$$\bar{\mathbf{N}}^{\sigma^T}\left[\bar{\boldsymbol{\sigma}}^E(\tau)+\bar{\boldsymbol{\sigma}}^S\right]+\bar{\mathbf{N}}^{p^T}\bar{\mathbf{p}}^E(\tau) < \bar{\mathbf{Y}}+\delta\bar{\mathbf{Y}}, \qquad \forall\, \tau \geq \tau^{**} \tag{62}$$

If self-stresses $\bar{\boldsymbol{\sigma}}^S$ are identified with those generated by the actual plastic strains after SD, then inequality (62) coincides with (59) for $\delta\bar{\mathbf{Y}} \to \mathbf{0}$.

(q.e.d.)

modelled fields (*b*) is identified with the (compatible and conservation abiding) kinematic variab rates in the complementary process (*C*). Thus, the virtual work equation (54) becomes:

$$\left(\bar{\boldsymbol{\sigma}}^C - \bar{\boldsymbol{\sigma}}^S\right)^T \dot{\bar{\boldsymbol{\varepsilon}}}^C + \left(\bar{\mathbf{p}}^C\right)^T \dot{\bar{\boldsymbol{\zeta}}}^C - \left(\bar{\mathbf{p}}^C\right)^T \bar{\mathbf{G}}^T \bar{\mathbf{q}}^C = 0 \tag{55}$$

In order to obtain Eq. (55) from Eq. (54) through the above interpretation of (*a*) and (*b*), account has been taken of the following circumstances: $\bar{\mathbf{p}}^S = \mathbf{0}$, Eq. (50); on Γ_t tractions are zero in both (*C*) and (*S*), so that $t_i^a = 0$; on Γ_u the constraints in the process (*C*) are fixed (i.e. $\dot{u}_i^b = 0$); on Γ_p pressures $\tilde{\bar{\mathbf{p}}}^C$ vanish in (*C*), while on Γ_q fluxes $\tilde{\bar{\mathbf{q}}}^C$ vanish in (*C*).

If Eq. (55) is now used to substitute the last term in the expression (52) of function $\dot{\chi}(t)$, this reads:

$$\dot{\chi}(\tau) = -\left(\bar{\boldsymbol{\sigma}}^C - \bar{\boldsymbol{\sigma}}^S\right)^T \dot{\bar{\boldsymbol{\varepsilon}}}^{pC} - \left(\bar{\mathbf{p}}^C\right)^T \dot{\bar{\boldsymbol{\zeta}}}^{pC} \tag{56}$$

Adding and subtracting $\bar{\boldsymbol{\sigma}}^E$ and $\bar{\mathbf{p}}^E$ in the former and latter brackets of Eq. (56), respectively, in view of the constitutive equations which formulate the normality rule, Eq. (56) becomes:

$$\dot{\chi}(\tau) = -\dot{\bar{\boldsymbol{\lambda}}}^T \left[\bar{\mathbf{N}}_\sigma^T \bar{\boldsymbol{\sigma}} + \bar{\mathbf{N}}_p^T \bar{\mathbf{p}}\right] + \dot{\bar{\boldsymbol{\lambda}}}^T \left[\bar{\mathbf{N}}_\sigma^T \left(\bar{\boldsymbol{\sigma}}^E + \bar{\boldsymbol{\sigma}}^S\right) + \bar{\mathbf{N}}_p^T \left(\bar{\mathbf{p}}^E + \bar{\mathbf{p}}^S\right)\right] \tag{57}$$

Now, let the former expression in square brackets be substituted into Eq. (57) by means of Eq. (33a) with $\bar{\mathbf{H}} = \mathbf{0}$, and let the complementarity relationship, Eq. (34c) be used. Thus, the time derivative $\dot{\chi}$ acquires the form:

$$\dot{\chi}(\tau) = \dot{\bar{\boldsymbol{\lambda}}}^T \bar{\boldsymbol{\varphi}}\left(\bar{\boldsymbol{\sigma}}^E + \bar{\boldsymbol{\sigma}}^S, \ \bar{\mathbf{p}}^E + \bar{\mathbf{p}}^S\right) \tag{58}$$

Suppose now that the following inequality holds at any time τ after a finite time τ':

$$\bar{\mathbf{N}}_\sigma^T \left(\bar{\boldsymbol{\sigma}}^E + \bar{\boldsymbol{\sigma}}^S\right) + \bar{\mathbf{N}}_p^T \left(\bar{\mathbf{p}}^E + \bar{\mathbf{p}}^S\right) < \bar{\mathbf{Y}}, \qquad \forall \tau \geq \tau' \tag{59}$$

This inequality entails that there exists some constant vector V with all negative components which bound from above the corresponding components of vector $\bar{\boldsymbol{\varphi}}$ in Eq. (58). Since $\dot{\bar{\boldsymbol{\lambda}}} \geq \mathbf{0}$, it follows that:

$$\dot{\chi}(\tau) \leq \mathbf{V}^T \dot{\bar{\boldsymbol{\lambda}}}(\tau), \qquad \text{whence:} \qquad \chi(\tau) - \chi(0) \leq \mathbf{V}^T \left[\bar{\boldsymbol{\lambda}}(\tau) - \bar{\boldsymbol{\lambda}}(0)\right] \tag{60}$$

5 Shakedown Analysis by Linear Programming

Straightforward application of statements I and II of Section 4 leads to assessments of structural safety centred on the following optimization problem:

$$s = \max_{\mu, \bar{\boldsymbol{\sigma}}^S, \tau'} \{\mu\} \qquad \text{subject to:}$$

$$\left[\bar{\mathbf{N}}^{\sigma^T} \bar{\boldsymbol{\sigma}}^E(\tau) + \bar{\mathbf{N}}^{p^T} \bar{\mathbf{p}}^E(\tau)\right]\mu + \bar{\mathbf{N}}^{\sigma^T} \bar{\boldsymbol{\sigma}}^S \leq \bar{\mathbf{Y}}, \qquad \bar{\mathbf{B}}^T \bar{\boldsymbol{\sigma}}^S = \mathbf{0} \qquad \forall\, \tau \geq \tau' \tag{63}$$

The (stability) assumption that infinitesimal perturbations of yield limits have consequences of the same order on the shakedown behaviour legitimates the transition from strict to weak inequalities in the sufficient SD condition, Eq. (59), and the consequent identification of this with the necessary condition (see e.g. Kaliszky, 1989; König, 1987).

Let the "envelope vector" $\bar{\mathbf{M}} \equiv \{\ldots \bar{M}_\alpha \ldots\}^T$ be defined as follows, index α running over the set of all n_y yield modes in the whole discrete (FE) model:

$$\bar{M}_\alpha(\tau') = \max_{\tau \geq \tau'} \left\{ \left(\bar{\mathbf{N}}_\alpha^\sigma\right)^T \bar{\boldsymbol{\sigma}}^E(\tau) + \left(\bar{\mathbf{N}}_\alpha^p\right)^T \bar{\mathbf{p}}^E(\tau) \right\} \tag{64}$$

It is worth noting that if a time τ' is chosen (i.e. no longer considered as a variable), optimization (63) reduces to linear programming (LP) and provides a lower bound of the safety factor s. The removal of time τ' from the set of the optimization variables can preserve s as optimal value in (63) with remarkable computational simplification when the history of external actions is periodic or can be regarded as periodic to engineering purposes. In fact, in this practically frequent case the poroelastic response $\bar{\boldsymbol{\sigma}}^E, \bar{\mathbf{p}}^E$ can be split into a periodic addend independent from the initial conditions and a transient addend dependent on them. The latter addend is damped off (i.e. is made lower than any accepted uncertainty margin) at a finite time τ' which therefore can be assumed as the time origin.

In this case, vector $\bar{\mathbf{M}}$, defined by Eq. (64), becomes constant (no longer function of τ') and Eq. (63) reduces to a linear programming (LP) problem. In structural engineering the external actions consist of variable "live loads" to be amplified by the load factor μ, and of time-constant "dead" load (tipically self-weight). Accordingly, vector $\bar{\mathbf{M}}$ will be split into two addends, namely $\bar{\mathbf{M}} = \bar{\mathbf{M}}' + \bar{\mathbf{M}}''$. Then the search for the safety factor s with respect to inadaptation materializes to the LP problem:

$$s = \max_{\mu, \bar{\boldsymbol{\sigma}}^S} \{\mu\}, \quad \text{subject to:} \quad \mu \bar{\mathbf{M}}' + \bar{\mathbf{N}}^{\sigma^T} \bar{\boldsymbol{\sigma}}^S \leq \bar{\mathbf{Y}} - \bar{\mathbf{M}}'', \quad \bar{\mathbf{B}}^T \bar{\boldsymbol{\sigma}}^S = \mathbf{0} \tag{65}$$

Clearly, like in plasticity, the self-equilibrium equations, Eq. (65b), can be given the form $\bar{\boldsymbol{\sigma}}^S = \bar{\mathbf{D}}\,\bar{\boldsymbol{\rho}}$ (by inverting a nonsingular submatrix of $\bar{\mathbf{B}}$ with the order equal to its column number) so that the "redundant" subvector $\bar{\boldsymbol{\rho}}$ of $\bar{\boldsymbol{\sigma}}^S$ become the only variables besides μ:

$$s = \max_{\mu,\,\bar{\boldsymbol{\rho}}} \{\mu\}, \quad \text{subject to:} \quad \mu\bar{\mathbf{M}}' + \bar{\mathbf{N}}^{\sigma^T}\bar{\mathbf{D}}\,\bar{\boldsymbol{\rho}} \le \bar{\mathbf{Y}} - \bar{\mathbf{M}}'' \tag{66}$$

Let the constant self-stresses $\bar{\boldsymbol{\sigma}}^S$ be conceived as (steady-state) poroelastic response (*S*) to plastic strains $\bar{\boldsymbol{\varepsilon}}^p$ in the solid skeleton. These, in turn, depend through Eq. (34a) on the plastic multipliers $\bar{\boldsymbol{\lambda}}$, which, therefore, can be used as independent variables. Therefore the following LP problem turns out to be an alternative to LP optimization (65):

$$s = \max_{\mu,\,\bar{\boldsymbol{\lambda}}} \{\mu\}, \quad \text{subject to:} \quad \mu\bar{\mathbf{M}}' + \bar{\mathbf{N}}^{\sigma^T}\bar{\mathbf{Z}}\bar{\mathbf{N}}^{\sigma}\bar{\boldsymbol{\lambda}} \le \bar{\mathbf{Y}} - \bar{\mathbf{M}}'', \quad \bar{\boldsymbol{\lambda}} \ge \mathbf{0} \tag{67}$$

having set:

$$\bar{\mathbf{Z}} = \bar{\mathbf{E}}_{dr}\bar{\mathbf{B}}\left(\bar{\mathbf{B}}^T\bar{\mathbf{E}}_{dr}\bar{\mathbf{B}}\right)^{-1}\bar{\mathbf{B}}^T\bar{\mathbf{E}}_{dr} - \bar{\mathbf{E}}_{\varepsilon}, \qquad \bar{\mathbf{E}}_{dr} = \left(\bar{\mathbf{E}}_{\varepsilon} - \bar{\mathbf{E}}_{\varepsilon\zeta}\bar{\mathbf{E}}_{\zeta}^{-1}\bar{\mathbf{E}}_{\zeta\varepsilon}\right) \tag{68}$$

The matrix $\bar{\mathbf{Z}}$ of influence coefficients (of imposed strains on consequent self-stresses) concerns only the solid skeleton regarded as linearly elastic and is symmetric and negative semidefinite in the present context of mixed FE discretization in Prager's generalized variables. The saturation liquid has an influence on SD merely through its pressure distribution in space and time $\bar{\mathbf{p}}^E(t)$ captured by vector $\bar{\mathbf{M}}$.

The "direct" methods above formulated for the assessment of the safety factor against inadaptation on the basis of statements I and II can be elucidated as for their practical and computational potentialities and limitations by the following comments.

(A) All three formulations (65), (66) and (67) require in general a two-stage computing procedure: a merely linear but still time-marching poroelastic analysis for the given history of external actions, and, subsequently, an optimization by conventional, routinely solvable linear programming. Linear programming (LP) has extensively been employed in computational in rigid-plastic limit analysis (see e.g.: Corradi and Zavelani, 1974; Sloan, 1988, 1989; Gross-Weege, 1997; Pastor et al., 2000). Nowadays, large scale LP problems are efficiently solvable by finite termination algorithms, which are modern variants of the classical Simplex Method, and, alternatively, by means of (generally superior) "interior point" techniques stemming from Karmarkar's 1985 seminal paper.

(B) In general, the "direct" approach does not provide great computational gains as long as it requires to follow the load history up to $\tau = \infty$. However, this requirement can be relaxed when the *loading history* can be interpreted as a sequence, in unknown order, of a small number of known loading histories which are separated in time by intervals long enough to permit a steady-state to intervene only between them under the dead loads. Then poroplastic SD analysis can be organized according to the same pattern as the "unrestricted" dynamic SD analysis presented in detail by C.Polizzotto et alii in another Chapter of this book and, hence, not expounded here (see Polizzotto et al., 1993). In the presence of loading periodicity, the computing burden of the SD analysis can be drastically reduced: in fact, the computation of the periodic response (*E*), can be carried out efficiently by integral transforms, or even analytically for simple geometries (see Cocchetti and Maier, 2000a, Appendix I).

(C) External actions which eventually become time-independent after the extinction of a transient regime, give rise to a time-independent poroelastic state. If the relevant (constant in time) stress and pressure vectors, say $\bar{\boldsymbol{\sigma}}^E$ and $\bar{\mathbf{p}}^E$, are introduced in it, Eq. (64) provides $\bar{M}_\alpha$ without maximization with respect to time, simply as the projection of vector $\left\{\left(\bar{\boldsymbol{\sigma}}^E\right)^T, \left(\bar{\mathbf{p}}^E\right)^T\right\}^T$ on the unit outward normal to the α-*th* yield plane. In this particular situation, the LP problems (65)-(67) yield the safety factor s with respect to plastic collapse, i.e. define the carrying capacity of the modelled poroplastic system asymptotically in time. In other terms, the present SD theorems (statements I and II and their synthesis into formulation (63) as a maximization problem) specialize to limit analysis by the static approach generalized to PWL poroplasticity.

(D) In view of engineering applications of the preceding direct methods, it is worth noting that the generation of PWL constitutive laws entails a computing cost, which is generally more than compensated by the savings entailed by the transition from (convex) nonlinear programming to LP. Moreover, like in plasticity, see e.g. Tin-Loi (1989), the LP problems (65)-(67) may be substantially reduced in size by tentatively ignoring every yield mode α for which $\bar{Y}_\alpha - \bar{M}_\alpha$ exceeds a pre-assigned tolerance (i.e. whose maximum projection $\bar{M}_\alpha$ is sufficiently far from the relevant yield plane, so that this is unlikely to be active). Those of the neglected yield inequalities which turn out to be violated by the stresses resulting from the solution of the trial LP are re-considered in a next LP problem.

(E) Recourse to the LP dual problem and its processing, if advantageous, is automatically made in most LP software and provides further savings. From the standpoint of the SD theory, formal dualization in the sense of convex analysis (see e.g. Gao, 1999) and mechanical interpretation of the dual (like in Maier, 1969, for plasticity) would lead to a generalization to poroplasticity of Koiter's kinematic theorem: a theoretically interesting issue not to be discussed herein.

6 Upper Bounds on Post-Shakedown Quantities

Shakedown analysis may be nonconservative when some of the idealizations on which it rests are violated. Consider in particular the following hypotheses: (*a*) "small" deformation and serviceability within SD ranges; (*b*) unlimited ductility; (*c*) in poroplasticity, linear diffusion equation, i.e. Darcy law with permeability constant in time.

Remedies to possible unconservative results of SD analysis, as a consequence of invalidation of some basic assumption, may be found still in the area of direct methods. Specifically, remedies may be provided by upper bounds on the post-SD values of meaningful quantities which depend on the whole history of the poroplastic system. Upper bounds on some meaningful residual displacements, plastic strains and volumetric plastic strains in some representative crucial points, if they turn out to be lower than critical thresholds, guarantee that the above hypotheses (*a*) (*b*) (*c*), respectively, are actually acceptable in the engineering situation considered; otherwise they may give guidelines for suitably adjusted re-formulations and iterated solutions of the SD analysis problem.

In the framework of the space discretization and constitutive piecewise linearization adopted in what precedes, a post-shakedown quantity of all three of the above mentioned kinds turns out to be a linear function of the actual post-shakedown generalized plastic multiplier vector $\bar{\boldsymbol{\lambda}}'$. In fact, account taken of Eq. (34a), we can write for any plastic strain component, say ε^p, in a point $\hat{\mathbf{x}}$ (e.g., in a Gauss point):

$$\varepsilon^p(\hat{\mathbf{x}}) = \mathbf{r}^T \boldsymbol{\Psi}_\varepsilon(\hat{\mathbf{x}})\, \bar{\mathbf{N}}^\sigma\, \bar{\boldsymbol{\lambda}}' \tag{69}$$

where $\mathbf{r}$ denotes a Boolean vector which extracts and sums the modelled normal strains in $\hat{\mathbf{x}}$ out of the modelled plastic strain field.

Similarly, using Eq. (42) solved with respect to displacements for $\bar{\mathbf{p}} = \mathbf{0}, \hat{\bar{\mathbf{b}}} = \mathbf{0}, \bar{\mathbf{t}} = \mathbf{0}$ and denoting by $\mathbf{s}$ the Boolean vector which selects, by its nonzero entries, the desired component, in $\hat{\mathbf{x}}$ (say, a FE node), of the modelled residual displacement field governed by vector $\bar{\mathbf{u}}^p$:

$$u_i^p(\hat{\mathbf{x}}) = \mathbf{s}^T \boldsymbol{\Psi}_u(\hat{\mathbf{x}})\, \bar{\mathbf{K}}^{-1} \bar{\mathbf{B}}^T\, \bar{\mathbf{E}} \bar{\mathbf{N}}^\sigma\, \bar{\boldsymbol{\lambda}}' \tag{70}$$

Clearly, a linear dependence on vector $\bar{\boldsymbol{\lambda}}'$ holds also for other quantities, such as residual stresses and permanent fluid content, in a suitably selected point $\hat{\mathbf{x}}$.

Henceforth, reference will be made to any of the above considered quantities, denoted by Q and expressed as a linear combination of plastic multipliers through a coefficient vector $\bar{\mathbf{V}}$:

$$Q = \bar{\mathbf{V}}^T\, \bar{\boldsymbol{\lambda}}' \tag{71}$$

It is understood that Q is part of the structural response to the variable repeated loads regarded as basic loading ($\mu = 1$) and that there is a safety factor $s > 1$ with respect to them.

Hardening is now allowed for through the matrix $\bar{\mathbf{H}}$ consistently with the adopted piecewiselinearization of the poroplastic material model. It is assumed that $\bar{\mathbf{H}}$ is positive semidefinite and symmetric. In a point $\check{\mathbf{x}}$ of the poroplastic body, let $\mathcal{D}(\check{\mathbf{x}})$ denote the value attained asymptotically in time by the time-cumulative dissipated energy per unit volume. In the present context of piecewiselinear poroplastic and finite element models, $\mathcal{D}(\check{\mathbf{x}})$ is a convex quadratic function of vector $\bar{\boldsymbol{\lambda}}'$ and that upper bounds on it can be achieved by straightforward generalization of the theoretical developments and conclusions concerning any quantity Q covered by Eq. (71).

Now the following statement (III in the present Chapter) might be proven, but the rather lengthy proof, not given here for space limitations, can be found in Cocchetti and Maier (2000b). This theorem and the subsequent one (IV) represent generalizations to poroplasticity of bounding inequalities established earlier in PWL plasticity, see e.g. Maier (1973), Maier and Vitiello (1974).

(III) Optimal Bound Theorem. The post shakedown value Q of a quantity linearly dependent, through the coefficient vector $\bar{\mathbf{V}}$, on the (unknown, history-dependent) post-shakedown residual plastic multiplier vector $\bar{\boldsymbol{\lambda}}'$, is bounded from above by the saddle-point value of the following min-max problem in the space of the variable vector $\left\{\bar{\boldsymbol{\lambda}}^{*T},\ \bar{\boldsymbol{\lambda}}^{T}\right\}^{T}$:

$$Q \le \tilde{Q}_1 = \min_{\bar{\boldsymbol{\lambda}}^*} \max_{\bar{\boldsymbol{\lambda}}} \left\{\bar{\mathbf{V}}^T \bar{\boldsymbol{\lambda}}\right\}, \quad \text{subject to:} \tag{72}$$

$$\bar{\mathbf{M}}' - \left(\bar{\mathbf{A}}_0 + \bar{\mathbf{H}}\right)\bar{\boldsymbol{\lambda}} \le \bar{\mathbf{Y}}^0 - \bar{\mathbf{M}}'', \qquad \bar{\boldsymbol{\lambda}} \ge \mathbf{0} \tag{73}$$

$$s\,\bar{\mathbf{M}}' - s\left(\bar{\mathbf{A}}_0 + \bar{\mathbf{H}}\right)\bar{\boldsymbol{\lambda}}^* \le \bar{\mathbf{Y}}^0 - \bar{\mathbf{M}}'', \qquad \bar{\boldsymbol{\lambda}}^* \ge \mathbf{0} \tag{74}$$

$$\frac{s-1}{s}\left[\left(\bar{\mathbf{Y}}^0 - \bar{\mathbf{M}}''\right)^T \bar{\boldsymbol{\lambda}} + \frac{1}{2}\bar{\boldsymbol{\lambda}}^T \bar{\mathbf{H}} \bar{\boldsymbol{\lambda}}\right] - \bar{\boldsymbol{\lambda}}^{*T} \bar{\mathbf{A}}_0\, \bar{\boldsymbol{\lambda}} + \frac{1}{2}\bar{\boldsymbol{\lambda}}^T \bar{\mathbf{A}}_0\, \bar{\boldsymbol{\lambda}} \le 0 \tag{75}$$

As a useful interpretation of the above statement, it is worth noting that the necessary static condition for SD leads to the constraints (73) for the (unknown) actual plastic multipliers generated in the modelled system along its evolution under the external actions. A maximization under these constraints alone of the quantity to bound as a function of $\bar{\boldsymbol{\lambda}}$ would lead to a usually very loose and useless upper bound on it.

In the proof here omitted (see Cocchetti and Maier, 2000b), an energy inequality in the form of Eq. (75) is established between the actual unknown evolution of the system and the fictitious situation generated by plastic multipliers $\bar{\boldsymbol{\lambda}}^*$ (constant in time) and such that the sufficient SD condition is fulfilled for the live loads amplified by the safety factor s. This vector $\bar{\boldsymbol{\lambda}}^*$, which influences the sought upper bound through the above energy inequality, is employed in the above formulation, Eqs. (72)-(75), as minimization variables in order to improve the upper bound.

It is generally difficult to numerically compute the optimal (minimal) bound achievable in principle by solving the minmax problem (72)-(75). By constraint relaxations in Eqs. (72)-(75) various suboptimal bounds can be generated and are easier to compute. The one expressed by the following statement, in the writers' experience, probably attains a satisfactory compromise between the conflicting requirements of accuracy and computing economy.

(IV) Suboptimal Bound. The post-shakedown value Q of a quantity linearly dependent, through the coefficient vector $\bar{\mathbf{V}}$, on the residual plastic multiplier vector $\bar{\boldsymbol{\lambda}}'$ is bounded from above by the optimal value $\tilde{Q}_2$ of the following maximization problems:

$$\tilde{Q}_2 = \max_{\bar{\boldsymbol{\lambda}}} \left\{ \bar{\mathbf{V}}^T \bar{\boldsymbol{\lambda}} \right\}, \quad \text{subject to:} \tag{76}$$

$$\bar{\mathbf{M}}' - \left(\bar{\mathbf{A}}_0 + \bar{\mathbf{H}}\right) \bar{\boldsymbol{\lambda}} \leq \bar{\mathbf{Y}}^0 - \bar{\mathbf{M}}'', \qquad \bar{\boldsymbol{\lambda}} \geq \mathbf{0} \tag{77}$$

$$\frac{s-1}{s} \left(\bar{\mathbf{Y}}^0 - \bar{\mathbf{M}}''\right)^T \bar{\boldsymbol{\lambda}} \leq \hat{\Pi} \tag{78}$$

where:

$$\hat{\Pi} = \min_{\bar{\boldsymbol{\lambda}}^*} \left\{ \frac{1}{2} \bar{\boldsymbol{\lambda}}^{*T} \bar{\mathbf{A}}_0 \bar{\boldsymbol{\lambda}}^* \right\}, \quad \text{subject to:} \tag{79}$$

$$s \bar{\mathbf{M}}' - s \left(\bar{\mathbf{A}}_0 + \bar{\mathbf{H}}\right) \bar{\boldsymbol{\lambda}}^* \leq \bar{\mathbf{Y}}^0 - \bar{\mathbf{M}}'', \qquad \bar{\boldsymbol{\lambda}}^* \geq \mathbf{0} \tag{80}$$

Proof. Note, by trivial algebra, that the sum of the last two terms on the l.h.s. of Eq. (80) satisfy the following inequality, since matrix $\bar{\mathbf{A}}_0$ is symmetric and positive semidefinite, namely:

$$\begin{aligned} -\bar{\boldsymbol{\lambda}}^{*T} \bar{\mathbf{A}}_0 \bar{\boldsymbol{\lambda}} + \frac{1}{2} \bar{\boldsymbol{\lambda}}^T \bar{\mathbf{A}}_0 \bar{\boldsymbol{\lambda}} &= \\ = -\frac{1}{2} \bar{\boldsymbol{\lambda}}^{*T} \bar{\mathbf{A}}_0 \bar{\boldsymbol{\lambda}}^* + \frac{1}{2} \left(\bar{\boldsymbol{\lambda}}^* - \bar{\boldsymbol{\lambda}}\right)^T \bar{\mathbf{A}}_0 \left(\bar{\boldsymbol{\lambda}}^* - \bar{\boldsymbol{\lambda}}\right) &\geq -\frac{1}{2} \bar{\boldsymbol{\lambda}}^{*T} \bar{\mathbf{A}}_0 \bar{\boldsymbol{\lambda}}^* \end{aligned} \tag{81}$$

As a consequence of (81) and of the nonnegativeness of the quadratic form associated to the hardening matrix $\bar{\mathbf{H}}$, inequality (75) is easily seen (by dropping nonnegative terms on the l.h.s.) to imply the weaker inequality (78) with the nonnegative quadratic form $\Pi = \frac{1}{2} \bar{\boldsymbol{\lambda}}^{*T} \bar{\mathbf{A}}_0 \bar{\boldsymbol{\lambda}}^*$ on the r.h.s. and a linear form in $\bar{\boldsymbol{\lambda}}^*$ on the l.h.s.. Therefore, a decoupling occurs, in the sense that, first, Π can be minimized under the constraints (74) and, second, its minimum $\hat{\Pi}$ can be introduced into the relaxed inequality (78), to be combined with Eq. (73), i.e. with constraints (77), for the subsequent maximization (76).

(q.e.d.)

It is worth noting that the upper bounds $\tilde{Q}_2$ are generally "worse" than the preceding one $\tilde{Q}_1$ ($\tilde{Q}_1 \leq \tilde{Q}_2$), but can be computed by solving (through widely available software), first, a convex quadratic programming problem, Eqs. (79)-(80), and subsequently a linear programming problem, Eqs. (76)-(78).

Of course, in order to compute bounds use can be made of provisions apt to reduce the number of yield planes and, hence, of the dimensionality of the variable vectors $\bar{\boldsymbol{\lambda}}$ and $\bar{\boldsymbol{\lambda}}^*$ (see e.g. Tin-Loi, 1989); these provisions are the same as those which have been pointed out in Section 5.

7 Conclusions

This Chapter has concisely presented recent extensions of shakedown theorems and "direct" methods for limit-state analysis of structures, namely for shakedown analysis, with its specialization to limit analysis, and procedures apt to provide upper bounds on post-shakedown history-dependent quantities.

The theoretical developments considered in precedes exhibit the following peculiarities: fully saturated two-phase (solid-liquid) poroplastic material models, stable in Drucker's sense; piecewiselinear approximation of the constitutive models; classical (Darcy's) linear filtration law; mixed finite element modelling, which preserves the essential features of the continuum formulations; linear programming as mathematical and computational context for assessing safety factors; quadratic and linear programming methods for computing those upper bounds which are expected to materialize a good compromise between accuracy and cost-effectiveness requirements.

The computational experiences achieved so far (expounded in the referenced publications on the present subject by the authors) show that the direct methods in poroplasticity presented herein turn out to be competitive with conventional time-stepping analyses in various engineering situations.

Future research is desirable for further generalizations of the present results. The basic shakedown theorem valid in the presence of the time-dependence due to both dynamics and filtration has been already established by Cocchetti and Maier (2001). Continuum damage might be allowed for in poroplasticity following the path of reasoning already developed in plasticity, e.g. by Hachemi and Weichert (1992, 1997). A still challenging open issue is at present the extension of shakedown theory and direct methods to partially saturated poroplasticity with variable permeability and, more in general, to the broad area of multiphase media with other constitutive nonlinearities besides inelastic strains in the solid skeleton and irreversible variations of liquid content envisaged in what precedes.

References

Capurso, M. (1979). Some upper bound principles for plastic strains in dynamic shakedown of elastoplastic structures. *J. Struct. Mech.*. 7:1–20.

Ceradini, G. (1969). Sull'adattamento dei corpi elastoplastici soggetti ad azioni dinamiche. *Giornale del Genio Civile*. 415:239–258.

Cocchetti, G. and Maier, G. (1998). Static shakedown theorems in piecewiselinearized poroplasticity. *Arch. Appl. Mech.*. 68:651–661.

Cocchetti, G. and Maier, G. (2000a). Shakedown analysis in poroplasticity by linear programming. *Int. J. Num. Meth. Engng.*. 47(1-3):141–168.

Cocchetti, G. and Maier, G. (2000b). Upper bounds on postshakedown quantities in poroplasticity. In Weichert, D. and Maier, G., eds., *Inelastic Analysis of Structures under Variable Repeated Loads*. Kluwer. 289–314.

Cocchetti, G. and Maier, G. (2001). A shakedown theorem in poroplastic dynamics. *Rend. Acc. Naz. Lincei*. (accepted for publication).

Corigliano, A., Maier, G. and Pycko, S. (1995). Dynamic shakedown analysis and bounds for elastoplastic structures with nonassociative, internal variable constitutive laws. *Int. J. Sol. Struct.*. 32:3145–3166.

Corradi, L. and Maier, G. (1973). Inadaptation theorems in the dynamics of elastic-work hardening structures. *Ingenieur-Archiv*. 43:44–57.

Corradi, L. and Maier, G. (1974). Dynamic non-shakedown theorem for elastic perfectly-plastic continua. *J. Mech. Phys. Sol.*. 22:401–413.

Corradi, L. and Zavelani, A. (1974). A linear programming approach to shakedown analysis of structures. *Comp. Meth. Appl. Mech. Eng.*. 3:37–53.

Coussy O. (1995). *Mechanics of porous continua*. Chichester: John Wiley & Sons.

Gao, D. Y. (1999). *Duality principles in nonconvex systems; theory, methods and applications*. Dordrecht: Kluwer Acad. Publ.

Genna, F. (1991). Bilateral bounds for structures under dynamic shakedown conditions. *Meccanica*. 26:37–46.

Gross-Weege, J. (1997). On the numerical assessment of the safety factor of elastic-plastic structures under variable loading. *Int. J. Mech. Sci.*. 39:417–433.

Hachemi, A. and Weichert, D. (1992). An extension of the static shakedown theorem to a certain class of inelastic materials with damage. *Arch. Mech.*. 44:491-498.

Hachemi, A. and Weichert, D. (1997). Application of shakedown theory to damaging inelastic material under mechanical and thermal loads. *Int. J. Mech. Sci.*. 39:1067-1076.

Kaliszky, S. (1989). Plasticity: theory and engineering applications. Amsterdam: Elsevier.

König, J. A. (1987). *Shakedown of elastic-plastic structures*. Amsterdam: Elsevier.

Lewis, R. W. and Schrefler, B. A. (1998). *The finite element method in the static and dynamic deformation and consolidation of porous media*. Chichester: John Wiley & Sons, 2nd edition.

Lloyd Smith, D., ed. (1990). *Mathematical programming methods in structural plasticity*. New York: Springer-Verlag.

Maier, G. (1969). Shakedown theory in perfect elastoplasticity with associated and nonassociated flow-laws: a finite element, linear programming approach. *Meccanica*. 4:250–260.

Maier, G. (1970). A matrix structural theory of piecewise-linear plasticity with interacting yield planes. *Meccanica*. 5:55–66.

Maier, G. (1973). Upper bounds on deformations of elastic-workhardening structures in the presence of dynamic and second-order effects. *J. Struct. Mech.*. 2:265–280.

Maier, G. (1976). Piecewise linearization of yield criteria in structural plasticity. *Solid Mechanics Archives*. 2/3:239–281.

Maier, G., Carvelli, V. and Cocchetti, G. (2000). On direct methods for shakedown and limit analysis. *European Journal of Mechanics A/Solids*. Special Issue. 19:79–100.

Maier, G. and Comi, C. (1997). Variational finite element modelling in poroplasticity. In Reddy, B. D., ed., *Recent Developments in Computational and Applied Mechanics*. Barcelona: CIMNE, 180–199.

Maier, G. and Vitiello, E. (1974). Bounds on plastic strains and displacements in dynamic shakedown of workhardening structures. *ASME, J. Appl. Mech.*. 41:434–440.

Pastor, J., Thai T.-H. and Francescato P. (2000). New bounds for the height limit of a vertical slope. *Int. J. Num. Analyt. Meth. Geomech.*. 24:165–182.

Polizzotto, C., Borino, G., Caddemi, S. and Fuschi, P. (1993). Theorems of restricted dynamic shakedown. *Int. J. Mech. Sci.*. 35:787–801.

Ponter, A. R. S. (1972). Deformation, displacement and work bounds for structures in a state of creep and subject to variable loading. *ASME, Journal of Applied Mechanics*. 39:953–959.

Ponter, A. R. S. (1975). General displacement and work bounds for dynamically loaded bodies. *J. Mech. Phy. Solids*. 23:151–163.

Ponter, A. R. S. and Williams, J. J. (1973). Work bounds and associated deformation of cyclically loaded creeping structures. *ASME, J. Appl. Mech.*. 40:921–927.

Sloan, S. W. (1988). Lower bound limit analysis using finite elements and linear programming. *Int. J. Num. Analytical Meth. Geomech.*. 12:61–77.

Sloan, S. W. (1989). Upper bound limit analysis using finite elements and linear programming. *Int. J. Num. Analytical Meth. Geomech.*. 13:263–282.

Tin-Loi, F. (1989). A constraint selection technique in limit analysis. *Appl. Math. Modelling*. 13:442–446.

Tin-Loi, F. (1990). A yield surface linearization procedure in limit analysis. *Mech. Struct. & Mach.*. 18:135–149.

Weichert, D. and Maier, G., eds. (2000). Inelastic analysis of structures under variable repeated loads. Dordrecht: Kluwer Academic Publishers.

A Kinematic Method for Shakedown and Limit Analysis of Periodic Heterogeneous Media

Giulio Maier and Valter Carvelli

Technical University (Politecnico) of Milan, Milan, Italy

Abstract. In this Chapter the kinematic (second, Koiter's) shakedown theorem is applied to the representative volume of periodic heterogeneous media with Huber-Mises local plastic behavior. The adopted formulation of shakedown analysis is based on periodicity boundary conditions, conventional finite element modeling and penalization enforcement of plastic incompressibility. A cost-effective iterative solution procedure is discussed and computationally tested. Numerical tests and engineering applications are presented with reference to perforated plates and metal-matrix unidirectional fiber-reinforced composites.

1 Introduction

In several engineering situations and advanced technologies, structural analysis and design concern heterogeneous systems which exhibit ductility of material behaviour and periodicity of texture as main basic features. Two typical categories of such systems are considered in what follows: perforated steel plates, often used in power plants; metal-matrix unidirectional fiber-reinforced composites, employed particularly by aerospace industries.

In view of the expected ductile inelastic behaviour, lack of shakedown (SD) under variable-repeated loads or, as a special case, plastic collapse under monotonically increasing loads can reasonably be regarded as the main critical events with respect to which the safety margins must be assessed.

Periodic space distribution of geometric and physical properties makes it possible and, clearly, computationally very advantageous, to select as the space domain of the limit state analysis problem the "representative volume" (*RV*). This is defined as the minimum volume which contains all geometrical and physical information about the heterogeneous medium and can be conceived as generating the whole solid by repeated translations, see e.g. Suquet (1982).

In the above circumstances a practically meaningful shakedown analysis (SDA) problem can be formulated in the spirit of homogenization theory as follows, (Carvelli et al., 1999b, 2000; Dvorak et al., 1994; Ponter and Leckie, 1998a). A "load domain" is assigned in the space of average (or "macroscopic") external actions, i.e. a region of that space within which they slowly (without inertia forces) fluctuate in time a supposedly unbounded number of times according to unknown time histories. The structural analysis problem is to achieve "directly" (i.e. not by step-by-step inelastic computations along loading histories, which are unknown) practically essential information on the "safety factor", i.e. the number s such that SD occurs if the "load multiplier" μ is below s ($\mu<s$), it does not if μ is above s ($\mu>s$). Collapse for $\mu>s$ means here that either incremental collapse or alternating plasticity occurs because yielding (dissipative) processes never cease locally (at the microscopic level).

The objectives and limitations of what follows in this Chapter are clarified by the following remarks.

(a) The external actions of practical interest on the RV are average stresses and/or temperature, but only the former will be considered here (as live loads, in the absence of dead loads). As long as they do not alter material properties, such as yield limit or/and elastic stiffness, like e.g. in Ponter and Leckie (1998b), thermal effects could be covered by straightforward extensions (and have no consequence on SD if they do not fluctuate in time).

(b) When the load domain shrinks to a point, i.e. no fluctuation of external actions is expected, then SDA reduces to limit analysis (LA) with respect to plastic collapse (however, SDA and LA have been implemented by the writers in separate computer codes, as it is advisable in general). It is worth noting that the boundary of the set of average stresses which can be carried by the heterogeneous material characterizes the homogenized material as its "strength", see e.g. Carvelli et al. (2000), de Buhan and Taliercio (1988), Francescato and Pastor (1997), Taliercio (1992); on the contrary any shakedown limit locus depends not only on material properties (constitutive and geometric) at the microscale, but also on the geometry of the considered basic load domains, of which that locus represents the envelop after amplification by the computed safety factors.

(c) Only a recently developed kinematic approach and the relevant solution technique, (Carvelli et al., 1999a, 1999b, 2000), will be presented herein. Alternative kinematic solutions and static methods are presented in other Chapters of this book and elsewhere, see e.g. Ponter and Leckie (1998a), Tarn et al. (1975), Weichert et al. (1999a, 1999b).

(d) Basically the same approach and computational procedure adopted herein for SDA of periodic media has been applied to homogeneous elastic-plastic solids, (Zhang, 1995), and to defective pressurized pipelines and pressure vessels with openings (Carvelli et al., 1999a). However, only heterogeneous bodies are focused here in view of the relative novelty, interesting peculiarities and technological importance of the kinematic SDA of periodic systems.

(e) Perfect plasticity with Mises-Huber yield criterion is assumed herein locally, for materials at the microscale. Hardening behaviour (linear or nonlinear with saturation) and/or damage might be allowed for by a generalization similar to the one presented e. g. in Corigliano et al. (1995), Druyanov and Roman (1999), Hachemi and Weichert (1997), Polizzotto et al. (1996). However, possible local unstable behaviours such as softening, fiber buckling and debonding on matrix-fiber interfaces, cannot be covered by the present approach and, hence, caution is needed in adopting it in practice when those phenomena cannot be ruled out a priori.

(f) This Chapter is in no way intended as a survey of the present knowledge on its subject: a survey and a fairly comprehensive bibliography can be founded e.g. in Maier et al. (2000, 2001), whereas the reference list at the end of this Chapter concerns almost exclusively contributions closely related to specific recent research results presented herein.

2 Kinematic Formulation of Shakedown Analysis with Periodicity Conditions

The core of SD theory, consisting of classical Melan's static and Koiter's kinematic theorems, is dealt with in the classical comprehensive essay by Koiter (1960), in more recent, widely available textbooks on plasticity, such as Kaliszky's (Kaliszky, 1989) and Lubliner's (Lubliner, 1990) and, with more details, in monographs such as König (1987) and Cohn and Maier (1979).

Therefore, only the following theorem is provided below without proof here as a conceptual basis for the kinematic approach to SD analysis adopted here. In view of applications, this unifying statement expresses a single condition, both necessary and sufficient, for lack of SD. It reads:

Statement. The body or structure will not shakedown if and only if, (at least) one "admissible plastic strain rate cycle" (APSRC) $\dot{\varepsilon}_{ij}^{p}(t)$ exists such that, over any time increment T, the work done because of this cycle by the external actions (here restricted to boundary tractions $\overline{p}_i$ for brevity) exceeds the internal dissipation due to it. By definition, an APSRC is characterized by the geometric compatibility ("kinematic admissibility") of the cumulative plastic strain field generated at its end (for t=T).

Such unified theorem, somewhat controversial in the past, rigorously requires a further, quite reasonable, hypothesis of material stability, besides Drucker's postulate: namely, the hypothesis that an infinitesimal perturbation of the yield limits, generates infinitesimal consequences on its behaviour, in particular in terms of the SD limit s.

Since all relations concerning APSRC are positively homogeneous, of first order, a common positive multiplier is available and can be exploited to normalize the work of external actions. This work in turn can be conveniently replaced, through a virtual equation, by the work performed by the stresses σ_{ij}^{e} due to the given loads in a hypothetical fictitious linear elastic process (the computation of these elastic stresses in the RV is outlined in the Appendix):

$$\int_0^T\int_S \overline{p}_i \dot{u}_i^p \, dS dt = \int_0^T\int_V \sigma_{ij}^{e} \dot{\varepsilon}_{ij}^{p} \, dV dt = 1 \tag{1}$$

where: V denotes the volume occupied by the body considered; S its boundary; $\dot{u}^p$ the displacement rates due to the APSRC conceived as external action acting on a linear elastic RV.

Focussing now on the specific class of solids in point, i.e. heterogeneous with periodic microscale, the following homogenization concepts have to be considered and employed in the formulation of SD analysis, see Figure 1:

Any RV in the heterogeneous material is associated with a point in a fictitious homogenized material (Figure 1c) where "macroscopic" stress Σ_{ij} and strain E_{ij} tensors are supposed to act. These variables are defined as the volume averages over the RV of the relevant microscopic stress $\sigma_{ij}(x_i)$ and strain tensors $\varepsilon_{ij}(x_i)$ $(i,j{=}1,2,3)$

$$\Sigma_{ij} \equiv \frac{1}{V}\int_V \sigma_{ij}\, dV ; \qquad E_{ij} \equiv \frac{1}{V}\int_V \varepsilon_{ij}\, dV \tag{2}$$

If the fields Σ_{ij} and E_{ij} are uniform over the homogenized solid, at sufficiently large distance from the boundary of the body the microscopic strain and stress fields conform to the periodicity of the microstructural geometry see e.g. Suquet (1982). This means that:

$$\sigma_{ij} n_j \quad \text{anti-periodic} \tag{3}$$

$$\varepsilon_{ij}(u_i) = E_{ij} + \tilde{\varepsilon}_{ij}(\tilde{u}_i) \quad \forall x_i \in V \qquad \left(\frac{1}{V}\int_V \tilde{\varepsilon}_{ij} dV = 0\right) \tag{4}$$

where $\tilde{\varepsilon}_{ij}$ is the space-fluctuating strain that conforms with the periodicity of the microstructure. Condition (3) shows that at two homologous points on opposite sides of Γ (where the outward normals n_i are also opposite (Figure 1b)), the stress vector $\sigma_{ij} n_j$ takes opposite values.

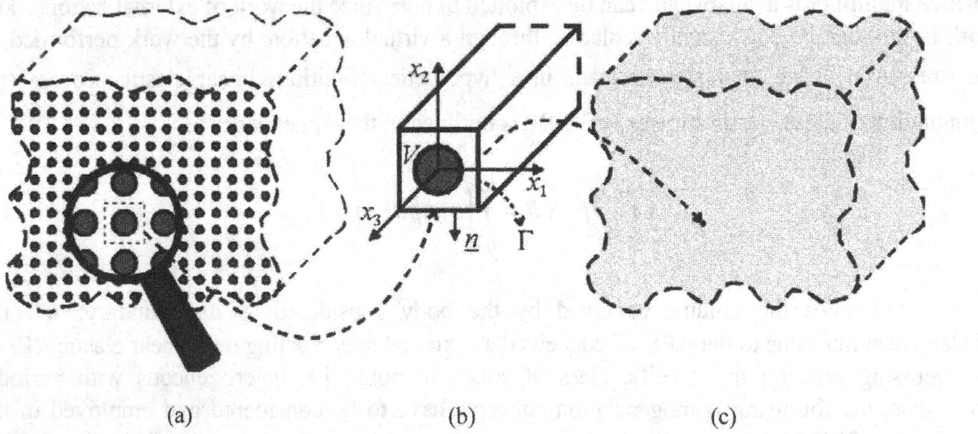

Figure 1. Homogenization schematically depicted: (a) periodic heterogeneous medium; (b) representative volume (RV); (c) homogenized material.

The relation (4) entails splitting the local displacement field into:

$$u_i = E_{ij}x_j + \tilde{u}_i \tag{5}$$

where $\tilde{u}_i(x_i)$ is the RV-periodic part of the microscopic displacement field u_i.

Let us assume a stress field σ_{ij} and a displacement field u_i satisfying conditions (3) and (4), respectively. Then the volume average over any RV of the microscopic work of σ_{ij} for the strain field ε_{ij} is equal to the macroscopic work $\Sigma_{ij}E_{ij}$:

$$\frac{1}{V}\int_V \sigma_{ij}\varepsilon_{ij}dV = \Sigma_{ij}E_{ij} \tag{6}$$

Relation (6) expresses the virtual work principle for a RV subjected to "loads" Σ_{ij}. This relation highlights the meaning of Σ_{ij} and E_{ij} as work-conjugate macroscopic variables.

On the basis of the preceding remarks (a)-(c), the SD analysis of periodic media can be formulated in the following general terms:

$$s = \min_{\dot{\varepsilon}_{ij}^p, \Delta\tilde{u}_i, \Delta E_{ij}^p} \int_0^T \int_V D(\dot{\varepsilon}_{ij}^p)dVdt, \quad \text{subject to:}$$

$$\begin{aligned}
&\int_0^T \int_V \sigma_{ij}^e \dot{\varepsilon}_{ij}^p \, dVdt = 1 \\
&\int_0^T \dot{\varepsilon}_{ij}^p dt = \Delta\tilde{\varepsilon}_{ij}^p + \Delta E_{ij}^p = \mathrm{R}(\Delta\tilde{u}_i) + \Delta E_{ij}^p \quad \text{in } V \\
&\Delta\tilde{u}_i \quad \text{periodic on } \Gamma
\end{aligned} \tag{7}$$

where $\mathrm{R}(\Delta\tilde{\underline{u}}) = \frac{1}{2}(\Delta\tilde{u}_{i,j} + \Delta\tilde{u}_{j,i})$ and s is the "safety factor" or shakedown limit of the average stress "loading" domain Ω (such that below it a state of adaptation takes place in the material in the sense that yielding processes eventually cease and above it they do not). In the present periodicity context, we define $\dot{\varepsilon}_{ij}^p(\dot{u}_i, t)$ as a microscopic admissible plastic strain rate cycle (APSRC) if the cumulative plastic strain increment over the time interval (0, T) is compatible (7c) in the peculiar sense of equation (4), namely if it consists of two addends, the former of which is compatible with a microscopic displacement field $\Delta\tilde{u}_i$ fulfilling the periodicity condition on the RV boundary Γ, equation (7d).

Now let the specialization be performed to Huber-Mises plastic constitutive law for the various material phases (matrix and fibers), after adopting matrix notation, namely:

$$\left.\begin{aligned} \Delta\mathbf{u}^T &= \{\Delta u_1 \quad \Delta u_2 \quad \Delta u_3\} \\ \boldsymbol{\sigma}^{eT} &= \{\sigma_{11}^e \quad \sigma_{22}^e \quad \sigma_{33}^e \quad \sigma_{12}^e \quad \sigma_{13}^e \quad \sigma_{23}^e\} \\ \boldsymbol{\varepsilon}^T &= \{\varepsilon_{11} \quad \varepsilon_{22} \quad \varepsilon_{33} \quad 2\varepsilon_{12} \quad 2\varepsilon_{13} \quad 2\varepsilon_{23}\} \end{aligned}\right\} \tag{8}$$

and similarly for vector representations of macroscopic stress Σ_{ij} and strain E_{ij}. Thus the shakedown analysis by the kinematic approach applied to the RV acquires the formulation:

$$s = \min_{\dot{\boldsymbol{\varepsilon}}^p, \Delta\tilde{\mathbf{u}}, \Delta\mathbf{E}^p} \sqrt{\tfrac{2}{3}} \int_0^T \int_V \sigma_0 \sqrt{\dot{\boldsymbol{\varepsilon}}^{pT} \mathbf{X} \, \dot{\boldsymbol{\varepsilon}}^p} \, dVdt \, , \quad \text{subject to:}$$

$$\begin{aligned} & \int_0^T \int_V \boldsymbol{\sigma}^{eT} \dot{\boldsymbol{\varepsilon}}^p dVdt = 1 \\ & \mathbf{Y}^T \dot{\boldsymbol{\varepsilon}}^p = 0 \quad \text{in } V, \ \forall t \\ & \int_0^T \dot{\boldsymbol{\varepsilon}}^p dt = \mathsf{R}(\Delta\tilde{\mathbf{u}}) + \Delta\mathbf{E}^p \quad \text{in } V \\ & \Delta\tilde{\mathbf{u}} \quad \text{periodic on } \Gamma \end{aligned} \tag{9}$$

In eqn. (9a) $\sigma_0(\mathbf{x})$ is the material yield limit and $\mathbf{X}$ is the usual symmetric matrix of numbers emerging from Mises yield function; in (9c) $\mathbf{Y}$ is a constant vector introduced to express in matrix form the volumetric strain.

3 Discretizations and Iterative Solution Technique

Both time and space integrals in formulation (9) have to be replaced by sums in view of numerical solutions. By a hypothesis which is very little restrictive in practice, let us assume that the load domain Ω (here in the space of the average stresses $\boldsymbol{\Sigma}$) is a polyhedron (or polygon or hyper-polyhedron, depending on the dimensionality), i.e. a convex region defined by a discrete number, say n, of vertices. Then the following theorem, due to König and Kleiber (1978) can be applied:

Shakedown occurs for the original polyhedrical load domain if it occurs for the load domain Ω^* reduced to the set of the n vertices of Ω.

Thus any APSRC reduces to a sequence of fields of plastic strain finite increments, say $\boldsymbol{\varepsilon}_k^p(\mathbf{x})$, (index k running over the vertex set), each one associated, through a dot product, to the preliminarily computed field (see Appendix) of local elastic stress response $\boldsymbol{\sigma}_k^e(\mathbf{x})$ to the k-th vertex external action, i.e. to $\boldsymbol{\Sigma}_k$.

As for the space integrals, they are reduced to a sum over the set I (run by index $i \in I$) of all Gauss points over the RV, when a conventional finite element (FE) model is adopted, as usual in computational mechanics, in order to algebrize boundary value problems.

Through both the discretizations (in time and space) which precede, SD analysis is cast into the following mathematical optimisation problem:

$$
\begin{aligned}
& s \cong \min_{\boldsymbol{\varepsilon}_{ki}^p, \Delta\tilde{\mathbf{U}}^*, \Delta\mathbf{E}^p} \sqrt{\tfrac{2}{3}} \sum_{k=1}^{n} \sum_{i \in I} \rho_i |\mathbf{J}|_i \sigma_0^i \sqrt{\boldsymbol{\varepsilon}_{ki}^{pT} \mathbf{X} \boldsymbol{\varepsilon}_{ki}^p}\,, \qquad \text{subject to:} \\
& \sum_{k=1}^{n} \sum_{i \in I} \rho_i |\mathbf{J}|_i \boldsymbol{\sigma}_{ki}^{eT} \boldsymbol{\varepsilon}_{ki}^p = 1 \\
& \sum_{k=1}^{n} \boldsymbol{\varepsilon}_{ki}^p = \mathbf{B}_i^* \Delta\tilde{\mathbf{U}}^* + \Delta\mathbf{E}^p \qquad \forall i \in I \\
& \mathbf{Y}^T \boldsymbol{\varepsilon}_{ki}^p = 0 \qquad \forall i \in I,\ k = 1,\dots,n
\end{aligned}
\tag{10}
$$

where I is the set of all Gauss integration points in the RV; ρ_i, $|\mathbf{J}|_i$ are the integration weights and the determinant of the Jacobian matrix (of the element-master mapping) computed at Gauss point i, respectively. Vector $\Delta\tilde{\mathbf{U}}^*$ gathers the nodal periodic displacement increments and matrix $\mathbf{B}^*$ assembles the compatibility matrices of each element. The periodicity conditions are accounted for both in $\Delta\tilde{\mathbf{U}}^*$ and $\mathbf{B}^*$.

The following remarks may elucidate the meaning of the above problem.

(i) As an alternative to the FE method, a boundary element method (BEM) and, in particular, the recently developed symmetric Galerkin BEM, see e.g. Bonnet et al. (1998), might be employed for the space discretization, e.g. Maier and Nappi (1984). BEM may imply computational savings, especially when plastic yielding processes can a priori be foreseen to be confined to a restricted region of the RV, so that domain discretization by cells can be limited to that region.

(ii) In the general formulation (7) and, hence, in the consequent formulations (9) and (10), the strain decomposition (4) dictated by the microscale periodicity is enforced only at the end of the APSRC as a kinematic compatibility requirement similar to the constraint (7c). Alternatively, by interpreting the strain splitting (4) as a constitutive requirement similar to plastic incompressibility (9c), it is enforced all along the APSRC and, hence, after time discretization, at each vertex k (k=1,…,n). In the latter case the following formulation, adopted in Carvelli et al. (1999b), would replace the preceding formulation (10):

$$s \cong \min_{\tilde{\boldsymbol{\varepsilon}}_{ki}^p, \mathbf{E}_k^p, \Delta\tilde{\mathbf{U}}^*, \Delta\mathbf{E}^p} \sqrt{\tfrac{2}{3}} \sum_{k=1}^{n} \sum_{i \in I} \rho_i |\mathbf{J}|_i \sigma_0^i \sqrt{\tilde{\boldsymbol{\varepsilon}}_{ki}^{pT} \mathbf{X} \tilde{\boldsymbol{\varepsilon}}_{ki}^p + 2\mathbf{E}_k^{pT} \mathbf{X} \tilde{\boldsymbol{\varepsilon}}_{ki}^p + \mathbf{E}_k^{pT} \mathbf{X} \mathbf{E}_k^p} \,,$$

subject to :

$$\sum_{k=1}^{n} \sum_{i \in I} \rho_i |\mathbf{J}|_i \boldsymbol{\sigma}_{ki}^{eT} (\tilde{\boldsymbol{\varepsilon}}_{ki}^p + \mathbf{E}_k^p) = 1 \tag{11}$$

$$\sum_{k=1}^{n} (\tilde{\boldsymbol{\varepsilon}}_{ki}^p + \mathbf{E}_k^p) = \mathbf{B}_i^* \Delta\tilde{\mathbf{U}}^* + \Delta\mathbf{E}^p \qquad \forall i \in I$$

$$\mathbf{Y}^T (\tilde{\boldsymbol{\varepsilon}}_{ki}^p + \mathbf{E}_k^p) = 0 \qquad \forall i \in I,\ k = 1,\ldots,n$$

Computational experience shows that the numerical results achieved by the two formulations are practically the same. However, the former (10), besides being more mechanically sound, involves less variables and, hence, is preferable in practice.

(iii) It is well known in computational mechanics that the displacement field may lead to “locking” phenomena in limit analysis and, for the same reason, in SD analysis as well. The reason of locking is that the kinematic constraints implicit in the displacement interpolations, combined with those arising from the constitutive model (e.g., in associative plasticity, the normality rule here reflected by the plastic incompressibility), reduce sometimes drastically the set of kinematically admissible mechanisms within which the analysis seeks the/a actual mechanism of collapse (or incremental collapse or alternating plasticity). The following remedies have been proposed in the literature:

(a) Higher order shape functions with reduced integration over FEs, which, however, may lead to the opposite trouble of kinematic indeterminacies (such as stressless “hourglass modes”, see e.g. Nagtegaal et al. (1974)).

(b) Mixed FEMs based on multifield (not only displacements) modelling: compatibility, like the other field equations, is enforced in an average approximate sense (rather than locally everywhere). In particular, for each pair of conjugate fields, the shape functions are related to each other, so that the dot products are preserved and global FE constitutive laws emerge in “Prager’s generalized variables”, see e.g. Comi et al. (1992), Corradi and Maier (1981). This special FEM is adopted in the poroplasticity Chapter of the present book.

(c) Penalization: a norm of the global violation of some constraints is defined, it is multiplied by one or more penalty factors and added to the objective function to minimize, so that thereafter it is pushed towards zero “in a soft manner” by the solution process, see e.g. Fiacco and McCornick (1968). The optimal choice of the penalty factor, say α, is problem-dependent and is crucial for the accuracy of the solution. In fact a too small factor α implies poor fulfilment of the relevant constraint (i.e. its relaxation). Hence, in the present context of SD and LA, a lower, possibly too conservative, safety factor s should be expected to result from relatively small penalty factor; too large α may lead to a higher nonconservative s or to numerical instability and lack of convergence.

In order to exploit its implementation implicitly, also in threedimensional problems, traditional FEM is adopted herein together with the penalization provision (c). Therefore, the problem to solve becomes:

$$
\begin{aligned}
& s \cong \min_{\boldsymbol{\varepsilon}_{ki}^{p},\Delta\tilde{\mathbf{U}}^{*},\Delta\mathbf{E}^{p}} \sqrt{\tfrac{2}{3}}\sum_{k=1}^{n}\sum_{i\in I}\rho_i|\mathbf{J}|_i\sigma_0^i\sqrt{\boldsymbol{\varepsilon}_{ki}^{pT}\mathbf{X}\boldsymbol{\varepsilon}_{ki}^{p}}+\alpha\sum_{k=1}^{n}\sum_{i\in I}\rho_i|\mathbf{J}|_i\boldsymbol{\varepsilon}_{ki}^{p^T}\mathbf{Y}\mathbf{Y}^T\boldsymbol{\varepsilon}_{ki}^{p}, \\
& \text{subject to:} \\
& \sum_{k=1}^{n}\sum_{i\in I}\rho_i|\mathbf{J}|_i\boldsymbol{\sigma}_{ki}^{eT}\boldsymbol{\varepsilon}_{ki}^{p}=1 \\
& \sum_{k=1}^{n}\boldsymbol{\varepsilon}_{ki}^{p}=\mathbf{B}_i^{*}\Delta\tilde{\mathbf{U}}^{*}+\Delta\mathbf{E}^{p} \qquad \forall i\in I
\end{aligned}
\tag{12}
$$

The peculiar features of problem (12) are: linear equality constraints, convex but nonsmooth objective function. The nonsmoothness is due to the fact that each (*k-th*) addend is not differentiable in the origin (for $\boldsymbol{\varepsilon}_{ki}^{p}=\mathbf{0}$). Differentiability might be restored in the generalized sense of "subdifferential" according to relatively recent developments in convex and nonconvex analysis, see e.g. Gao et al. (2001). Recourse will not be made here to the elegant and novel formalism of nonsmooth mathematics.

Numerical solutions are pursued in what follow by the classical Lagrange multiplier method and by an "ad hoc" version of an iterative algorithm originally proposed by Zhang et alii (1993, 1995), and Liu et alii (1995).

The Lagrangian function of problem (12) reads:

$$
\begin{aligned}
\mathsf{L}(\boldsymbol{\varepsilon}_{ki}^{p},\Delta\tilde{\mathbf{U}}^{*},\Delta\mathbf{E}^{p},\lambda,\mathbf{l}_i)=\Bigg\{&\sqrt{\tfrac{2}{3}}\sum_{k=1}^{n}\sum_{i\in I}\rho_i|\mathbf{J}|_i\sigma_0^i\sqrt{\boldsymbol{\varepsilon}_{ki}^{pT}\mathbf{X}\boldsymbol{\varepsilon}_{ki}^{p}}+\lambda\left[1-\sum_{k=1}^{n}\sum_{i\in I}\rho_i|\mathbf{J}|_i\boldsymbol{\sigma}_{ki}^{eT}\boldsymbol{\varepsilon}_{ki}^{p}\right] \\
&+\sum_{i\in I}\mathbf{l}_i\left[\sum_{k=1}^{n}\boldsymbol{\varepsilon}_{ki}^{p}-\mathbf{B}_i^{*}\Delta\tilde{\mathbf{U}}^{*}-\Delta\mathbf{E}^{p}\right]+\tfrac{1}{2}\alpha\sum_{k=1}^{n}\sum_{i\in I}\rho_i|\mathbf{J}|_i\boldsymbol{\varepsilon}_{ki}^{pT}\mathbf{Y}\mathbf{Y}^{T}\boldsymbol{\varepsilon}_{ki}^{p}\Bigg\}
\end{aligned}
\tag{13}
$$

where λ and $\mathbf{l}_i$ denote Lagrange multipliers.

The stationarity conditions of the Lagrangian lead to a set of nonlinear equations. These are solved herein by means of the above mentioned algorithm which reduces to solving, at the iteration h+1, the following system of linear equations:

$$
\begin{aligned}
&\sqrt{\tfrac{2}{3}}\rho_i|\mathbf{J}|_i\sigma_0^i \frac{\mathbf{X}\boldsymbol{\varepsilon}_{ki}^{p\,h+1}}{D_{ki}^h} - \lambda^{h+1}\rho_i|\mathbf{J}|_i\boldsymbol{\sigma}_{ki}^e + \mathbf{l}_i^{h+1} \\
&\qquad + \alpha\rho_i|\mathbf{J}|_i\mathbf{C}\,\boldsymbol{\varepsilon}_{ki}^{p\,h+1} = \mathbf{0} \qquad (\forall i \in I,\ k = 1,\ldots,n) \\
&\sum_{i\in I}\mathbf{B}_i^{*T}\mathbf{l}_i^{h+1} = \mathbf{0} \\
&\sum_{i\in I}\mathbf{l}_i^{h+1} = \mathbf{0} \\
&\sum_{k=1}^{n}\sum_{i\in I}\rho_i|\mathbf{J}|_i\boldsymbol{\sigma}_{ki}^{eT}\boldsymbol{\varepsilon}_{ki}^{p\,h+1} = 1 \\
&\sum_{k=1}^{n}\boldsymbol{\varepsilon}_{ki}^{p\,h+1} - \mathbf{B}_i^*\Delta\tilde{\mathbf{U}}^{*h+1} - \Delta\mathbf{E}^{p\,h+1} = \mathbf{0} \qquad \forall i \in I
\end{aligned}
\tag{14}
$$

where: $\mathbf{C} = \mathbf{Y}\mathbf{Y}^T$; $\boldsymbol{\varepsilon}_{ki}^{p\,h+1}$, $\Delta\tilde{\mathbf{U}}^{*h+1}$, $\Delta\mathbf{E}^{p\,h+1}$, λ^{h+1}, $\mathbf{l}_i^{h+1}$ are the unknowns at the current iteration; $D_{ki}^h = \sqrt{\boldsymbol{\varepsilon}_{ki}^{pT\,h}\mathbf{X}\,\boldsymbol{\varepsilon}_{ki}^{p\,h}}$ is known since it is computed on the basis of values at the preceding iteration.

The iterative procedure cannot proceed when D_{ki}^h is evaluated in the non-yielding zone as a manifestation of the nonsmoothness of the objective function (12a) in the mathematical programming formulation (12). To avoid this difficulty the Gauss point set I, at corner k, is split into the subset R_k^{h+1} of the points where, in the light of the results from the preceding iteration h, no plastic strains develop and the complementary subset P_k^{h+1} of the points where dissipation occurs, namely:

$$
\begin{aligned}
P_k^{h+1} &= \{i \in I \ \text{ such that } \ D_{ki}^h \neq 0\} \\
R_k^{h+1} &= \{i \in I \ \text{ such that } \ D_{ki}^h = 0\}
\end{aligned}
\tag{15}
$$

By another application of the penalty method, β being a suitably chosen small number, account taken of eqns. (15) the denominators D_{ki}^h in eqns. (14a) are replaced by:

$$
\overline{D}_{ki}^h = \begin{cases} D_{ki}^h & \forall i \in P_k^{h+1} \\ \beta << 1 & \forall i \in R_k^{h+1} \end{cases}
\tag{16}
$$

Let us set:

$$\mathbf{H}^h_{ki} = \sqrt{\tfrac{2}{3}}\sigma^i_0 \frac{\mathbf{X}}{\overline{D}^h_{ki}} + \alpha \mathbf{C} \tag{17}$$

$$\left.\begin{aligned} \Delta\widetilde{\mathbf{U}}^{*h+1} &= \lambda^{h+1}\Delta\widetilde{\mathbf{v}}^{*h+1} \\ \Delta\mathbf{E}^{p^{h+1}} &= \lambda^{h+1}\Delta\mathbf{e}^{p^{h+1}} \end{aligned}\right\} \tag{18}$$

At iteration h+1, some algebraic manipulations lead to the following equivalent form of problem (14), modified according to eqns. (16, 17, 18):

$$\begin{aligned} &\Big(\sum_{i\in I}\rho_i|\mathbf{J}|_i\mathbf{B}^{*T}_i\big(\sum_{k=1}^{n}(\mathbf{H}^h_{ki})^{-1}\big)^{-1}\mathbf{B}^*_i\Big)\Delta\widetilde{\mathbf{v}}^{*h+1} + \Big(\sum_{i\in I}\rho_i|\mathbf{J}|_i\mathbf{B}^{*T}_i\big(\sum_{k=1}^{n}(\mathbf{H}^h_{ki})^{-1}\big)^{-1}\Big)\Delta\mathbf{e}^{p^{h+1}} \\ &= \sum_{i\in I}\rho_i|\mathbf{J}|_i\mathbf{B}^{*T}_i\big(\sum_{k=1}^{n}(\mathbf{H}^h_{ki})^{-1}\big)^{-1}\big(\sum_{k=1}^{n}(\mathbf{H}^h_{ki})^{-1}\boldsymbol{\sigma}^e_{ki}\big) \end{aligned} \tag{19}$$

$$\begin{aligned} &\Big(\sum_{i\in I}\rho_i|\mathbf{J}|_i\big(\sum_{k=1}^{n}(\mathbf{H}^h_{ki})^{-1}\big)^{-1}\mathbf{B}^*_i\Big)\Delta\widetilde{\mathbf{v}}^{*h+1} + \Big(\sum_{i\in I}\rho_i|\mathbf{J}|_i\big(\sum_{k=1}^{n}(\mathbf{H}^h_{ki})^{-1}\big)^{-1}\Big)\Delta\mathbf{e}^{p^{h+1}} \\ &= \sum_{i\in I}\rho_i|\mathbf{J}|_i\big(\sum_{k=1}^{n}(\mathbf{H}^h_{ki})^{-1}\big)^{-1}\big(\sum_{k=1}^{n}(\mathbf{H}^h_{ki})^{-1}\boldsymbol{\sigma}^e_{ki}\big) \end{aligned} \tag{20}$$

$$\lambda^{h+1} = \left\{\sum_{k=1}^{n}\sum_{i\in I}\rho_i|\mathbf{J}|_i\boldsymbol{\sigma}^{eT}_{ki}(\mathbf{H}^h_{ki})^{-1}\big(\sum_{m=1}^{n}(\mathbf{H}^h_{mi})^{-1}\big)^{-1}\Big[\mathbf{B}^*_i\Delta\widetilde{\mathbf{v}}^{*h+1} + \Delta\mathbf{e}^{p^{h+1}} + \sum_{m=1}^{n}(\mathbf{H}^h_{mi})^{-1}(\boldsymbol{\sigma}^e_{ki}-\boldsymbol{\sigma}^e_{mi})\Big]\right\}^{-1} \tag{21}$$

$$\boldsymbol{\varepsilon}^{p^{h+1}}_{ki} = (\mathbf{H}^h_{ki})^{-1}\big(\sum_{m=1}^{n}(\mathbf{H}^h_{mi})^{-1}\big)^{-1}\left[\mathbf{B}^*_i\Delta\widetilde{\mathbf{U}}^{*h+1} + \Delta\mathbf{E}^{p^{h+1}} + \lambda^{h+1}\sum_{m=1}^{n}(\mathbf{H}^h_{mi})^{-1}(\boldsymbol{\sigma}^e_{ki}-\boldsymbol{\sigma}^e_{mi})\right] \tag{22}$$

At each iteration (say h+1), $\Delta\widetilde{\mathbf{v}}^{*h+1}$ and $\Delta\mathbf{e}^{p^{h+1}}$ are solution of the linear system eqns. (19, 20). Substituting them into (21) and using eqns. (18) and (21), one obtains the microscopic plastic strain sub-increment $\boldsymbol{\varepsilon}^{p^{h+1}}_{ki}$ at corner k (k=1,...,n) and at each Gauss point i.

The iterative procedure centered on eqns. (19)-(22) starts (for h=0) by assuming the body in a fully plastic state and setting $D^0_{ki} = 1 \ \forall i \in I,\ k = 1,...,n$.

The above iterative procedure is suitable to computer implementation and turned out to converge fairly fast, with appropriate empirically selected penalty factors α in all the tests performed so far, including the applications outlined in the next Section.

4 Applications to Perforated Plates

In this Section the kinematic method, presented in what precedes, is employed to compute macroscopic shakedown domains of perforated mild steel (Mises) thin plates with a periodic distribution of circular holes.

Reference is made first to a thin specimen with a regular array of circular holes with volumetric ratio (void over total volume) c_h=0.3, such that a square representative volume (*RV*) can be singled out as shown in Figure 3a (where also the adopted mesh of 4-nodes finite element is visualized). Analyses are carried out assuming plane stress states and considering the square *RV* subjected to principal macroscopic stresses Σ_1 and Σ_2 (see Figure 3a). The stresses (i.e. "the loads" for the *RV*) Σ_1 and Σ_2 fluctuate in time independently between zero and a prescribed maximum value. The assigned load domains are depicted in Figure 2 in the plane of the macroscopic stresses Σ_1, Σ_2 normalized with respect to the yield stress σ_0 of the material.

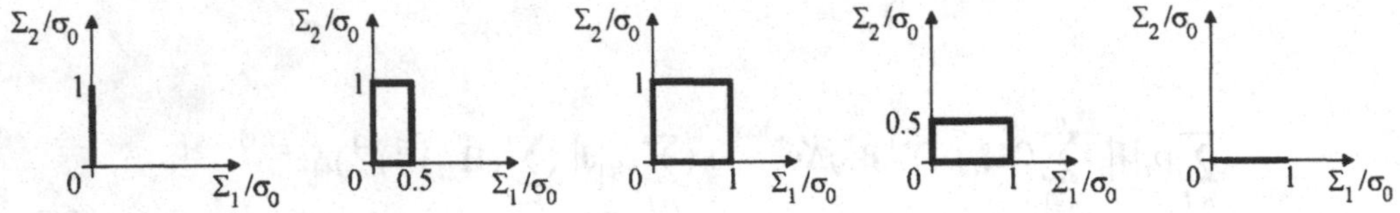

Figure 2. Assigned basic macroscopic stress domains.

For each assigned macroscopic stress domain, the shakedown limit factor is obtained by performing a shakedown analysis by the proposed method. Every macroscopic stress history inside an amplified limit domain leads to adaptation of the homogenized material. The envelope of the limit domains defines "shakedown domain" of the heterogeneous material for the class of rectangular load domains specified in Figure 2.

Figure 3b shows the shakedown domain of the considered square *RV* for the orientations θ=45° of the principal stress Σ_2 with respect to axis x_2. Comparing the shakedown domain with the plastic collapse locus obtained in Carvelli et al. (2000), a significant reduction can be observed (Figure 3b).

The shakedown domain of the homogenized material is now computed for a thin plate perforated by a hexagonal pattern of circular holes, with volume fraction c_h=0.5, by applying the proposed method to the hexagonal *RV* under biaxial macroscopic stress conditions shown in Figure 4a. The resulting shakedown domain envelopes are presented in Figure 4b,c for two orientations θ. Considering uniaxial macroscopic stress histories along the direction x_1 or x_2, shakedown analysis predicts the same strength limits provided by limit analysis. On the contrary, biaxial macroscopic stress paths lead to a shakedown domain smaller than the plastic collapse one. The above results show that, if the considered perforated plate with circular hole and hexagonal

RV is subjected to variable macroscopic stress states according to unpredictable time histories within known intervals, the strength limit of the homogenized (plane stress) plate can undergo a significant reduction. This kind of practically important information can be provided in a computationally cost-effective way by a direct method like the kinematic one exposed herein, based on the shakedown theory and able to take into consideration the texture periodicity.

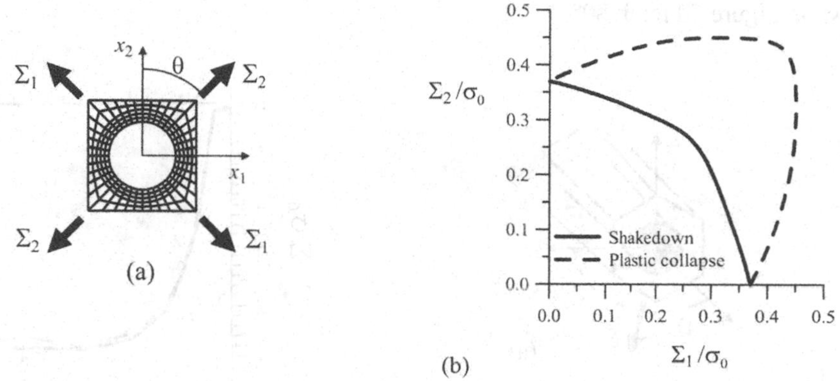

Figure 3. (a) square representative volume (*RV*) of a perforated plate with c_h=0.3 and FE mesh. (b) Plastic collapse locus and shakedown domain for θ=45° and for loading domains of Figure 2.

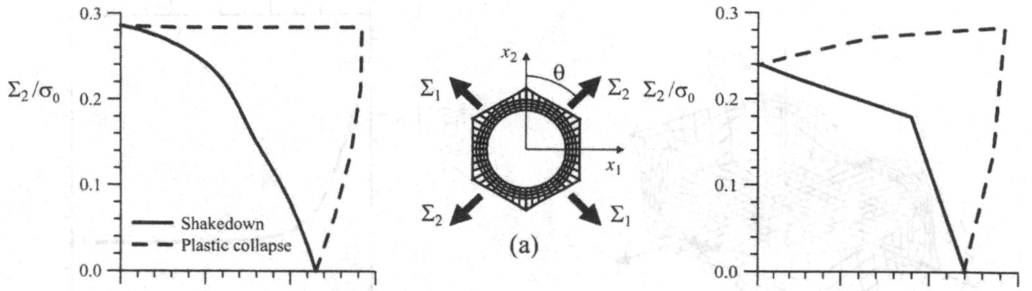

Figure 4. (a) hexagonal *RV* with c_h=0.5 and FE mesh. Plastic collapse locus and shakedown domains for θ=0° (b) and θ=45° (c).

5 Applications to Metal-Matrix Composites

The *RV* shown in Figure 5a characterizes the periodic heterogeneous geometric texture at the microscale of a ductile metal-matrix composite (MMC) with fiber volume fraction c_f=0.65. Both materials are attributed elastic-perfectly plastic von Mises behavior defined by yield limits with fiber (σ_0^f) vs. matrix (σ_0^m) ratio equal to 8.7. The homogenized SD limits for uniaxial average stress states fluctuating between 0 and Σ have been computed by the FE procedure summarized in Section 3, for several angles θ of the stress axis from the fiber axis. These results are visualized in Figure 5b and compared to the uniaxial plastic collapse limits presented in Carvelli et alii (2000). Figure 5c depicts the incremental collapse mechanism for θ=10° and the adopted mesh (432 FE with bilinear interpolation). The convergence of the iteration sequence with penalty factor $\alpha=10^6$, is shown in Figure 5d for θ=80°.

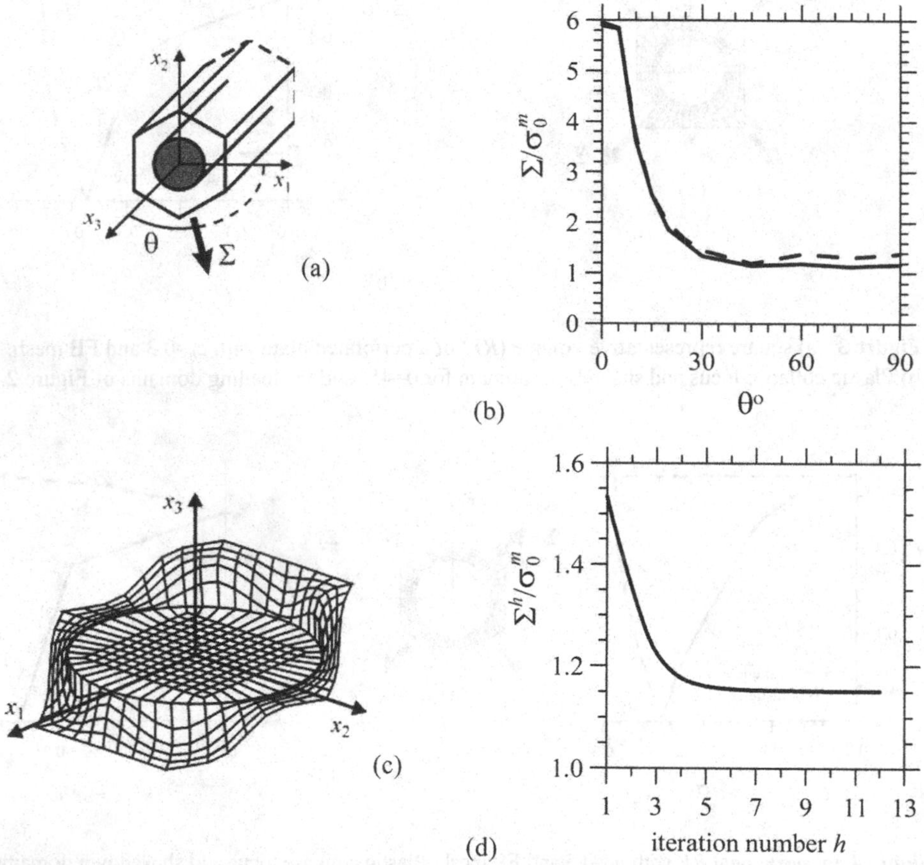

Figure 5. (a) representative volume; (b) uniaxial plastic collapse limit (dashed line) and shakedown limit (solid line) vs. direction θ of uniaxial average stress states; (c) FE mesh and incremental collapse mechanism for θ=10°; (d) convergence of the iteration algorithm for θ=80°.

Acknowledgements. This Chapter contains results achieved in the frame of a Research Project on advanced materials sponsored by CNR (Italian Research Council).

References

Bonnet, M. Maier, G. and Polizzotto, C. (1998). Symmetric Galerkin boundary element method. *Appl. Mech. Reviews ASME* 51: 669-704.

Carvelli, V., Cen, Z., Liu, Y. and Maier, G. (1999a). Shakedown analysis of defective pressure vessels by a kinematic approach. *Arch. Appl. Mech.* 69: 751-764.

Carvelli, V., Maier, G. and Taliercio, A. (1999b). Shakedown analysis of periodic heterogeneous materials by a kinematic approach. *J. Mech. Eng.* (Strojnicky Časopis) 50: 229-240.

Carvelli, V., Maier, G. and Taliercio, A. (2000). Kinematic limit analysis of periodic heterogeneous media. *Comp. Mod. Eng. Sci.* 1: 15-26.

Cohn, M. Z., Maier, G. and Grierson, D. eds. (1979). *Engineering plasticity by mathematical programming*. Pergamon Press, New York.

Comi, C., Maier, G., Perego, U. (1992). Generalized variable finite element modelling and extremum theorems in stepwise holonomic elastoplasticity with internal variables. *Comput. Meth. Appl. Mech. Eng.* 96: 133-171.

Corigliano, A., Maier, G. and Pycko, S. (1995). Dynamic shakedown analysis and bounds for elastoplastic structures with nonassociative, internal variable constitutive laws. *Int. J. Solids and Structures* 32: 3145-3166.

Corradi, L. and Maier, G. (1981). Finite element elastoplastic and limit analysis: some consistency criteria and their implications. In Wunderlich W., Stein E. and Bathe K .J., eds., *Nonlinear Finite Element Analysis in Structural Mechanics*. Springer Verlag, 290-306.

de Buhan, P. and Taliercio, A. (1988). Critère de résistance macroscopique pour les matériaux composites à fibres. *C. R. Acad. Sci. Paris* Série II 307: 227-232.

Druyanov, B. and Roman, I. (1999). Conditions for shakedown of damaged elastic plastic bodies. *European J. Mech. A/Solids* 18: 641-651.

Dvorak, G. J., Lagoudas, D. C. and Huang, C. M. (1994). Fatigue damage and shakedown in metal matrix composite laminates. *Mech. Composite Mat. Struct.* 1: 171-202.

Fiacco, A, V. and McCormick, G. P. (1968). Nonlinear programming: sequential unconstrained minimization techniques. Wiley Publ. Co., New York. Reprinted by SIAM Publications in 1990.

Francescato, P. and Pastor, J. (1997). Lower and upper numerical bounds to the off-axis strength of unidirectional fiber-reinforced composites by limit analysis methods. *European J. Mech. A/Solids* 16: 213-234.

Gao, D. Y., Ogden, R. W. and Stavroulakis, G.E. (2001). Nonsmooth-nonconvex mechanics. Kluwer Akademic Publ., Dordrecht.

Hachemi, A. and Weichert, D. (1997). Application of shakedown theory to damaging inelastic material under mechanical and thermal loads. *Int. J. Mech. Sci.* 39: 1067-1076.

Kaliszky, S. (1989). *Plasticity: theory and engineering applications*. Elsevier, Amsterdam, North Holland.

Koiter, W. T. (1960). General theorems for elastic-plastic solids. In Sneddon, J. N. and Hill, R., eds., *Progress in solid mechanics*. North Holland, Amsterdam. 165-221.

König, J. A. (1987). *Shakedown of elastic-plastic structures*. Elsevier, Amsterdam.

König, J. A. and Kleiber, M. (1978). On a new method of shakedown analysis. *Bulletin de l'Academie Polonaise des Sciences - Sér. des Scien. Techn.* 26: 165-171.

Liu, Y. H., Cen, Z. Z. and Xu, B. Y. (1995). A numerical method for plastic limit analysis of 3-D structures. *Int. J. Sol. Struct.* 32: 1645-1658.

Lubliner, J. (1990). *Plasticity theory*. Mc Millan Publ., New York.

Maier, G., Carvelli, V., Cocchetti, G. (2000). On direct methods for shakedown and limit analysis, *Eur. J. Mech. A/Solids*. 19: S79-S100.

Maier, G., Carvelli, V., Taliercio, A. (2001). Limit and shakedown analysis of periodic heterogeneous media. In Lemaitre, J, ed., *Handbook of Material Behaviour Models*, Academic Press, San Diego, 10.8: 1017-1028.

Maier, G. and Nappi, A. (1984). On bounding post-shakedown quantities by the boundary element method. *Engineering Analysis* 1: 223-229.

Nagtegaal, J. C., Parks, D. M. and Rice, J. R. (1974). On numerically accurate finite element solutions in the fully-plastic range. *Comp. Meth. Appl. Mech. Eng.* 4: 153-177.

Polizzotto, C., Borino, G. and Fuschi, P. (1996) An extended shakedown theory for elastic-plastic-damage material models. *European J. Mech. - A/Solids* 15: 825-858.

Ponter, A. R. S. and Leckie, F. A. (1998a). Bounding properties of metal-matrix composites subjected to cyclic loading. *J. Mech. Phys. Solids* 46: 697-717.

Ponter, A. R. S. and Leckie, F. A. (1998b). On the behaviour of metal matrix composites subjected to cyclic thermal loading. *J. Mech. Phys. Solids* 46: 2183-2199.

Suquet, P. (1982). Analyse limite et homogénéisation. *C. R. Acad. Sci. Paris* Série II 296: 1355-1358.

Taliercio, A. (1992). Lower and upper bounds to the macroscopic strength domain of a fiber-reinforced composite material. *Int. J. Plasticity* 8: 741-762.

Tarn, J. Q., Dvorak, G. J. and Rao, M. S. M. (1975). Shakedown of unidirectional composites. *Int. J. Sol. Struct.* 11: 751-764.

Weichert, D., Hachemi, A. and Schwabe, F. (1999a). Application of shakedown analysis to the plastic design of composites. *Arch. Appl. Mech.* 69: 623-633.

Weichert, D., Hachemi, A. and Schwabe, F. (1999b). Shakedown analysis of composites. *Mech. Res. Comm.* 26: 309-318.

Zhang, Y.G., Zhang, P. and Lu, M. W. (1993). Computational limit analysis of rigid-plastic bodies in plane strain. *Acta Mech. Sol. Sinica* 6: 341-348.

Zhang, Y.G. (1995). An iterative algorithm for kinematic shakedown analysis. *Comp. Meth. Appl. Mech. Eng.* 127: 217-226.

Zhang, Y.G. and Lu, M. W. (1995). An algorithm for plastic limit analysis. *Comp. Meth. Appl. Mech. Eng.* 126: 333-341.

Appendix: On Linear Elastic Computations Preliminary to Shakedown Analysis of Periodic Media

In Section 3 it has been seen that, for each (k-th) vertex $\mathbf{\Sigma}_k$ (k=1,...,n) of the loading domain Ω in the space of the average stresses $\mathbf{\Sigma}$, the consequent linear elastic stress field $\boldsymbol{\sigma}_k^e(\underline{x})$ must be determined in each (i-th) Gauss point ($i \in I$) of the FE model of the representative volume (RV) and entered as data into the minimization problem to solve for shakedown analysis purposes. A homogenization procedure apt to provide the above results is outlined below in matrix notation.

Each (r-th) constituent material (or "phase") at the microscale is characterized by its elastic stiffness matrix $\mathbf{D}_r$: e.g., for the metal matrix composites $\mathbf{D}_m$ is the stiffness of the matrix and $\mathbf{D}_f$ that of the fibres, perfect adhesion being assumed between them; for perforated plates $\mathbf{D}_f$=$\mathbf{0}$.

The total potential energy of the representative volume (RV), in general terms, reads:

$$\Pi(\mathbf{u}) = \int_V \tfrac{1}{2}\boldsymbol{\sigma}^T[\boldsymbol{\varepsilon}(\mathbf{u})]\,\boldsymbol{\varepsilon}(\mathbf{u})\,dV - \int_\Gamma \mathbf{p}^T\mathbf{u}\,d\Gamma \tag{A1}$$

The virtual work principle implies the equalities:

$$\int_\Gamma \mathbf{p}^T\mathbf{u}\,d\Gamma = \int_V \boldsymbol{\sigma}^T\boldsymbol{\varepsilon}\,dV = \mathbf{\Sigma}^T\mathbf{E} \tag{A2}$$

Periodicity, see eqn. (4), and kinematic compatibility, expressed for the strain field modelled in accordance with the adopted FE space discretization, require that:

$$\boldsymbol{\varepsilon}(\mathbf{x}) = \mathbf{E} + \tilde{\boldsymbol{\varepsilon}}(\mathbf{x}) = \mathbf{E} + \mathbf{B}(\mathbf{x})\tilde{\mathbf{U}}^* \tag{A3}$$

where: $\tilde{\mathbf{U}}^*$ denotes the vector of independent nodal displacements, account taken of the periodicity boundary conditions on displacements; $\mathbf{B}$ is the compatibility matrix, consistent with the adopted shape functions for displacements.

Making use of (A2) and (A3) and of Gaussian integration over the discrete FE model of the RV, the potential energy (A1) can be expressed as follows:

$$\Pi(\mathbf{E},\tilde{\mathbf{U}}^*) = \frac{1}{2V}\sum_{i\in I}\rho_i|\mathbf{J}|_i(\mathbf{E}^T\mathbf{D}_i\mathbf{E} + 2\mathbf{E}^T\mathbf{D}_i\mathbf{B}_i\tilde{\mathbf{U}}^* + \tilde{\mathbf{U}}^{*T}\mathbf{B}_i^T\mathbf{D}_i\mathbf{B}_i\tilde{\mathbf{U}}^*) - \mathbf{\Sigma}^T\mathbf{E} \tag{A4}$$

where: index i runs over the set I of all Gauss points $\mathbf{x}_i$ in the RV; ρ_i and $\mathbf{J}$ denote Gauss weights and Jacobian matrix of the master element mapping onto the FE; $\mathbf{D}_i$ represent the elastic stiffness matrix of the (homogeneous) material surrounding point i; $\mathbf{B}_i$=$\mathbf{B}(\mathbf{x}_i)$.

The stationarity conditions of Π read, with obvious meaning of the new matrix symbols:

$$\frac{\partial \Pi}{\partial \widetilde{\mathbf{U}}^*} = \mathbf{K}\widetilde{\mathbf{U}}^* + \mathbf{H}^T\mathbf{E} = \mathbf{0} \tag{A5}$$

$$\frac{\partial \Pi}{\partial \mathbf{E}} = \mathbf{H}\widetilde{\mathbf{U}}^* + \mathbf{G}\mathbf{E} = \mathbf{\Sigma} \tag{A6}$$

Let the equation (A5) be solved with respect to vector $\widetilde{\mathbf{U}}^*$ and be substituted into eqn. (A6) (matrix $\mathbf{K}=\mathbf{K}^T$ is not singular, since all rigid body motions are meant as usual to be prevented by statically determinate, non-redundant constraints). Thus the elastic stiffness **D** of the homogenised medium is obtained:

$$\mathbf{\Sigma} = \mathbf{D}\mathbf{E}, \quad \text{with} \quad \mathbf{D} = \mathbf{G} - \mathbf{H}\mathbf{K}^{-1}\mathbf{H}^T \tag{A7}$$

Equation (A5) now provides the displacement vector due to a given vertex $\mathbf{\Sigma}_k$ of the loading domain in the average stress space $\mathbf{\Sigma}$:

$$\widetilde{\mathbf{U}}_k^* = -\mathbf{K}^{-1}\mathbf{H}^T\mathbf{D}^{-1}\mathbf{\Sigma}_k \tag{A8}$$

whence, finally, through (A3) the elastic stress $\boldsymbol{\sigma}_r^e$ is obtained in each phase r and in each Gauss points i pertaining to it:

$$[\boldsymbol{\sigma}_k^e(\mathbf{x}_i)]_r = \mathbf{D}_r(\mathbf{I} - \mathbf{B}(\mathbf{x}_i)\mathbf{K}^{-1}\mathbf{H}^T)\mathbf{D}^{-1}\mathbf{\Sigma}_k \tag{A9}$$

Shakedown of Structures Subjected to Dynamic External Actions and Related Bounding Techniques

Castrenze Polizzotto[1], Guido Borino[1] and Paolo Fuschi[2]

[1] University of Palermo, Palermo, Italy
[2] University of Reggio Calabria, Reggio Calabria, Italy

Abstract. The shakedown theory for dynamic external actions is expounded considering elastic-plastic internal-variable material models endowed with hardening saturation surface and assuming small displacements and strains as long with negligible effects of temperature variations on material data. Two sorts of dynamic shakedown theories are presented, i.e.: i) *Unrestricted* dynamic shakedown, in which the structure is subjected to (unknown) sequences of short-duration excitations belonging to a known excitation domain, with no-load no-motion time periods in between and for which a unified framework with quasi-static shakedown is presented; and ii) *Restricted* dynamic shakedown, in which the structure is subjected to a specified infinite-duration load history. Two general bounding principles are also presented, one is non evolutive in nature and holds for repeated loads below the shakedown limit, the other is evolutive and holds for a specified load history either below and above the shakedown limit. Both principles are applicable in either statics and dynamics to construct bounds to the actual plastic deformation parameters. A continuum solid mechanics approach is used throughout, but a class of discrete models (finite elements with piecewise linear yield and saturation surfaces and plastic deformability lumped at Gauss points) are also considered. Extensions of shakedown theorems to materials with temperature dependent yield and saturation functions are also presented.

1 Introduction

The origins of dynamic shakedown can be traced back to the pioneering works of Ceradini (1969) and Corradi and Maier (1974) who extended to dynamics, respectively, the (Melan) static-type and (Koiter) kinematic-type theorems of classical shakedown theory, which is concerned with quasi-static loads and perfect plasticity. These theorems were then generalized to cope with more realistic material models, as linearly hardening materials (Corradi and Maier, 1973), second-order geometric effects (Corradi and De Donato, 1975), internal-variable material models with nonlinear hardening (Maier and Novati, 1990; Comi and Corigliano, 1991).

All these referenced papers consider the structure subjected to a load history specified in the infinite time interval $t \geq 0$. However, in practice, time-variable loads are known only for finite time intervals, and the infinite-duration load histories sometimes considered, as e.g. the periodic loads, are just extrapolations of finite-duration ones. Furthermore, as pointed out by Polizzotto (1984a), the dynamic shakedown mentioned above turns out to be conceptually different from the classical one. This difference stems from the circumstance that, in the presence of a specified load history, no load repetitions are allowed in addition to those (if any) foreseen by the loading programme itself. On the contrary, the classical (quasi-static) shakedown is deeply rooted on the concept of *load repetition*; that is, any load belonging to a given domain of potentially active loads can in principle be repeated any number of times.

For the above reasons, according to a proposal by Polizzotto (1984a), dynamic shakedown dealing with a specified load history is referred to as *restricted*, whereas *unrestricted* is shakedown with load schemes allowing for load repetitions. The shakedown being a long term structural phenomenon, a consequent main difference between these two types of shakedown —so far not explicitly mentioned in the literature— is the following: with unrestricted shakedown, all load conditions are equally responsible in determining the shakedown occurrence, since every individual load is always potentially active during the structure's life; on the contrary, with unrestricted shakedown, only the load conditions acting after a long term (i.e. those of the asymptotic load history) are responsible in determining the shakedown occurrence, whereas the loads acting during the initial (finite) time period (i.e. those of the initial transient load history) are irrelevant in respect to the shakedown occurrence, since they constitute isolated irrepeatable load conditions that in any case may produce only finite plastic deformations. These features of restricted dynamic shakedown make it less adequate as safety criterion than unrestricted dynamic shakedown.

Often in practice, the loads can be considered as an unknown sequence of short-duration excitations of relatively high intensities, with periods of no loads (or of residual static loads) between two subsequent excitations, during which the structure can be considered motionless due to damping. All these excitations can be thought of to belong to a specified (excitation) domain, each excitation being allowed to be repeated any number of times. In this way, a loading scheme with repeated loads, conceptually similar to that of quasi-static shakedown, is envisaged. More precisely: in statics, we deal with load points within a load domain; in dynamics, we deal with excitation points within an excitation domain. Shakedown as based on the repeated excitation load scheme belongs to the category of unrestricted dynamic shakedown. Ceradini (1969) considered special load histories as sequences of short-duration excitations like in the above and showed that his dynamic shakedown theorem simplified to take a form similar to the analogous quasi-static theorem, however he did not account for load repetition.

The purpose of the present chapter is to present the essentials of the theory of dynamic shakedown for structures with an elastic-plastic rate-independent material endowed with internal variables and hardening saturation surface, stable in the Drucker sense (Drucker, 1960). The unrestricted dynamic shakedown will be presented first together with a few applications; then the restricted dynamic shakedown will be addressed. The related bounding techniques will be also presented, in a unified framework for quasi-static and dynamic loads. All the topics are developed with a continuum solid mechanics approach, but finite element discrete models will also be considered. Extensions of the shakedown theorems to materials endowed with yield and saturation functions depending on temperature will be also presented with reference to quasi-static external actions.

Notation. A compact notation is used throughout, with bold face letters for vectors and tensors. The 'dot' and the 'colon' products of vectors and tensors denote simple and double index contraction operations, respectively. For instance, for the vectors $\boldsymbol{u} = \{u_i\}$, $\boldsymbol{n} = \{n_i\}$, the tensors $\boldsymbol{\sigma} = \{\sigma_{ij}\}$, $\boldsymbol{\varepsilon} = \{\varepsilon_{ij}\}$, $\boldsymbol{E} = \{E_{ijhk}\}$, the following operations hold: $\boldsymbol{u} \cdot \boldsymbol{n} = u_i n_i$, $\boldsymbol{\sigma} \cdot \boldsymbol{n} = \{\sigma_{ij} n_j\}$, $\boldsymbol{\sigma} : \boldsymbol{\varepsilon} = \sigma_{ij} \varepsilon_{ji}$, $\boldsymbol{E} : \boldsymbol{\varepsilon} = \{E_{ijhk} \varepsilon_{kh}\}$, where the subscripts denote Cartesian components and the repeated index summation rule is applied. The upper dot denotes time derivative, i.e. $\dot{\boldsymbol{u}} = \partial \boldsymbol{u}/\partial t$. An orthogonal Cartesian axes system is used, with coordinates $\boldsymbol{x} = (x_1, x_2, x_3)$. The symbol ∇^s = symmetric part of the gradient operator, e.g. $\nabla^s \boldsymbol{u} = \{(\partial u_i/\partial x_j + \partial u_j/\partial x_i)/2\}$; also, 'div' = divergence operator, e.g. div $\boldsymbol{\sigma} = \{\partial \sigma_{ji}/\partial x_j\}$. Other symbols will be defined in the text where they appear for the first time.

2 The Material Model

For the purposes of the present chapter, an elastic-plastic rate-independent associative material model is considered. This material behaves as a standard material (Halphen and Nguyen, 1975) as far as its hardening state is below the saturation threshold, but like a perfect plastic one when its hardening state reaches a critical saturation threshold (Fuschi and Polizzotto, 1998). The reversible constitutive behaviour of this material is described by the state equations:

$$\boldsymbol{\sigma} = \boldsymbol{E} : \boldsymbol{\varepsilon}^e, \qquad \boldsymbol{\chi} = \boldsymbol{\chi}(\boldsymbol{\xi}) \equiv \frac{\partial \Psi_{\text{in}}}{\partial \boldsymbol{\xi}}. \tag{1}$$

Here, $\boldsymbol{E}$ is the usual (symmetric, positive definite) fourth-order moduli tensor of linear elasticity, whereas $\Psi_{\text{in}}(\boldsymbol{\xi})$ is the (convex) hardening potential, which represents the amount of energy stored within the microstructure as a consequence of the slip adjustments described by the internal variables $\boldsymbol{\xi}$ and by the related thermodynamic forces $\boldsymbol{\chi}$ (static internal variables). Both $\boldsymbol{\xi}$ and $\boldsymbol{\chi}$ may be scalars, or vectors, or tensors in practice, but here we formally treat them as vectors for simplicity. The additive strain decomposition holds, i.e.

$$\boldsymbol{\varepsilon} = \boldsymbol{\varepsilon}^e + \boldsymbol{\varepsilon}^p \tag{2}$$

where $\boldsymbol{\varepsilon}^e$ = elastic strain, $\boldsymbol{\varepsilon}^p$ = plastic strain.

The material irreversible constitutive behaviour is described as follows:

$$f(\boldsymbol{\sigma}, \boldsymbol{\chi}) \leq 0, \qquad \dot{\lambda} \geq 0, \qquad \dot{\lambda} f(\boldsymbol{\sigma}, \boldsymbol{\chi}) = 0 \tag{3}$$

$$\phi(\boldsymbol{\chi}) \leq 0, \qquad \dot{\mu} \geq 0, \qquad \dot{\mu}\, \phi(\boldsymbol{\chi}) = 0 \tag{4}$$

$$\dot{\boldsymbol{\varepsilon}}^p = \dot{\lambda} \frac{\partial f}{\partial \boldsymbol{\sigma}}, \qquad -\dot{\boldsymbol{\xi}} = \dot{\lambda} \frac{\partial f}{\partial \boldsymbol{\chi}} + \dot{\mu} \frac{\partial \phi}{\partial \boldsymbol{\chi}} \tag{5}$$

where f is the yield function and ϕ the saturation function, both being smooth and convex by hypothesis. Equations (3)-(5) provide the plastic mechanism ($\dot{\boldsymbol{\varepsilon}}^p$, $-\dot{\boldsymbol{\xi}}$) through the normality and the usual load/unload rules. The intrinsic dissipation density associated with this mechanism reads

$$D = \boldsymbol{\sigma} : \dot{\boldsymbol{\varepsilon}}^p - \boldsymbol{\chi} \cdot \dot{\boldsymbol{\xi}} \tag{6}$$

where the dot product denotes scalar product also between non-Cartesian vectors (like $\boldsymbol{\chi}$ and $\dot{\boldsymbol{\xi}}$). Equation (6) provides D as a nonnegative one-degree homogeneous function, $D = D(\dot{\boldsymbol{\varepsilon}}^p, \dot{\boldsymbol{\xi}})$, which for nonsaturated hardening states coincides with the dissipation function $D_0(\dot{\boldsymbol{\varepsilon}}^p, \dot{\boldsymbol{\xi}})$ related to the standard material model. The following conditions hold, i.e.

$$\boldsymbol{\sigma} = \frac{\partial D}{\partial \dot{\boldsymbol{\varepsilon}}^p}, \qquad \boldsymbol{\chi} = -\frac{\partial D}{\partial \dot{\boldsymbol{\xi}}} \tag{7}$$

for any nontrivial mechanism $(\dot{\boldsymbol{\varepsilon}}^p, \dot{\boldsymbol{\xi}})$.

A peculiarity of this model is that $\dot{\boldsymbol{\xi}}$ can vanish even with both $\dot{\lambda}$ and $\dot{\mu}$ being positive. When this is the case (*critical saturation state*), we can write:

$$\frac{\partial f_\sigma}{\partial \boldsymbol{\chi}} = -q \frac{\partial \phi}{\partial \boldsymbol{\chi}} \qquad (q = \dot{\mu}/\dot{\lambda} > 0) \tag{8}$$

which states that the yield surface $f_\sigma(\boldsymbol{\chi}) = f(\boldsymbol{\sigma}, \boldsymbol{\chi}) = 0$ imbedded in the $\boldsymbol{\chi}$-space (hence with $\boldsymbol{\sigma}$ considered fixed) finds itself in a position of external contact with the saturation surface $\phi(\boldsymbol{\chi}) = 0$, Fig. 1(a).

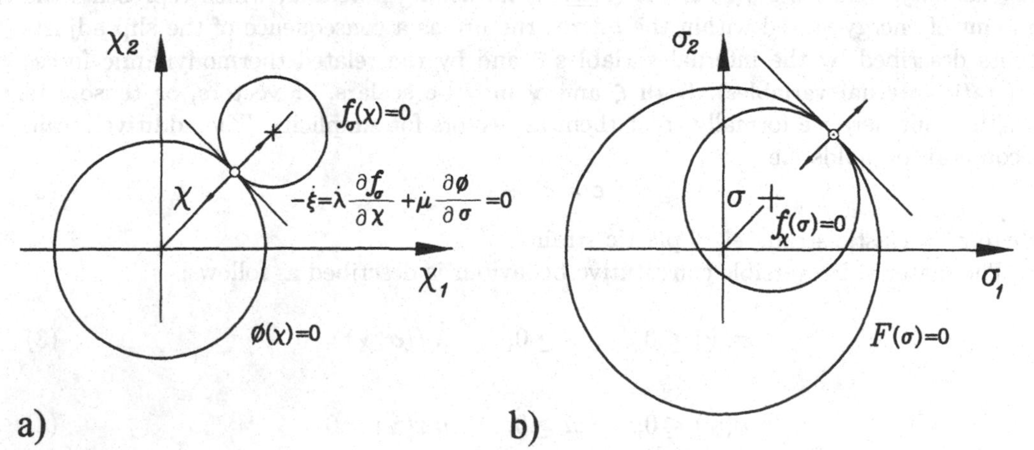

Figure 1. Saturation surface and related bounding surface: a) in the $\boldsymbol{\chi}$-space, the saturation and the yield surface are in a mutual external contact position, such that $\dot{\boldsymbol{\xi}} = \boldsymbol{0}$; b) In the $\boldsymbol{\sigma}$-space, the bounding surface $F(\boldsymbol{\sigma}) = 0$ envelops all possible positions of the yield surface, $f_\chi(\boldsymbol{\sigma}) = 0$, on changing $\boldsymbol{\chi}$ on the saturation surface $\phi(\boldsymbol{\chi}) = 0$.

Equation (8), together with $\phi(\chi) = 0$, provides the relation $\chi = \chi(\sigma)$ between the contact point χ and the related stress σ, and this relation, substituted in the equation $f(\sigma, \chi) = 0$, leads to the equation $F(\sigma) \equiv f(\sigma, \chi(\sigma)) = 0$. $F(\sigma) = 0$ is the envelope of the yield functions $f_\chi(\sigma) = 0$ generated by allowing χ to change in all possible ways on the saturation surface. It thus results that in the critical saturation state the yield surface $f_\chi(\sigma) = 0$ is in a position of internal contact with the envelope surface $F(\sigma) = 0$, which therefore acts as a limit (or bounding) surface, Fig. 1(b).

In a critical saturation state, being $\dot{\xi} = 0$, we have: $D = \sigma : \dot{\varepsilon}^p$, $\dot{\varepsilon}^p = \dot{\lambda}\, \partial f / \partial \sigma = \dot{\lambda}\, \partial F / \partial \sigma$; that is, the material behaves as a perfectly plastic material, the only changes allowed being neutral changes of hardening from a critical state to a similar one (Fuschi and Polizzotto, 1998).

It is worth noting that the above material model admits a maximum dissipation theorem that reads as in the following:

$$\left.\begin{aligned} D(\dot{\varepsilon}^p, \dot{\xi}) &= \max_{(\sigma,\chi)} \; (\sigma : \dot{\varepsilon}^p - \chi \cdot \dot{\xi}) \\ \text{s.t.} \;\; & f(\sigma, \chi) \leq 0, \qquad \phi(\chi) \leq 0 \end{aligned}\right\} \tag{9}$$

where 's.t.' stands for 'subject to' and $\dot{\varepsilon}^p, \dot{\xi}$ are arbitrarily given. It would be easy to show that the Kuhn-Tucker conditions of (9) coincide with the constitutive equations (3)–(5).

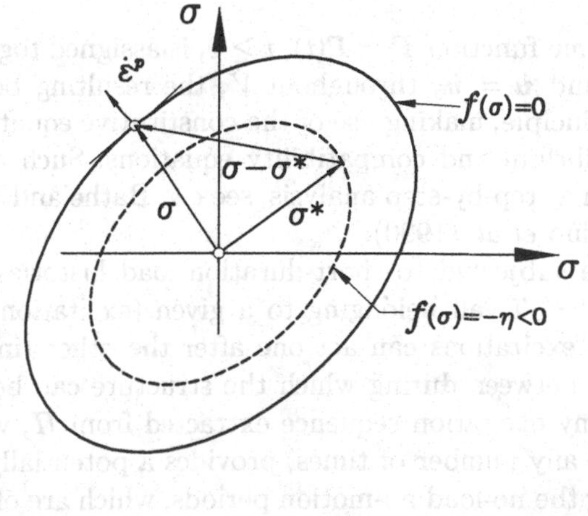

Figure 2. Stronger Druckerian inequality for a perfectly plastic material: for any σ^* not exceeding the interior surface $f(\sigma^*) = -\eta < 0$, it is $(\sigma - \sigma^*) : \dot{\varepsilon}^p > 0$ if $\dot{\varepsilon}^p \neq 0$, and $(\sigma - \sigma^*) : \dot{\varepsilon}^p = 0$ if $\dot{\varepsilon}^p = 0$.

The following Druckerian inequality can be shown to hold as a consequence of (9); that is:

$$(\boldsymbol{\sigma}-\boldsymbol{\sigma}^*):\dot{\boldsymbol{\varepsilon}}^p-(\boldsymbol{\chi}-\boldsymbol{\chi}^*)\cdot\dot{\boldsymbol{\xi}}\geq 0 \tag{10}$$

where the pairs $(\boldsymbol{\sigma},\boldsymbol{\chi})$ and $(\dot{\boldsymbol{\varepsilon}}^p,\dot{\boldsymbol{\xi}})$ correspond to each other through the constitutive equations (3)–(5), whereas the pair $(\boldsymbol{\sigma}^*,\boldsymbol{\chi}^*)$ is any plastically admissible statical state, i.e. satisfying $f(\boldsymbol{\sigma}^*,\boldsymbol{\chi}^*)\leq 0$ and $\phi(\boldsymbol{\chi}^*)\leq 0$. The equality sign holds true in (10) only if either $\dot{\boldsymbol{\varepsilon}}^p=\boldsymbol{0}$ and $\dot{\boldsymbol{\xi}}=\boldsymbol{0}$, or $\boldsymbol{\sigma}=\boldsymbol{\sigma}^*$ and $\boldsymbol{\chi}=\boldsymbol{\chi}^*$, or both.

A stronger form of the inequality (10) can be derived by allowing $(\boldsymbol{\sigma}^*,\boldsymbol{\chi}^*)$ to lie *inside* the yield and saturation surfaces, i.e. on letting $f(\boldsymbol{\sigma}^*,\boldsymbol{\chi}^*)<0$ and $\phi(\boldsymbol{\chi}^*)<0$; in which case (10) holds with the positive sign whenever the mechanism $(\dot{\boldsymbol{\varepsilon}}^p,\dot{\boldsymbol{\xi}})$ is not a trivial one, but with the equality sign if, and only if, the mechanism is a trivial one. We refer to this property of (10) as ***stronger Druckerian inequality*** in the following. In Fig. 2 this concept is illustrated for a perfectly plastic material.

3 The Dynamic Shakedown Problem

Let a solid body occupy the (open) domain V in its undeformed initial state and let it be referred to a Cartesian orthogonal system of co-ordinates (x_1,x_2,x_3). The body is subjected to body forces in V, tractions on the free part, S_t, of its boundary surface $S=\partial V$, imposed displacements on the constrained part, S_u, of S, $(S_t\cup S_u=S,\ S_t\cap S_u=\emptyset)$, and to imposed thermal-like strains in V. All these actions are by hypothesis governed by a vector $\boldsymbol{P}$ of independent parameters, which means that the vector $\boldsymbol{P}$ generates the entire load distribution for the considered structure. By hypothesis the material is as described in Section 2. Furthermore, the temperature variations, if any, do not affect the material data; also, the displacements $\boldsymbol{u}$ and strains $\boldsymbol{\varepsilon}$ induced by the external actions are infinitesimal.

Once the vectorial time function $\boldsymbol{P}=\boldsymbol{P}(t)$, $t\geq 0$, is assigned together with the initial conditions as $\boldsymbol{u}=\boldsymbol{u}_0$ and $\dot{\boldsymbol{u}}=\dot{\boldsymbol{u}}_0$ throughout V, the resulting body's motion can be evaluated, at least in principle, making use of the constitutive equations (1)–(5) together with the relevant equilibrium and compatibility equations. Such an evolutive analysis can be achieved through a step-by-step analysis, see e.g. Bathe and Wilson (1976), Owen and Hinton (1980), Borino *et al.* (1990).

Let the structure be subjected to short-duration load histories referred to as excitations, say $\boldsymbol{P}(t)$, $0\leq t\leq T$, all belonging to a given (excitation) domain, Π, of the functional space. These excitations can act one after the other (in an arbitrary order) with no-load periods in between during which the structure can be considered motionless due to damping. Any excitation sequence extracted from Π, with every excitation being possibly repeated any number of times, provides a potentially active load history $\boldsymbol{P}(t)$, $t\geq 0$. Eliminating the no-load no-motion periods, which are of no relevance for the present purposes, such a load history can be represented as a sequence, say:

$$\boldsymbol{P}(t)=\{\boldsymbol{P}_n(t)\in\Pi,\quad t_{n-1}\leq t\leq t_n,\quad n=1,2,\ldots\} \tag{11}$$

where $\boldsymbol{P}_n(t)=\boldsymbol{P}_n(t_{n-1}+\tau)$, $0\leq\tau\leq T_n=t_n-t_{n-1}$ and $t_0=0$. $\boldsymbol{P}(t)$ of (11) constitutes an ***admissible load history*** (ALH) for the given structure. A loading scheme like the above is referred to as a ***repeated excitation load scheme.***

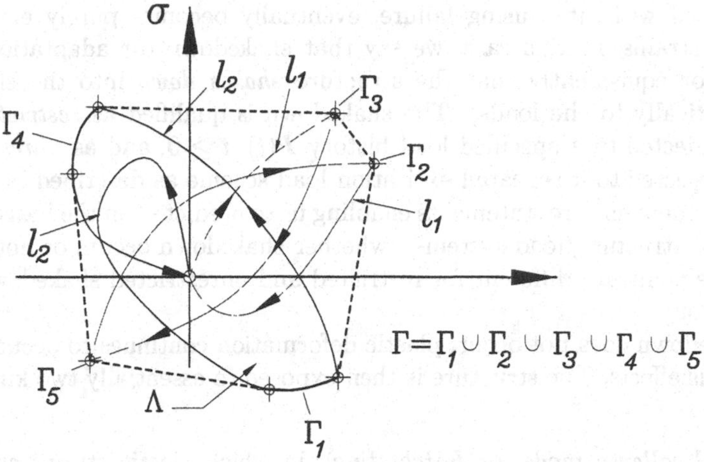

Figure 3. Typical elastic stress paths ℓ_1, ℓ_2 at a point $\boldsymbol{x} \in V$, related elastic stress domain Λ and (reduced) envelope Γ (the dashed flat regions are dropped to obtain the reduced envelope).

Let $\boldsymbol{\sigma}^E(\boldsymbol{x}, \tau)$ denote the structure's elastic (dynamic) stress response to the excitation $\boldsymbol{P}(\tau)$, $0 \le \tau \le T$ with zero initial conditions, and let ℓ_x denote the related stress path at $\boldsymbol{x} \in V$. Considering all the excitations of Π, a set $\{\ell_x\}$ of elastic stress paths is generated at every $\boldsymbol{x} \in V$. Denoting by Λ_x the convex hull of $\{\ell_x\}$, the set

$$\Lambda = \{\Lambda_x : \forall \boldsymbol{x} \in V\} \tag{12}$$

constitutes the ***elastic stress (response) domain*** of the structure/load system, whereas the set

$$\partial\Lambda = \{\partial\Lambda_x : \forall \boldsymbol{x} \in V\} \tag{13}$$

constitutes the related ***elastic stress (response) envelope***. The smallest subset of $\partial\Lambda$, say $\Gamma \subseteq \partial\Lambda$, the convex hull of which coincides with Λ, is the ***reduced elastic stress envelope***. Γ is obtained as the set $\{\Gamma_x\}$, by elimination, from every $\partial\Lambda_x$, of the (open) flat regions and edges (if any); and it may thus include isolated stress points (Fig. 3). Had Λ_x a polyhedral shape, Γ_x would consist of only isolated points, i.e. the vertices of Λ_x. For simplicity, Γ is simply referred to as the elastic stress envelope of the system, Γ_x is its intersection at $\boldsymbol{x}$. In case of quasi-static loads, ℓ_x reduces to a single stress point.

The dynamic elastic stress response $\boldsymbol{\sigma}^E(\boldsymbol{x}, t)$ to any ALH (with zero initial conditions) has a stress path L_x formed up by a sequence of elementary paths ℓ_x (related to the constituting excitations) and thus L_x lies entirely in the convex hull (Λ_x) of Γ_x at all $\boldsymbol{x} \in V$. The elastic stress envelope Γ characterizes the structure/load system in relation to shakedown. The evaluation of Γ requires a preliminary elastic analysis of the structure under a sufficient number of loading conditions.

The (dynamic) shakedown problem for the given structure/load system characterized by the elastic stress envelope Γ, is posed to assess whether the structural response, after a transient elastic-plastic phase during which some finite amount of plastic deformations may be produced without causing failure, eventually becomes purely elastic with no further plastic strains. In this case, we say that shakedown (or adaptation) occurs in the structure, or equivalently that the structure ***shakes down*** into the elastic regime (or ***adapts*** elastically to the loads). The shakedown is qualified as ***restricted*** when the structure is subjected to a specified load history $\boldsymbol{P}(t)$, $t \geq 0$, and as ***unrestricted*** if the structure is subjected to a repeated excitation load scheme as described in this section. The shakedown theorems are statements enabling us to predict —on the basis of some prerequisites of the structure/load system— whether shakedown occurs or not; it happens that these statements are different for restricted and unrestricted shakedown, as it will be shown later on.

When shakedown does not occur, plastic deformation continues to occur with cumulating dangerous effects. The structure is then exposed to essentially two kinds of failure modes, that is:

- *Incremental collapse mode* (or *Ratchetting*), in which plastic strains cumulate progressively, soon becoming intolerably large;
- *Alternating plasticity collapse mode* (or *Plastic shakedown*), in which plastic strains remain small, but exhibit opposite signs in subsequent time intervals, with consequent excessive increase of plastic dissipation and fatigue failure.

Shakedown occurrence is a necessary condition for structural safety, not a sufficient one, because plastic deformations produced during the transient elastic-plastic response may result excessive with respect to some relevant safety criteria and cause (local or global) failure. A shakedown analysis must thus include an additional work to estimate the above plastic deformations, in general by bounding techniques.

In the following, the unrestricted dynamic shakedown is referred to simply as dynamic shakedown whenever no misconfusion can arise.

4 Dynamic Shakedown Theorems

Theorems of (unrestricted) dynamic shakedown can be cast in a form quite similar to that of theorems of quasi-static shakedown, as shown in this section.

4.1 Static-Type Theorem

The following static-type theorem can be proven for a structure subjected to repeated excitations and characterized by an elastic stress envelope Γ.

- *Static Theorem for Unrestricted Dynamic Shakedown.*

A necessary and sufficient condition for shakedown to occur in a structure subjected to repeated excitations and with the elastic stress envelope Γ, is that there exist load-independent fields of self-stresses, $\hat{\boldsymbol{\sigma}}^r$, and of static internal variables, $\hat{\boldsymbol{\chi}}$, such as to satisfy the conditions

$$f(\boldsymbol{s} + \hat{\boldsymbol{\sigma}}^r, \hat{\boldsymbol{\chi}}) \leq 0, \qquad \forall \boldsymbol{s} \in \Gamma_{\boldsymbol{x}}, \quad \forall \boldsymbol{x} \in V \tag{14}$$

$$\phi(\hat{\boldsymbol{\chi}}) \leq 0, \qquad \forall \boldsymbol{x} \in V. \tag{15}$$

Here, $\boldsymbol{s}$ denotes stress co-ordinates of points in Γ_x = intersection of Γ at $\boldsymbol{x} \in V$. The fields $\widehat{\boldsymbol{\sigma}}^r$ and $\widehat{\boldsymbol{\chi}}$ will on occasion be referred to as the (*static*) *shakedown parameters*, the conditions (14) and (15) as the (*static*) *shakedown criterion*. $\mathcal{S}$ denotes the set of all such shakedown parameters.

Proof. The necessity is proved first. If, by hypothesis, shakedown occurs, the actual post-transient response of the structure to any ALH, say $\boldsymbol{P}(t)$, $t \geq 0$, has the form $\boldsymbol{\sigma}(\boldsymbol{x},t) = \boldsymbol{\sigma}^E(\boldsymbol{x},t) + \boldsymbol{\sigma}_a^r(\boldsymbol{x})$, $\boldsymbol{\chi}(\boldsymbol{x},t) = \boldsymbol{\chi}_a(\boldsymbol{x})$ for all $t \geq t_a$, where t_a is the instant of the shakedown occurrence, and $\boldsymbol{\sigma}_a^r(\boldsymbol{x})$, $\boldsymbol{\chi}_a(\boldsymbol{x})$ are both consequences of the plastic strains occurred before shakedown. Obviously, it is $f(\boldsymbol{\sigma}^E + \boldsymbol{\sigma}_a^r, \boldsymbol{\chi}_a) \leq 0$, $\phi(\boldsymbol{\chi}_a) \leq 0$ in V, $\forall t \geq t_a$. The latter inequalities hold good for whatsoever subsequent load history, hence for any stress point $\boldsymbol{\sigma}^E = \boldsymbol{s}$ ranging within the (closed) domain Λ. As Λ includes Γ, it results that (14) and (15) are satisfied, but $\widehat{\boldsymbol{\sigma}}^r$ and $\widehat{\boldsymbol{\chi}}$ replaced by $\boldsymbol{\sigma}_a^r$ and $\boldsymbol{\chi}_a$, respectively.

As to the sufficiency, let $\boldsymbol{P}(t)$, $t \geq 0$, be any ALH and let $\boldsymbol{\sigma}$, $\boldsymbol{\varepsilon}$, $\boldsymbol{u}$, $\boldsymbol{\chi}$, $\boldsymbol{\xi}, \ldots$ denote the related actual elastic-plastic response. Inequalities (14) and (15) are by hypothesis fulfilled with some fields $\widehat{\boldsymbol{\sigma}}^r$ and $\widehat{\boldsymbol{\chi}}$ of $\mathcal{S}$ in the following more stringent form

$$f(\boldsymbol{s} + \widehat{\boldsymbol{\sigma}}^r, \widehat{\boldsymbol{\chi}}) \leq -\eta, \qquad \forall \boldsymbol{s} \in \Gamma_x, \quad \forall \boldsymbol{x} \in V \tag{16}$$

$$\phi(\widehat{\boldsymbol{\chi}}) \leq -\eta, \qquad \forall \boldsymbol{x} \in V \tag{17}$$

where $\eta > 0$ is a small constant scalar. Due to the convexity of f and ϕ, the latter inequalities hold true also when, at every $\boldsymbol{x} \in V$, the stress point $\boldsymbol{s}$ ranges within the (closed) convex hull of Γ_x, i.e. $\forall \boldsymbol{s} \in \Lambda_x$. Because $\boldsymbol{P}(t)$ is an ALH, the elastic stress response $\boldsymbol{\sigma}^E$ at $\boldsymbol{x}$ belongs to Λ_x and thus the fictitious stress

$$\widehat{\boldsymbol{\sigma}}(\boldsymbol{x},t) = \boldsymbol{\sigma}^E(\boldsymbol{x},t) + \widehat{\boldsymbol{\sigma}}^r(\boldsymbol{x}) \tag{18}$$

and $\widehat{\boldsymbol{\chi}}$ satisfy the inequalities:

$$f(\widehat{\boldsymbol{\sigma}}, \widehat{\boldsymbol{\chi}}) \leq -\eta, \qquad \phi(\widehat{\boldsymbol{\chi}}) \leq -\eta, \qquad \text{in } V, \ \forall t \geq 0. \tag{19}$$

Making use of the Druckerian inequality (10) (in its stronger form), we can write

$$J \equiv (\boldsymbol{\sigma} - \widehat{\boldsymbol{\sigma}}) : \dot{\boldsymbol{\varepsilon}}^p - (\boldsymbol{\chi} - \widehat{\boldsymbol{\chi}}) \cdot \dot{\boldsymbol{\xi}} \geq 0 \qquad \text{in } V, \ \forall t \geq 0 \tag{20}$$

where $\boldsymbol{\sigma}$, $\boldsymbol{\chi}$, $\dot{\boldsymbol{\varepsilon}}^p$, $\dot{\boldsymbol{\xi}}$ pertain to the actual response and $\widehat{\boldsymbol{\sigma}}$, $\widehat{\boldsymbol{\chi}}$ are evaluated at the same point $\boldsymbol{x}$ and time t as $\boldsymbol{\sigma}$ and $\boldsymbol{\chi}$. Note that the equality sign holds in (20) if, and only if, the plastic mechanism $(\dot{\boldsymbol{\varepsilon}}^p, -\dot{\boldsymbol{\xi}})$ is vanishing. Using the identity

$$\dot{\boldsymbol{\varepsilon}}^p = \dot{\boldsymbol{\varepsilon}} - \dot{\boldsymbol{\varepsilon}}^E - \boldsymbol{E}^{-1} : (\dot{\boldsymbol{\sigma}} - \dot{\widehat{\boldsymbol{\sigma}}}), \tag{21}$$

and by an integration over V, we write:

$$\int_V (\boldsymbol{\sigma} - \widehat{\boldsymbol{\sigma}}) : (\dot{\boldsymbol{\varepsilon}} - \dot{\boldsymbol{\varepsilon}}^E) \, \mathrm{d}V - \int_V (\boldsymbol{\sigma} - \widehat{\boldsymbol{\sigma}}) : \boldsymbol{E}^{-1} : (\dot{\boldsymbol{\sigma}} - \dot{\widehat{\boldsymbol{\sigma}}}) \, \mathrm{d}V$$
$$- \int_V (\boldsymbol{\chi} - \widehat{\boldsymbol{\chi}}) \cdot \dot{\boldsymbol{\xi}} \, \mathrm{d}V = \int_V J \, \mathrm{d}V > 0 \qquad \forall t \geq 0, \tag{22}$$

where the positive sign is due to the fact that $\dot{\varepsilon}^p$ and $\dot{\boldsymbol{\xi}}$ cannot be identically vanishing (if so, the real deformation process would be a purely elastic one, which is excluded). Applying the virtual work principle we can write:

$$\int_V (\boldsymbol{\sigma}-\widehat{\boldsymbol{\sigma}}) : (\dot{\boldsymbol{\varepsilon}}-\dot{\boldsymbol{\varepsilon}}^E)\,\mathrm{d}V = -\int_V \rho(\ddot{\boldsymbol{u}}-\ddot{\boldsymbol{u}}^E)\cdot(\dot{\boldsymbol{u}}-\dot{\boldsymbol{u}}^E)\,\mathrm{d}V - \int_V \mu_d(\dot{\boldsymbol{u}}-\dot{\boldsymbol{u}}^E)\cdot(\dot{\boldsymbol{u}}-\dot{\boldsymbol{u}}^E)\,\mathrm{d}V \quad (23)$$

where ρ is the mass density and μ_d the damping coefficient. Let us note that (Halphen, 1978; Maier, 1987; Polizzotto *et al.*, 1991):

$$\left.\begin{aligned} (\boldsymbol{\chi}-\widehat{\boldsymbol{\chi}})\cdot\dot{\boldsymbol{\xi}} &= \frac{\mathrm{d}}{\mathrm{d}t}\, Z(\boldsymbol{\chi},\widehat{\boldsymbol{\chi}}), \\ Z(\boldsymbol{\chi},\widehat{\boldsymbol{\chi}}) &\equiv \Omega(\widehat{\boldsymbol{\chi}}) - \Omega(\boldsymbol{\chi}) - \boldsymbol{\xi}(\boldsymbol{\chi})\cdot(\widehat{\boldsymbol{\chi}}-\boldsymbol{\chi}) \end{aligned}\right\} \quad (24)$$

where $\Omega(\boldsymbol{\chi})$ is the Legendre transform of $\Psi_{in}(\boldsymbol{\xi})$, i.e. $\Omega(\boldsymbol{\chi}) = \boldsymbol{\chi}\cdot\boldsymbol{\xi} - \Psi_{\text{in}}(\boldsymbol{\xi})$, such that $\boldsymbol{\xi} = \partial\Omega/\partial\boldsymbol{\chi}$ and $Z(\boldsymbol{\chi},\widehat{\boldsymbol{\chi}}) \geq 0$ due to the convexity of Ψ_{in}, hence of Ω. Then, introducing the positive-definite functional

$$\begin{aligned} L(t) = &\int_V \frac{1}{2}\rho(\dot{\boldsymbol{u}}-\dot{\boldsymbol{u}}^E)\cdot(\dot{\boldsymbol{u}}-\dot{\boldsymbol{u}}^E)\,\mathrm{d}V + \int_0^t\int_V \mu_d(\dot{\boldsymbol{u}}-\dot{\boldsymbol{u}}^E)\cdot(\dot{\boldsymbol{u}}-\dot{\boldsymbol{u}}^E)\,\mathrm{d}V\,\mathrm{d}\tau \\ &+ \int_V \frac{1}{2}(\boldsymbol{\sigma}-\widehat{\boldsymbol{\sigma}}) : \boldsymbol{E}^{-1} : (\boldsymbol{\sigma}-\widehat{\boldsymbol{\sigma}})\,\mathrm{d}V + \int_V Z(\boldsymbol{\chi},\widehat{\boldsymbol{\chi}})\,\mathrm{d}V \end{aligned} \quad (25)$$

eq. (22) can be rewritten as

$$\frac{\mathrm{d}}{\mathrm{d}t}L(t) = -\int_V J\,\mathrm{d}V < 0 \qquad \forall t \geq 0. \quad (26)$$

The function $L(t)$, which cannot be negative any time in the deformation process, is a scalar measure of the discrepancy between the actual response and the fictitious one (i.e. $\widehat{\boldsymbol{\sigma}} = \boldsymbol{\sigma}^E + \widehat{\boldsymbol{\sigma}}^r$ in terms of stresses). The real deformation process requires that these two responses approach to each other as much as possible, that is with decreasing values of L at increasing t, at a nonzero rate as far as plastic strains occur. Therefore, there must exist some (finite) time t_a such that $\mathrm{d}L/\mathrm{d}t = 0\ \forall t \geq t_a$, hence $J = 0$ in $V\ \forall t \geq t_a$, which —by the stronger Druckerian inequality— implies that plastic strains stop being produced, i.e. $\dot{\boldsymbol{\varepsilon}}^p = \boldsymbol{0}$, $\dot{\boldsymbol{\xi}} = \dot{\boldsymbol{\chi}} = \boldsymbol{0}$ everywhere in V for $t \geq t_a$. As this result holds for any ALH, we can conclude that shakedown certainly occurs under the assumed hypotheses. We can also state that, except perhaps in special situations not contemplated here, shakedown also occurs at the limit for $\eta \to 0$. QED.

4.2 Kinematic-Type Theorem

Let us introduce, at every point $\boldsymbol{x} \in V$, a set of *fictitious* plastic strains and kinematic internal variables, say $\boldsymbol{\varepsilon}^{pc}$ and $\boldsymbol{\xi}^c$ which are in one-to-one correspondence with the stress points $\boldsymbol{s} \in \Gamma_{\boldsymbol{x}}$. We write $\boldsymbol{\varepsilon}^{pc}(\boldsymbol{x},\boldsymbol{s})$, $\boldsymbol{\xi}^c(\boldsymbol{x},\boldsymbol{s})$ to make evident this correspondence. These $\boldsymbol{\varepsilon}^{pc}$ and $\boldsymbol{\xi}^c$ are thus fields over $V \times \Gamma$. These fields are required to satisfy the condition:

$$\int_V\int_{\Gamma_{\boldsymbol{x}}} [\boldsymbol{\sigma}^r(\boldsymbol{x}) : \boldsymbol{\varepsilon}^{pc}(\boldsymbol{x},\boldsymbol{s}) - \boldsymbol{\chi}(\boldsymbol{x})\cdot\boldsymbol{\xi}^c(\boldsymbol{x},\boldsymbol{s})]\,\mathrm{d}\Gamma(\boldsymbol{s})\,\mathrm{d}V = 0 \quad (27)$$

identically for any pair $(\sigma^r, \chi) \in \mathcal{S}$. It can be shown, by a procedure which is skipped here for brevity, (Polizzotto *et al.*, 1993), that the fields ε^{pc}, ξ^c satisfying (27) are those that satisfy the following equations:

$$\Delta\varepsilon^{pc}(x) = \int_{\Gamma_x} \varepsilon^{pc}(x, s)\, \mathrm{d}\Gamma(s), \qquad \Delta\xi^c(x) = \int_{\Gamma_x} \xi^c(x, s)\, \mathrm{d}\Gamma(s), \quad \text{in } V \tag{28}$$

$$\Delta\varepsilon^{pc} = \nabla^s u^c \ \text{ in } V, \quad u^c = 0 \text{ on } S_u \tag{29}$$

$$\Delta\xi^c = 0 \text{ in } V. \tag{30}$$

Equation (28) gives definitions of the *ratchet* fields $\Delta\varepsilon^{pc}, \Delta\xi^c$ as cumulated values over the elastic stress envelope Γ_x; (29) and (30) are conditions that restrict the ratchet fields, that is, $\Delta\varepsilon^{pc}$ must be compatible with the displacements u^c vanishing on S_u, whereas $\Delta\xi^c$ must vanish everywhere in V.

If Γ_x includes a set of discrete points, say s_k $(k = 1, 2, \ldots, N)$ then (28) must be written as follows:

$$\Delta\varepsilon^{pc}(x) = \int_{\Gamma_x} \varepsilon^{pc}(x, s)\, \mathrm{d}\Gamma(s) + \sum_{k=1}^{N} \varepsilon^{pc}_k(x) \tag{31}$$

$$\Delta\xi^c(x) = \int_{\Gamma_x} \xi^c(x, s)\, \mathrm{d}\Gamma(s) + \sum_{k=1}^{N} \xi^c(x). \tag{32}$$

When Γ_x contains only discrete points, the integrals on the r.h.s. of (31) and (32) disappear. Note that ε^{pc} and ξ^c defined over Γ_x are referred to the unit dimension of Γ_x. Considered that the extended definitions (31) and (32) can be directly derived from (28) by attributing to every isolated point s_k of Γ_x values of ε^{pc} and ξ^c as Dirac deltas, i.e. $\varepsilon^{pc}(x, s) = \varepsilon^{pc}_k(x)\,\delta_D(s - s_k)$, $\xi^c(x, s) = \xi^c_k(x)\,\delta_D(s - s_k)$, in the following the general definition (28) will be referred to.

A set K of fields ε^{pc}, ξ^c, that is

$$K = \{\varepsilon^{pc}(x, s), \xi^c(x, s) \,:\, \forall\, s \in \Gamma_x,\ \forall\, x \in V\}, \tag{33}$$

complying with (28)–(30) constitutes, by definition, a ***plastic accumulation mechanism*** (PAM), which is a generalization of the ***kinematically admissible plastic strain cycle*** of Koiter (1960). The complete set of PAMs is denoted by $\mathcal{M}$, which is thus dual of $\mathcal{S}$. The definition of PAM given here is an extension of that proposed by Polizzotto *et al.* (1991) within quasi-static shakedown for internal variable constitutive models.

After these premises, the following kinematic-type theorem can be proven, in which the PAMs are used as ***kinematic shakedown parameters***:

- *Kinematic Theorem for Unrestricted Dynamic Shakedown.*

A necessary and sufficient condition for shakedown to occur in a structure subjected to repeated excitations and with an elastic stress envelope Γ, is that the inequality

$$\int_V \int_{\Gamma_x} [D(\varepsilon^{pc}(x, s), \xi^c(x, s)) - s : \varepsilon^{pc}(x, s)]\, \mathrm{d}\Gamma(s)\, dV \geq 0 \tag{34}$$

be satisfied for all PAMs in $\mathcal{M}$.

Proof. Assuming that shakedown occurs, by the static theorem there exist some static shakedown parameters, say $\widehat{\boldsymbol{\sigma}}^r$ and $\widehat{\boldsymbol{\chi}}$, such as to satisfy (14) and (15). Thus, the pair $(\boldsymbol{s}+\widehat{\boldsymbol{\sigma}}^r, \widehat{\boldsymbol{\chi}})$ constitutes a feasible solution to problem (9) related to the maximum dissipation theorem, but written with $\dot{\boldsymbol{\varepsilon}}^p$ and $\dot{\boldsymbol{\xi}}$ replaced by $\boldsymbol{\varepsilon}^{pc}$ and $\boldsymbol{\xi}^c$, the latter being evaluated at the same $\boldsymbol{s}$ and $\boldsymbol{x}$ which the mentioned pair refers to. Thus we can write:

$$D(\boldsymbol{\varepsilon}^{pc}, \boldsymbol{\xi}^c) \geq (\boldsymbol{s} + \widehat{\boldsymbol{\sigma}}^r) : \boldsymbol{\varepsilon}^{pc} - \widehat{\boldsymbol{\chi}} \cdot \boldsymbol{\xi}^c, \tag{35}$$

which holds for all $\boldsymbol{s} \in \Gamma_x$ and for all $\boldsymbol{x} \in V$. Integrating (35) over Γ_x first and then over V gives

$$\int_V \int_{\Gamma_x} D(\boldsymbol{\varepsilon}^{pc}, \boldsymbol{\xi}^c) \, \mathrm{d}\Gamma(\boldsymbol{s}) \, \mathrm{d}V \geq \int_V \int_{\Gamma_x} \boldsymbol{s} : \boldsymbol{\varepsilon}^{pc} \, \mathrm{d}\Gamma(\boldsymbol{s}) \, \mathrm{d}V, \tag{36}$$

since the integrals

$$\int_V \widehat{\boldsymbol{\sigma}}^r : \int_{\Gamma_x} \boldsymbol{\varepsilon}^{pc} \, \mathrm{d}\Gamma = \int_V \widehat{\boldsymbol{\sigma}}^r : \Delta\boldsymbol{\varepsilon}^{pc} \, \mathrm{d}V = 0 \tag{37}$$

$$\int_V \widehat{\boldsymbol{\chi}} \cdot \int_{\Gamma_x} \boldsymbol{\xi}^c \, \mathrm{d}\Gamma = \int_V \widehat{\boldsymbol{\chi}} \cdot \Delta\boldsymbol{\xi}^c \, \mathrm{d}V = 0 \tag{38}$$

vanish by (28)–(30). Equation (36) is recognized to coincide with the kinematic shakedown criterion (34) and thus the necessity part of the theorem is proven.

In order to prove the sufficiency part, we start from (34), but in the following more restrictive form

$$\mathcal{E}_m[\boldsymbol{\varepsilon}^{pc}, \boldsymbol{\xi}^c] \equiv \int_V \int_{\Gamma_x} [D(\boldsymbol{\varepsilon}^{pc}, \boldsymbol{\xi}^c) - m\boldsymbol{s} : \boldsymbol{\varepsilon}^{pc}] \, \mathrm{d}\Gamma(\boldsymbol{s}) \, \mathrm{d}V \geq 0 \tag{39}$$

where $m > 1$ is a scalar parameter (slightly greater than 1). $\mathcal{E}_m$ as given by (39) is a functional defined over the set $\mathcal{M}$ of PAMs. Because $\mathcal{M}$ includes the trivial PAM (i.e. $\boldsymbol{\varepsilon}^{pc} = \boldsymbol{0}$, $\boldsymbol{\xi}^c = \boldsymbol{0}$ in $V \times \Gamma$, for which $\mathcal{E}_m = 0$), it can be stated that the problem

$$\min_K \mathcal{E}_m[\boldsymbol{\varepsilon}^{pc}, \boldsymbol{\xi}^c] \qquad \text{s.t.} \;\; K \in \mathcal{M} \;\;\; (m \text{ fixed}) \tag{40}$$

has an absolute (vanishing) minimum, and that therefore the related Euler-Lagrange equations must have a solution.

We proceed with the Lagrange multiplier method in order to derive these Euler-Lagrange equations. To this purpose, the constrained minimization problem (40) is transformed into an unconstrained one by appending the constraint equations (28)-(30) to the objective functional of (40). So we can write the Lagrangian functional:

$$\begin{aligned} \mathcal{E}_m^L = \mathcal{E}_m[\boldsymbol{\varepsilon}^{pc}, \boldsymbol{\xi}^c] &+ \int_V m\widehat{\boldsymbol{\sigma}}^r : \left[\nabla^s \boldsymbol{u}^c - \int_{\Gamma_x} \boldsymbol{\varepsilon}^{pc} \, \mathrm{d}\Gamma(\boldsymbol{s})\right] \mathrm{d}V \\ &+ \int_V \int_{\Gamma_x} m\widehat{\boldsymbol{\chi}} \cdot \boldsymbol{\xi}^c \, \mathrm{d}\Gamma(\boldsymbol{s}) \, \mathrm{d}V - \int_{S_u} \boldsymbol{n} \cdot (m\widehat{\boldsymbol{\sigma}}^r) \cdot \boldsymbol{u}^c \, \mathrm{d}S \end{aligned} \tag{41}$$

where $\boldsymbol{n}$ is the unit external normal to $S = \partial V$ and $m\widehat{\boldsymbol{\sigma}}^r$, $m\widehat{\boldsymbol{\chi}}$ denote $\boldsymbol{s}$-independent Lagrange multipliers (the factor $m > 1$ is introduced for convenience). The first variation of (41), with respect to the primal variables, after some mathematics including the divergence theorem where appropriate, reads as follows:

$$\delta \mathcal{E}^L_m = \int_V \int_{\Gamma_x} \delta \boldsymbol{\varepsilon}^{pc} : \left[\frac{\partial D}{\partial \boldsymbol{\varepsilon}^{pc}} - m(\boldsymbol{s} + \widehat{\boldsymbol{\sigma}}^r) \right] \mathrm{d}\Gamma(\boldsymbol{s})\, \mathrm{d}V + \int_V \int_{\Gamma_x} \delta \boldsymbol{\xi}^c \cdot \left[\frac{\partial D}{\partial \boldsymbol{\xi}^c} + m \widehat{\boldsymbol{\chi}} \right] \mathrm{d}\Gamma(\boldsymbol{s})\, \mathrm{d}V$$
$$- m \int_V \delta \boldsymbol{u}^c \cdot \operatorname{div} \widehat{\boldsymbol{\sigma}}^r \, \mathrm{d}V + m \int_{S_t} \delta \boldsymbol{u}^c \cdot \widehat{\boldsymbol{\sigma}}^r \cdot \boldsymbol{n} \, \mathrm{d}S. \tag{42}$$

Because $\delta \mathcal{E}^L_m$ must vanish for arbitrary choices of the variation fields, the stationarity conditions turn out to read as follows:

$$m\,(\boldsymbol{s} + \widehat{\boldsymbol{\sigma}}^r) = \frac{\partial D}{\partial \boldsymbol{\varepsilon}^{pc}}, \qquad m\,\widehat{\boldsymbol{\chi}} = -\frac{\partial D}{\partial \boldsymbol{\xi}^c}, \qquad \forall \boldsymbol{s} \in \Gamma_x \quad \forall \boldsymbol{x} \in V \tag{43}$$

$$\operatorname{div} \widehat{\boldsymbol{\sigma}}^r = \boldsymbol{0} \;\text{ in } V, \qquad \widehat{\boldsymbol{\sigma}}^r \cdot \boldsymbol{n} = \boldsymbol{0} \;\text{ on } S_t \tag{44}$$

which hold together with (28)–(30). Equations (43) and (44) qualify the Lagrange multipliers $\widehat{\boldsymbol{\sigma}}^r$ and $\widehat{\boldsymbol{\chi}}$ as, respectively, self-stress and static internal variable fields both being load-independent. Furthermore, from (43) the pair $(m\,(\boldsymbol{s}+\widehat{\boldsymbol{\sigma}}^r), m\,\widehat{\boldsymbol{\chi}})$, being derived from the relevant intrinsic dissipation potential D, turns out to be plastically admissible, that is

$$f(m\,(\boldsymbol{s}+\widehat{\boldsymbol{\sigma}}^r), m\,\widehat{\boldsymbol{\chi}}) \le 0 \qquad \forall \boldsymbol{s} \in \Gamma_x \quad \forall \boldsymbol{x} \in V \tag{45}$$

$$\phi(m\,\widehat{\boldsymbol{\chi}}) \le 0 \qquad \text{in } V. \tag{46}$$

Since $f(m\,\boldsymbol{\sigma}, m\,\boldsymbol{\chi}) > f(\boldsymbol{\sigma}, \boldsymbol{\chi})$ and $\phi(m\,\boldsymbol{\chi}) > \phi(\boldsymbol{\chi})$ for any $\boldsymbol{\sigma}$, $\boldsymbol{\chi}$ and $m > 1$, (45) and (46) imply that

$$f(\boldsymbol{s} + \widehat{\boldsymbol{\sigma}}^r, \widehat{\boldsymbol{\chi}}) < 0, \quad \phi(\widehat{\boldsymbol{\chi}}) < 0 \qquad \forall \boldsymbol{s} \in \Gamma_x \quad \forall \boldsymbol{x} \in V \tag{47}$$

which, by the static theorem, means that shakedown certainly occurs. We arrive at the same conclusion on letting m decrease; at the limit for $m \to 1$, (45) and (46) coincide with (14) and (15). QED.

On concluding this section, we remark that the above theorems are extensions to materials with internal variables and hardening saturation surface of analogous theorems given by Borino and Polizzotto (1995, 1996) for perfect plasticity, though the latter theorems were there expressed in terms of time-dependent elastic stresses. The use of the (reduced) elastic stress envelope made here is the implementation of a procedure proposed by Polizzotto (1985b), see also Panzeca *et al.* (1990). Similar theorems were given by Pham (1996) for perfect plasticity. Also, the above static-type theorem is an extension to dynamics of a theorem given by Fuschi (1999) for quasi-static loads and material models as the present one.

5 The Shakedown Safety Factor

For the structure/load system subjected to time variable loads, a safety factor, β_a say, can be defined as the maximum value of a scalar multiplier β affecting all the external actions, such that shakedown occurs for any $\beta < \beta_a$, whereas for $\beta = \beta_a$ a shakedown limit state is produced. Classically, there are two ways to compute β_a, one based on the static shakedown theorem, another based on the kinematic one (Koiter, 1960).

Considering the case of repeated excitations of the previous sections, amplifying the external actions by β implies an equal amplification of Λ and Γ, such that the static shakedown criterion (14) and (15) can be restated as:

$$f(\beta \boldsymbol{s}+\widehat{\boldsymbol{\sigma}}^r,\widehat{\boldsymbol{\chi}})\leq 0 \qquad \forall \boldsymbol{s}\in\Gamma_x \quad \forall \boldsymbol{x}\in V \tag{48}$$

$$\phi(\widehat{\boldsymbol{\chi}})\leq 0 \qquad \forall \boldsymbol{x}\in V. \tag{49}$$

Therefore, β_a can be obtained by solving the following problem:

—*Safety Factor Problem by the Static Approach*

$$\left.\begin{array}{l} \beta_a = \max\limits_{(\beta,\widehat{\boldsymbol{\sigma}}^r,\widehat{\boldsymbol{\chi}})} \beta \qquad \text{subject to :} \\ \\ \text{—Constraints (48) and (49)} \\ \text{—Equilibrium equations on } \widehat{\boldsymbol{\sigma}}^r. \end{array}\right\} \tag{50}$$

The limit state generated for $\beta=\beta_a$ can be derived from (50) by studying the related Euler-Lagrange equations. To this purpose, let the Lagrangian functional be written as:

$$\begin{aligned}\mathcal{L} = -\beta &+ \int_V\int_{\Gamma_x} \lambda^c f(\beta\boldsymbol{s}+\widehat{\boldsymbol{\sigma}}^r,\widehat{\boldsymbol{\chi}})\,\mathrm{d}\Gamma(\boldsymbol{s})\,\mathrm{d}V + \int_V\int_{\Gamma_x} \mu^c\phi(\widehat{\boldsymbol{\chi}})\,\mathrm{d}\Gamma(\boldsymbol{s})\,\mathrm{d}V \\ &+ \int_V \boldsymbol{u}^c\cdot\operatorname{div}\widehat{\boldsymbol{\sigma}}^r\,\mathrm{d}V - \int_{S_t}\boldsymbol{u}^c\cdot\widehat{\boldsymbol{\sigma}}^r\cdot\boldsymbol{n}\,\mathrm{d}S, \end{aligned} \tag{51}$$

where $\lambda^c(\boldsymbol{x},\boldsymbol{s})\geq 0$, $\mu^c(\boldsymbol{x},\boldsymbol{s})\geq 0$ and $\boldsymbol{u}^c(\boldsymbol{x})$ are Lagrange multipliers. The first variation of $\mathcal{L}$, after some easy mathematical transformations can be written as follows:

$$\begin{aligned}\delta\mathcal{L} = \delta\beta&\left[-1+\int_V\int_{\Gamma_x}\boldsymbol{s}:\left(\frac{\partial f}{\partial\widehat{\boldsymbol{\sigma}}}\lambda^c\right)\mathrm{d}\Gamma(\boldsymbol{s})\,\mathrm{d}V\right] \\ &+\int_V\delta\widehat{\boldsymbol{\sigma}}^r:\left[\int_{\Gamma_x}\left(\lambda^c\frac{\partial f}{\partial\widehat{\boldsymbol{\sigma}}}\right)\mathrm{d}\Gamma(\boldsymbol{s})-\nabla^s\boldsymbol{u}^c\right]\mathrm{d}V+\int_{S_u}\boldsymbol{n}\cdot\delta\widehat{\boldsymbol{\sigma}}^r\cdot\boldsymbol{u}^c\,\mathrm{d}S \\ &+\int_V\delta\widehat{\boldsymbol{\chi}}\cdot\int_{\Gamma_x}\left[\lambda^c\frac{\partial f}{\partial\widehat{\boldsymbol{\chi}}}+\mu^c\frac{\partial\phi}{\partial\widehat{\boldsymbol{\chi}}}\right]\mathrm{d}\Gamma(\boldsymbol{s})\,\mathrm{d}V \\ &+\int_V\int_{\Gamma_x}\delta\lambda^c f(\widehat{\boldsymbol{\sigma}},\widehat{\boldsymbol{\chi}})\,\mathrm{d}\Gamma(\boldsymbol{s})\,\mathrm{d}V+\int_V\int_{\Gamma_x}\delta\mu^c\phi(\widehat{\boldsymbol{\chi}})\,\mathrm{d}\Gamma(\boldsymbol{s})\,\mathrm{d}V \\ &+\int_V\delta\boldsymbol{u}^c\cdot\operatorname{div}\widehat{\boldsymbol{\sigma}}^r\,\mathrm{d}V-\int_{S_t}\delta\boldsymbol{u}^c\cdot\widehat{\boldsymbol{\sigma}}^r\cdot\boldsymbol{n}\,\mathrm{d}S, \end{aligned} \tag{52}$$

where we have set $\widehat{\boldsymbol{\sigma}}=\beta\boldsymbol{s}+\widehat{\boldsymbol{\sigma}}^r$. Considering that $\mathcal{L}$ must take a minimum with respect to the primal variables $\beta,\widehat{\boldsymbol{\sigma}}^r,\widehat{\boldsymbol{\chi}}$, but a maximum with respect to the Lagrange multipliers, the stationarity conditions read:

$$\widehat{\boldsymbol{\sigma}}=\beta\boldsymbol{s}+\widehat{\boldsymbol{\sigma}}^r \qquad \forall\boldsymbol{s}\in\Gamma_x \quad \forall\boldsymbol{x}\in V \tag{53}$$

$$\varepsilon^{pc} = \lambda^c \frac{\partial f}{\partial \widehat{\sigma}} \qquad -\xi^c = \lambda^c \frac{\partial f}{\partial \widehat{\chi}} + \mu^c \frac{\partial \phi}{\partial \widehat{\chi}}, \qquad \forall\, s \in \Gamma_x \quad \forall\, x \in V \tag{54}$$

$$f(\widehat{\sigma}, \widehat{\chi}) \leq 0, \qquad \lambda^c \geq 0, \qquad \lambda^c f(\widehat{\sigma}, \widehat{\chi}) = 0 \qquad \forall\, s \in \Gamma_x \quad \forall\, x \in V \tag{55}$$

$$\phi(\widehat{\chi}) \leq 0, \qquad \mu^c \geq 0, \qquad \mu^c \phi(\widehat{\chi}) = 0 \qquad \forall\, s \in \Gamma_x \quad \forall\, x \in V \tag{56}$$

$$\int_V \int_{\Gamma_x} s : \varepsilon^{pc} \, \mathrm{d}\Gamma(s) \, \mathrm{d}V = 1 \tag{57}$$

$$\Delta\varepsilon^{pc} = \int_{\Gamma_x} \varepsilon^{pc} \, \mathrm{d}\Gamma(s), \qquad \Delta\xi^c = \int_{\Gamma_x} \xi^c \, \mathrm{d}\Gamma(s) = 0 \quad \text{in } V \tag{58}$$

$$\Delta\varepsilon^{pc} = \nabla^s u^c \quad \text{in } V, \qquad u^c = 0 \quad \text{on } S_u \tag{59}$$

$$\operatorname{div} \widehat{\sigma}^r = 0 \quad \text{in } V, \qquad \widehat{\sigma}^r \cdot n = 0 \quad \text{on } S_t. \tag{60}$$

Before making any effort to interpret (53)–(60), we observe that the latter equation system expresses necessary and sufficient conditions for problem (50), the latter being convex. This implies that the/a solution to (53)–(60), say $\beta^*, \widehat{\sigma}^{r*}, \ldots$, is such that the triple $(\beta^*, \widehat{\sigma}^{r*}, \widehat{\chi}^*)$ extracted from it solves problem (50) and thus, in particular, $\beta^* = \beta_a$.

Another point to remark is the uniqueness of solution to (50) and thus of (53)–(60). In fact, let $(\cdot)'$ and $(\cdot)''$ denote two distinct solutions of the problem. By the Druckerian inequality (10) we can write:

$$(\widehat{\sigma}' - \widehat{\sigma}'') : (\varepsilon^{pc\,\prime} - \varepsilon^{pc\,\prime\prime}) - (\widehat{\chi}' - \widehat{\chi}'') \cdot (\xi^{c\,\prime} - \xi^{c\,\prime\prime}) \geq 0 \qquad \text{in } \Gamma \times V \tag{61}$$

where $\widehat{\sigma}' = \beta_a s + \widehat{\sigma}^{r'}$, $\widehat{\sigma}'' = \beta_a s + \widehat{\sigma}^{r''}$. Integrating (61) over $\Gamma \times V$ gives

$$\int_V (\widehat{\sigma}^{r'} - \widehat{\sigma}^{r''}) : (\Delta\varepsilon^{pc\,\prime} - \Delta\varepsilon^{pc\,\prime\prime}) \, \mathrm{d}V - \int_V (\widehat{\chi}' - \widehat{\chi}'') \cdot (\Delta\xi^{c\,\prime} - \Delta\xi^{c\,\prime\prime}) \, \mathrm{d}V \geq 0. \tag{62}$$

But both integrals are vanishing, the first one by the virtual work principle and also by (59) and (60), the second one by (58). Therefore (61) holds with the equality sign in the whole $\Gamma \times V$, which implies that, necessarily, $\widehat{\sigma}^{r'} = \widehat{\sigma}^{r''}$, $\widehat{\chi}' = \widehat{\chi}''$, $\varepsilon^{pc\,\prime} = \varepsilon^{pc\,\prime\prime}$, $\xi^{c\,\prime} = \xi^{c\,\prime\prime}$ in the region of V where the mechanism $(\varepsilon^{pc}, -\xi^c)$ is nontrivial, but in the remaining region it may be $\widehat{\sigma}^{r'} \neq \widehat{\sigma}^{r''}$ and $\widehat{\chi}' \neq \widehat{\chi}''$. Because ε^{pc} and ξ^c are uniquely determined in V, from (59) follows that also u^c is unique. So, in conclusion, the solution to the above problem is unique for all, except for $\widehat{\sigma}^r$ and $\widehat{\chi}$ in the region (if any) not suffering plastic deformations.

Let us apply the kinematic-type shakedown theorem by writing (34) with s replaced by βs, i.e.

$$\int_V \int_{\Gamma_x} D(\boldsymbol{\varepsilon}^{pc}, \boldsymbol{\xi}^c)\, \mathrm{d}\Gamma(\boldsymbol{s})\, \mathrm{d}V \geq \beta \int_V \int_{\Gamma_x} \boldsymbol{s} : \boldsymbol{\varepsilon}^{pc}\, \mathrm{d}\Gamma(\boldsymbol{s})\, \mathrm{d}V. \tag{63}$$

The latter inequality is satisfied for arbitrary choices of the PAM and for $\beta \leq \beta_a$. Thus, rewriting (63) with $\beta = \beta_a$, we have

$$\beta_a \leq \frac{\int_V \int_{\Gamma_x} D(\boldsymbol{\varepsilon}^{pc}, \boldsymbol{\xi}^c)\, \mathrm{d}\Gamma(\boldsymbol{s})\, \mathrm{d}V}{\int_V \int_{\Gamma_x} \boldsymbol{s} : \boldsymbol{\varepsilon}^{pc}\, \mathrm{d}\Gamma(\boldsymbol{s})\, \mathrm{d}V} \tag{64}$$

provided the double integral on the r.h.s. of (63) be positive, as assumed. Inequality (64) means that the functional

$$\beta_{\text{kin}}[\boldsymbol{\varepsilon}^{pc}, \boldsymbol{\xi}^c] = \frac{\int_V \int_{\Gamma_x} D(\boldsymbol{\varepsilon}^{pc}, \boldsymbol{\xi}^c)\, \mathrm{d}\Gamma(\boldsymbol{s})\, \mathrm{d}V}{\int_V \int_{\Gamma_x} \boldsymbol{s} : \boldsymbol{\varepsilon}^{pc}\, \mathrm{d}\Gamma(\boldsymbol{s})\, \mathrm{d}V} \tag{65}$$

provides an upper bound to β_a. Since β_{kin} is bounded from below, it must have a minimum in the set $\mathcal{M}$ of all PAMs. The PAMs being defined within a positive scalar multiplier, we can normalize the PAMs by writing a condition like (57), after which the minimum of β_{kin} is obtained by solving the problem:

—*Safety Factor Problem by the Kinematic Approach:*

$$\left.\begin{aligned} &\beta_a \leq \widehat{\beta} = \min_{(\boldsymbol{\varepsilon}^{pc}, \boldsymbol{\xi}^c)} \quad \int_V \int_{\Gamma_x} D(\boldsymbol{\varepsilon}^{pc}, \boldsymbol{\xi}^c)\, \mathrm{d}\Gamma(\boldsymbol{s})\, \mathrm{d}V \\ &\text{subject to:} \qquad (\boldsymbol{\varepsilon}^{pc}, \boldsymbol{\xi}^c) \in \mathcal{M}_{\text{Nor}} \end{aligned}\right\} \tag{66}$$

where $\mathcal{M}_{\text{Nor}}$ denotes the set of all normalized PAMs, i.e. satisfying (57).

First of all we prove that $\beta_a = \widehat{\beta}$. In fact, let the PAM resulting from solving problem (53)–(60) be used, that is $\boldsymbol{\varepsilon}^{pc*}$, $\boldsymbol{\xi}^{c*}$, which is normalized by (57). For the latter PAM, we can write:

$$\beta_a \leq \widehat{\beta} \leq \beta^*, \tag{67}$$

β^* being the value correspondingly taken on by the objective functional of (66). On the other hand, using (53)–(60), we can write:

$$\begin{aligned} \int_V \int_{\Gamma_x} D(\boldsymbol{\varepsilon}^{pc*}, \boldsymbol{\xi}^{c*})\, \mathrm{d}\Gamma(\boldsymbol{s})\, \mathrm{d}V &= \int_V \int_{\Gamma_x} [\widehat{\boldsymbol{\sigma}}^* : \boldsymbol{\varepsilon}^{pc*} - \widehat{\boldsymbol{\chi}}^* \cdot \boldsymbol{\xi}^{c*}]\, \mathrm{d}\Gamma(\boldsymbol{s})\, \mathrm{d}V \\ &= \beta_a \int_V \int_{\Gamma_x} \boldsymbol{s} : \boldsymbol{\varepsilon}^{pc*}\, \mathrm{d}\Gamma(\boldsymbol{s})\, \mathrm{d}V + \int_V \widehat{\boldsymbol{\sigma}}^{r*} : \Delta\boldsymbol{\varepsilon}^{pc*} \mathrm{d}V - \int_V \widehat{\boldsymbol{\chi}}^* \cdot \Delta\boldsymbol{\xi}^{c*} \mathrm{d}V \end{aligned} \tag{68}$$

which, the last two integrals on the r.h.s. being vanishing, reduces to

$$\int_V \int_{\Gamma_x} D(\boldsymbol{\varepsilon}^{pc*}, \boldsymbol{\xi}^{c*}) \, \mathrm{d}\Gamma(\boldsymbol{s}) \, \mathrm{d}V = \beta_a. \tag{69}$$

Since the l.h.s. of (69) equals β^*, we have $\beta^* = \beta_a$ and thus, by (67), it must be $\beta_a = \widehat{\beta}$. Therefore, the minimization problem (66) gives the shakedown safety factor β_a. QED.

It can be proved that the Euler-Lagrange equations of problem (66) substantially coincide with eqs. (53)–(60), but this point is skipped for brevity.

The shakedown limit state, to which the structure is able to report itself under loads amplified by β_a, is described by eqs. (53)–(60). In this state, a (fictitious) plastic strain mechanism $(\boldsymbol{\varepsilon}^{pc}, -\boldsymbol{\xi}^c)$ is produced at every stress point $\boldsymbol{s} \in \Gamma_x$, $\forall \boldsymbol{x} \in V$. This deformation process is not evolutive in nature, as no stress redistribution occurs. The set of these plastic mechanisms constitutes a normalized PAM. The latter PAM represents the incipient inadaptation collapse mechanism, which may be either an incremental one if $\Delta\boldsymbol{\varepsilon}^{pc}$ is nonvanishing at least somewhere in V, or an alternating plasticity one if $\Delta\boldsymbol{\varepsilon}^{pc} = \boldsymbol{0}$ everywhere in V.

It is worth remarking that, with the present constitutive model endowed with a saturation surface, either kinds of inadaptation modes are rendered possible. In fact, let for simplicity of reasoning the yield function be taken as $f = f_0(\boldsymbol{\sigma} - \boldsymbol{\chi}_K) - \chi_I \leq 0$ where $\boldsymbol{\chi}_K$ is the back-stress tensor associated with kinematic hardening and χ_I a scalar internal variable (drag-stress) associated with isotropic hardening. By (54) and (58), the conditions $\Delta\boldsymbol{\xi}_K^c = \boldsymbol{0}$, $\Delta\xi_I^c = 0$ in V give:

$$\Delta\boldsymbol{\varepsilon}^{pc} = \Delta\mu^c \frac{\partial\phi}{\partial\widehat{\boldsymbol{\chi}}_K} \qquad \text{in } V \tag{70}$$

$$\Delta\lambda^c = \Delta\mu^c \frac{\partial\phi}{\partial\widehat{\chi}_I} \qquad \text{in } V \tag{71}$$

where, by definition,

$$\Delta\lambda^c = \int_{\Gamma_x} \lambda^c \, \mathrm{d}\Gamma(\boldsymbol{s}), \qquad \Delta\mu^c = \int_{\Gamma_x} \mu^c \, \mathrm{d}\Gamma(\boldsymbol{s}). \tag{72}$$

It results that, in the limit state, $\Delta\boldsymbol{\varepsilon}^{pc}$ is different from zero if the r.h.s. of (70) is nonvanishing, i.e. the saturation surface is actually attained ($\Delta\mu^c \neq 0$) and $\partial\phi/\partial\widehat{\boldsymbol{\chi}}_K \neq \boldsymbol{0}$. Otherwise, if the r.h.s. of (70) is vanishing because $\partial\phi/\partial\widehat{\boldsymbol{\chi}}_K = \boldsymbol{0}$ (the normal to the saturation surface is parallel to the χ_I axis), but $\Delta\mu^c \neq 0$, it is $\Delta\boldsymbol{\varepsilon}^{pc} = \boldsymbol{0}$, $\Delta\lambda^c \neq 0$, i.e. alternating plasticity occurs (as it would be always the case for unlimited kinematic hardening). Also, from (71) we have that, in case of no saturation surface, ($\Delta\mu^c = 0$), it is $\Delta\lambda^c = 0$ identically, that is shakedown would always occur, whereas this is not the case with the present material model. See Fuschi (1999) for a similar discussion within quasi-static shakedown.

6 Two Simple Applications

Two simple applications to steel frame structures subjected to repeated excitations are presented in this Section.

6.1 Frame Structure Subjected to Seismic Loads

As shown by Borino and Polizzotto (1995), seismic loads can be modelled as a repeated-excitation load scheme. To this purpose, a seismic ground wave is represented as one with a discrete spectrum, i.e. denoting by $\ddot{u}_g$ the ground acceleration,

$$\ddot{u}_g(\tau) = E(\tau) \sum_{i=1}^{N} (\alpha_i \cos \omega_i \tau + \beta_i \sin \omega_i \tau) \tag{73}$$

where the ω_i are the wave circular frequencies, the scalars α_i, β_i are the wave parameters and $E(\tau)$ is a shape function to model the peaks' envelope. A simple choice for $E(\tau)$ is

$$E(\tau) = \frac{\tau}{T_0} \exp\left(1 - \frac{\tau}{T_0}\right) \tag{74}$$

where $E(T_0) = 1 = \max E(\tau)$.

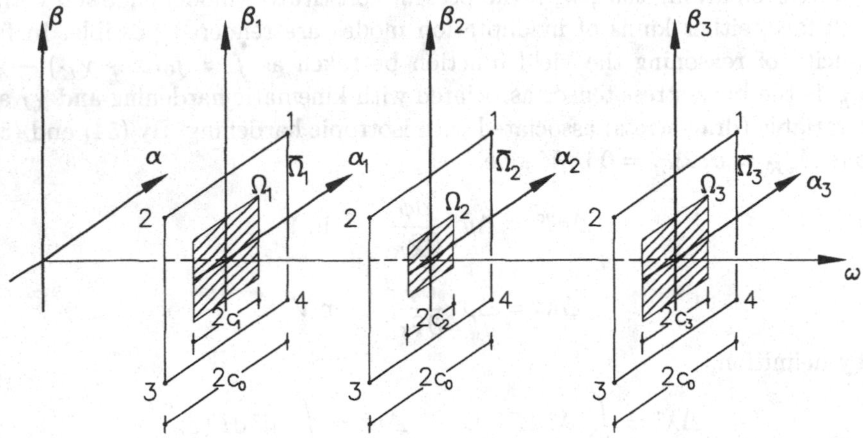

Figure 4. Three-dimensional space (ω, α, β) of the single-frequency seismic waves.

Equation (73) represents a two-parameter family of waves. A subset Π of this wave family is obtained by allowing α_i, β_i to range each within the (closed) interval $(-c_i, +c_i)$, where the N scalars c_i are given. This means that the point (α_i, β_i) of the (α_i, β_i)-plane range within a square Ω_i of sides $2c_i$ and centred at the origin, Fig. 4. The subset Π is then composed of N-frequency waves as (73), each of which is represented by N points (α_i, β_i), one for every square Ω_i and possesses a maximum power not exceeding $2(c_1^2 + c_2^2 + \ldots + c_N^2)$.

The frequencies ω_i are chosen coincident with the first N natural frequencies of the structure whereas the parameters c_i are taken as

$$c_i = \sqrt{8\, \zeta_i\, \omega_i\, S_g(\omega_i)}, \quad (i = 1, 2, \ldots, N) \tag{75}$$

where the ζ_i coefficients denote the damping ratios of the structure and $S_g(\omega_i)$ is given by

$$S_g(\omega_i) = \frac{S_0\left[1 + 4\zeta_g^2(\omega_i/\omega_g)^2\right]}{\left[1-(\omega_i/\omega_g)^2\right]^2 + 4\zeta_g(\omega_i/\omega_g)^2}. \tag{76}$$

Here, $S_g(\omega)$ denotes the power spectral density function of a simple oscillator, with natural frequency ω_g and damping ratio ζ_g, which according to the Kanai and Tajimi ground filter model (Tajimi, 1960) simulates the ground response to seismic waves from the deep rock bed with uniform power density S_0.

A perfectly plastic steel frame structure as in Fig. 5 has been considered. It has been discretized into seven beam elements, with 10 critical sections where bending plastic hinges can develop. The frame cross section is of the type HE300B, with an area $A = 149cm^2$ and second area moment $I = 25,166cm^4$. Other data are: Young modulus $E = 21,000kN/cm^2$, yield bending moment $M^y = 78,760kNcm$. The frame is loaded by its own weight plus a permanent load $q^0 = 400N/cm$ uniformly distributed over the horizontal beams, and concentrated loads $F^0 = 100KN$ at the three nodes, as shown in Fig. 5. The total permanent load gives the mass of the frame for the dynamic analysis. A damping ratio $\zeta = 0.05$ has been considered.

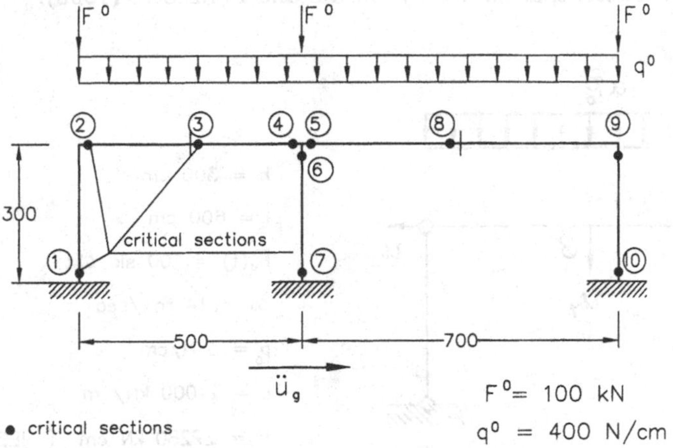

Figure 5. Steel frame under repeated seismic loads and permanent vertical loads.

The frame is subjected to repeated excitations (the seismic waves) as described previously, with common duration $T = 20s$, $T_0 = T/5$, $\zeta_g = 0,5$, $\omega_g = 6\pi$, $S_0 = 10$. Only the first three natural frequencies of the frame have been considered for the analysis. Table 1 collects details related to the computation of the constants c_1, c_2, c_3.

A series of dynamic elastic analyses, each for a particular seismic wave, enabled the maximum and minimum bending moments to be evaluated. Then, the shakedown safety factor has been determined solving a linear programming problem obtaining $\beta_a = 4.425$. For more details of this procedure, see Borino and Polizzotto (1995).

i	ω_i	$\Delta\omega_i = 2\zeta_i\omega_i$	$S_g(\omega_i)$	$c_i = 2\sqrt{S_g(\omega_i)\,\Delta\omega_i}$
1	23.17	2.317	14.169	11.460
2	37.31	3.731	3.955	7.683
3	71.04	7.104	0.806	4.787
				$c_0 = 23.930$

Table 1. Piecewise discretization of the power spectral density function and computation of the constants c_1, c_2, c_3.

6.2 Frame Structure With Variable Appended Mass

Often structures are subjected to variable (dynamic) loads and to permanent (static) ones, the latter being constituted by the weight of the structure and of the other masses appended to it. Typical examples are the ware-house structures, the water reservoirs and the like. In such cases, the mass of the system changes parametrically with the permanent (static) load. It follows that the shakedown load domain in the space of the dynamic and the static loads is *nonconvex*, as shown by Borino and Polizzotto (1996).

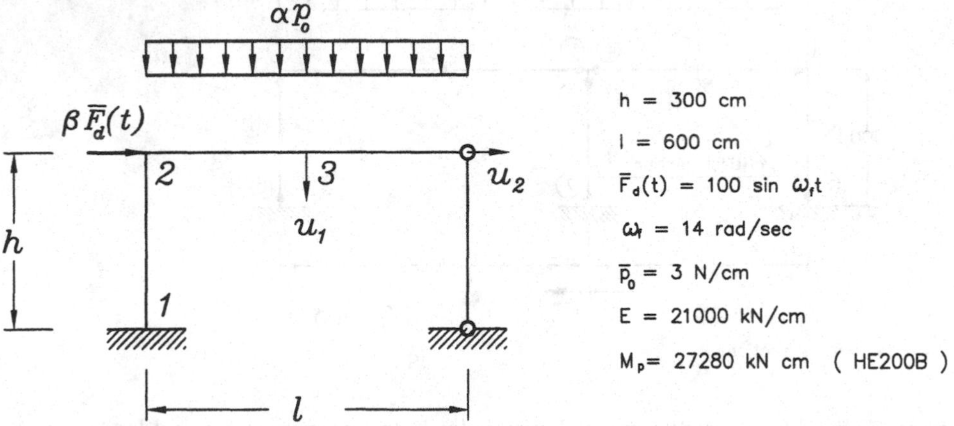

Figure 6. Bending frame of constant cross section subjected to a static load $\alpha\bar{p}_0$ and a variable load $\beta\bar{F}_d(t)$ as indefinite repetition of the sinusoidal excitation $f(\tau)$.

In the latter paper, a simple perfectly-plastic steel frame as in Fig. 6 has been considered, subjected to a permanent vertical load $\alpha\bar{p}_0$ and a sinusoidally time variable load $\beta\bar{F}_d(t)$. The mass matrix of the system is of the form

$$\boldsymbol{M} = \alpha\,\boldsymbol{M}_0 + \boldsymbol{M}_d. \tag{77}$$

As the dynamic response of the structure depends on α, the maximum and minimum bending moments also depend on α. The shakedown load boundary is reported in Fig. 7 as the curve $\beta = g(\alpha)$, together with the elastic load boundary, as well as the shakedown load boundary for the case $M_0 = 0$. The nonconvexity of the shakedown load boundary is a notable feature to be taken into account for safety purposes.

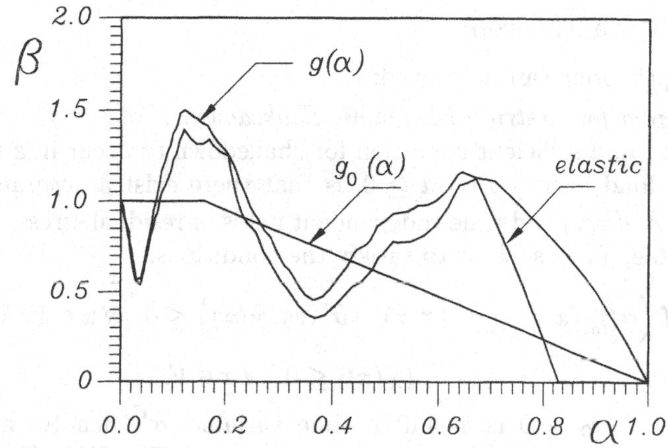

Figure 7. Shakedown boundary curves of the frame of Fig. 6: with appended mass, $\beta = g(\alpha)$, and without appended mass, $\beta = g_o(\alpha)$; elastic limit curve for the frame with appended mass.

7 Restricted Dynamic Shakedown

As stated in the Introduction Section, restricted dynamic shakedown is shakedown of structures subjected to dynamic external actions specified for all times, from $t = 0$ to $t = \infty$. A typical example often considered is a periodic load, i.e. $P(t) = P(t + \Delta t)$ with period Δt, but other examples may be considered. Though these examples are likely theoretical extrapolations of real situations in which loads are known only for finite time intervals, however the study of shakedown for such infinite-duration loads is considered relevant for design purposes since it may provide useful information on the asymptotic behaviour of the system.

Historically, the form of dynamic shakedown first studied is restricted shakedown. Ceradini (1969) formulated his static-type theorem as an extension of the classical Melan theorem to dynamics under a specified load history, and considered the case of sequences of short-duration excitations as a special case of the former, pointing out how the theorem then simplifies; but he however did not consider the possibility for every excitation to be repeated indefinitely. The categorization of dynamic shakedown as restricted and unrestricted was proposed by Polizzotto, (1984a).

In this Section we report (without proof) two theorems of restricted dynamic shakedown, one of static type, the other of kinematic type. To this purpose, we assume that the structure is subjected to a specified load history (LH), say $\boldsymbol{P}(t)$, $t \geq 0$. We denote by $\boldsymbol{\sigma}^E(\boldsymbol{x}, t)$ the related elastic (dynamic) stress response (with arbitrary but fixed initial conditions, e.g. zero initial conditions), by $\boldsymbol{\sigma}^F(\boldsymbol{x}, t)$ a free-motion stress (i.e. the stresses arising in the system considered elastic as a consequence of some imposed initial conditions, say $\boldsymbol{u}^F = \boldsymbol{u}_0^F$ and $\dot{\boldsymbol{u}}^F = \dot{\boldsymbol{u}}_0^F$ in V at $t = 0$). The material is the same as that considered in Section 2.

7.1 Static-Type Theorem

The following theorem can be proved:

- *Static Theorem for Restricted Dynamic Shakedown.*

A necessary and sufficient condition for shakedown to occur in a structure subjected to a specified load history $\boldsymbol{P}(t)$, $t \geq 0$, is that there exist a *separation* time t_0, a free-motion stress $\widehat{\boldsymbol{\sigma}}^F(\boldsymbol{x}, \tau)$ and time-independent fields of residual stress, $\widehat{\boldsymbol{\sigma}}^r(\boldsymbol{x})$, and of static internal variable, $\widehat{\boldsymbol{\chi}}(\boldsymbol{x})$, such as to satisfy the conditions:

$$f\left(\boldsymbol{\sigma}^E_{(t_0)}(\boldsymbol{x}, \tau) + \widehat{\boldsymbol{\sigma}}^F(\boldsymbol{x}, \tau) + \widehat{\boldsymbol{\sigma}}^r(\boldsymbol{x}), \widehat{\boldsymbol{\chi}}(\boldsymbol{x})\right) \leq 0 \quad \forall \boldsymbol{x} \in V, \; \forall \tau \geq 0 \tag{78}$$

$$\phi\left(\widehat{\boldsymbol{\chi}}(\boldsymbol{x})\right) \leq 0 \quad \forall \boldsymbol{x} \in V. \tag{79}$$

Here $\tau = t - t_0 \geq 0$ is a shifted time variable, $\widehat{\boldsymbol{\sigma}}^F$ denotes a free-motion stress generated by initial conditions imposed at time $\tau = 0$. The fields $(\widehat{\boldsymbol{\sigma}}^r, \widehat{\boldsymbol{\chi}}, \widehat{\boldsymbol{\sigma}}^F)$ constitute *(static) shakedown parameters* and $\mathcal{S}_d = \{\widehat{\boldsymbol{\sigma}}^r, \widehat{\boldsymbol{\chi}}, \widehat{\boldsymbol{\sigma}}^F\}$ is the set of such fields. Also, $\boldsymbol{\sigma}^E_{(t_0)}$ denotes the elastic stress response back-ward truncated at the separation time t_0, i.e.

$$\boldsymbol{\sigma}^E_{(t_0)}(\boldsymbol{x}, \tau) = \boldsymbol{\sigma}^E(\boldsymbol{x}, t_0 + \tau), \quad \forall \tau \geq 0. \tag{80}$$

Since the proof of the above theorem can be achieved in as much a similar way as for the unrestricted dynamic shakedown of Section 4.1, we skip it for simplicity. The above theorem is an extension to the present context of Ceradini's theorem (1969) and of its generalization to internal variable material models given by Maier and Novati (1990). However, the present theorem has a quite different formulation as it accounts for the fictitious initial conditions of Ceradini's theorem by the free-motion stress $\widehat{\boldsymbol{\sigma}}^F$, as in Polizzotto (1984a, 1984b). A few further comments are appropriate here.

a) The free-motion stresses, $\widehat{\boldsymbol{\sigma}}^F$, are representative of the dynamic effects produced by plastic strains over the body's motion during the transient phase up to the adaptation time, t_a; the analogous static effects are represented by $\widehat{\boldsymbol{\sigma}}^r$. In Ceradini's (1969) approach, these dynamic effects are simulated by means of fictitious initial conditions fixed at the remote time $t = 0$.

b) The actual initial conditions associated with the given LH have no influence on the capacity of the structure to adapt to the given LH, nor on the related shakedown safety factor. In fact, assuming that in (78) $\boldsymbol{\sigma}^E$ is associated with zero initial conditions, the analogous response in case of nonzero initial conditions, say $\boldsymbol{\sigma}^{E\prime}$, can be expressed as $\boldsymbol{\sigma}^{E\prime} = \boldsymbol{\sigma}^E + \boldsymbol{\sigma}^{EF}$, with $\boldsymbol{\sigma}^{EF}$ being the free-motion stress generated by the mentioned initial conditions. Thus, inequality (78) becomes

$$f\left(\boldsymbol{\sigma}^E_{(t_0)} + \boldsymbol{\sigma}^{EF}_{(t_0)} + \widehat{\boldsymbol{\sigma}}^F + \widehat{\boldsymbol{\sigma}}^r, \widehat{\boldsymbol{\chi}}\right) \leq 0, \tag{81}$$

which coincides with (78) because the addend $(\sigma^{EF}_{(t_0)} + \hat{\sigma}^F)$ does represent a free-motion stress like $\hat{\sigma}^F$. This fact is beneficial because the actual initial conditions, like the initial residual stresses, are hardly known in practice.

c) If the LH has finite duration, say $0 \leq t \leq T$, the above theorem turns out to be defective, as in fact we can always choose a time $t_0 > T$, such that $\sigma^E_{(t_0)}$ is a free-motion stress, and thus inequalities (78) and (79) can be always satisfied by taking $\hat{\sigma}^F = -\sigma^E_{(t_0)}$. In other words, for T finite, shakedown always occurs because after $t_0 > T$ no further plastic strains can be produced, but the theorem gives no indication about whether shakedown occurs before the given LH extinguishes.

d) According to a definition given by Polizzotto (1984a, 1984b, 1985) and by Polizzotto *et al.* (1993) for undamped structures, the minimum value of $t_0 \geq 0$ for which (78) and (79) can be satisfied with some set $(\hat{\sigma}^F, \hat{\sigma}^r, \hat{\chi})$, identifies with the ***minimum adaptation time*** (MAT) of the structure/load system, that is the minimum time required for shakedown to occur, provided suitable initial conditions are imposed to the system. Denoting by t^* the MAT, shakedown occurs before the LH's termination if $t^* < T$. An example is provided by a periodic load, in which case $t^* = 0$, and the optimal initial conditions are the periodicant ones (Maier and Novati, 1990), whenever the load is below the shakedown limit, but t^* does not exist in the opposite case.

e) For a LH with a very long time duration T, the damping effects cannot be disregarded in real structures. Since, then, any free vibration motion rapidly extinguishes, the yield inequality (78) should be better enforced with the free-motion stress $\hat{\sigma}^F$ being disregarded. This would produce a notable simplification in the shakedown criterion since the latter can then be expressed only in terms of the shakedown parameters $(\hat{\sigma}^r, \hat{\chi}) \in \mathcal{S} \subset \mathcal{S}_d$.

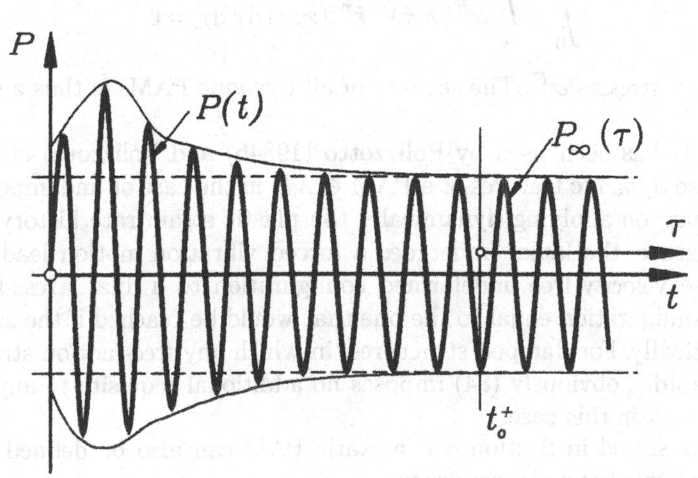

Figure 8. Loading history $P(t)$, $t \geq 0$, becoming periodic for $t \geq t_0^+$, i.e. $P_\infty(\tau) \equiv P(t_0^+ + \tau) = P_\infty(\tau + \Delta t)$.

f) The separation time t_0 is an essential ingredient of the shakedown criterion (78) and (79); it possesses the paramount role of cutting off, from the given LH $\boldsymbol{P}(t)$, $t \geq 0$, an initial piece of finite duration as irrelevant for shakedown, but leaving the subsequent LH as responsible for the shakedown occurrence. Considered that shakedown is a long term structural phenomenon, it follows that the right separation time to consider in the shakedown criterion is $t_0 = \infty$; or, in other words, the subsequent LH to consider is the asymptotic one, i.e. $\boldsymbol{P}_{(\infty)}(\tau) = \lim \boldsymbol{P}(t_0 + \tau)$ for $t_0 \to \infty$. For instance, Fig. 8, the LH has as asymptotic counterpart a periodic LH, $\boldsymbol{P}_{(\infty)}(\tau)$, $\tau \geq 0$, coinciding with $\boldsymbol{P}(t_0^+ + \tau)$; then, the shakedown occurrence in the structure is determined by $\boldsymbol{P}_{(\infty)}(\tau)$, the initial LH up to t_0^+ being irrelevant.

7.2 Kinematic-Type Theorem

As for unrestricted dynamic shakedown, the concept of PAM plays a crucial role in the formulation of kinematic-type theorems. In the present context, we distinguish two kinds of such PAMs, one *static*, the other *dynamic*. The static one is identical to that defined in Section 4.2, but it is here redefined to make it fit to the present load scheme. Namely, a static PAM is constituted by fields as $\dot{\boldsymbol{\varepsilon}}^{pc}(\boldsymbol{x},\tau)$, $\dot{\boldsymbol{\xi}}^{c}(\boldsymbol{x},\tau)$, $\tau \geq 0$, satisfying the following conditions:

$$\Delta \boldsymbol{\varepsilon}^{pc} = \int_0^\infty \dot{\boldsymbol{\varepsilon}}^p \, \mathrm{d}\tau = \nabla^s \boldsymbol{u}^c \quad \text{in } V, \qquad \boldsymbol{u}^c = \boldsymbol{0} \quad \text{on } S_u, \tag{82}$$

$$\Delta \boldsymbol{\xi}^c = \int_0^\infty \dot{\boldsymbol{\xi}}^c \, \mathrm{d}\tau = \boldsymbol{0} \quad \text{in } V. \tag{83}$$

The set of all such PAMs is denoted $\mathcal{M}$.

A dynamic PAM is a static PAM satisfying the additional (dynamic) requisite of being orthogonal to any free-motion stress $\boldsymbol{\sigma}^F$, i.e. such that

$$\int_0^\infty \int_V \boldsymbol{\sigma}^F(\boldsymbol{x},\tau) : \dot{\boldsymbol{\varepsilon}}^{pc}(\boldsymbol{x},\tau) \, \mathrm{d}V \mathrm{d}\tau = 0 \tag{84}$$

for such arbitrary stresses $\boldsymbol{\sigma}^F$. The set $\mathcal{M}_d$ of all dynamic PAMs is thus a subset of $\mathcal{M}$, i.e. $\mathcal{M}_d \subseteq \mathcal{M}$.

Equation (84) has been used by Polizzotto (1984b) and Polizzotto *et al.* (1993) to derive the precise dynamic features of a PAM $\in \mathcal{M}_d$ in the case of undamped structures. It was found that, on applying dynamically the plastic strain rate history $\dot{\boldsymbol{\varepsilon}}^{pc}(\boldsymbol{x},\tau)$ on the elastic structure, the latter undergoes a forced vibration motion leading from the initial stress-free velocity-free undeformed configuration to a final stress-free velocity-free deformed configuration equal to the one that would be reached if the same $\dot{\boldsymbol{\varepsilon}}^{pc}(\boldsymbol{x},\tau)$ acted quasi-statically. For damped structures, in which any free-motion stress $\boldsymbol{\sigma}^F$ tends to extinguish rapidly, obviously (84) imposes no additional requisite to any static PAM such that $\mathcal{M}_d = \mathcal{M}$ in this case.

Note that, as stated in Section 4.2, a static PAM can also be defined as the fields $\dot{\boldsymbol{\varepsilon}}^{pc}(\boldsymbol{x},\tau)$, $\dot{\boldsymbol{\xi}}^{c}(\boldsymbol{x},\tau)$ satisfying the condition:

$$\int_0^\infty \int_V \left(\boldsymbol{\sigma}^r : \dot{\boldsymbol{\varepsilon}}^{pc} - \boldsymbol{\chi} \cdot \dot{\boldsymbol{\xi}}^c \right) \mathrm{d}V \mathrm{d}\tau = 0 \tag{85}$$

for all *time-independent* fields of self-stresses, σ^r, and static internal variable, χ, i.e. $\forall\,(\sigma^r, \chi) \in \mathcal{S}$. This definition was shown to be equivalent to (82) and (83) by Polizzotto *et al.* (1993). The condition (84) must be added to (85) in order to define a dynamic PAM for undamped systems. Thus, whereas $\mathcal{M}$ is dual of $\mathcal{S} = \{\sigma^r, \chi\}$, $\mathcal{M}_d$ is dual of $\mathcal{S}_d = \{\sigma^r, \chi, \sigma^F\}$.

With the latter definitions in mind, the following kinematic theorem can be proved:

- *Kinematic Theorem for Restricted Dynamic Shakedown.*

A necessary and sufficient condition for shakedown to occur in a structure subjected to an infinite-duration LH, say $\boldsymbol{P}(t)$, $t \geq 0$, is that there exists some (separation) time $t_0 \geq 0$ such that the inequality

$$\mathcal{E}\left[\dot{\varepsilon}^{pc}, \dot{\xi}^c, t_0\right] \equiv \int_0^\infty \int_V \left[D\left(\dot{\varepsilon}^{pc}, \dot{\xi}^c\right) - \sigma^E_{(t_0)} : \dot{\varepsilon}^{pc}\right] \mathrm{d}V\,\mathrm{d}\tau \geq 0 \tag{86}$$

be satisfied for all PAMs (taken from either $\mathcal{M}_d$ in case of undamped structures, or $\mathcal{M}$ in case of damped ones).

The proof of this theorem is omitted for simplicity, also because it is quite similar to that of Section 4.2. See Polizzotto *et al.* (1993) for more details on this point. The theorem above is different from that of Corradi and Maier (1973, 1974), since the latter makes use of PAMs $\in \mathcal{M}$ in all cases and exhibits a mixed-type feature, whereas the theorems given here are dual of each other.

7.3 The Shakedown Safety Factor

As observed in Section 7.1, remark e), for LHs of very long time duration, damping cannot be ignored and thus the static shakedown theorem of Section 7.1 applies with $\widehat{\sigma}^F$ dropped, whereas the kinematic theorem of Section 7.2 applies with PAMs $\in \mathcal{M}$. Correspondingly, the shakedown safety factor can be evaluated as in the following (see Section 5 for comparison), either by the ***static approach*** by solving the problem:

$$\left.\begin{aligned} &\bar{\beta}(t_0) = \max_{(\beta, \widehat{\sigma}^r, \widehat{\chi})} \beta \qquad \text{subject to :} \\ &f(\beta\sigma^E_{(t_0)} + \widehat{\sigma}^r, \widehat{\chi}) \leq 0, \quad \phi(\widehat{\chi}) \leq 0 \quad \text{in } V, \quad \forall \tau \geq 0 \\ &\operatorname{div} \widehat{\sigma}^r = \boldsymbol{0} \quad \text{in } V, \qquad \widehat{\sigma}^r \cdot \boldsymbol{n} = \boldsymbol{0} \quad \text{on } S_t \end{aligned}\right\} \tag{87}$$

or, by the ***kinematic approach*** by solving the problem:

$$\left.\begin{aligned} &\bar{\beta}(t_0) = \min_{(\dot{\varepsilon}^{pc}, \dot{\xi}^c)} \int_0^\infty \int_V D(\dot{\varepsilon}^{pc}, \dot{\xi}^c)\,\mathrm{d}V\mathrm{d}\tau \qquad \text{subject to :} \\ &\int_0^\infty \int_V \sigma^E_{(t_0)} : \dot{\varepsilon}^{pc}\,\mathrm{d}V\mathrm{d}\tau = 1, \qquad (\dot{\varepsilon}^{pc}, \dot{\xi}^c) \in \mathcal{M} \end{aligned}\right\} \tag{88}$$

and then searching for the asymptotic value of the nondecreasing function $\bar{\beta}(t_0)$, that is

$$\beta_a = \lim_{t_0 \to \infty} \bar{\beta}(t_0) = \max_{t_0 \geq 0} \bar{\beta}(t_0). \tag{89}$$

The function $\bar{\beta}(t_0)$ is nondecreasing due to the fact that an increase of t_0 produces a reduction of constraints in (87). The following holds:

$$\frac{\mathrm{d}\bar{\beta}}{\mathrm{d}t_0} = -\bar{\beta}(t_0) \int_0^\infty \int_V \dot{\boldsymbol{\sigma}}^E_{(t_0)} : \dot{\boldsymbol{\varepsilon}}^{pc} \, \mathrm{d}V \mathrm{d}\tau \geq 0, \tag{90}$$

as it can be shown by the Euler-Lagrange equations related to either (87) and (88).

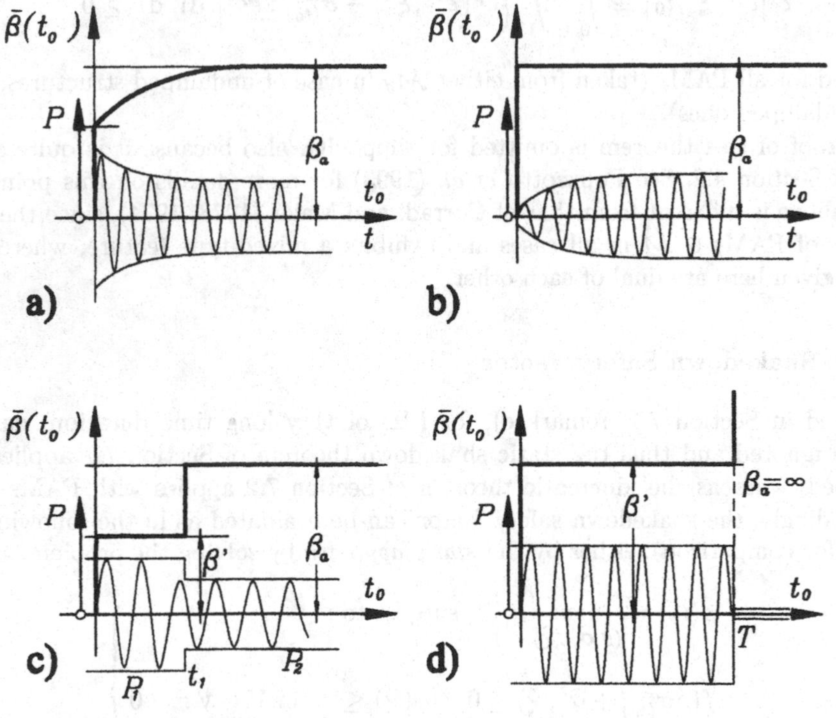

Figure 9. Typical examples of load histories and related curves $\bar{\beta}(t_o)$.

Using the considerations developed at point f) in Section 7.1, let $\boldsymbol{\sigma}^E_{(\infty)}(\boldsymbol{x},\tau)$ be the elastic stress response related to the asymptotic LH, $\boldsymbol{P}_{(\infty)}(\tau)$. Obviously, the asymptotic elastic stress response can be directly obtained as the limit:

$$\boldsymbol{\sigma}^E_{(\infty)}(\boldsymbol{x},\tau) = \lim_{t_0 \to \infty} \boldsymbol{\sigma}^E(\boldsymbol{x}, t_0 + \tau) \qquad \forall \tau \geq 0. \tag{91}$$

Then, we realize that β_a can be directly obtained as the optimal objective value of either (87) and (88), but with this $\boldsymbol{\sigma}^E_{(\infty)}$ in place of $\boldsymbol{\sigma}^E_{(t_0)}$. Whenever the limit (91) is difficult to obtain, it may be sufficient to solve either (87) or (88), with a t_0 value sufficiently large to have $\bar{\beta}(t_0) \simeq \beta_a$.

Possible shapes of the function $\bar{\beta}(t_0)$ were discussed by Polizzotto (1984a) for perfectly plastic materials, but they remain valid also in the present context. In Figs. 9(a-d) are sketched a few typical shapes of the curve $\bar{\beta}(t_0)$ that may be encountered in practice.

Fig. 9(a) refers to a load of the type $\boldsymbol{P}(t) = \boldsymbol{P}_0[1+b\exp(-ct)]\sin\omega t$, that is sinusoidal with amplitudes exponentially varying from $1 + b > 1$ for $t = 0$ to 1 for $t \to \infty$. It is intuitively evident that in this case the curve $\bar{\beta}(t_0)$ provided by either (87) and (88) is a monotonically increasing one, as shown in Fig. 9(a), with the asymptotic value $\bar{\beta}(\infty) = \beta_a$. The latter value can be directly obtained from (87), or (88), considering the asymptotic periodic load $\boldsymbol{P}_0 \sin\omega t$. For $b = 0$ a periodic load is generated, in which case $\bar{\beta}(t_0) = \text{const} = \beta_a$.

Fig. 9(b) refers to the case of a load as in the previous case, but $b < 0$. In this case, obviously it is $\bar{\beta}(t_0) = \text{const} = \beta_a$. The latter value β_a can be directly obtained from (87), or (88), considering the asymptotic load $\boldsymbol{P}_0 \sin\omega t$.

Fig. 9(c) is related to a load $\boldsymbol{P}(t)$ consisting of a periodic load $\boldsymbol{P}_1(t)$ for $0 \le t \le t_1$ and a periodic load $\boldsymbol{P}_2(t)$ with a smaller amplitude for $t \ge t_1$. In this case, the curve $\bar{\beta}(t_0)$ is piecewise constant, with $\bar{\beta} = \beta'$ for $0 \le t_0 \le t_1$ and $\bar{\beta} = \beta'' > \beta'$ for $t_0 \ge t_1$. Obviously, $\beta_a = \beta''$ and this value can be directly obtained from (87), or (88), for the asympotic load $\boldsymbol{P} = \boldsymbol{P}_2(t)$. Finally, Fig. 9(d) considers the case of a load $\boldsymbol{P}(t)$ with a finite duration $0 \le t \le T$. The curve $\bar{\beta} = \bar{\beta}(t_0)$ has a finite value β' for $0 \le t_0 \le T$, but an infinite value for $t_0 > T$. Thus, $\beta_a = \infty$, accordingly to the fact that (restricted) shakedown always occurs in this case.

Note that, were the loads capable of repetitions in all the considered cases, the shakedown safety factor would result as follows: $\beta_a = \bar{\beta}(0)$ in Fig. 9(a), β_a unaltered in Fig. 9(b), $\beta_a = \beta'$ in Fig. 9(c), $\beta_a = \beta'$ in Fig. 9(d).

8 Bounding Techniques

In the present context, bounding techniques are meant as appropriate analytical-numerical procedures leading to the evaluation of a scalar quantity, say U, which is an upper bound to a specific scalar measure, say η, of the inelastic deformation experienced by a structure subjected to some loading programme, such that the following inequality holds:

$$\eta \le U. \tag{92}$$

η can in practice have a variety of meanings, as component of the inelastic strain and of the residual displacement, inelastic dissipation work and the like. Additionally, η can have local or global significance according to whether it refers to a single point of the structure, or to a region which is a finite portion of, or even coincides with, the entire volume of it.

In consideration that η depends on the actual structural response, and thus on the actual load history, η is to be considered as a time function, $\eta(t)$, which can in principle be evaluated by an evolutive analysis. However, such an analysis being in general computationally expensive, in certain circumstances (e.g. early stages of the design process) it may be considered sufficient to evaluate U, likely with less computational efforts than η. U gives a piece of information on η, whose usefulness obviously increases with the closeness of U to η.

The bounding techniques are in general presented in association of shakedown theory. This is due to the fact that shakedown theorems tell nothing about the amount of plastic deformations produced in the transient phase before the shakedown occurrence and that the bounding techniques are most appropriate tools to estimate such deformations.

A bounding technique can be characterized in two ways: i) the simplicity of the related analytical-numerical procedure and ii) the stringentness of the upper bound provided. It is intuitively evident that less stringent upper bounds are to be expected with procedures of increasing simplicity. So, in practice, what is important is to find a satisfactory compromise between the simplicity of the procedure and closeness of the bound.

Such a compromise is perhaps realized with a class of bounding techniques holding for repeated load schemes and widely used in practice. We refer to Leckie (1974), Capurso *et al.* (1978), König and Maier (1981), Polizzotto (1982a, 1984a), Panzeca *et al.* (1990), as review papers; see also Ponter (1972, 1975), Capurso (1979), Polizzotto *et al.* (1991). With the latter class of bounding techniques, the upper bound U does not depend on the actual load history which the deformation parameter η refers to, nor on the time at which η is computed. As a consequence of this fact, U actually represents an upper bound to the peak value of η within the considered time interval. A drawback of this sort of bounding techniques is that these techniques can be applied only for loads below the shakedown limit and that the upper bound rapidly deteriorates when the load approaches the latter limit.

Another class of bounding techniques collects the so-called *convergent* bounding techniques (Polizzotto, 1982b, 1984a, 1984c, 1986), which apply to a specified load history, either below and above the shakedown limit. With these bounding techniques, upper and lower bounds U can be constructed, the computation of which requires a suitably simplified *evolutive analysis* over the given load history; furthermore, U can be rendered as close to η as desired, but obviously with adequately reduced simplification degrees and thus at increasing computational costs. Such procedures are particularly suitable for creeping materials, but they can also be applied to elastic-plastic ones, for load ranges within which the repeated-load techniques are useless. We will refer to these bounding techniques as *evolutive* ones.

The bounding techniques were mainly developed during the years 1970-1985; then there has been a considerable reduction in the related research activity, probably because meantime the numerical methods for elastic plastic evolutive analysis improved notably together with the computer capabilities. A revival of interest is observed at present (Maier *et al.*, 2000), likely due to the more complex material constitutive models considered in nowadays engineering practice, such that simplified analysis methods extended to such materials deserve being developed.

In this section, a bounding principle for repeated loads is first presented such as to cope with material models as described in Section 2. The related bounding techniques are implemented with the use of the so-called *Perturbation Method* devised by Polizzotto (1982a, 1982b, 1984a, 1984c) and constituting a generalization of the Dummy Load Method of Ponter (1972, 1975) and pursued by others (Capurso *et al.*, 1978; Leckie, 1974).

In consideration of the unified character of the shakedown theorems for quasi-static loads and for repeated excitations, the above bounding principle with related bounding techniques are presented in a similar unified form for statics and dynamics. An evolutive-type bounding principle is presented in Section 9.

8.1 General Bounding Principle for Repeated Loads

The structure/load system introduced in Section 3 is here reconsidered characterized by the (reduced) elastic stress envelope $\Gamma = \{\Gamma_x, \forall \boldsymbol{x} \in V\}$. Let us introduce the scalar field $\gamma = \gamma(\boldsymbol{x}) \geq 0$ and the stress field $\boldsymbol{p} = \boldsymbol{p}(\boldsymbol{x})$, which are referred to as *perturbation functions* in the following, and let the quantity

$$\eta = \int_V \left[\gamma \int_0^{t_1} D\left(\dot{\boldsymbol{\varepsilon}}^p, \dot{\boldsymbol{\xi}}\right) \mathrm{d}t + \boldsymbol{p} : \int_0^{t_1} \dot{\boldsymbol{\varepsilon}}^p \, \mathrm{d}t \right] \mathrm{d}V \tag{93}$$

be assumed as a suitable scalar measure of the actual plastic deformation produced in the structure in the time interval $(0, t_1)$ by the application of some ALH, say $\boldsymbol{P}(t), t \geq 0$. Note that the fields γ and $\boldsymbol{p}$ are mutually exclusive, that is one of them vanishes identically when the other does not. A pair $(\gamma, \boldsymbol{p})$ constitutes a *perturbation mode*.

Also, let us define the following *modified* fictitious stress and static internal variable fields, i.e.

$$\widehat{\boldsymbol{\sigma}}^* = (\boldsymbol{s} + \widehat{\boldsymbol{\sigma}}^r)\, g + \alpha \boldsymbol{p}, \quad \widehat{\boldsymbol{\chi}}^* = \widehat{\boldsymbol{\chi}} g \qquad \forall \boldsymbol{s} \in \Gamma_x, \ \forall \boldsymbol{x} \in V \tag{94}$$

where $\alpha > 0$ is the *perturbation multiplier* and g is defined as

$$g = 1 + \alpha \gamma(\boldsymbol{x}) \qquad \text{in } V, \tag{95}$$

whereas $\widehat{\boldsymbol{\sigma}}^r$ and $\widehat{\boldsymbol{\chi}}$ are shakedown parameters, i.e. time independent fields of self-stresses and internal variables, respectively.

Let us assume that, for a given perturbation mode $(\gamma, \boldsymbol{p})$, there exist some shakedown parameters $\widehat{\boldsymbol{\sigma}}^r$, $\widehat{\boldsymbol{\chi}}$ and a scalar $\alpha > 0$, such that the related modified fictitious stress and static internal variable fields, $\widehat{\boldsymbol{\sigma}}^*$ and $\widehat{\boldsymbol{\chi}}^*$, satisfy the yield and hardening saturation conditions, i.e.

$$f(\widehat{\boldsymbol{\sigma}}^*, \widehat{\boldsymbol{\chi}}^*) \leq 0, \quad \phi(\widehat{\boldsymbol{\chi}}^*) \leq 0 \qquad \forall \boldsymbol{s} \in \Gamma_x, \ \forall \boldsymbol{x} \in V. \tag{96}$$

Since the latter inequalities can also be written as

$$f(\widehat{\boldsymbol{\sigma}} g + \alpha \boldsymbol{p}, \widehat{\boldsymbol{\chi}} g) \leq 0, \quad \phi(\widehat{\boldsymbol{\chi}} g) \leq 0 \qquad \forall \boldsymbol{s} \in \Gamma_x, \forall \boldsymbol{x} \in V, \tag{97}$$

where, by definition, $\widehat{\boldsymbol{\sigma}} = \boldsymbol{s} + \widehat{\boldsymbol{\sigma}}^r$, we are allowed to consider the perturbation mode as producing a modification over the yield and saturation functions, instead over the fields $\widehat{\boldsymbol{\sigma}}$ and $\widehat{\boldsymbol{\chi}}$. With this interpretation, (97) can also be written as

$$f^*_\alpha(\widehat{\boldsymbol{\sigma}}, \widehat{\boldsymbol{\chi}}) \leq 0, \quad \phi^*_\alpha(\widehat{\boldsymbol{\chi}}) \leq 0 \qquad \forall \boldsymbol{s} \in \Gamma_x, \ \forall \boldsymbol{x} \in V, \tag{98}$$

where f^*_α and ϕ^*_α, are the modified (or perturbed) yield and saturation functions. f^*_α and ϕ^*_α can be thought of as being derived from f and ϕ, respectively, by applying the given perturbation mode. The latter produces, at the generic point $\boldsymbol{x} \in V$, a ***homothetic contraction*** with ratio $1 : g$ of both f and ϕ, along with a ***rigid translation*** $\alpha \boldsymbol{p}$ of the yield function, Fig. 10.

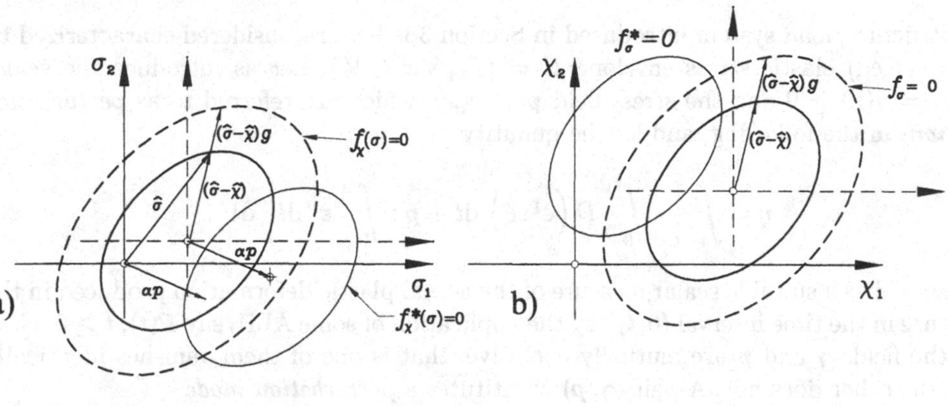

Figure 10. Geometrical sketch illustrating the effects of a perturabation on the yield surface: a) in the $\boldsymbol{\sigma}$-space and b) in the $\boldsymbol{\chi}$-space.

The (either static, or dynamic) elastic stress response, $\boldsymbol{\sigma}^E(\boldsymbol{x}, t)$, to any ALH $\boldsymbol{P}(t)$, $t \geq 0$, (with zero initial conditions) constitutes a path not exceeding Λ_x everywhere in V. Since, due to the convexity of f and ϕ, (97) and (98) are satisfied also for $\boldsymbol{s}$ ranging within the convex hull of Γ, i.e. Λ, we can state that (97) holds good also in the form

$$f(g\widehat{\boldsymbol{\sigma}} + \alpha \boldsymbol{p}, g\widehat{\boldsymbol{\chi}}) \leq 0, \quad \phi(g\widehat{\boldsymbol{\chi}}) \leq 0 \qquad \forall t \geq 0 \text{ in } V \tag{99}$$

where $\widehat{\boldsymbol{\sigma}} = \boldsymbol{\sigma}^E + \widehat{\boldsymbol{\sigma}}^r$ and $\boldsymbol{s} = \boldsymbol{\sigma}^E$ is the elastic stress response to the considered ALH. Then, denoting by $\boldsymbol{\sigma}$, $\boldsymbol{\varepsilon}$, $\boldsymbol{\varepsilon}^p$, $\boldsymbol{\chi}$,... the actual elastic plastic response to this ALH and making use of the Druckerian inequality (10), we can write

$$(\boldsymbol{\sigma} - g\widehat{\boldsymbol{\sigma}} - \alpha \boldsymbol{p}) : \dot{\boldsymbol{\varepsilon}}^p - (\boldsymbol{\chi} - g\widehat{\boldsymbol{\chi}}^r) : \dot{\boldsymbol{\xi}} \geq 0, \quad \forall t \geq 0 \text{ in } V. \tag{100}$$

This, summing and subtracting the terms $g\,\boldsymbol{\sigma}$ and, respectively, $g\,\boldsymbol{\chi}$ within the parentheses and using (95) gives

$$\alpha\,\gamma\, D(\dot{\boldsymbol{\varepsilon}}^p, \dot{\boldsymbol{\xi}}) + \alpha \boldsymbol{p} : \dot{\boldsymbol{\varepsilon}}^p \leq g\left[(\boldsymbol{\sigma} - \widehat{\boldsymbol{\sigma}}) : \dot{\boldsymbol{\varepsilon}}^p - (\boldsymbol{\chi} - \widehat{\boldsymbol{\chi}}) \cdot \dot{\boldsymbol{\xi}}\right] \qquad \forall t \geq 0 \text{ in } V. \tag{101}$$

from where multiplying by $\bar{g}/g$ with $\bar{g} = \max g$ in V (and noting that $g \equiv \bar{g} = 1$ whenever $\boldsymbol{p} \not\equiv \boldsymbol{0}$ such that $\boldsymbol{p} = \bar{g}\,\boldsymbol{p}/g$) we have:

$$\int_V \left[\frac{\bar{g}\gamma}{g} D + \boldsymbol{p} : \dot{\boldsymbol{\varepsilon}}^p\right] \mathrm{d}V \leq \frac{\bar{g}}{\alpha}\left[\int_V (\boldsymbol{\sigma} - \widehat{\boldsymbol{\sigma}}) : \dot{\boldsymbol{\varepsilon}}^p\, \mathrm{d}V - \int_V (\boldsymbol{\chi} - \widehat{\boldsymbol{\chi}}) \cdot \dot{\boldsymbol{\xi}}\, \mathrm{d}V\right] \tag{102}$$

where we have set:

$$\bar{g} = 1 + \alpha\,\bar{\gamma} \qquad \bar{\gamma} = \max_{\boldsymbol{x}\in V} \gamma(\boldsymbol{x}). \tag{103}$$

The first integral on the r.h.s. of (102) can be transformed by means of the identity (21) —here also holding— as follows:

$$\int_V (\boldsymbol{\sigma} - \widehat{\boldsymbol{\sigma}}) : \dot{\boldsymbol{\varepsilon}}^p \, \mathrm{d}V = \int_V (\boldsymbol{\sigma} - \widehat{\boldsymbol{\sigma}}) : (\dot{\boldsymbol{\varepsilon}} - \dot{\boldsymbol{\varepsilon}}^E) \, \mathrm{d}V - \int_V (\boldsymbol{\sigma} - \widehat{\boldsymbol{\sigma}}) : \boldsymbol{E}^{-1} : \left(\dot{\boldsymbol{\sigma}} - \dot{\widehat{\boldsymbol{\sigma}}}\right) \mathrm{d}V \tag{104}$$

in which the first integral on the r.h.s., by the virtual work principle, is found to be either vanishing for quasi-static loads, or equivalent to

$$\int_V (\boldsymbol{\sigma} - \widehat{\boldsymbol{\sigma}}) : (\dot{\boldsymbol{\varepsilon}} - \dot{\boldsymbol{\varepsilon}}^E) \, \mathrm{d}V = -\int_V \rho \,(\ddot{\boldsymbol{u}} - \ddot{\boldsymbol{u}}^E) \cdot (\dot{\boldsymbol{u}} - \dot{\boldsymbol{u}}^E) \, \mathrm{d}V - \int_V \mu_d \,(\dot{\boldsymbol{u}} - \dot{\boldsymbol{u}}^E) \cdot (\dot{\boldsymbol{u}} - \dot{\boldsymbol{u}}^E) \, \mathrm{d}V \tag{105}$$

in the more general case of dynamic loads. Thus, substituting (105) in (104) and then in (102), integrating the latter in time over $(0, t_1)$ and with the aid of the functional $L(t)$ of (25), we have

$$\int_0^{t_1} \int_V \left[\frac{\bar{g}\gamma}{g} D + \boldsymbol{p} : \dot{\boldsymbol{\varepsilon}}^p \right] \mathrm{d}V \mathrm{d}t \leq \frac{\bar{g}}{\alpha} \left[L(0) - L(t_1) \right] \tag{106}$$

which holds for any $t_1 > 0$. Since $D \geq 0$ and $\bar{g}/g \geq 1$, hence $\bar{g}D/g \geq D$ in V, remembering (93) and (103) we have

$$\eta(t_1) \leq \int_0^{t_1} \int_V \left[\frac{\bar{g}\gamma}{g} D + \boldsymbol{p} : \dot{\boldsymbol{\varepsilon}}^p \right] \mathrm{d}V \mathrm{d}t; \tag{107}$$

also, since $L(t)$ is positive definite, the subtractive term on the r.h.s. of (106) can be disregarded and thus (107) can be enforced by writing:

$$\eta(t_1) \leq U \equiv \frac{1 + \alpha\bar{\gamma}}{\alpha} L(0), \tag{108}$$

where $L(0)$ is the functional

$$L(0) = \frac{1}{2} \int_V \widehat{\boldsymbol{\sigma}}^r : \boldsymbol{E}^{-1} : \widehat{\boldsymbol{\sigma}}^r \, \mathrm{d}V + \int_V \Omega\,(\widehat{\boldsymbol{\chi}}) \, \mathrm{d}V. \tag{109}$$

Equation (108) is a general bounding inequality in which the upper bound U depends only on α, $\widehat{\boldsymbol{\sigma}}^r$ and $\widehat{\boldsymbol{\chi}}$, whereas the actual deformation parameter η depends on the considered ALH and the time t_1. As a consequence, inequality (108) is actually equivalent to

$$\eta_{\max} \equiv \max_{\{\text{ALHs}\}} \max_{t_1 \geq 0} \eta\,[t_1, \text{ALH}] \leq U\,(\alpha, \widehat{\boldsymbol{\sigma}}^r, \widehat{\boldsymbol{\chi}}). \tag{110}$$

Inequality (110) is the mathematical expression of a general bounding principle for repeated loads, which can be phrased as in the following.

— ***Bounding Principle for Repeated Loads.*** For a given structure subjected to repeated (either quasi-static, or dynamic) loads, characterized by a (reduced) elastic stress envelope Γ, and for a given perturbation mode $(\gamma, \boldsymbol{p})$ and some perturbation multiplier $\alpha > 0$, if shakedown occurs in the structure obeying the yield and saturation functions modified by the considered perturbation (i.e. if (98) is satisfied) then the peak value of the actual plastic deformation measure, η, produced by any ALH turns out to be bounded from above, i.e. (110) holds with $\hat{\boldsymbol{\sigma}}^r$ and $\hat{\boldsymbol{\chi}}$ being the shakedown parameters.

Formally, the bound inequality (110) is the same for either quasi-static and dynamic loads; what changes with the two types of external actions is the size and shape of the elastic stress envelope Γ.

The above bounding principle is an extension to the present context of an analogous principle given by Polizzotto *et al.* (1991) for quasi-static loads and internal-variable plastic constitutive models without hardening saturation surface.

8.2 Assessment of Plastic Deformation Via Bounding Techniques

As stated before, the deformation measure η can take different meanings with different suitable choices of the perturbation mode. To show this point, the following cases are considered.

Bounds to the Intrinsic Dissipation. Considering a perturbation mode with $\boldsymbol{p} \equiv \boldsymbol{0}$, we have

$$\eta = \int_V \gamma(\boldsymbol{x}) \left(\int_0^{t_1} D(\dot{\boldsymbol{\varepsilon}}^p, \dot{\boldsymbol{\xi}})\, \mathrm{d}t \right) \mathrm{d}V = \int_V \gamma(\boldsymbol{x}) W(\boldsymbol{x}, t_1)\, \mathrm{d}V \tag{111}$$

where $W(\boldsymbol{x}, t_1)$ is the intrinsic dissipation energy per unit volume wasted as heat from $t = 0$ to $t = t_1$. Moreover, if $\gamma(\boldsymbol{x}) = 1$ for all $\boldsymbol{x} \in V$, (110) becomes

$$\eta = \int_V W(\boldsymbol{x}, t_1)\, \mathrm{d}V \tag{112}$$

which is the global intrinsic dissipation, whereas if $\gamma(\boldsymbol{x}) = 1$ for $\boldsymbol{x} \in \Delta V \subset V$ and $\gamma(\boldsymbol{x}) = 0$ for $\boldsymbol{x} \notin \Delta V$ (hence $\bar{g} = 1 + \alpha$), (111) becomes

$$\eta = \int_{\Delta V} W(\boldsymbol{x}, t_1)\, \mathrm{d}V \tag{113}$$

which is the intrinsic dissipation developed in some (small) region ΔV of V.

Bounds to the Plastic Strains. On setting $\gamma \equiv 0$ (hence $g = \bar{g} = 1$) and $\boldsymbol{p} \neq \boldsymbol{0}$ only within some subregion ΔV, where $\boldsymbol{p}$ possesses a single nonvanishing component, say $p_{hk} = 1/\Delta V$, then η becomes

$$\eta = \frac{1}{\Delta V} \int_{\Delta V} \varepsilon^p_{hk}(\boldsymbol{x}, t_1)\, \mathrm{d}V, \tag{114}$$

that is, it coincides with the mean value of the hk-component of the tensor $\boldsymbol{\varepsilon}^p$ within the small subregion ΔV at time t_1.

Bounds to Residual Displacements. The perturbation mode is one with $\gamma \equiv 0$ and $\boldsymbol{p}$ identified as the elastic stress response to some static force $\boldsymbol{F}^*$ smoothly distributed over a small subregion $\Delta V \in V$ (Dummy Load Method, Ponter (1972, 1975)). Then, denoting by $\boldsymbol{\varepsilon}^r(\boldsymbol{x}, t_1)$ and $\boldsymbol{\sigma}^r(\boldsymbol{x}, t_1)$ the residual strains and stresses associated with $\boldsymbol{\varepsilon}^p(\boldsymbol{x}, t_1)$, i.e. $\boldsymbol{\varepsilon}^r(\boldsymbol{x}, t_1) = \boldsymbol{E}^{-1} : \boldsymbol{\sigma}^r(\boldsymbol{x}, t_1) + \boldsymbol{\varepsilon}^p(\boldsymbol{x}, t_1)$, the quantity η becomes

$$\eta = \int_V \boldsymbol{p}(\boldsymbol{x}) : \boldsymbol{\varepsilon}^p(\boldsymbol{x}, t_1)\, \mathrm{d}V = \int_V \boldsymbol{p} : \boldsymbol{\varepsilon}^r\, \mathrm{d}V - \int_V \boldsymbol{p} : \boldsymbol{E}^{-1} : \boldsymbol{\sigma}^r\, \mathrm{d}V. \tag{115}$$

Since the last integral of (115) vanishes ($\boldsymbol{E}^{-1} : \boldsymbol{p}$ is compatible with zero displacements on S_u, $\boldsymbol{\sigma}^r$ is self-equilibrated), it follows from (115), applying the virtual work principle and assuming that $\boldsymbol{F}^*$ has a single nonvanishing component in ΔV, say $F_k^* = 1/\Delta V$:

$$\eta = \int_V \boldsymbol{F}^* \cdot \boldsymbol{u}^r\, \mathrm{d}V = \frac{1}{\Delta V} \int_{\Delta V} u_k^r\, \mathrm{d}V, \tag{116}$$

that is, η coincides with the mean value of the k-component of the residual displacement vector $\boldsymbol{u}^r$ within ΔV at time t_1.

Summarizing, we can state that the general bounding principle previously given can be particularized in several ones by suitably choosing the perturbation mode. In this way it is possible to derive bounding inequalities for overall or local plastic dissipation work, plastic strain components and residual displacement components at specific points of the body. This bounding principle is thus a unified extended formulation of a number of particular principles given separately by different authors, see e.g. Capurso *et al.*, (1978); Ponter (1972, 1975), Capurso (1979).

8.3 The Bound Computation Problems

The bounding principle of the previous section is a means for constructing upper bounds to the actual plastic deformations. To this purpose, once the perturbation mode $(\gamma, \boldsymbol{p})$ has been chosen in relation to the particular deformation measure of interest, we have to solve a search problem to find some $\alpha > 0$ and some shakedown parameters $\widehat{\boldsymbol{\sigma}}^r$, $\widehat{\boldsymbol{\chi}}$ such as to satisfy the inequalities:

$$f_\alpha^*(\boldsymbol{s} + \widehat{\boldsymbol{\sigma}}^r, \widehat{\boldsymbol{\chi}}) \leq 0, \qquad \phi_\alpha^*(\widehat{\boldsymbol{\chi}}) \leq 0 \qquad \forall \boldsymbol{s} \in \Gamma \quad \forall \boldsymbol{x} \in V \tag{117}$$

where f_α^* and ϕ_α^* are the relevant yield and saturation function modified by the chosen perturbation mode. f_α^* and ϕ_α^* are defined as

$$f_\alpha^* = \frac{1}{g} f\left((\boldsymbol{s} + \widehat{\boldsymbol{\sigma}}^r) g + \alpha \boldsymbol{p}, \widehat{\boldsymbol{\chi}} g\right), \qquad \phi_\alpha^* = \frac{1}{g} \phi\left(\widehat{\boldsymbol{\chi}} g\right) \tag{118}$$

with g given by (95). For $\alpha > 0$ fixed, the above search problem identifies with the shakedown problem for the given structure/load system, but with the yield and saturation functions f and ϕ replaced by f_α^* and ϕ_α^* of (118).

For a fixed $\alpha > 0$, any feasible solution $(\widehat{\boldsymbol{\sigma}}^r, \widehat{\boldsymbol{\chi}})$ to (117) guarantees that shakedown occurs in the modified structure subjected to the given loads, as well as that the actual plastic deformation, produced in the real structure under the given loads, has a measure η bounded from above by the upper bound $U(\alpha, \widehat{\boldsymbol{\sigma}}^r, \widehat{\boldsymbol{\chi}})$ of (108) and (109). It follows that, for every $\alpha > 0$, a bound U to η actually exists only within the shakedown limit of the modified structure. This limit can thus be assessed by solving the shakedown problem for the modified structure.

To this purpose, let $\boldsymbol{s}$ be replaced by $\beta\boldsymbol{s}$ in (117) and let problem (50) be formulated with f and ϕ replaced by f^*_α and ϕ^*_α, respectively. We thus can write:

$$\left.\begin{array}{l} \beta^*_a(\alpha) = \max\limits_{(\beta,\widehat{\boldsymbol{\sigma}}^r,\widehat{\boldsymbol{\chi}})} \beta \qquad \text{subject to :} \\[2ex] f^*_\alpha(\beta\boldsymbol{s}+\widehat{\boldsymbol{\sigma}}^r,\widehat{\boldsymbol{\chi}})\leq 0, \quad \phi^*_\alpha(\widehat{\boldsymbol{\chi}})\leq 0 \ \ \forall\boldsymbol{s}\in\Gamma \ \ \forall\boldsymbol{x}\in V \\[2ex] \operatorname{div}\widehat{\boldsymbol{\sigma}}^r = \boldsymbol{0} \ \text{ in } V, \qquad \widehat{\boldsymbol{\sigma}}^r\cdot\boldsymbol{n} = \boldsymbol{0} \ \text{ on } S_t. \end{array}\right\} \tag{119}$$

The optimality conditions of (119) read as (53)-(60), but we report them here for more clarity:

$$\boldsymbol{\varepsilon}^{p*} = \lambda^*\frac{\partial f^*_\alpha}{\partial\widehat{\boldsymbol{\sigma}}}, \qquad -\boldsymbol{\xi}^* = \lambda^*\frac{\partial f^*_\alpha}{\partial\widehat{\boldsymbol{\chi}}} + \mu^*\frac{\partial\phi^*_\alpha}{\partial\widehat{\boldsymbol{\chi}}} \qquad \forall\boldsymbol{s}\in\Gamma_x, \quad \forall\boldsymbol{x}\in V \tag{120}$$

$$f^*_\alpha(\widehat{\boldsymbol{\sigma}},\widehat{\boldsymbol{\chi}})\leq 0, \qquad \lambda^*\geq 0, \qquad \lambda^* f^*_\alpha(\widehat{\boldsymbol{\sigma}},\widehat{\boldsymbol{\chi}}) = 0 \qquad \forall\boldsymbol{s}\in\Gamma_x, \quad \forall\boldsymbol{x}\in V \tag{121}$$

$$\phi^*_\alpha(\widehat{\boldsymbol{\chi}})\leq 0, \qquad \mu^*\geq 0, \qquad \mu^*\phi^*_\alpha(\widehat{\boldsymbol{\chi}}) = 0 \qquad \forall\boldsymbol{s}\in\Gamma_x, \quad \forall\boldsymbol{x}\in V \tag{122}$$

$$\int_V\int_{\Gamma_x}\boldsymbol{s}:\boldsymbol{\varepsilon}^{p*}\,\mathrm{d}\Gamma(\boldsymbol{s})\,\mathrm{d}V = 1 \tag{123}$$

$$\Delta\boldsymbol{\varepsilon}^{p*} \equiv \int_{\Gamma_x}\boldsymbol{\varepsilon}^{p*}\,\mathrm{d}\Gamma(\boldsymbol{s}) = \nabla^s\boldsymbol{u}^* \ \text{ in } V, \qquad \boldsymbol{u}^* = \boldsymbol{0} \quad \text{on } S_u \tag{124}$$

$$\Delta\boldsymbol{\xi}^* \equiv \int_{\Gamma_x}\boldsymbol{\xi}^*\,\mathrm{d}\Gamma(\boldsymbol{s}) = \boldsymbol{0} \qquad \text{in } V \tag{125}$$

with $\widehat{\boldsymbol{\sigma}} = \boldsymbol{s}+\widehat{\boldsymbol{\sigma}}^r$, besides the equilibrium equations for the self-stress $\widehat{\boldsymbol{\sigma}}^r$. The meaning of the above equation set is as specified in Section 5 for (53)-(60). After obtainment of the solution to the modified shakedown problem (119), the related shakedown parameters $\widehat{\boldsymbol{\sigma}}^r$, $\widehat{\boldsymbol{\chi}}$ together with the selected α, can be used for the computation of $U = U(\alpha, \widehat{\boldsymbol{\sigma}}^r, \widehat{\boldsymbol{\chi}})$, which is an upper bound to the actual plastic deformation produced by loads amplified by $\beta \leq \beta^*_a(\alpha)$. Since in general the perturbation weakens the structure (but this is not always true), and thus $\beta^*_a(\alpha) < \beta^*_a(0) \equiv \beta_a$, the derived bound holds for any $\beta < \beta_a$ —but it diverges as $\beta \to \beta_a$.

We can thus state that, with the above procedure, the construction of an upper bound to the actual plastic deformation requires the solution of the shakedown safety factor problem for the modified structure. Because α is taken fixes during the latter computation, the upper bound $U(\alpha, \widehat{\sigma}^r, \widehat{\chi})$ so obtained may be minimizated by a suitable choice of α.

It may be of interest to evaluate a load factor β_L such that for any $\beta \leq \beta_L$ the actual deformation measure η satisfies $\eta \leq \eta_L$, where η_L is a fixed upper limit to η. This problem can be solved by a procedure (Polizzotto, 1984b; Panzeca *et al.*, 1990) as follows. First we solve problem (119) for varying values of α, so to obtain the functions $\beta^*_a(\alpha)$, $\widehat{\sigma}^r(\boldsymbol{x}, \alpha)$, $\widehat{\chi}(\boldsymbol{x}, \alpha)$ and thus $U = U(\alpha, \widehat{\sigma}^r, \widehat{\chi}) = \bar{U}(\alpha)$. Then, the equality $\bar{U}(\alpha) = \eta_L$ provides the value $\alpha = \alpha_L$, by which we have $\beta_L = \beta^*_a(\alpha_L)$. In this way, it is guaranteed that, for any $\beta \leq \beta_L = \beta^*_a(\alpha_L)$, the inequality $\eta_{\max} \leq \bar{U}(\alpha_L) = \eta_L$ is certainly satisfied.

9 General Evolutive-Type Bounding Principle

The bounding principle of Section 8 is not evolutive in nature and leads to upper bounds, U, expressed in terms of time-independent quantities, but these upper bounds become too large as the load approaches the shakedown limit load. The bounding principle given in this section is evolutive in nature, that is, it requires an evolutive elastic-plastic analysis all along a given load history (LH), but in some way ***simpler*** than the incremental one required for a complete knowledge of the actual elastic-plastic response to the same LH. It can be applied for any LH, either below and above the shakedown limit, and leads to upper and to lower bounds U that can be made as stringent as desired to the actual deformation, obviously at increasing computational costs. One way to accomplish simplified evolutive analyses is offered by the so-called deformation-theory, or holonomic, plasticity. The restrictive hypothesis of linear hardening is here introduced, i.e. it is $\Psi_{\text{in}} = (1/2)\boldsymbol{\xi} \cdot \boldsymbol{B} \cdot \boldsymbol{\xi}$, hence $\chi = \boldsymbol{B} \cdot \boldsymbol{\xi}$ with $\boldsymbol{B}$ positive definite.

9.1 Derivation of the Principle

Let the structure considered in Section 3 be subjected to a given LH with initial conditions $\boldsymbol{u} = \boldsymbol{u}_0$, $\dot{\boldsymbol{u}} = \dot{\boldsymbol{u}}_0$ in V at $t = 0$. Assuming a perturbation mode $(\gamma, \boldsymbol{p})$ and a perturbation multiplier $\alpha > 0$, the modified, or perturbed, yield and saturation functions f^*_α, ϕ^*_α of (118) are defined, i.e.

$$f^*_\alpha(\sigma, \chi) \equiv \frac{1}{g} f(g\sigma + \alpha\boldsymbol{p}, g\chi), \qquad \phi^*_\alpha(\chi) \equiv \frac{1}{g}\phi(g\chi). \tag{126}$$

where $g = 1 + \alpha\gamma$. We consider two deformation processes: the *real* deformation process, described by unstarred variables, which is governed by the material yield and saturation functions $f(\sigma, \chi)$ and $\phi(\chi)$; the *fictitious* deformation process, described by starred variables, which is governed by the modified yield and saturation functions $f^*_\alpha(\sigma^*, \chi^*)$ and $\phi^*_\alpha(\chi^*)$. The real process complies with constitutive equations established in Section 2. The fictitious process complies with the same constitutive equations, but with the complementarity conditions relaxed, i.e.

$$f^*_\alpha(\sigma^*, \chi^*) \leq 0, \qquad \dot{\lambda}^* \geq 0 \tag{127}$$

$$\phi^*_\alpha(\boldsymbol{\chi}^*) \leq 0, \qquad \dot{\mu}^* \geq 0 \tag{128}$$

$$\dot{\boldsymbol{\varepsilon}}^{p*} = \dot{\lambda}^* \frac{\partial f^*_\alpha}{\partial \boldsymbol{\sigma}^*}, \qquad -\dot{\boldsymbol{\xi}}^* = \dot{\lambda}^* \frac{\partial f^*_\alpha}{\partial \boldsymbol{\chi}^*} + \dot{\mu}^* \frac{\partial \phi^*_\alpha}{\partial \boldsymbol{\chi}^*}, \tag{129}$$

all the other equations remaining unaltered.

For a given LH and initial conditions, whereas the actual deformation process is (in general) unique, the fictitious one is not unique. Let $\boldsymbol{u}^*, \boldsymbol{\sigma}^*, \boldsymbol{\varepsilon}^{p*}, \boldsymbol{\chi}^*, \ldots$ denote a fictitious response to the given LH with related initial conditions, say $\boldsymbol{u}^* = \boldsymbol{u}^*_0$, $\dot{\boldsymbol{u}}^* = \dot{\boldsymbol{u}}^*_0$ in V at $t = 0$. Then, because the yield and saturation conditions in (127) and (128) imply that

$$f\,(g\boldsymbol{\sigma}^* + \alpha \boldsymbol{p}, g\boldsymbol{\chi}^*) \leq 0, \qquad \phi\,(g\boldsymbol{\chi}^*) \leq 0 \qquad \forall t \geq 0 \text{ in } V, \tag{130}$$

the Druckerian inequality (10) can be applied considering the actual deformation process in association with the pair $(\boldsymbol{\sigma}^*, \boldsymbol{\chi}^*)$ as plastically allowable. So we can write the inequality:

$$(\boldsymbol{\sigma} - g\boldsymbol{\sigma}^* - \alpha \boldsymbol{p}) : \dot{\boldsymbol{\varepsilon}}^p - (\boldsymbol{\chi} - g\boldsymbol{\chi}^*) \cdot \dot{\boldsymbol{\xi}} \geq 0 \qquad \forall t \geq 0 \text{ in } V, \tag{131}$$

which is similar to (100) and thus (102) and (108) can be used here to write

$$\alpha\,\dot{\eta}(t) \leq \bar{g}\left[(\boldsymbol{\sigma} - \boldsymbol{\sigma}^*) : \dot{\boldsymbol{\varepsilon}}^p - (\boldsymbol{\chi} - \boldsymbol{\chi}^*) \cdot \dot{\boldsymbol{\xi}}\right] \qquad \forall t \geq 0 \text{ in } V. \tag{132}$$

Because of the convexity of f and ϕ, we can write the following:

$$f(\boldsymbol{\sigma}, \boldsymbol{\chi}) - f(\widehat{\boldsymbol{\sigma}}^*, \widehat{\boldsymbol{\chi}}^*) \geq (\boldsymbol{\sigma} - \widehat{\boldsymbol{\sigma}}^*) : \frac{\partial f}{\partial \widehat{\boldsymbol{\sigma}}^*} + (\boldsymbol{\chi} - \widehat{\boldsymbol{\chi}}^*) \cdot \frac{\partial f}{\partial \widehat{\boldsymbol{\chi}}^*} \tag{133}$$

$$\phi(\boldsymbol{\chi}) - \phi(\widehat{\boldsymbol{\chi}}^*) \geq (\boldsymbol{\chi} - \widehat{\boldsymbol{\chi}}^*) \cdot \frac{\partial \phi}{\partial \widehat{\boldsymbol{\chi}}^*} \tag{134}$$

which hold for any $\boldsymbol{\sigma}$, $\boldsymbol{\chi}$ and $\widehat{\boldsymbol{\sigma}}^* = g\boldsymbol{\sigma}^* + \alpha \boldsymbol{p}$, $\widehat{\boldsymbol{\chi}}^* = g\boldsymbol{\chi}^*$. Since

$$\frac{\partial f}{\partial \widehat{\boldsymbol{\sigma}}^*} = \frac{\partial f^*_\alpha}{\partial \boldsymbol{\sigma}^*}, \qquad \frac{\partial \phi}{\partial \widehat{\boldsymbol{\chi}}^*} = \frac{\partial \phi^*_\alpha}{\partial \boldsymbol{\chi}^*}, \tag{135}$$

the inequalities (133) and (134), after multiplication by $\dot{\lambda}^* > 0$ and $\dot{\mu}^* > 0$, respectively, and then summing with each other, and also noting that $\dot{\lambda}^* f(\boldsymbol{\sigma}, \boldsymbol{\chi}) \leq 0$, $\dot{\mu}^* \phi(\boldsymbol{\chi}) \leq 0$, turn out to be equivalent to:

$$(\widehat{\boldsymbol{\sigma}}^* - \boldsymbol{\sigma}) : \dot{\boldsymbol{\varepsilon}}^{p*} - (\widehat{\boldsymbol{\chi}}^* - \boldsymbol{\chi}) \cdot \dot{\boldsymbol{\xi}}^* \geq \dot{\lambda}^* f(\widehat{\boldsymbol{\sigma}}^*, \widehat{\boldsymbol{\chi}}^*) + \dot{\mu}^* \phi(\widehat{\boldsymbol{\chi}}^*). \tag{136}$$

The latter inequality constitutes a generalization of the Druckerian inequality (10), see also Polizzotto (1982b, 1984b, 1985). Inequality (136) can then be transformed as

$$\alpha\,\bar{g}\,\gamma\,D^* + \alpha \boldsymbol{p} : \dot{\boldsymbol{\varepsilon}}^{p*} - \bar{g}\left[(\boldsymbol{\sigma} - \boldsymbol{\sigma}^*) : \dot{\boldsymbol{\varepsilon}}^{p*} - (\boldsymbol{\chi} - \boldsymbol{\chi}^*) \cdot \dot{\boldsymbol{\xi}}^*\right] \geq \bar{g}\,g[\dot{\lambda}^* f^*_\alpha + \dot{\mu}^* \phi^*_\alpha] \tag{137}$$

where $D^* = \boldsymbol{\sigma}^* : \dot{\boldsymbol{\varepsilon}}^{p*} - \boldsymbol{\chi}^* \cdot \dot{\boldsymbol{\xi}}^*$. On setting

$$\dot{\eta}^* = \gamma D^* + \boldsymbol{p} : \dot{\boldsymbol{\varepsilon}}^{p*} \tag{138}$$

and summing (132) and (137) with each other gives

$$\dot{\eta}(t) \leq \dot{\eta}^*(t) + \frac{\bar{g}}{\alpha}\left[(\boldsymbol{\sigma}-\boldsymbol{\sigma}^*) : (\dot{\boldsymbol{\varepsilon}}^p - \dot{\boldsymbol{\varepsilon}}^{p*}) - (\boldsymbol{\chi}-\boldsymbol{\chi}^*)\cdot(\dot{\boldsymbol{\xi}} - \dot{\boldsymbol{\xi}}^*) - g\,(\dot{\lambda}^* f_\alpha^* + \dot{\mu}^* \phi_\alpha^*)\right]. \tag{139}$$

Then, since

$$\dot{\boldsymbol{\varepsilon}}^p - \dot{\boldsymbol{\varepsilon}}^{p*} = \dot{\boldsymbol{\varepsilon}} - \dot{\boldsymbol{\varepsilon}}^* - \boldsymbol{E}^{-1} : (\dot{\boldsymbol{\sigma}} - \dot{\boldsymbol{\sigma}}^*)\,, \tag{140}$$

we can write

$$\int_V (\boldsymbol{\sigma}-\boldsymbol{\sigma}^*) : (\dot{\boldsymbol{\varepsilon}}^p - \dot{\boldsymbol{\varepsilon}}^{p*})\,\mathrm{d}V = \int_V (\boldsymbol{\sigma}-\boldsymbol{\sigma}^*) : (\dot{\boldsymbol{\varepsilon}} - \dot{\boldsymbol{\varepsilon}}^*)\,\mathrm{d}V - \int_V (\boldsymbol{\sigma}-\boldsymbol{\sigma}^*) : \boldsymbol{E}^{-1} : (\dot{\boldsymbol{\sigma}} - \dot{\boldsymbol{\sigma}}^*)\,\mathrm{d}V \tag{141}$$

where, by the virtual work principle,

$$\int_V (\boldsymbol{\sigma}-\boldsymbol{\sigma}^*) : (\dot{\boldsymbol{\varepsilon}}-\dot{\boldsymbol{\varepsilon}}^*)\,\mathrm{d}V = -\int_V \rho\,(\ddot{\boldsymbol{u}}-\ddot{\boldsymbol{u}}^*)\cdot(\dot{\boldsymbol{u}}-\dot{\boldsymbol{u}}^*)\,\mathrm{d}V - \int_V \mu_d\,(\dot{\boldsymbol{u}}-\dot{\boldsymbol{u}}^*)\cdot(\dot{\boldsymbol{u}}-\dot{\boldsymbol{u}}^*)\,\mathrm{d}V. \tag{142}$$

Thus, introducing the positive definite functional (similar to L in (25)):

$$\begin{aligned} L^*(t) = &\int_V \frac{1}{2}\rho(\dot{\boldsymbol{u}}-\dot{\boldsymbol{u}}^*)\cdot(\dot{\boldsymbol{u}}-\dot{\boldsymbol{u}}^*)\mathrm{d}V + \int_0^t\int_V \mu_d(\dot{\boldsymbol{u}}-\dot{\boldsymbol{u}}^*)\cdot(\dot{\boldsymbol{u}}-\dot{\boldsymbol{u}}^*)\mathrm{d}V\mathrm{d}\tau \\ &+ \int_V \frac{1}{2}(\boldsymbol{\sigma}-\boldsymbol{\sigma}^*) : \boldsymbol{E}^{-1} : (\boldsymbol{\sigma}-\boldsymbol{\sigma}^*)\mathrm{d}V + \int_V \frac{1}{2}(\boldsymbol{\chi}-\boldsymbol{\chi}^*)\cdot\boldsymbol{B}^{-1}\cdot(\boldsymbol{\chi}-\boldsymbol{\chi}^*)\mathrm{d}V, \end{aligned} \tag{143}$$

inequality (139) (which holds for all $t \geq 0$ and for all $\boldsymbol{x} \in V$), after integration over V and $(0, t_1)$ and substitution of (142) in (141) and then in (139), yields the following bounding inequality

$$\eta(t_1) \leq \eta^*(t_1) + \frac{\bar{g}}{\alpha}\left\{L^*(0) - L^*(t_1) - \int_0^{t_1}\int_V g\left[\dot{\lambda}^* f_\alpha^*(\boldsymbol{\sigma}^*, \boldsymbol{\chi}^*) + \dot{\mu}^* \phi_\alpha^*(\boldsymbol{\chi}^*)\right]\mathrm{d}V\mathrm{d}t\right\}. \tag{144}$$

Dropping the nonnegative subtractive term $L^*(t_1)$, so enforcing the inequality and posing

$$U^* = \frac{\bar{g}}{\alpha}\left\{L^*(0) - \int_0^{t_1}\int_V g\left[\dot{\lambda}^* f_\alpha^*(\boldsymbol{\sigma}^*, \boldsymbol{\chi}^*) + \dot{\mu}^* \phi_\alpha^*(\boldsymbol{\chi}^*)\right]\mathrm{d}V\mathrm{d}t\right\}. \tag{145}$$

we finally obtain the following bounding inequality

$$\eta(t_1) \leq U \equiv \eta^*(t_1) + U^* \tag{146}$$

which holds for any fictitious deformation process associated with the given LH and initial conditions. $L^*(0)$ of (145) is expressed as

$$L^*(0) = \int_V \frac{1}{2}\rho(\dot{\boldsymbol{u}}_0 - \dot{\boldsymbol{u}}_0^*)\cdot(\dot{\boldsymbol{u}}_0 - \dot{\boldsymbol{u}}_0^*)\mathrm{d}V + \int_V \frac{1}{2}(\boldsymbol{\sigma}_0 - \boldsymbol{\sigma}_0^*):\boldsymbol{E}^{-1}:(\boldsymbol{\sigma}_0 - \boldsymbol{\sigma}_0^*)\mathrm{d}V$$
$$+ \int_V \frac{1}{2}(\boldsymbol{\chi}_0 - \boldsymbol{\chi}_0^*)\cdot\boldsymbol{B}^{-1}\cdot(\boldsymbol{\chi}_0 - \boldsymbol{\chi}_0^*)\mathrm{d}V \quad (147)$$

and thus depends only on the known initial conditions of the actual and fictitious deformation processes. As the upper bound U depends only on the fictitious deformation process, (146) represents a general bounding principle of evolutive type which can be phrased as follows:

Evolutive Bounding Principle. For a given elastic plastic structure with linear hardening and saturation surface and for a given perturbation mode $(\gamma, \boldsymbol{p})$ with a perturbation multiplier $\alpha > 0$, the related measure of the plastic deformation induced by a given LH in a specified time interval $(0, t_1)$ can be bounded from above in terms of a fictitious deformation process, i.e.

$$\eta\,[\gamma,\, \boldsymbol{p},\, t_1] \leq U\;[\text{ fictitious deformation process within } (0, t_1)]\,. \quad (148)$$

Note that, with $g = 1 - \alpha\gamma$, $\gamma \geq 0$ and $-\boldsymbol{p}$ in place of $\boldsymbol{p}$, the bound (146) becomes

$$\eta(t_1) \geq \eta^*(t_1) - U^* \quad (149)$$

which is a lower bound to η, provided that $1 - \alpha\bar{\gamma} > 0$.

The functional U in (146) exhibits a convergence property. Intuitively, we can observe that the set of the fictitious responses includes the actual response, which corresponds to $\alpha = 0$. Thus, any computational strategy leading to the actual response with $\alpha = 0$ produces the minimum value of U which identifies with η. But, the more stringent U to η, the more costly the computational effort required. At the limit, for $U = \eta$, the computational cost is that of an incremental, step-by-step, analysis.

9.2 Particularizations

The bounding principle given in section (9.1) can be specialized in various ways, first by selecting the perturbation mode $(\gamma, \boldsymbol{p})$ in relation to the particular deformation measure of interest, secondly by selecting a particular type of fictitious response. A possibility in the latter sense consists in subdividing the time interval $(0, t_1)$ in relatively large steps (i.e. larger than such intervals are taken for a step-by-step analysis) and in performing holonomic plasticity analysis for everyone of them. Inequality (146) then provides a notable bounding comparison between the two types of analysis.

Other possibilities are realized by making the fictitious responses to be elastic, i.e. taking $\dot{\lambda}^* = \dot{\mu}^* = 0$ for the entire fictitious deformation process, and fixing the initial conditions in the form

$$\boldsymbol{u}_0^* = \boldsymbol{u}_0 + \hat{\boldsymbol{u}}_0, \quad \dot{\boldsymbol{u}}_0^* = \dot{\boldsymbol{u}}_0 + \hat{\dot{\boldsymbol{u}}}_0 \qquad \text{in } V$$
$$\boldsymbol{\sigma}_0^* = \boldsymbol{\sigma}_0 + \hat{\boldsymbol{\sigma}}_0^F + \hat{\boldsymbol{\sigma}}^r, \quad \boldsymbol{\chi}_0^* = \boldsymbol{\chi}_0 + \hat{\boldsymbol{\chi}} \qquad \text{in } V \quad (150)$$

where $\widehat{\sigma}^F$ is the free-motion stress generated by the initial conditions $\widehat{u}_0$, $\dot{\widehat{u}}_0$ at $t = 0$ and $\widehat{\sigma}^r$, $\widehat{\chi}$ are time-independent stresses and static internal variables associated with some initial plastic strains, ϵ_0^{p*} in V. As, then, $\sigma^* = \sigma^E + \widehat{\sigma}^F + \widehat{\sigma}^r$ and $\chi^* = \widehat{\chi}$ with σ^E being the elastic stress response to the LH, (127) and (128) reduce to

$$f_\alpha^* = \frac{1}{g} f\left((\sigma^E + \widehat{\sigma}^F + \widehat{\sigma}^r)g + \alpha p, \widehat{\chi} g\right) \leq 0, \quad \phi_\alpha^* = \frac{1}{g}\phi(\widehat{\chi} g) \leq 0, \qquad \forall t \geq 0 \text{ in } V \tag{151}$$

whereas (129) drops and the upper bound U reduces to

$$U = \frac{\bar{g}}{\alpha} L^*(0) = \frac{\bar{g}}{\alpha}\left[\frac{1}{2}\int_V \rho \dot{\widehat{u}}_0 \cdot \dot{\widehat{u}}_0 \, \mathrm{d}V + L(0)\right] \tag{152}$$

where $L(0)$ is given by (109). We have so derived a non-evolutive type of bounding principle holding for a specified LH, either dynamic, or quasi-static (in which case, obviously, $\dot{\widehat{u}}_0 \equiv 0$ and $\widehat{\sigma}^F \equiv 0$). It can be phrased as follows.

Non-Evolutive Bounding Principle for a Specified LH. For a given structure subjected to a given LH and initial conditions and for a specified perturbation mode, (γ, p), if there exists some $\alpha > 0$ and some shakedown parameters $(\widehat{\sigma}^F, \widehat{\sigma}^r, \widehat{\chi})$ such as to satisfy (151), then the actual plastic deformation measure, η, produced by the given LH in a specified interval $(0, t_1)$ can be bounded from above by the upper bound U of (152).

Note that $L(0)$ in (152) has been taken as in (109), where a nonlinear hardening was considered, whereas in this section linear hardening has been considered. In effect, (152) holds with nonlinear hardening, as it would be possible to show by direct proof, but this point is given for granted here for brevity.

Another specialization of the evolutive bounding principle is obtained in the case in which the LH is an ALH with respect to some load domain of repeated excitations. In this case, as in the previous one, we consider fictitious elastic responses (i.e. $\dot{\lambda}^* = \dot{\mu}^* = 0$ identically) and moreover

$$u_0^* = u_0, \quad \dot{u}_0^* = \dot{u}_0 \qquad \text{in } V$$
$$\sigma_0^* = \sigma_0 + \widehat{\sigma}^r, \quad \chi_0^* = \widehat{\chi} \qquad \text{in } V. \tag{153}$$

As, correspondingly, $\sigma^* = \sigma^E + \widehat{\sigma}^r$, $\chi^* = \widehat{\chi}$ with $\sigma^E = s$ allowed to range within the (reduced) elastic stress response envelope Γ, (127) and (128) reduce to

$$f_\alpha^* = \frac{1}{g} f((s + \widehat{\sigma}^r)g + \alpha p, \widehat{\chi}) \leq 0, \quad \phi_\alpha^* = \frac{1}{g}\phi(\widehat{\chi} g) \leq 0, \qquad \forall s \in \Gamma_x, \; \forall x \in V \tag{154}$$

whereas (129) drops and the upper bound U becomes

$$U = \frac{\bar{g}}{\alpha} L(0). \tag{155}$$

The same remark on $L(0)$ made before can be repeated here. Since (154), multiplied by $g \geq 1$, coincides with (97), and (155) with (108), we conclude stating that the bounding principle for repeated loads can be derived as a specialization of the evolutive bounding principle.

10 Discrete Models

For computational purposes, the continuum models must be discretized, either by the finite element method (FEM), or by the boundary element method (BEM). This implies, for instance in the more familiar case of the FEM, that the elastic analysis can be addressed by solving a differential equation as

$$\boldsymbol{M}\ddot{\boldsymbol{u}} + \boldsymbol{V}\dot{\boldsymbol{u}} + \boldsymbol{K}\boldsymbol{u} = \boldsymbol{F}(t), \qquad t \geq 0 \tag{156}$$

with the related initial conditions in the dynamic case, or the equation

$$\boldsymbol{K}\boldsymbol{u} = \boldsymbol{F}(t), \qquad t \geq 0 \tag{157}$$

in the quasi-static case. In the above equations, $\boldsymbol{F}(t)$ is a given nodal load history, $\boldsymbol{K}$ is the system's stiffness matrix, $\boldsymbol{M}$ the mass matrix and $\boldsymbol{V}$ the damping matrix. Denoting by $\boldsymbol{u}^E = \boldsymbol{u}^E(t)$ the elastic displacement response of the discrete system (solution of either (156), or (157), the nodal displacements $\boldsymbol{q}^E_{(e)}$ of the element (e) are related to $\boldsymbol{u}^E$ by

$$\boldsymbol{q}^E_{(e)} = \boldsymbol{C}_{(e)}\boldsymbol{u}^E \tag{158}$$

where $\boldsymbol{C}_{(e)}$ is the compatibility matrix, whereas the nodal forces $\boldsymbol{Q}^E_{(e)}$ are expressed as

$$\boldsymbol{Q}^E_{(e)} = \boldsymbol{K}_{(e)}\boldsymbol{q}^E_{(e)} \tag{159}$$

where $\boldsymbol{K}_{(e)}$ is the element stiffness matrix. From $\boldsymbol{Q}^E_{(e)}$ the local stress $\boldsymbol{\sigma}^E$ can be obtained by writing

$$\boldsymbol{\sigma}^E = \boldsymbol{S}_{(e)}\boldsymbol{Q}^E_{(e)} \tag{160}$$

where $\boldsymbol{S}_{(e)}$ is a suitable stress-shape function, or stress-influence matrix, of the element. From the equality

$$\boldsymbol{Q}^{E^T}_{(e)}\boldsymbol{q}^E_{(e)} = \int_{V_{(e)}} \boldsymbol{\sigma}^{E^T}\boldsymbol{\varepsilon}^E \,\mathrm{d}V \tag{161}$$

follows that

$$\boldsymbol{q}^E_{(e)} = \int_{V_{(e)}} \boldsymbol{S}^T_{(e)}\boldsymbol{\varepsilon}^E \,\mathrm{d}V \tag{162}$$

where $(\cdot)^T$ means transpose of $(\cdot)$. In general, the local stress $\boldsymbol{\sigma}^E$ is computed —for every element— at a discrete number (n_g) of points (usually the Gauss points), where the plasticity conditions are solely enforced.

Within shakedown theory, the use of piecewise linearly discretized yield functions is of common practice. Though this may lead to drastic changes in the shape of the yield surface, there are however undoubtful advantages with using piecewise linear yield functions (Maier, 1973). Considered that for a wide class of structures the yield functions are piecewise linear in their own nature, as for instance for frame and truss models, we consider here a material with yield and saturation functions in the form

$$\boldsymbol{f} = \boldsymbol{N}^T\boldsymbol{\sigma} - \boldsymbol{G}^T\boldsymbol{\chi} - \boldsymbol{k} \leq \boldsymbol{0}, \qquad \boldsymbol{\phi} = \boldsymbol{J}^T\boldsymbol{\chi} - \boldsymbol{h} \leq \boldsymbol{0} \tag{163}$$

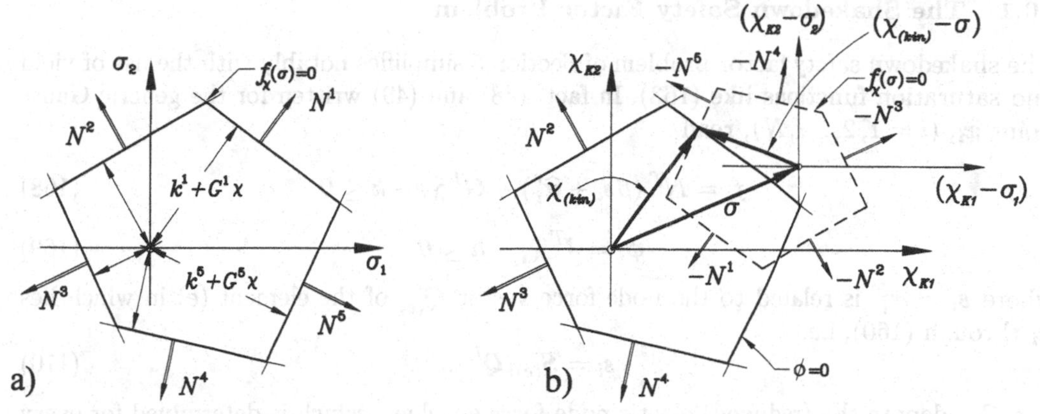

Figure 11. Piecewise linear yield and saturation surfaces (in two dimensions) for kinematic and isotropic hardening: a) yield surface in the stress space; b) yield and saturation surfaces in the back-stress χ_K-space.

where $\boldsymbol{f} = \{f_j\}$, $\boldsymbol{\phi} = \{\phi_j\}$ collect the yield and saturation potentials of the faces of the polyhedral surfaces in the $\boldsymbol{\sigma}$-space and the $\boldsymbol{\chi}$-space, $\boldsymbol{N}$ is a matrix that collects the unit external normals to the latter faces in the $\boldsymbol{\sigma}$-space, $\boldsymbol{k} > \boldsymbol{0}$ is a vector that collects their distances from the origin in the virgin state, Fig. 11(a). The matrix $\boldsymbol{J}$ and the vector $\boldsymbol{h} > \boldsymbol{0}$ are to be interpreted analogously in relation to the saturation polyhedron, Fig. 11(b). The matrix $\boldsymbol{G}$ is representative of the external normals to the faces of the yield polyhedron in the $\boldsymbol{\chi}$-space. For (linear) kinematic and isotropic hardening, $\boldsymbol{G}$ and $\boldsymbol{J}$ can be taken in the form

$$\boldsymbol{G} = \begin{bmatrix} \boldsymbol{N} \\ \boldsymbol{k}^T \end{bmatrix}, \qquad \boldsymbol{J} = \begin{bmatrix} \boldsymbol{N} & \boldsymbol{0} \\ \boldsymbol{k}^T & k_0 \end{bmatrix} \tag{164}$$

where $k_0 > 0$ is a scalar constant whereas correspondingly $\boldsymbol{\chi}$ is decomposed as

$$\boldsymbol{\chi} = \begin{Bmatrix} \chi_K \\ \chi_I \end{Bmatrix}, \tag{165}$$

where χ_K is the back stress tensor and χ_I is the drag stress, related to the isotropic hardening. Often the yield functions are directly expressed in terms of node-forces $\boldsymbol{Q}_{(e)}$, but this case is here treated as a special case obtained by taking $\boldsymbol{S}_{(e)} =$ Identity.

With the yield and saturation functions as in (163), the plastic mechanism $(\dot{\boldsymbol{\varepsilon}}^p, -\dot{\boldsymbol{\xi}})$ of (5) takes on the form

$$\dot{\boldsymbol{\varepsilon}}^p = \boldsymbol{N}\dot{\boldsymbol{\lambda}}, \qquad \dot{\boldsymbol{\xi}} = \boldsymbol{G}\dot{\boldsymbol{\lambda}} - \boldsymbol{J}\dot{\boldsymbol{\mu}} \tag{166}$$

where $\dot{\boldsymbol{\lambda}}$ and $\dot{\boldsymbol{\mu}}$ are yield and saturation coefficient vectors, whereas the intrinsic dissipation D of (6) reads

$$D = \boldsymbol{k}^T\dot{\boldsymbol{\lambda}} + \boldsymbol{h}^T\dot{\boldsymbol{\mu}}. \tag{167}$$

10.1 The Shakedown Safety Factor Problem

The shakedown safety factor problem of Section 5 simplifies notably with the use of yield and saturation functions like (163). In fact, (48) and (49) written for the generic Gauss point $\boldsymbol{x}_i$, $(i = 1, 2, \ldots, N)$, read

$$\boldsymbol{f}_i = \boldsymbol{N}^T (\beta \boldsymbol{s}_i + \widehat{\boldsymbol{\sigma}}_i^r) - \boldsymbol{G}^T \widehat{\boldsymbol{\chi}}_i - \boldsymbol{k} \leq \boldsymbol{0} \tag{168}$$

$$\boldsymbol{\phi}_i = \boldsymbol{J}^T \widehat{\boldsymbol{\chi}}_i - \boldsymbol{h} \leq \boldsymbol{0} \tag{169}$$

where $\boldsymbol{s}_i = \boldsymbol{\sigma}_i^E$ is related to the node force vector $\boldsymbol{Q}_{(e)}^E$ of the element (e) in which lies $\boldsymbol{x}_i$ through (160), i.e.

$$\boldsymbol{s}_i = \boldsymbol{S}_{(e)i} \boldsymbol{Q}_{(e)}^E. \tag{170}$$

Let $\Gamma_{(e)}$ denote the (reduced) elastic node-force envelope, which is determined for every element by considering the elastic responses $\boldsymbol{Q}_{(e)}^E(t)$ to all ALHs $\boldsymbol{P}(t)$. Substituting (170) in (168), the latter becomes

$$\boldsymbol{f}_i = \beta \boldsymbol{N}^T \boldsymbol{S}_{(e)i} \boldsymbol{Q}_{(e)}^E + \boldsymbol{N}^T \widehat{\boldsymbol{\sigma}}_i^r - \boldsymbol{G}^T \widehat{\boldsymbol{\chi}}_i - \boldsymbol{k} \leq \boldsymbol{0} \tag{171}$$

where $\boldsymbol{Q}_{(e)}^E$ must range within the envelope $\Gamma_{(e)}$ of the element (e) including $\boldsymbol{x}_i$. Thus, denoting by $N_g = n_g N_e$ the total number of Gauss points in all N_e elements, let us set:

$$\boldsymbol{R}_i = \max_{\boldsymbol{Q}_{(e)}^E \in \Gamma_{(e)i}} \boldsymbol{N}^T \boldsymbol{S}_{(e)i} \boldsymbol{Q}_{(e)}^E \qquad \forall i \in I(N_g) \tag{172}$$

where $\Gamma_{(e)i}$ is related to the element (e) in which $\boldsymbol{x}_i$ lies and the maximization operation is performed component-wise; then (171) turns out to be equivalent to

$$\boldsymbol{f}_i = \beta \boldsymbol{R}_i + \boldsymbol{N}^T \widehat{\boldsymbol{\sigma}}_i^r - \boldsymbol{G}^T \widehat{\boldsymbol{\chi}}_i - \boldsymbol{k} \leq \boldsymbol{0}. \tag{173}$$

The vectors $\boldsymbol{R}_i$, $(i = 1, 2, \ldots, N_g)$, are referred to as the *elastic node-force envelope projections* (on the yielding modes) at the Gauss points.

With this definition in mind, problem (50) transforms into the following

$$\left.\begin{array}{l} \beta_a = \max\limits_{(\beta, \widehat{\boldsymbol{\chi}}_i, \widehat{\boldsymbol{Q}}_{(e)}^r)} \beta \qquad \text{subject to :} \\[2ex] \beta \boldsymbol{R}_i + \boldsymbol{N}^T \boldsymbol{S}_{(e)i} \widehat{\boldsymbol{Q}}_{(e)}^r - \boldsymbol{G}^T \widehat{\boldsymbol{\chi}}_i - \boldsymbol{k} \leq \boldsymbol{0}, \qquad \forall i \in I(N_g) \\[2ex] \boldsymbol{J}^T \widehat{\boldsymbol{\chi}}_i - \boldsymbol{h} \leq \boldsymbol{0} \qquad \forall i \in I(N_g) \\[2ex] \boldsymbol{C}^T \left\{ \widehat{\boldsymbol{Q}}_{(e)}^r \right\} = \boldsymbol{0} \end{array}\right\} \tag{174}$$

where we have set $\widehat{\boldsymbol{\sigma}}_i^r = \boldsymbol{S}_{(e)i} \widehat{\boldsymbol{Q}}_{(e)}^r$. $\left\{ \widehat{\boldsymbol{Q}}_{(e)}^r \right\}$ collects all node force vectors $\widehat{\boldsymbol{Q}}_{(e)}^r$, which by the last constraint equation is required to constitute a self-equilibrated stress state. Problem (174) is a linear programming (LP) problem, which can thus easily solved using routine computer programmes (Simplex Method). However, (174) is an approximation of the actual continuum problem, first because of the approximations inherent to the discretization procedure, second because the equilibrium equations are in general not satisfied inside the individual element. Nevertheless, the results so obtained are the exact solution for structural models as bending-hinge frame and truss models and generally a good approximation to the exact values for other types of structural models.

Analogous simplifications are obtained for the kinematic problem (66). The simplified version of the latter problem can be obtained either directly from (66) with the aid of an appropriate reasoning, (in which the fields $\dot{\boldsymbol{\varepsilon}}^p$ and $\boldsymbol{\xi}$ must be substituted by vectors $\boldsymbol{\epsilon}_i^p$, $\boldsymbol{\xi}_i$ lumped at the Gauss points $\boldsymbol{x}_i$) or, also, from (174) as dual LP problem. We can thus obtain:

$$\left.\begin{array}{l} \beta_a = \min\limits_{\left(\boldsymbol{\lambda}_i^c,\boldsymbol{\mu}_i^c,\boldsymbol{q}_{(e)}^c,\boldsymbol{u}^c\right)} \sum\limits_{i=1}^{N_g}\left(\boldsymbol{k}^T\,\boldsymbol{\lambda}_i^c + \boldsymbol{h}^T\,\boldsymbol{\mu}_i^c\right) \qquad \text{subject to :} \\[2ex] \boldsymbol{G}\boldsymbol{\lambda}_i^c - \boldsymbol{J}\boldsymbol{\mu}_i^c = \boldsymbol{0}, \qquad \boldsymbol{\lambda}_i^c \geq \boldsymbol{0}, \quad \boldsymbol{\mu}_i^c \geq \boldsymbol{0} \qquad \forall i \in I(N_g) \\[2ex] \sum\limits_{i=1}^{n_{g\,(e)}} \boldsymbol{S}_{(e)i}^T\,\boldsymbol{N}\boldsymbol{\lambda}_i^c = \boldsymbol{q}_{(e)}^c \qquad \forall e \in I(N_e) \\[2ex] \left\{\boldsymbol{q}_{(e)}^c\right\} = \boldsymbol{C}\boldsymbol{u}^c \\[2ex] \sum\limits_{i=1}^{N_g} \boldsymbol{R}_i\,\boldsymbol{\lambda}_i^c = 1. \end{array}\right\} \tag{175}$$

Remembering (164), the first constraint of (175) implies $\Delta\boldsymbol{\varepsilon}_i^{pc} = \boldsymbol{N}\boldsymbol{\lambda}_i^c = \boldsymbol{N}\boldsymbol{\mu}_{Ki}^c$ and $\boldsymbol{k}^T\,\boldsymbol{\lambda}_i^c = \boldsymbol{k}^T\,\boldsymbol{\mu}_{Ki}^c + k_0\,\mu_{Ii}^c$. The incipient collapse mode is alternating plasticity if in the optimal solution of (175): $\boldsymbol{N}\boldsymbol{\lambda}_i^c = \boldsymbol{0}$ for all is, but $\boldsymbol{\lambda}_i^c \neq \boldsymbol{0}$, $\boldsymbol{\mu}_{Ki}^c = \boldsymbol{0}$ and $\mu_{Ii}^c \neq 0$.

10.2 The Bounding Principle for Repeated Loads

The bounding principle of Section 8 can be straightforwardly transformed as follows. Let the upper bound U of (108) be considered first, and let to this purpose $\Omega(\widehat{\boldsymbol{\chi}})$ be taken as a quadratic form, i.e. $\Omega = \widehat{\boldsymbol{\chi}}^T\,\boldsymbol{B}^{-1}\,\widehat{\boldsymbol{\chi}}/2$, with $\widehat{\boldsymbol{\chi}} = \sum_{i=1}^{N_g}\widehat{\boldsymbol{\chi}}_i\,\delta_D(\boldsymbol{x}-\widehat{\boldsymbol{x}}_i)$. Since $\widehat{\boldsymbol{\sigma}}^r = \boldsymbol{S}_{(e)}\,\widehat{\boldsymbol{Q}}_{(e)}^r$ in the element (e), (109) becomes

$$L(0) = \sum_{(e)} \frac{1}{2}\,\widehat{\boldsymbol{Q}}_{(e)}^{r\,T}\,\boldsymbol{A}_{(e)}\,\widehat{\boldsymbol{Q}}_{(e)}^r + \sum_{i=1}^{N_g}\frac{1}{2}\,\widehat{\boldsymbol{\chi}}_i^T\,\boldsymbol{B}^{-1}\,\widehat{\boldsymbol{\chi}}_i \tag{176}$$

where $\boldsymbol{A}_{(e)}$ is the elastic compliance of the eth element, given by

$$\boldsymbol{A}_{(e)} = \int_{V_e} \boldsymbol{S}_{(e)}^T\,\boldsymbol{E}^{-1}\,\boldsymbol{S}_{(e)}\,\mathrm{d}V. \tag{177}$$

The modified yield and saturation conditions, remembering (118), can be written as (note: $\boldsymbol{p} = \boldsymbol{p}/g_i$):

$$\boldsymbol{f}_i{}^* = \frac{1}{g_i}\,\boldsymbol{f}_i = \boldsymbol{N}^T(\boldsymbol{s}_i + \widehat{\boldsymbol{\sigma}}_i^r) - \boldsymbol{G}^T\,\widehat{\boldsymbol{\chi}}_i - \frac{1}{g_i}\,\boldsymbol{k} + \alpha\,\boldsymbol{N}^T\,\boldsymbol{p}_i \leq \boldsymbol{0} \tag{178}$$

$$\boldsymbol{\phi}_i{}^* = \frac{1}{g_i}\,\boldsymbol{\phi}_i = \boldsymbol{J}^T\,\widehat{\boldsymbol{\chi}}_i - \frac{\boldsymbol{h}}{g_i} \leq \boldsymbol{0} \tag{179}$$

which, introducing the *perturbation vectors* $\boldsymbol{k}_i^*$ and $\boldsymbol{h}_i^*$ given by

$$\boldsymbol{k}_i^* = \frac{\gamma_i}{g_i}\,\boldsymbol{k} + \boldsymbol{N}^T\,\boldsymbol{p}_i, \qquad \boldsymbol{h}_i^* = \frac{\gamma_i}{g_i}\,\boldsymbol{h} \qquad \forall i \in I(N_g) \tag{180}$$

can alternatively be written in the form

$$\boldsymbol{f}_i{}^* = \boldsymbol{N}^T(\boldsymbol{s}_i + \widehat{\boldsymbol{\sigma}}_i^r) - \boldsymbol{G}^T\,\widehat{\boldsymbol{\chi}}_i + \alpha\,\boldsymbol{k}_i^* - \boldsymbol{k} \leq \boldsymbol{0} \tag{181}$$

$$\boldsymbol{\phi}_i^* = \boldsymbol{J}^T\,\widehat{\boldsymbol{\chi}}_i + \alpha\,\boldsymbol{h}_i^* - \boldsymbol{h} \leq \boldsymbol{0}. \tag{182}$$

In (181) and (182) the perturbation vectors $\boldsymbol{k}_i^*$ and $\boldsymbol{h}_i^*$ constitute themselves a perturbation mode and can be directly chosen without passing through the continuum perturbation mode $(\gamma, \boldsymbol{p})$.

The l.h.s. of (106), using (167) with the fields

$$\dot{\boldsymbol{\lambda}}(\boldsymbol{x}) = \sum_{i=1}^{N_g} \dot{\boldsymbol{\lambda}}_i\,\delta_D(\boldsymbol{x} - \boldsymbol{x}_i), \qquad \dot{\boldsymbol{\mu}}(\boldsymbol{x}) = \sum_{i=1}^{N_g} \dot{\boldsymbol{\mu}}_i\,\delta_D(\boldsymbol{x} - \boldsymbol{x}_i), \tag{183}$$

and remembering (179) and (182), transforms as in the following:

$$\begin{aligned}
\sum_{i=1}^{N_g}\left[\frac{\bar{g}\,\gamma_i}{g_i}\,D(\boldsymbol{\epsilon}_i^p, \boldsymbol{\xi}_i) + \boldsymbol{p}_i : \boldsymbol{\epsilon}_i^p\right]_{t_1} &= \\
&= \bar{g}\sum_{i=1}^{N_g}\left[\frac{\gamma_i}{g_i}\,\boldsymbol{k}^T\,\boldsymbol{\lambda}_i + \frac{\gamma_i}{g_i}\,\boldsymbol{h}^T\,\boldsymbol{\mu}_i + \boldsymbol{p}_i^T\,\boldsymbol{N}\boldsymbol{\lambda}_i\right]_{t_1} \\
&= \bar{g}\sum_{i=1}^{N_g}\left[\boldsymbol{k}_i^{*T}\,\boldsymbol{\lambda}_i(t_1) + \boldsymbol{h}_i^{*T}\,\boldsymbol{\mu}_i(t_1)\right]
\end{aligned} \tag{184}$$

such that inequality (106) becomes

$$\sum_{i=1}^{N_g}\left(\boldsymbol{k}_i^{*T}\,\boldsymbol{\lambda}_i + \boldsymbol{h}_i^{*T}\,\boldsymbol{\mu}_i\right)\Big|_{t_1} \leq \frac{1}{\alpha}\,[L(0) - L(t_1)]\,. \tag{185}$$

Then, choosing as deformation measure the quantity

$$\eta(t_1) = \sum_{i=1}^{N_g}\left[\boldsymbol{k}_i^{*T}\,\boldsymbol{\lambda}_i(t_1) + \boldsymbol{h}_i^{*T}\,\boldsymbol{\mu}_i(t_1)\right], \tag{186}$$

the bounding principle finally takes on the form:

$$\eta(t_1) \le U = \frac{1}{\alpha} L(0) \tag{187}$$

with $L(0)$ given by (176).

Note that, for the validity of the bound inequality (187), the yield and saturation conditions (181) and (182) must be satisfied at all Gauss points and for all stress coordinates $\boldsymbol{s}_i = \boldsymbol{S}_{(e)i} \boldsymbol{Q}^E_{(e)}$, that is, for all $\boldsymbol{Q}^E_{(e)} \in \Gamma_{(e)}$. This implies that (181) is equivalent to

$$\boldsymbol{f}_i{}^* = \boldsymbol{R}_i + \boldsymbol{N}^T \boldsymbol{S}_{(e)i} \widehat{\boldsymbol{Q}}^r_{(e)} - \boldsymbol{G}^T \widehat{\boldsymbol{\chi}}_i + \alpha \boldsymbol{k}^*_i - \boldsymbol{k} \le \boldsymbol{0} \tag{188}$$

to be satisfied for $\forall i \in I(N_g)$, with vectors $\widehat{\boldsymbol{Q}}^r_{(e)}$ complying with the equilibrium equation on the last line of (174).

In conclusion, for a finite element/Gauss point elastic-plastic structural system (that is, a finite element discretized structure with plastic deformability lumped at the element Gauss points) with piecewise linear yield function and saturation surfaces, the following statement can be phrased:

For a given structure of the above sort, subjected to repeated loads and characterized by elastic node-force envelope projection vectors $\{\boldsymbol{R}_i\}$ at the system's Gauss points $\boldsymbol{x}_i$, $(i = 1, 2, \ldots, N_g)$, and for a given perturbation mode $\{\boldsymbol{k}^*_i, \boldsymbol{h}^*_i\}$, if correspondingly a scalar $\alpha > 0$ and some self-equilibrated node force vectors $\{\widehat{\boldsymbol{Q}}^r_{(e)}\}$ and static internal variable vectors $\{\widehat{\boldsymbol{\chi}}_i\}$ can be found, such as to satisfy the yield and saturation conditions modified by the given perturbation mode, then the actual plastic deformation, η, produced during any ALH is bounded from above, i.e.

$$\eta_{\max} [\boldsymbol{k}^*_i, \boldsymbol{h}^*_i] \le U \left(\alpha, \widehat{\boldsymbol{Q}}^r_{(e)}, \widehat{\boldsymbol{\chi}}_i\right) \tag{189}$$

where U is given by (187) and (176).

Considered that the conditions required for $\{\widehat{\boldsymbol{\sigma}}^r_i\}$ and $\{\widehat{\boldsymbol{\chi}}_i\}$ are the same as for shakedown to occur in the structure with modified yield and saturation conditions, it follows that the bounding principle of Section 8 —of which that given above is a specialization— still holds (with due straightforward changes), as it would be easy to prove directly.

10.3 Possible Choices of the Perturbation Vectors

The same special bound cases contemplated in Section 8 can now be obtained by suitable choices of the perturbation vectors $\{\boldsymbol{k}^*_i, \boldsymbol{h}^*_i\}$. Bounds to the intrinsic dissipation work are obtained by choosing $\boldsymbol{k}^*_i = \gamma_i \boldsymbol{k}$, $\boldsymbol{h}^*_i = \gamma_i \boldsymbol{h}$ with $\gamma_i \ge 0$ for all $i \in I(N_g)$. With this choice $\eta(t_1)$ becomes

$$\eta(t_1) = \sum_{i=1}^{N_g} \gamma_i \left(\boldsymbol{k}^T \boldsymbol{\lambda}_i + \boldsymbol{h}^T \boldsymbol{\mu}_i\right)\Big|_{t_1}, \tag{190}$$

which can be either the global dissipation if $\gamma_i = 1$ for all $i \in I(N_g)$, or the eth element value if $\gamma_i = 1$ for $i = 1, 2, \ldots, n_{g(e)}$ and $\gamma_i = 0$ for all other is.

Bounds to plastic strains are obtained by choosing $\boldsymbol{k}_i^* = \boldsymbol{N}^T \boldsymbol{p}_i$, $\boldsymbol{h}_i^* = \boldsymbol{0}$ with $\boldsymbol{p}_i$ arbitrary stress vectors. Then

$$\eta(t_1) = \sum_{i=1}^{N_g} \boldsymbol{p}_i^T \boldsymbol{N} \boldsymbol{\lambda}_i \tag{191}$$

which is a measure of plastic strains. If additionally $\boldsymbol{p}_i = \boldsymbol{S}_{(e)} \boldsymbol{Q}_{(e)}^*$ with $\left\{ \boldsymbol{Q}_{(e)}^* \right\}$ being the elastic node force response to some static (dummy) load $\boldsymbol{F}^*$, it can be shown that (191) becomes

$$\eta(t_1) = \boldsymbol{F}^{*^T} \boldsymbol{u}^r(t_1) \tag{192}$$

where $\boldsymbol{u}^r(t_1)$ is the residual displacement vector corresponding to the plastic strains $\{\boldsymbol{N}\boldsymbol{\lambda}_i\}$.

11 Extensions to Materials with Temperature-Dependent Yield Stress

When temperature changes are modest, the usual hypothesis of temperature-independent material data is quite adequate, and thermal effects can be accounted for only through the consequent thermal strains induced in the structure. For more important temperature variations, the above hypothesis is no longer acceptable. Then, the yield function is to be considered dependent on temperature, θ, but convex in the stress space for every θ value. With a yield function of this sort, Prager (1956) and König (1982) extended, respectively, the classical Melan and Koiter theorems to account for the thermal effects on the yield surface. The extended theorems are rather straightforward, as they maintain the same conceptual and mathematical structures of the classical ones, but with the requirement that the shakedown criterion must be enforced ***for every possible combination of the loads with temperature***, both varying in their respective ranges.

As in general dynamic loads vary more rapidly than temperatures,we consider here only quasi-static loads. The above extended theorems are hereafter reported (without proofs) for a structural system as that of Section 3, subjected to repeated loads and temperature variations ranging within given domains, the yield and saturation functions being both temperature dependent, but convex for every θ value. Then, denoting by $\boldsymbol{\sigma}^E = \boldsymbol{\sigma}^E(\boldsymbol{x})$ the thermo-elastic stress response to any admissible thermo-mechanical load condition within the given (load and temperature) domains, the above extended shakedown theorems can be rephrased as follows.

11.1 Static-Type Theorem

A necessary and sufficient condition for shakedown to occur in a structure subjected to repeated loads $\boldsymbol{P} \in \Pi$ and to temperature variations within some range, is that there exist some fields of self stresses, $\hat{\boldsymbol{\sigma}}^r$, and of statical internal variables, $\hat{\boldsymbol{\chi}}$, both independent of loads and temperature, such as to satisfy the conditions:

$$f(\sigma^E + \widehat{\sigma}^r, \widehat{\chi}^r, \theta) \leq 0, \qquad \phi(\widehat{\chi}^r, \theta) \leq 0 \quad \text{in } V \tag{193}$$

for all possible load/temperature combinations.

Aside the simplicity of the above statement, there is a complexity in the likely high number of load/temperature combinations to consider, since in fact all possible temperature distributions must be combined with every load condition, due to the fact that f and ϕ may be nonconvex in the (σ, χ, θ)-space.

11.2 Kinematic-Type Theorem

The extended kinematic shakedown theorem given by König (1972) can be phrased as follows:

A necessary and sufficient condition for shakedown to occur in a structure subjected to repeated loads $P \in \Pi$ and to temperature variations within some range, is that the inequality

$$\int_0^{t_1}\int_V \left[D(\dot{\varepsilon}^{pc}, \dot{\xi}^c, \theta) - \sigma^E : \dot{\varepsilon}^{pc}\right] dV\, dt \geq 0 \tag{194}$$

be satisfied for any PAM, i.e. for any pair of fields $(\dot{\varepsilon}^{pc}, \dot{\xi}^c)$ such that

$$\Delta\varepsilon^{pc} = \int_0^{t_1} \dot{\varepsilon}^{pc}\, dt, \qquad \Delta\xi^c = \int_0^{t_1} \dot{\xi}^c\, dt = 0 \qquad \text{in } V \tag{195}$$

$$\Delta\varepsilon^{pc} = \nabla^s u^c \quad \text{in } V, \qquad u^c = 0 \quad \text{on } S_u, \tag{196}$$

and for any admissible thermo-mechanical load path $Q(t) = \{P(t), \theta(t)\}$.

As the dissipation function D, associated with the yield and saturation function, f and ϕ, may be a nonconvex function of θ, the same remarks as for the static-type theorem can be repeated here in regards to the likely excessive number of load/temperature combinations to consider.

An additional drawback exhibited by the above kinematic theorem is that it in general cannot be used as an upper bound statement for the shakedown safety factor to the load/temperature combinations, see Borino and Polizzotto (1997) and Borino (2000). In fact, inequality (197), rewritten with $\beta_a\sigma^E$ in place of σ^E and $\beta_a\theta$ in place of θ, gives

$$\beta_a \leq \int_0^{t_1}\int_V D(\dot{\varepsilon}^{pc}, \dot{\xi}^c, \beta_a\theta)\, dV\, dt \Bigg/ \int_0^{t_1}\int_V \sigma^E : \dot{\varepsilon}^{pc}\, dV\, dt, \tag{197}$$

provided that, as assumed,

$$\int_0^{t_1}\int_V \sigma^E : \dot{\varepsilon}^{pc}\, dV\, dt > 0. \tag{198}$$

Obviously (197) does not provide a proper upper bound to β_a. Exceptions to this can occur when D is a linear function θ, as for instance in the case of perfect plasticity and

$f = f_0(\boldsymbol{\sigma}) - k_0(1 - c\theta) \leq 0$, for which $D = (1 - c\theta)D_0$, D_0 being the dissipation function for $\theta = 0$ (König, 1982); then (197) can be replaced by

$$\beta_a \leq \int_0^{t_1}\int_V D_0(\dot{\boldsymbol{\varepsilon}}^{pc})\, dV\, dt \Bigg/ \int_0^{t_1}\int_V \left[c\theta D_0(\dot{\boldsymbol{\varepsilon}}^{pc}) + \boldsymbol{\sigma}^E : \dot{\boldsymbol{\varepsilon}}^{pc}\right] dV\, dt, \qquad (199)$$

which is a proper upper bound to β_a.

11.3 A Consistent Approach

In practice, f is a convex function of θ for a rather wide range of temperatures (from -10 ^{0}C to 600 ^{0}C for many metals and alloys) and can be treated as a linear one for less wide ranges of temperatures. Assuming a similar behaviour also for ϕ, as shown by Borino and Polizzotto (1997) and Borino (2000), a consistent, thermodynamically coherent, approach to shakedown can be formulated, in which the temperature θ is treated as an additional state variable to which the *plastic entropy* rate, $\dot{\eta}^p$, is associated as an additional evolutive variable.

The material constitutive behaviour is governed by thermoplastic evolutive laws as follows:

$$\dot{\boldsymbol{\varepsilon}}^p = \dot{\lambda}\frac{\partial f}{\partial \boldsymbol{\sigma}}, \quad -\dot{\boldsymbol{\xi}} = \dot{\lambda}\frac{\partial f}{\partial \boldsymbol{\chi}} + \dot{\mu}\frac{\partial \phi}{\partial \boldsymbol{\chi}} \qquad (200)$$

$$\dot{\eta}^p = \dot{\lambda}\frac{\partial f}{\partial \theta} + \dot{\mu}\frac{\partial \phi}{\partial \theta} \qquad (201)$$

$$f(\boldsymbol{\sigma}, \boldsymbol{\chi}, \theta) \leq 0, \quad \dot{\lambda} \geq 0, \quad \dot{\lambda}\, f(\boldsymbol{\sigma}, \boldsymbol{\chi}, \theta) = 0 \qquad (202)$$

$$\phi(\boldsymbol{\chi}, \theta) \leq 0, \quad \dot{\mu} \geq 0, \quad \dot{\mu}\, \phi(\boldsymbol{\chi}, \theta) = 0 \qquad (203)$$

which have been shown (Borino and Polizzotto, 1997; Borino, 2000) to be thermodynamically consistent. The plastic entropy turns out to be an addition to the internal entropy production of the material. The above thermo-plastic material model admits a maximum intrinsic dissipation principle, that is:

$$\left.\begin{array}{l} D(\dot{\boldsymbol{\varepsilon}}^p, \dot{\boldsymbol{\xi}}, \dot{\eta}^p) = \max\limits_{(\boldsymbol{\sigma}, \boldsymbol{\chi}, \theta)} \left(\boldsymbol{\sigma} : \dot{\boldsymbol{\varepsilon}}^p - \boldsymbol{\chi} \cdot \dot{\boldsymbol{\xi}} + \theta \dot{\eta}^p\right) \\ \text{subject to} \quad f(\boldsymbol{\sigma}, \boldsymbol{\chi}, \theta) \leq 0, \quad \phi(\boldsymbol{\chi}, \theta) \leq 0 \end{array}\right\} \qquad (204)$$

From (204) the following Druckerian inequality can be derived:

$$(\boldsymbol{\sigma} - \boldsymbol{\sigma}^*) : \dot{\boldsymbol{\varepsilon}}^p - (\boldsymbol{\chi} - \boldsymbol{\chi}^*) \cdot \dot{\boldsymbol{\xi}} + (\theta - \theta^*)\, \dot{\eta}^p \geq 0 \qquad (205)$$

which holds for any $\boldsymbol{\sigma}^*$, $\boldsymbol{\chi}^*$, θ^* such that $f(\boldsymbol{\sigma}^*, \boldsymbol{\chi}^*, \theta^*) \leq 0$ and $\phi(\boldsymbol{\chi}^*, \theta^*) \leq 0$, but $(\boldsymbol{\sigma}, \boldsymbol{\chi}, \theta)$ and $(\dot{\boldsymbol{\varepsilon}}^p, \dot{\boldsymbol{\xi}}, \dot{\eta}^p)$ related to each other through (200)–(203).

Let us assume a repeated load domain Π of polyhedral shape, with *basic loads* $\boldsymbol{P}_i$ $(i = 1, 2, \ldots, N_P)$. Analogously, let every admissible temperature field $\theta(\boldsymbol{x})$ be expressed as a linear combination of the basic temperature fields $\theta_j(\boldsymbol{x})$, $(j = 1, 2, \ldots, N_\theta)$. Then, combining every basic load $\boldsymbol{P}_i$ with the basic temperature fields θ_j, we obtain $N = N_P N_\theta$ basic thermo-mechanical loads, say $\boldsymbol{Q}_k$, $(k = 1, 2, \ldots, N)$. Let $\boldsymbol{\sigma}_k^E = \boldsymbol{\sigma}_k^E(\boldsymbol{x})$ be the elastic stress response to $\boldsymbol{Q}_k$.

With the load scheme as described above, the static shakedown criterion takes on the simpler form:

$$f(\boldsymbol{\sigma}_k^E + \hat{\boldsymbol{\sigma}}^r, \hat{\boldsymbol{\chi}}, \theta_k) \leq 0, \qquad \phi(\hat{\boldsymbol{\chi}}, \theta_k) \leq 0 \quad \text{in } V, \quad \forall k \in I(N) \tag{206}$$

which replaces (193) in the static shakedown theorem.

Analogously, the kinematic shakedown criterion reads

$$\sum_{k=1}^{N} \int_V \left[D\left(\boldsymbol{\varepsilon}_k^{pc}, \boldsymbol{\xi}_k^c, \eta_k^{pc}\right) - \boldsymbol{\sigma}_k^E : \boldsymbol{\varepsilon}_k^{pc} - \theta_k \eta_k^{pc} \right] \mathrm{d}V \geq 0 \tag{207}$$

where the fields $\boldsymbol{\varepsilon}_k^{pc}$, $\boldsymbol{\xi}_k^c$, η_k^{pc} constitute a Plastic Accumulation Mechanism (PAM), that is, they satisfy the following (compatibility) equations:

$$\Delta \boldsymbol{\varepsilon}^{pc} = \sum_{k=1}^{N} \boldsymbol{\varepsilon}_k^{pc} = \nabla^s \boldsymbol{u}^c \quad \text{in } V, \ \boldsymbol{u}^c = \boldsymbol{0} \quad \text{on } S_u \tag{208}$$

$$\Delta \boldsymbol{\xi}^c = \sum_{k=1}^{N} \boldsymbol{\xi}_k^c = \boldsymbol{0} \quad \text{in } V \tag{209}$$

$$\eta_k^{pc} \geq 0 \quad \text{in } V. \tag{210}$$

Equations (207) to (210) replace (194) and (195) in the kinematic shakedown theorem.

The shakedown safety factor can accordingly be expressed as follows:

– *By the static approach*

$$\left.\begin{aligned} &\beta_a = \max_{(\beta, \hat{\boldsymbol{\sigma}}^r, \hat{\boldsymbol{\chi}})} \beta \quad \text{s.t.} \\ &f(\boldsymbol{\sigma}_k^E + \hat{\boldsymbol{\sigma}}^r, \hat{\boldsymbol{\chi}}, \theta_k) \leq 0, \quad \phi(\hat{\boldsymbol{\chi}}, \theta_k) \leq 0 \quad \text{in } V, \quad \forall k \in I(N) \\ &\operatorname{div} \hat{\boldsymbol{\sigma}}^r = \boldsymbol{0} \quad \text{in } V, \qquad \hat{\boldsymbol{\sigma}}^r \cdot \boldsymbol{n} = \boldsymbol{0} \quad \text{on } S_t \end{aligned}\right\} \tag{211}$$

– *By the kinematic approach*

$$\left.\begin{array}{l}
\beta_a = \max\limits_{(\varepsilon_k^{pc},\xi_k^c,\eta_k^{pc})} \sum\limits_{k=1}^{N_{B\theta}} \int_V D(\varepsilon_k^{pc},\xi_k^c,\eta_k^{pc})\, dV \qquad \text{s.t.} \\
\sum\limits_{k=1}^{N} \int_V (\sigma_k^E : \varepsilon_k^{pc} + \theta_k \eta_k^{pc})\, dV = 1 \\
\Delta\varepsilon^{pc} = \sum\limits_{k=1}^{N} \varepsilon_k^{pc} = \nabla^s \boldsymbol{u}^c \quad \text{in } V,\ \boldsymbol{u}^c = \boldsymbol{0} \ \text{on } S_u \\
\Delta\xi^c = \sum\limits_{k=1}^{N} \xi_k^c = \boldsymbol{0} \quad \text{in } V \\
\eta_k^{pc} \geq 0 \quad \text{in } V.
\end{array}\right\} \tag{212}$$

12 Conclusions

We have presented two categories of dynamic shakedown, i.e. unrestricted and restricted. In the unrestricted dynamic shakedown, repeated excitations load schemes are considered, that is unknown sequences of short-duration excitations all belonging to a given excitation domain, with no-load no-motion periods between two subsequent excitations. Two theorems have been given, extensions of the classical Melan and Koiter theorems to dynamics and to internal variable constitutive models with saturation surface. Both theorems have been expressed in terms of elastic stress response envelope, thus without involving the time variable, in a way making the two theorems to take a unified format for quasi-static and dynamic loads. Such a procedure, already attempted in previous works (Polizzotto, 1985; Panzeca *et al.*, 1990; Pham, 1996), deserves further study, also because the unrestricted dynamic shakedown with its repeated excitations load scheme is believed to be most appropriate for a wider class of engineering situations of practice (e.g. seismic loads, wind and sea wave actions, rapidly transient vehicles on a bridge, etc.).

We have also discussed the shakedown limit state of the structure to which the latter is able to report itself under loads at the shakedown limit. The study of the equations governing this limit state is useful in order to understand the phenomena occurring in the structure on transition from the safe shakedown state to the unsafe one towards exposure to inadaptation collapse mechanisms. The role of the saturation surface has been pointed out within this context, but further study on this point is hoped for.

In the restricted dynamic shakedown, the structure is subjected to an infinite duration load history. Two theorems, dual of each other, have been given (without proofs), i.e. a static-type one with shakedown parameters including free-motion stresses (instead of fictitious initial conditions), and a kinematic-type one expressed in terms of dynamic plastic accumulation mechanisms (instead of static ones), the latter being a generalization of the Koiter's kinematically admissible plastic strain cycles. Peculiarities of this type of shakedown, herein pointed out are:

i) the independence of the structure's shakedown capacity and of the related shakedown safety factor from the initial conditions (displacements and velocities, as well as residual stresses);

ii) the separation time, which cuts out from the given load history the initial (finite) transient piece as irrelevant for the shakedown occurrence;

iii) the asymptotic load history, which only is responsible for the shakedown occurrence, which leads to notable simplifications of the related shakedown analysis;

iv) damping, which cannot be ignored in this case of load histories with very long time duration, by which the free-motion stresses can be dropped as a shakedown parameter, with further notable simplifications in the shakedown analysis.

The restricted dynamic shakedown theory turns out to be less adequate as safety criterion than the unrestricted one.

Two bounding principles have been presented. One, nonevolutive in nature, is applicable for either quasi-static or dynamic repeated loads below the shakedown limit, and is useful to construct upper bounds to a number of plastic deformation measures, as intrinsic dissipation, plastic strain and residual displacement components. These upper bounds diverge as the load approaches to the shakedown limit.

An evolutive bounding principle has been also presented, which is applicable for a (either quasi-static, or dynamic) given load history below and above the shakedown limit. This principle requires an evolutive analysis, but simpler than a step-by-step one (e.g. an holonomic analysis with larger time steps) and leads to either lower and upper bounds to the actual deformation parameters. These bounds can be rendered as stringent as desired, but at more costly computational efforts. The bounding principle for repeated loads is shown to be a particular case of the evolutive one.

Discrete models, with finite elements, piecewise linear yield and saturation surfaces and plastic deformability lumped at the Gauss points, have been considered for computational purposes.

Extensions to materials with temperature dependent yield stress have been also presented together with a thermodynamically consistent formulation of shakedown theory in which the temperature and the plastic entropy play a crucial role as additional state and evolutive variables.

References

Bathe, K.-J., and Wilson, E. L. (1976). *Numerical Methods in Finite Element Analysis.* Englewood Cliffs, N.J.: Prentice Hall.

Borino, G. (2000). Consistent shakedown theorems for materials with temperature dependent yield functions. *International Journal of Solids and Structures.* 37:3121–3147.

Borino, G., Caddemi, S., and Polizzotto, C. (1990). Mathematical programming methods for the evaluation of dynamic plastic deformations. In Lloyd Smith, D., ed., *Mathematical Programming Methods in Structural Plasticity.* Wien: Springer-Verlag. 349–372.

Borino, G., and Polizzotto, C. (1995). Dynamic shakedown of structures under repeated seismic loads. *Journal of Engineering Mechanics* 121:1306–1314.

Borino, G., and Polizzotto, C. (1996). Dynamic shakedown of structures with variable appended masses and subjected to repeated excitations. *Int. Journal of Plasticity* 12:215–228.

Borino, G, and Polizzotto, C. (1997). Shakedown theorems for a class of materials with temperature-dependent yield stress. In Owen, D. R. J., Oñate, E. and Hinton, E., eds., *Computational Plasticity*. Barcelona, Spain, CIMNE. Part 1, 475–480.

Capurso, M., Corradi, L., and Maier, G. (1978). Bounds on deformations and displacements in shakedown theory. In Ecole Politechnique Palaiseau, ed., *Materials and Structures under Cyclic Loads*. Paris: Association Amicale des Ingénieurs Anciens Elèves, Ecóle National Ponts et Chaussees. 231–244.

Capurso, M. (1979). Some upper bounds principles for plastic strains in dynamic shakedown of elastoplastic structures. *Journal of Structural Mechanics* 7:1–20.

Ceradini, G. (1969). On shakedown of elastic-plastic solids under dynamic actions. *Giornale del Genio Civile* 4-5:239–258 (in Italian).

Comi, C., and Corigliano, A. (1991). Dynamic shakedown in elastoplastic structures with general internal variable constitutive laws. *Int. Journal of Plasticity* 7:679–692.

Corradi, L., and Maier, G. (1973). Inadaptation theorems in the dynamics of elastic-workhardening structures. *Ingenieur Archiv* 43:44–57.

Corradi, L., and Maier, G. (1974). On non-shakedown theorems for elastic perfectly plastic continua. *Journal of Mechanics and Physics of Solids* 22:401–413.

Corradi, L., and De Donato, O. (1975). Dynamic shakedown theory allowing for second-order geometric effects. *Meccanica* 10:93–98.

Drucker, D. C. (1960). Plasticity. In Goodier, J. N., and Hoff, J. H., eds., *Structural Mechanics*. London: Pergamon Press. 407–455.

Fuschi, P., and Polizzotto, C. (1998). Internal-variable constitutive model for rate-independent plasticity with hardening saturation surface. *Acta Mechanica* 129:73–95.

Fuschi, P. (1999). Structural shakedown for elastic-plastic materials with hardening saturation surface. *Int. Journal of Solids and Structures* 36:219–240.

Halphen, B. and Nguyen, Q. S. (1975). Sur les matériaux standard généralisées. *Journal de Mécanique* 14:39–63.

Halphen, B. (1979). Accomodation et adaptation des structures élasto-visco-plastiques et plastiques. In Ecole Politechnique Palaiseau, ed., *Matériaux et Structures Sous Chargement Cyclique*. Paris: Association Amicale des Ingénieurs Anciens Elèves, Ecóle National Ponts et Chaussées. 203–229.

Koiter, W. J. (1960). General theorems for elastic-plastic solids. In Hill, R., and Sneddon, I., eds., *Progress in Solid Mechanics*. North-Holland: vol.I. 167–221.

König, A, and Maier, G. (1981). Shakedown analysis of elastoplastic structures: a review of recent developments. *Nuclear Engineering and Design* 66:81–95.

König, A. (1982). Shakedown criteria in the case of loading and temperature variations. *Journal de Mécanique Appliquée* Special Issue, 99–108.

Leckie, F. A. (1974). *Review of bounding techniques in shakedown and ratchetting at elevated temperature*. Welding Research Council Bullettin, 195:1–11.

Maier, G. (1973). A shakedown matrix theory allowing for workhardening and second-order geometric effects. In Sawczuk, A., eds., *Foundations of Plasticity*. Leyden, Noordhoff. 417–433.

Maier, G. (1987). A generalization to nonlinear hardening of the first shakedown theorem for discrete elastic-plastic models. *Atti Accademia dei Lincei, Rendiconti di Fisica* 8, LXXXI:161–174.

Maier, G., and Novati, G. (1990). Dynamic shakedown and bounding theory for a class of nonlinear hardening discrete structural models. *Int. Journal of Plasticity* 7:679–692.

Maier, G., Carvelli, V., and Cocchetti, G. (2000). On direct methods for shakedown and limit analysis. *European Journal of Mechanics A/Solids* (to appear).

Owen, D. R. J., and Hinton, E. (1980). *Finite Element in Plasticity.* Swansea, U.K.: Pineridge Press Ltd.

Panzeca, T., Polizzotto, C., and Rizzo, S. (1990). Bounding techniques and their application to simplified plastic analysis of structures. In Lloyd Smith, D., ed., *Mathematical Programming Methods in Structural Plasticity.* Wien: Springer-Verlag. 315–348.

Pham, D. C. (1996). Dynamic shakedown and reduced kinematic theorems. *Int. Journal of Plasticity* 12:1055–1068.

Polizzotto, C. (1982a). A unified treatment of shakedown theory and related bounding techniques. *Solid Mechanics Archives* 7:19–75.

Polizzotto, C. (1982b). Bounding principles for elastic-plastic-creeping solids loaded below and above the shakedown limit. *Meccanica* 17:143–148.

Polizzotto, C. (1984a). On shakedown of structures under dynamic agencies. In Sawczuk, A., and Polizzotto, C., eds., *Inelastic Analysis Under Variable Loads.* Palermo: Cogras. 5–29.

Polizzotto, C. (1984b). Dynamic shakedown by modal analysis. *Meccanica* 19:133–144.

Polizzotto, C. (1984c). Deformation bounds for elastic-plastic solids within and out of the creep range. *Nuclear Engineering and Design* 83:293–301.

Polizzotto, C. (1985a). New developments in the theory of dynamic shakedown. In Sawczuk, A., and Bianchi, G., eds., *Plasticity Today.* Amsterdam: Elsevier. 343–360.

Polizzotto, C. (1985b). Bounding techniques and their application to the analysis of elastic-plastic structures. In Zarka, J., ed., *Simplified Analysis of Inelastic Structures Subjected to Statical and Dynamical loadings.* Palaiseau, France: Ecole Polytechnique .

Polizzotto, C. (1986). A convergent bounding principle for a class of elastoplastic strain-hardening solids. *Int. Journal of Plasticity* 2:359–370.

Polizzotto, C., Borino, G., Caddemi, S., and Fuschi, P. (1991). Shakedown problems for material models with internal variables. *European Journal of Mechanics A/Solids* 10:621–639.

Polizzotto, C., Borino, G., Caddemi, S., and Fuschi, P. (1993). Theorems of restricted dynamic shakedown. *Int. Journal of Mechanical Sciences* 35:787–801.

Ponter, A. R. S. (1972). An upper bound on the small displacements of elastic perfectly plastic structures. *Journal of Applied Mechanics* 39:959–963.

Ponter, A. R. S. (1975). General displacement and work bounds for dynamically loaded bodies. *Journal of Mechanics and Physics of Solids* 23:157–163.

Prager, W. (1956). Shakedown in elasticplastic media subjected to cycles of load and temperature In *Proceedings of A. Danusso Symposium on: La Plasticità nella Scienza delle Costruzioni*, Bologna, Nicola Zanichelli Editore, 239–244.

Tajimi, H. (1960). A statistical method to determining the maximum response of a building structure during an earthquake. In *Proceedings of the 2nd World Conference on Earthquake Engineering*, Tokyo, 2:781–797.

Shakedown of Structures Accounting for Damage Effects

Castrenze Polizzotto[1], Guido Borino[1] and Paolo Fuschi[2]

[1] University of Palermo, Palermo, Italy
[2] University of Reggio Calabria, Reggio Calabria, Italy

Abstract. Shakedown theory for elastic-plastic-damage materials is exposed. Two kinds of shakedown are considered: i) Enlarged shakedown (or simply shakedown), in which both plastic deformations and damage eventually cease, after which the structural response is purely elastic; ii) Weak-form shakedown, in which plastic deformations eventually cease together with their consequences (including ductile damage), not necessarily damage from other sources (which are however escluded by assumption). An (enlarged) shakedown static-type theorem is given for a class of D-stable structures. Sufficient theorems of weak-form shakedown are provided, i.e. a static-type one (quite similar to that of Hachemi and Weichert (1992)), and a kinematic-type one, and possible applications of the latter theorems to ductile-damage structures are also indicated.

1 Introduction

In the presence of a material exhibiting damage with consequent deterioration of its stiffness and resistance properties, the notion of shakedown must take on a wider meaning than the classic one. That is, for a structure of elastic-plastic-damage material, shakedown means that, after a certain transient period of loading in which plastic deformations and damage may occur without causing the structure to fail, finally the structure reacts to the subsequent loads as an elastic structure with no further plastic strains, nor damage. With the latter enlarged shakedown notion, the plastic and damage mechanisms obey each a specific constitutive law, with possible interactions between them; also, the related shakedown criterion remains valid for whatsoever relative incidence of the two mechanisms on the material behaviour, including the two (extreme) situations of elastic-plastic materials (no damage), for which the classical shakedown theory is recovered, and of elastic-brittle materials (no plastic strains) for which shakedown is only concerned with damage.

For many materials damage is a direct consequence of (large) plastic strains (ductile damage), in which case it is reasonable to expect that damage ceases to occur in a structure *as soon as plastic deformations cease independently of the actual evolutive law of damage*. This is a weak-form of shakedown, which actually turns out to be equivalent to the classical shakedown accounting for plastic deformations only, but the relevant shakedown criterion applied to a structure prone to damage induced by plastic strains (the damage evolutive law being left unspecified).

Shakedown for ductile damage as mentioned above will be referred to as *weak-form shakedown* in the following. It is of practical importance because the majority of metals and alloys behave like that. It is quite natural to approach shakedown for this sort of materials by applying the classical shakedown theorems to a damageable structure, in order to find out the pre-requisite criteria to guarantee that plastic strains and damage altogether eventually cease under a given loading programme, without the necessity to specify the damage evolutive law.

This research route was open by Hachemi and Weichert (1992), who applied the static shakedown theorem to generalized standard materials which exhibit ductile isotropic damage according to a model devised by Lemaitre and Chaboche (1985) and by Ju (1989). This work was then pursued by the same authors (Hachemi and Weichert, 1997, 1998) and by others (Feng and Yu, 1994, 1995; Druyanov and Roman, 1995, 1998; Polizzotto *et al.*, 2001) to study various aspects of the issue. The work by Siemaszko (1993) can be included in the same research line as it provides —through a numerical step-by-step analysis— the influence on shakedown loads of various structural and constitutive factors, as geometrical nonlinearities, nonlinear hardening and ductile damage, the latter being described by a porosity parameter according to a model of Perzyna (1984). It is worth noting that, whereas the weak-form shakedown theorems given by Hachemi and Weichert (1992, 1997, 1998) and by Druyanov and Roman (1995, 1998) are expressed in terms of the unknown actual damage process produced under the given loading program, those given by Polizzotto *et al.* (2001) are expressed in terms of ***trial damage***, i.e. an arbitrary damage within the domain of allowable damage (which in principle includes the actual one).

Shakedown in its wider meaning was addressed by Polizzotto et al. (1996) for a material model exhibiting nonlinear hardening and a multiparameter damage law. The material is endowed with a general free energy potential and obeys plasticity and damage associative evolutive laws with distinct yielding and damage surfaces. The coupling between damage and plasticity is achieved by the use of suitable internal variables appearing as arguments of both the yield and damage functions, instead of the effective stress concept of Lemaitre and Chaboche (1985). The main result of this paper is to have shown the equivalence of the addressed shakedown problem to a stability problem (D-stability, as it is called there) and to have provided, for a D-stable structure, a shakedown theorem which is an extension of the classical Melan one.

In the following, we first report the essentials of the enlarged shakedown theory assuming, for simplicity, materials that are linear elastic at constant damage. Then, we present a weak-form shakedown theory, which accounts only for plastic deformations, hence also for ductile damage, but is unable to account for other types of damage (anyway excluded by the assumed material model). The latter theory includes static-type and kinematic-type shakedown theorems and a procedure for the computation of the shakedown safety factor (Polizzotto *et al.* 2001. The static-type theorem is similar to that proposed by Hachemi and Weichert (1992).

Notation. A compact notation is used throughout, with bold face letters for vectors and tensors. The 'dot' and the 'colon' products of vectors and tensors denote simple and double index contraction operations, respectively. For instance, for the vectors $\boldsymbol{u} = \{u_i\}$, $\boldsymbol{n} = \{n_i\}$, the tensors $\boldsymbol{\sigma} = \{\sigma_{ij}\}$, $\boldsymbol{\varepsilon} = \{\varepsilon_{ij}\}$, $\boldsymbol{E} = \{E_{ijhk}\}$, the following operations hold: $\boldsymbol{u} \cdot \boldsymbol{n} = u_i n_i$, $\boldsymbol{\sigma} \cdot \boldsymbol{n} = \{\sigma_{ij} n_j\}$, $\boldsymbol{\sigma} : \boldsymbol{\varepsilon} = \sigma_{ij} \varepsilon_{ji}$, $\boldsymbol{E} : \boldsymbol{\varepsilon} = \{E_{ijhk} \varepsilon_{kh}\}$, where the subscripts denote Cartesian components and the repeated index summation rule is applied. The upper dot denotes time derivative, i.e. $\dot{\boldsymbol{u}} = \partial \boldsymbol{u}/\partial t$. An orthogonal Cartesian axes system is used, with coordinates $\boldsymbol{x} = (x_1, x_2, x_3)$. The symbol ∇^s = symmetric part of the gradient operator, e.g. $\nabla^s \boldsymbol{u} = \{(\partial u_i/\partial x_j + \partial u_j/\partial x_i)/2\}$; also, 'div' = divergence operator, e.g. div$\boldsymbol{\sigma} = \{\partial \sigma_{ji}/\partial x_j\}$. Other symbols will be defined in the text where they appear for the first time.

2 Enlarged Shakedown

In this section we address (enlarged) shakedown for elastic-plastic-damage material models which obey a particular evolutive law as described in the following.

2.1 The Material Model

We consider a material model endowed with a free energy potential

$$\Psi = \frac{1}{2}\boldsymbol{\varepsilon}^e : \boldsymbol{E}(\boldsymbol{\omega}) : \boldsymbol{\varepsilon}^e + \Psi_{\text{in}}(\boldsymbol{\xi}) \tag{1}$$

where $\boldsymbol{\varepsilon}^e$ = elastic strain tensor, $\boldsymbol{\omega}$ = damage tensor and $\boldsymbol{\xi}$ = (kinematic) internal variable tensor. In practice, $\boldsymbol{\omega}$ and $\boldsymbol{\xi}$ may be scalars, or vectors, or tensors, but both are formally treated here as vectors. $\boldsymbol{E}(\boldsymbol{\omega})$ is positive definite and has the usual symmetries of the fourth-order elastic moduli tensor for all values of $\boldsymbol{\omega}$; $\Psi_{\text{in}}(\boldsymbol{\xi})$ is convex and smooth. Temperature does not affect (1). The vector $\boldsymbol{\xi}$, which accounts for the changes of the material state at the micro-structural level as a consequence of the plastic and damage mechanisms, is split as

$$\boldsymbol{\xi} = \boldsymbol{\xi}^p + \boldsymbol{\xi}^d \tag{2}$$

where $\boldsymbol{\xi}^p$ is the part of $\boldsymbol{\xi}$ associated with the plastic mechanism, $\boldsymbol{\xi}^d$ that associated with the damage mechanism.

By suitable thermodynamics considerations, the state equations of the material are:

$$\boldsymbol{\sigma} = \boldsymbol{E}(\boldsymbol{\omega}) : \boldsymbol{\varepsilon}^e, \quad \boldsymbol{Y} = -\frac{1}{2}\boldsymbol{\varepsilon}^e : \frac{\partial \boldsymbol{E}}{\partial \boldsymbol{\omega}} : \boldsymbol{\varepsilon}^e, \quad \boldsymbol{\chi} = \frac{\partial \Psi_{\text{in}}}{\partial \boldsymbol{\xi}} \tag{3}$$

whereas the dissipation density reads

$$D = D_p + D_d \tag{4}$$

$$D_p = \boldsymbol{\sigma} : \dot{\boldsymbol{\varepsilon}}^p - \boldsymbol{\chi} \cdot \dot{\boldsymbol{\xi}}^p \geq 0, \qquad D_d = \boldsymbol{Y} \cdot \dot{\boldsymbol{\omega}} - \boldsymbol{\chi} \cdot \dot{\boldsymbol{\xi}}^d \geq 0. \tag{5}$$

The irreversible material behaviour is governed by a yield function, $f_p(\boldsymbol{\sigma}, \boldsymbol{\chi})$, and a damage function, $f_d(\boldsymbol{Y}, \boldsymbol{\chi})$, and obey the following constitutive equations:

$$\dot{\boldsymbol{\varepsilon}}^p = \dot{\lambda}_p \frac{\partial f_p}{\partial \boldsymbol{\sigma}}, \quad -\dot{\boldsymbol{\xi}}^p = \dot{\lambda}_p \frac{\partial f_p}{\partial \boldsymbol{\chi}} \tag{6}$$

$$f_p(\boldsymbol{\sigma}, \boldsymbol{\chi}) \leq 0, \quad \dot{\lambda}_p \geq 0, \quad \dot{\lambda}_p f_p(\boldsymbol{\sigma}, \boldsymbol{\chi}) = 0 \tag{7}$$

$$\dot{\omega} = \dot{\lambda}_d \frac{\partial f_d}{\partial \boldsymbol{Y}}, \qquad -\dot{\boldsymbol{\xi}}^d = \dot{\lambda}_d \frac{\partial f_d}{\partial \boldsymbol{\chi}} \tag{8}$$

$$f_d(\boldsymbol{Y}, \boldsymbol{\chi}) \leq 0, \qquad \dot{\lambda}_d \geq 0, \qquad \dot{\lambda}_d f_d(\boldsymbol{Y}, \boldsymbol{\chi}) = 0 \tag{9}$$

where $\dot{\lambda}_p$ and $\dot{\lambda}_d$ are the plastic and damage consistency coefficients. The fact that $\boldsymbol{\chi}$ (or at least a subset of components of $\boldsymbol{\chi}$) appears as argument of either f_p and f_d interprets the physical interaction between damage and plasticity.

Since the evolutive laws are associative, the following Druckerian inequality holds:

$$(\boldsymbol{\sigma} - \bar{\boldsymbol{\sigma}}) : \dot{\boldsymbol{\varepsilon}}^p + (\boldsymbol{Y} - \bar{\boldsymbol{Y}}) \cdot \dot{\omega} - (\boldsymbol{\chi} - \bar{\boldsymbol{\chi}}) \cdot \dot{\boldsymbol{\xi}} \geq 0 \tag{10}$$

in which the triples $(\boldsymbol{\sigma}, \boldsymbol{Y}, \boldsymbol{\chi})$ and $(\dot{\boldsymbol{\varepsilon}}^p, \dot{\omega}, \dot{\boldsymbol{\xi}} = \dot{\boldsymbol{\xi}}^p + \dot{\boldsymbol{\xi}}^d)$ correspond to each other through (6) to (9), whereas $(\bar{\boldsymbol{\sigma}}, \bar{\boldsymbol{Y}}, \bar{\boldsymbol{\chi}})$ is any triple such that $f_p(\bar{\boldsymbol{\sigma}}, \bar{\boldsymbol{\chi}}) \leq 0$ and $f_d(\bar{\boldsymbol{Y}}, \bar{\boldsymbol{\chi}}) \leq 0$. For further details see Polizzotto *et al.* (1996), Borino *et al.* (1996).

2.2 The Shakedown Problem

A solid body (or structure), composed of material as described previously, is referred to Cartesian orthogonal co-ordinates (x_1, x_2, x_3) in its undeformed initial state. It is subjected to external quasi-static actions as volume forces $\boldsymbol{b}$ distributed in the domain V, surface forces $\boldsymbol{T}$ distributed on the free part, S_t, of its boundary surface S, imposed displacements $\boldsymbol{g}$ on the constrained part, $S_u = S \backslash S_T$, and imposed strains $\boldsymbol{\varepsilon}^\theta$ in V. These actions are linearly expressed in terms of a vector $\boldsymbol{P}$ of independent parameters, which is allowed to range within a prescribed domain Π of the Euclidean space. Any load path $\boldsymbol{P} = \boldsymbol{P}(t)$, $0 \leq t \leq t_f$ in Π is an Admissible Load History (ALH) for the structure. The resulting displacements and strains are, by hypothesis, infinitesimal. The influence of temperature changes on material data is negligible.

For a specified load path $\boldsymbol{P} = \boldsymbol{P}(t)$, $0 \leq t \leq t_f$ with assigned initial conditions (e.g. $\boldsymbol{\varepsilon}^p = \boldsymbol{0}$, $\boldsymbol{\xi}^p = \boldsymbol{\xi}^d = \boldsymbol{0}$, $\omega = \boldsymbol{0}$ in V at $t = 0$), the related structural response can in principle be evaluated making use of the constitutive equations (2), (3) and (6)–(9), as well as of the equilibrium and compatibility equations, i.e.

$$\operatorname{div} \boldsymbol{\sigma} + \boldsymbol{b} = \boldsymbol{0} \quad in \ V, \qquad \boldsymbol{\sigma} \cdot \boldsymbol{n} = \boldsymbol{T} \quad on \ S_t \tag{11}$$

$$\boldsymbol{\varepsilon} = \nabla^s \boldsymbol{u} \quad in \ V, \qquad \boldsymbol{u} = \boldsymbol{g} \quad on \ S_u. \tag{12}$$

This response constitutes an ***equilibrium path*** of the structure. Suitable step-by-step procedures must be applied to this purpose, but here we are not interested in this analysis problem.

In fact, we want to find out —on the basis of some pre-requisite of the structure/load system— the conditions under which (enlarged) shakedown occurs in the structure subjected to repeated loads $\boldsymbol{P} \in \Pi$. Obviously, if shakedown occurs, the structure may suffer plastic deformations and damage during a transient initial period without failure, then no further plastic deformations, nor damage, occur and the actual response will coincide with the elastic response to the loads with initial plastic strains $\boldsymbol{\varepsilon}^p = \bar{\boldsymbol{\varepsilon}}^p$, damage, $\omega = \bar{\omega}$

and kinematic internal variables, $\boldsymbol{\xi} = \bar{\boldsymbol{\xi}} = \bar{\boldsymbol{\xi}}^p + \bar{\boldsymbol{\xi}}^d$, the overbarred symbols denoting values cumulated at the end of the transient phase.

In other words, if shakedown occurs, there exists an initial state of plastic strains, $\bar{\boldsymbol{\varepsilon}}^p(\boldsymbol{x})$, of damage, $\bar{\omega}(\boldsymbol{x})$, and of kinematic internal variables, $\bar{\boldsymbol{\xi}}(\boldsymbol{x})$, such that the elastic response to the loads with these initial conditions, say $\widehat{\boldsymbol{\sigma}}, \widehat{\boldsymbol{Y}}, \widehat{\boldsymbol{\chi}}$, does not violate the yield and damage conditions for any load $\boldsymbol{P} \in \Pi$. Is this condition also sufficient ? Here we try to give an answer to this question.

By hypothesis, there exists in the structure an initial state described by the fields $\bar{\boldsymbol{\varepsilon}}^p$, $\bar{\omega}$, $\bar{\boldsymbol{\xi}} = \bar{\boldsymbol{\xi}}^p + \bar{\boldsymbol{\xi}}^d$, such that the ensuing elastic response satisfies

$$f_p(\widehat{\boldsymbol{\sigma}}, \widehat{\boldsymbol{\chi}}) < 0, \qquad f_d(\widehat{\boldsymbol{Y}}, \widehat{\boldsymbol{\chi}}) < 0 \qquad \text{in } V \quad \forall \boldsymbol{P} \in \Pi_B \tag{13}$$

where Π_B is the set of the *basic loads*, i.e. the smallest subset of $\partial\Pi$, whose convex hull coincides with the convex hull of Π. Because of the convexity of f_p and f_d, (13) is also satisfied for any $\boldsymbol{P} \in \Pi$. So, considering an ALH, $\boldsymbol{P}(t)$, $0 \le t \le t_f$, and denoting by $\boldsymbol{\sigma}$, $\boldsymbol{\varepsilon}$, $\boldsymbol{\chi}, \ldots$ the related elastic-plastic-damage response of the structure, we can use the Druckerian inequality (10) to write:

$$j \equiv (\boldsymbol{\sigma} - \widehat{\boldsymbol{\sigma}}) : \dot{\boldsymbol{\varepsilon}}^p + \left(\boldsymbol{Y} - \widehat{\boldsymbol{Y}}\right) \cdot \dot{\omega} - (\boldsymbol{\chi} - \widehat{\boldsymbol{\chi}}) \cdot \dot{\boldsymbol{\xi}} \ge 0 \quad \text{in } V \tag{14}$$

which holds for $0 \le t \le t_f$ (with the equality sign only if $\dot{\boldsymbol{\varepsilon}}^p = \boldsymbol{0}$, $\dot{\omega} = 0$, $\dot{\boldsymbol{\xi}} = \boldsymbol{0}$). Noting that

$$\dot{\boldsymbol{\varepsilon}}^p = \dot{\boldsymbol{\varepsilon}} - \dot{\widehat{\boldsymbol{\varepsilon}}} - (\dot{\boldsymbol{\varepsilon}}^e - \dot{\widehat{\boldsymbol{\varepsilon}}}^e) \tag{15}$$

integrating (14) over V and substituting from (15), the integral

$$\int_V (\boldsymbol{\sigma} - \widehat{\boldsymbol{\sigma}}) : (\dot{\boldsymbol{\varepsilon}} - \dot{\widehat{\boldsymbol{\varepsilon}}})\, \mathrm{d}V = 0 \tag{16}$$

being vanishing by the virtual work principle, gives

$$\int_V \left[(\boldsymbol{\sigma} - \widehat{\boldsymbol{\sigma}}) : \left(\dot{\boldsymbol{\varepsilon}}^e - \dot{\widehat{\boldsymbol{\varepsilon}}}^e\right) - \left(\boldsymbol{Y} - \widehat{\boldsymbol{Y}}\right) \cdot \dot{\omega} + (\boldsymbol{\chi} - \widehat{\boldsymbol{\chi}}) \cdot \dot{\boldsymbol{\xi}} \right] \mathrm{d}V = - \int_V j\, \mathrm{d}V < 0. \tag{17}$$

By the position

$$W(t) = W_0 + \int_0^t \int_V \left[(\boldsymbol{\sigma} - \widehat{\boldsymbol{\sigma}}) : \left(\dot{\boldsymbol{\varepsilon}}^e - \dot{\widehat{\boldsymbol{\varepsilon}}}^e\right) - \left(\boldsymbol{Y} - \widehat{\boldsymbol{Y}}\right) \cdot \dot{\omega} + (\boldsymbol{\chi} - \widehat{\boldsymbol{\chi}}) \cdot \dot{\boldsymbol{\xi}} \right] \mathrm{d}V \mathrm{d}\tau \tag{18}$$

where W_0 is a constant, (17) can be written as

$$\dot{W} = - \int_V j\, \mathrm{d}V < 0. \tag{19}$$

The above procedure is typical of shakedown theory, in which an inequality like (19) is always encountered with the functional W exhibiting evident its feature of being positive definite. Here, a direct proof that W of (18) is positive definite seems not achievable depending on the mathematical structure of the first two products in the integrand on the r.h.s. of (18). In the next subsection, we will try an alternative indirect procedure. Now we conclude stating that, by (19), W decreases during the deformation process at a finite rate as far as some plastic and/or damage mechanism are active and thus, if W is positive definite —or in any case bounded from below—, then it must stop decreasing at a certain time, t_a say, after which $\dot{W} = 0$, hence $j = 0$ everywhere in V, i.e. (14) is satisfied as an equality, which is possible only if no further plastic and damage mechanisms occur, i.e. $\dot{\boldsymbol{\varepsilon}}^p = \boldsymbol{0}$, $\dot{\boldsymbol{\omega}} = \boldsymbol{0}$, $\dot{\boldsymbol{\xi}} = \boldsymbol{0}$ in V for all $t \geq t_a$. That is, shakedown occurs provided the functional W in (18) is positive definite, or anyway bounded from below. This condition can be assumed to hold also in the limit case in which the inequality sign $<$ of (13) is replaced by $\leq$.

2.3 D–Stability

Let the body considered before be subjected to an arbitrary (reference) load history, say $\boldsymbol{P}^{(a)}(t)$, $t \geq 0$, and related initial conditions $\bar{\boldsymbol{\varepsilon}}^{p\,(a)}$, $\bar{\boldsymbol{\omega}}^{(a)}$, $\bar{\boldsymbol{\xi}}^{(a)}$. An external agency slowly applies a load perturbance $\Delta\boldsymbol{P}$ consisting of volume forces $\Delta\boldsymbol{b}$ in V and surface forces $\Delta\boldsymbol{T}$ on S_t, specified in the time interval $0 \leq t \leq t_1$, possibly with changes on the initial conditions, say $\Delta\bar{\boldsymbol{\varepsilon}}^p$, $\Delta\bar{\boldsymbol{\omega}}$, $\Delta\bar{\boldsymbol{\xi}}$ in V. As a result of this perturbance, the body will be subjected to a new (neighbour) load history, say $\boldsymbol{P}^{(b)}(t) = \boldsymbol{P}^{(a)}(t) + \Delta\boldsymbol{P}(t)$ and new initial conditions, say $\bar{\boldsymbol{\varepsilon}}^{(b)} = \bar{\boldsymbol{\varepsilon}}^{(a)} + \Delta\bar{\boldsymbol{\varepsilon}}$, etc. Let the structural responses to $\boldsymbol{P}^{(a)}(t)$ and to $\boldsymbol{P}^{(b)}(t)$ be labeled with symbols as $(\cdot)^{(a)}$ and $(\cdot)^{(b)}$ and let $\Delta(\cdot) = (\cdot)^{(b)} - (\cdot)^{(a)}$ denote the related difference response. According to Drucker (1960), the structure is stable-in-the-large if the energy L_{ext} supplied by the external agency is nonnegative, i.e.

$$L_{\text{ext}} = \bar{W} + \int_0^{t_1} \left\{ \int_V \Delta\boldsymbol{b} \cdot \Delta\dot{\boldsymbol{u}} \mathrm{d}V + \int_{S_t} \Delta\boldsymbol{T} \cdot \Delta\dot{\boldsymbol{u}} \mathrm{d}S \right\} \mathrm{d}t \geq 0 \tag{20}$$

where $\bar{W}$ denotes the energy spent to modify the initial conditions. Inequality (20) must be satisfied for arbitrary perturbances, as well as for any load history $\boldsymbol{P}^{(a)}(t)$ belonging to some domain Π_s.

Since $\Delta\dot{\boldsymbol{\varepsilon}} = \Delta\dot{\boldsymbol{\varepsilon}}^e + \Delta\dot{\boldsymbol{\varepsilon}}^p$, we can write the identity

$$\begin{aligned} \Delta\boldsymbol{\sigma} : \Delta\dot{\boldsymbol{\varepsilon}} = & \left(\Delta\boldsymbol{\sigma} : \Delta\dot{\boldsymbol{\varepsilon}}^e - \Delta\boldsymbol{Y} \cdot \Delta\dot{\boldsymbol{\omega}} + \Delta\boldsymbol{\chi} \cdot \Delta\dot{\boldsymbol{\xi}} \right) \\ & + \left(\Delta\boldsymbol{\sigma} : \Delta\dot{\boldsymbol{\varepsilon}}^p + \Delta\boldsymbol{Y} \cdot \Delta\dot{\boldsymbol{\omega}} - \Delta\boldsymbol{\chi} \cdot \Delta\dot{\boldsymbol{\xi}} \right) \end{aligned} \tag{21}$$

from where, after integration over V, we obtain

$$\int_V \Delta\boldsymbol{\sigma} : \Delta\dot{\boldsymbol{\varepsilon}}\, \mathrm{d}V = \int_V \left[\Delta\boldsymbol{\sigma} : \Delta\dot{\boldsymbol{\varepsilon}}^e - \Delta\boldsymbol{Y} \cdot \Delta\dot{\boldsymbol{\omega}} + \Delta\boldsymbol{\chi} \cdot \Delta\dot{\boldsymbol{\xi}} \right] \mathrm{d}V + \int_V \Delta^2 D\, \mathrm{d}V \tag{22}$$

where $\Delta^2 D$ is the second variation of the dissipation function, i.e.

$$\Delta^2 D = \Delta\boldsymbol{\sigma} : \Delta\dot{\boldsymbol{\varepsilon}}^p + \Delta\boldsymbol{Y} \cdot \Delta\dot{\boldsymbol{\omega}} - \Delta\boldsymbol{\chi} \cdot \Delta\dot{\boldsymbol{\xi}} \geq 0. \tag{23}$$

Considering that the integral on the l.h.s. of (22) equals the integral on the r.h.s. of (20) by the virtual work principle, and with the position

$$W(t) = \bar{W} + \int_0^t \int_V \left[\Delta\boldsymbol{\sigma} : \Delta\dot{\boldsymbol{\varepsilon}}^e - \Delta \boldsymbol{Y} \cdot \Delta\dot{\boldsymbol{\omega}} + \Delta\boldsymbol{\chi} \cdot \Delta\dot{\boldsymbol{\xi}}\right] \mathrm{d}V \mathrm{d}\tau, \tag{24}$$

we can finally write (20) in the form

$$L_{\text{ext}} = W(t_1) + \int_0^{t_1} \int_V \Delta^2 D \, \mathrm{d}V \mathrm{d}t \geq 0. \tag{25}$$

As the second addend on the r.h.s. of (25) is nonnegative and monotonically increasing, a necessary condition of stability in the large in Drucker's sense is that $W(t)$ be bounded from below, i.e.

$$W(t) > -\infty \quad \forall t \geq 0 \tag{26}$$

(or, to within an additive constant, nonnegative). A structure for which (26) is satisfied is *D–stable* and the *D–Stability Principle* states that any structure that is stable-in-the-large in the Drucker sense is also D–stable (but the converse may not be true). Denoting Π_D the D–stability domain, it is $\Pi_s \subseteq \Pi_D$, Polizzotto *et al.* (1996).

Let the two structural responses of the previous subsections be considered again, i.e. the actual elastic-plastic-damage response $\boldsymbol{\sigma}, \boldsymbol{\varepsilon}, \ldots$ to the loads $\boldsymbol{P}(t)$ with zero initial conditions, as well as the fictitious elastic response $\hat{\boldsymbol{\sigma}}, \hat{\boldsymbol{\varepsilon}}, \ldots$ to the same loads with initial conditions $\bar{\boldsymbol{\varepsilon}}^p, \bar{\boldsymbol{\omega}}, \bar{\boldsymbol{\xi}}$. Thus, since the latter response, because of (13), is the actual equilibrium path of the structure under the loads $\boldsymbol{P}(t)$ and initial conditions $\bar{\boldsymbol{\varepsilon}}^p, \bar{\boldsymbol{\omega}}, \bar{\boldsymbol{\xi}}$, it can be taken as the reference response labeled as $(\cdot)^{(a)}$ (with $\dot{\boldsymbol{\varepsilon}}^{p\,(a)} = \boldsymbol{0}$, $\dot{\boldsymbol{\omega}}^{(a)} = \boldsymbol{0}$, $\dot{\boldsymbol{\xi}}^{(a)} = \boldsymbol{0}$ identically), whereas the perturbance consists in removing these initial conditions to obtain the (neighbour) elastic-plastic-damage equilibrium path (with $\dot{\boldsymbol{\varepsilon}}^{p\,(b)} = \dot{\boldsymbol{\varepsilon}}^p$, $\dot{\boldsymbol{\omega}}^{(b)} = \dot{\boldsymbol{\omega}}$, $\dot{\boldsymbol{\xi}}^{(b)} = \dot{\boldsymbol{\xi}}$ in V). It follows that the functional (24), applied to these two equilibrium paths, becomes as (18) with $\bar{W} = W_0$. Furthermore, if the structure is D–stable within a sufficiently large domain $\Pi_D \supset \Pi$, then (26) holds and the following static-type enlarged shakedown theorem can be stated:

For a given D–stable structure of elastic-plastic-damage material subjected to loads $\boldsymbol{P}$ ranging within a given domain, a necessary and sufficient condition for (enlarged) shakedown is that there exist some time-independent fields of plastic strains, $\bar{\boldsymbol{\varepsilon}}^p$, of damage, $\bar{\boldsymbol{\omega}}$, and kinematic internal variables, $\bar{\boldsymbol{\xi}}$, such that the ensuing fictitious elastic response satisfies the conditions

$$f_p(\hat{\boldsymbol{\sigma}}, \hat{\boldsymbol{\chi}}) \leq 0, \qquad f_d\left(\hat{\boldsymbol{Y}}, \hat{\boldsymbol{\chi}}\right) \leq 0 \quad \text{in } V, \ \forall \boldsymbol{P} \in \Pi_B. \tag{27}$$

The above statement considers D–stability as a pre-requisite of the structure, but unfortunately no criterion is available yet to recognize such a pre-requisite with reasonable *simplicity.*

3 Shakedown for Ductile Damage

3.1 Generals

In this section, we address the weak-form shakedown, that is,, shakedown for ductile damage, (i.e. for damage produced by plastic deformations). The material obeys the state equations (3) and the evolutive plasticity laws:

$$\dot{\varepsilon}^p = \dot{\lambda}\,\frac{\partial f_p}{\partial \sigma}, \qquad -\dot{\xi} = \dot{\lambda}\,\frac{\partial f_p}{\partial \chi} \tag{28}$$

$$f_p(\sigma,\chi,\omega) \le 0, \qquad \dot{\lambda} \ge 0, \qquad \dot{\lambda}\, f_p(\sigma,\chi,\omega) = 0. \tag{29}$$

The evolutive law for damage is not specified, but the related dissipation must be always nonnegative, i.e.

$$\boldsymbol{Y}\cdot\dot{\boldsymbol{\omega}} = -\frac{1}{2}\,\boldsymbol{\varepsilon}^e : \dot{\boldsymbol{E}} : \boldsymbol{\varepsilon}^e = \frac{1}{2}\,\boldsymbol{\sigma} : \dot{\boldsymbol{C}} : \boldsymbol{\sigma} \ge 0 \tag{30}$$

where $\boldsymbol{C}(\boldsymbol{\omega}) = \boldsymbol{E}^{-1}(\boldsymbol{\omega})$ and the equality sign holds if, and only if, either $\boldsymbol{\sigma} = \mathit{0}$, or $\dot{\boldsymbol{\omega}} = \mathit{0}$, or both.

If shakedown, in the weak sense meant here, actually occurs in the structure subjected to repeated loads $\boldsymbol{P} \in \Pi$, the structure exhibits plastic deformations and ductile damage during an initial transient period without failure, then plastic strains and their consequences (including ductile damage) cease, and the structure will respond to the subsequent loads as an elastic structure, with constant ductile damage,. So a necessary condition of shakedown is that there exists some initial plastic deformation state $(\bar{\varepsilon}^p, \bar{\xi})$ such that the elastic response of the structure to any ALH, this structure being damaged like in the actual post-transient process, does not violate the yield condition.

In the following subsections, two sufficient weak-form shakedown theorems are presented, one static and another kinematic (Polizzotto *et al.*, 2001). To this purpose the notion of ***trial damage domain***, $\mathcal{D}$, is introduced, which is the set of all sufficiently regular fields of allowable damage in V:

$$\mathcal{D} = \{\boldsymbol{\omega} = \boldsymbol{\omega}(\boldsymbol{x}) : 0 \le \boldsymbol{\omega} \le \boldsymbol{\omega}^a \quad \forall \boldsymbol{x} \in V\}. \tag{31}$$

Here, $\boldsymbol{\omega}^a$ is the vector of the maximum allowable damage values (smaller than the critical values). In principle, $\mathcal{D}$ includes the actual damage produced by any ALH.

3.2 Sufficient Static-Type Theorem

A sufficient condition for the weak-form shakedown can be stated as follows:

For a given elastic-plastic-damage structure subjected to repeated loads $\boldsymbol{P} \in \Pi$, a sufficient condition for weak-form shakedown to occur is that there exist in the structure some initial fields of self-stress, $\widehat{\boldsymbol{\sigma}}^r$, and of static internal variable, $\widehat{\boldsymbol{\chi}}$, such that the structure —being in an arbitrarily trial damag, say $\widehat{\boldsymbol{\omega}} \in \mathcal{D}$— provides elastic stress responses not violating the (equally damaged) yield condition, that is

$$f_p\left(\widehat{\boldsymbol{\sigma}}, \widehat{\boldsymbol{\chi}}, \widehat{\boldsymbol{\omega}}\right) < 0 \quad \text{in } V,\ \forall\, \boldsymbol{P} \in \Pi_B \tag{32}$$

where , by definition $\widehat{\sigma} = \sigma^E_{\widehat{\omega}} + \widehat{\sigma}^r$.

Proof. — Due to the convexity of f_p, (32) is also satisfied for all $P \in \Pi$ and thus also for any ALH, say $P(t)$, $t \geq 0$, even if $\widehat{\omega}$ changes in time, i.e. $\widehat{\omega}(t) \in \mathcal{D}\ \forall t \geq 0$. Let σ, ε, ω, ξ, ... denote the actual elastic-plastic-damage response to some ALH (with zero initial conditions). Since (32) holds also with $\widehat{\omega}$ coincident with the actual damage ω, (which, by assumption, does not exceed ω^a), the Druckerian inequality can be written as

$$j_1 \equiv (\sigma - \widehat{\sigma}) : \dot{\varepsilon}^p - (\chi - \widehat{\chi}) \cdot \dot{\xi} \geq 0 \quad \text{in } V, \forall t \geq 0 \tag{33}$$

in which the equality sign holds if, and only if, $\dot{\varepsilon}^p = 0$, $\dot{\xi} = 0$. Also, because of (30), we can write

$$j_2 \equiv \frac{1}{2}(\sigma - \widehat{\sigma}) : \dot{C} : (\sigma - \widehat{\sigma}) \geq 0 \quad \text{in } V, \forall t \geq 0, \tag{34}$$

where $\dot{C}(\omega)$ is also related to the actual damage. Summing (33) and (34) with each other, then integrating over V and using the identity

$$\dot{\varepsilon}^p = \dot{\varepsilon} - \dot{\widehat{\varepsilon}} - C : \left(\dot{\sigma} - \dot{\widehat{\sigma}}\right) - \dot{C} : (\sigma - \widehat{\sigma}) \tag{35}$$

gives

$$\int_V (j_1 + j_2)\, dV = \int_V (\sigma - \widehat{\sigma}) : \left(\dot{\varepsilon} - \dot{\widehat{\varepsilon}}\right) dV - \int_V (\chi - \widehat{\chi}) \cdot \dot{\xi}\, dV$$
$$- \int_V (\sigma - \widehat{\sigma}) : \left[C : \left(\dot{\sigma} - \dot{\widehat{\sigma}}\right) + \frac{1}{2}\dot{C} : (\sigma - \widehat{\sigma})\right] dV. \tag{36}$$

The first integral on the r.h.s. of (36) vanishes by the virtual work principle, whereas the remaining integral terms are the time derivative of the positive definite functional

$$W(t) = \int_V \frac{1}{2}(\sigma - \widehat{\sigma}) : C(\omega) : (\sigma - \widehat{\sigma})\, dV + \int_V Z(\chi, \widehat{\chi})\, dV \tag{37}$$

where $Z(\chi, \widehat{\chi})$, after Halphen (1978), Maier (1987) and Polizzotto *et al.* (1991), is defined as

$$Z(\chi, \widehat{\chi}) = \Omega(\widehat{\chi}) - \Omega(\chi) - \xi(\chi) \cdot (\widehat{\chi} - \chi) \geq 0 \tag{38}$$

and $\Omega(\chi)$ is the Legendre transform of $\Psi_{in}(\xi)$, i.e. $\Omega(\chi) = \chi \cdot \xi - \Psi_{in}(\xi)$, such that $\xi = \partial\Omega(\chi)/\partial\chi$. (Due to the convexity of Ψ_{in} and hence of Ω, Z is nonnegative.) Therefore, (36) becomes:

$$\frac{dW}{dt} = -\int_v (j_1 + j_2)\, dV < 0 \tag{39}$$

where the negative sign is due to the fact that (33) and (34) cannot be identically vanishing —otherwise the real process would be elastic since the beginning, what is excluded. By (39), W must decrease at a finite rate during the plastic-damage process and thus, by a classical argument, at a certain finite time t_a necessarily $\dot{W} = 0$, hence $j_1 = j_2 = 0$ everywhere in V, for all $t \geq t_a$, i.e. (33) and (34) hold as equalities. The former condition, i.e. $j_1 \equiv 0$, can be satisfied if, and only if, $\dot{\varepsilon}^p = 0$, $\dot{\xi} = 0$ everywhere in V and $\forall t \geq t_a$; whereas the latter one, i.e. $j_2 \equiv 0$, can be satisfied in two ways: either with $\dot{\omega} = \mathbf{0}$ in V for all $t \geq t_a$, what certainly occurs if damage is driven by plastic strains (ductile damage, as assumed), or because $\boldsymbol{\sigma} = \widehat{\boldsymbol{\sigma}}$ in V for all $t \geq t_a$, what certainly would occur for nonductile damage (brittle damage, or from other sources, anyway excluded by the assumed constitutive model). Thus weak-form shakedown occurs at $t = t_a$, meaning that plastic strains cease together with ductile damage. The theorem is so proven. QED.

Note that the above theorem reduces to that of Hachemi and Weichert (1992) if the damage is postulated to coincide with the actual one in the theorem's statement itself.

3.3 Sufficient Kinematic-Type Theorem

Another sufficient condition for the weak-form shakedown is the following:

For a given elastic-plastic-damage structure subjected to repeated loads $\boldsymbol{P} \in \Pi$, a sufficient condition for the weak-form shakedown to occur is that there exist a scalar $m > 1$ such that the inequality

$$\mathcal{E}_{\omega}[\dot{\boldsymbol{\varepsilon}}^{pc}, \dot{\boldsymbol{\xi}}^{c}] \equiv \int_0^{t_f} \int_V \left[D(\dot{\boldsymbol{\varepsilon}}^{pc}, \dot{\boldsymbol{\xi}}^{c}, \widehat{\omega}) - m \boldsymbol{\sigma}^E_{\widehat{\omega}} : \dot{\boldsymbol{\varepsilon}}^{pc} \right] \mathrm{d}V \, \mathrm{d}t \geq 0 \tag{40}$$

be satisfied for all PAMs (plastic accumulation mechanisms) and for all ALHs (admissible load histories), say $\boldsymbol{P}(t)$, $0 \leq t \leq t_f$, the damage field $\widehat{\omega} = \widehat{\omega}(\boldsymbol{x})$ being an arbitrary trial damage, i.e $\widehat{\omega} \in \mathcal{D}$.

Proof.— Let us consider an arbitrary, but fixed, trial damage, say $\widehat{\omega} = \widehat{\omega}(\boldsymbol{x}) \in \mathcal{D}$, and let $\boldsymbol{P}(t)$, $0 \leq t \leq t_f$, be a chosen ALH. Under these conditions, (40) implies that the functional $\mathcal{E}_{\widehat{\omega}}[\dot{\boldsymbol{\varepsilon}}^{pc}, \dot{\boldsymbol{\xi}}^{c}]$ has a vanishing absolute minimum in the set of all PAMs, that is in the set of fields $\dot{\boldsymbol{\varepsilon}}^{pc}(\boldsymbol{x}, t)$, $\dot{\boldsymbol{\xi}}^{c}(\boldsymbol{x}, t)$ such that

$$\Delta \boldsymbol{\varepsilon}^{pc}(\boldsymbol{x}) = \int_0^{t_f} \dot{\boldsymbol{\varepsilon}}^{pc}(\boldsymbol{x}, t) \, \mathrm{d}t = \nabla^s \boldsymbol{u}^c \text{ in } V, \qquad \boldsymbol{u}^c = \mathbf{0} \text{ on } S_u. \tag{41}$$

$$\Delta \boldsymbol{\xi}^{c}(\boldsymbol{x}) = \int_0^{t_f} \dot{\boldsymbol{\xi}}^{c}(\boldsymbol{x}, t) \, \mathrm{d}t = \mathbf{0} \quad \text{in } V \tag{42}$$

Because the problem

$$\min \mathcal{E}_{\widehat{\omega}}[\dot{\boldsymbol{\varepsilon}}^{pc}, \dot{\boldsymbol{\xi}}^{c}] \qquad \text{in the set } \{\dot{\boldsymbol{\varepsilon}}^{pc}, \dot{\boldsymbol{\xi}}^{c}\} \text{ of PAMs} \tag{43}$$

admits a solution (note that $\mathcal{E}_{\widehat{\omega}}[\mathbf{0}, \mathbf{0}] = 0$), then a solution must have the related Euler-Lagrange equations.

These equations can be obtained by the Lagrange multiplier method starting from the relevant Lagrangian functional, i.e.

$$\begin{aligned} \mathcal{E}^L_{\widehat{\omega}} = \mathcal{E}_{\widehat{\omega}}[\dot{\boldsymbol{\varepsilon}}^{pc}, \dot{\boldsymbol{\xi}}^{c}] &+ \int_V (m \widehat{\boldsymbol{\sigma}}^r) : \left[\nabla^s \boldsymbol{u}^c - \int_0^{t_f} \dot{\boldsymbol{\varepsilon}}^{pc} \, \mathrm{d}t \right] \mathrm{d}V \\ &- \int_{S_u} \boldsymbol{u}^c \cdot (m \widehat{\boldsymbol{\sigma}}^r) \cdot \boldsymbol{n} \, \mathrm{d}S + \int_0^{t_f} \int_V (m \widehat{\boldsymbol{\chi}}) \cdot \dot{\boldsymbol{\xi}}^{c} \, \mathrm{d}V \, \mathrm{d}t \end{aligned} \tag{44}$$

where $(m\widehat{\sigma}^r)$ and $(m\widehat{\chi})$ denote Lagrange multipliers. Then, writing the first variation of (44), we have, after some mathematics:

$$\begin{aligned}\delta\mathcal{E}^L_{\widehat{\omega}} &= \int_0^{t_f}\int_V \delta\dot{\varepsilon}^{pc} : \left[\frac{\partial D}{\partial\dot{\varepsilon}^{pc}} - m(\sigma^E_{\widehat{\omega}} + \widehat{\sigma}^r)\right] \mathrm{d}V\,\mathrm{d}t \\ &+ \int_0^{t_f}\int_V \delta\dot{\xi}^c \cdot \left[\frac{\partial D}{\partial\dot{\xi}^c} + m\widehat{\chi}\right] \mathrm{d}V\,\mathrm{d}t - m\int_V \delta u^c \cdot \mathrm{div}\,\widehat{\sigma}^r\,\mathrm{d}V \\ &+ m\int_{S_t} \delta u^c \cdot \widehat{\sigma}^r \cdot n\mathrm{d}S + m\int_V \delta\widehat{\sigma}^r : \left[\nabla^s u^c - \int_0^{t_f} \dot{\varepsilon}^{pc}\,\mathrm{d}t\right] \mathrm{d}V \\ &- m\int_{S_u} u^c \cdot \delta\widehat{\sigma}^r \cdot n\mathrm{d}S + m\int_V \delta\widehat{\chi} \cdot \int_0^{t_f} \dot{\xi}^c\,\mathrm{d}t\,\mathrm{d}V. \end{aligned} \tag{45}$$

Since $\delta\mathcal{E}^L_{\widehat{\omega}}$ must vanish for all variation fields, it results that the stationarity conditions read:

$$m(\sigma^E_{\widehat{\omega}} + \widehat{\sigma}^r) = \frac{\partial D}{\partial\dot{\varepsilon}^{pc}}, \qquad m\widehat{\chi} = -\frac{\partial D}{\partial\dot{\xi}^c} \quad \text{in } V,\ \forall t \in (0,t_f) \tag{46}$$

$$\mathrm{div}\,\widehat{\sigma}^r = \boldsymbol{0} \ \text{ in } V, \qquad \widehat{\sigma}^r \cdot n = \boldsymbol{0} \ \text{ on } S_t, \tag{47}$$

besides (41) and (42). The Lagrange multipliers are thus qualified as self-stresses and static internal variables, both being time-independent. From (46), the stress fields $m(\sigma^E_{\widehat{\omega}} + \widehat{\sigma}^r)$ and $m\widehat{\chi}$, being derived from the dissipation function, are plastically admissible, i.e.

$$f_p(m\widehat{\sigma}, m\widehat{\chi}, \widehat{\omega}) \leq 0 \quad \text{in } V,\ \forall t \in (0,t_f) \tag{48}$$

where $\widehat{\sigma} = \sigma^E_{\widehat{\omega}} + \widehat{\sigma}^r$ by definition. Since by the convexity of f_p, $f_p(m\sigma, m\chi, \widehat{\omega}) > f_p(\sigma, \chi, \widehat{\omega})\ \forall m > 1$, from (48) follows that

$$f_p(\widehat{\sigma}, \widehat{\chi}, \widehat{\omega}) < 0 \quad \text{in } V,\ \forall t \in (0,t_f). \tag{49}$$

The latter inequality holds good if, taking fixed the trial damage field, the ALH is changed in all possible ways and thus for all $\boldsymbol{P} \in \Pi \supset \Pi_B$, therefore (49) can also be written as

$$f_p(\sigma^E_{\widehat{\omega}} + \widehat{\sigma}^r, \widehat{\chi}, \widehat{\omega}) < 0 \quad \text{in } V,\ \forall \boldsymbol{P} \in \Pi_B, \tag{50}$$

which holds for arbitrary choices of the trial damage $\widehat{\omega} \in \mathcal{D}$. This is the sufficient static shakedown criterion of the previous subsection and thus the weak-form shakedown occurs. QED.

3.4 Shakedown Safety Factor

The above sufficient static and kinematic theorems are recognizable as the necessary and sufficient shakedown theorems for the structure considered elastic-plastic with a fixed damage distribution $\widehat{\omega}(\boldsymbol{x})$ in V. It follows that the shakedown safety factor of this damaged structure can be evaluated using one or another of the two theorems.

We can thus write, assuming $\Pi_B = \{\boldsymbol{P}_i, i = 1, 2, \ldots, N_B\}$ and with ω in place of $\widehat{\omega}$:

—*By the static-type approach*

$$\beta_s[\omega] = \max_{(\beta,\widehat{\sigma}^r,\widehat{\chi})} \beta \tag{51a}$$

subject to :

$$\left.\begin{array}{l} f_p\left(\beta\sigma^E_{\omega\,(i)} + \widehat{\sigma}^r, \widehat{\chi}, \omega\right) \leq 0 \quad \text{in } V, \forall i \in I(N_B) \\ \text{— Equilibrium equations on } \widehat{\sigma}^r \end{array}\right\} \tag{51b}$$

—*By the kinematic-type approach*

$$\beta_s[\omega] = \min_{(\varepsilon^{pc}_{(i)},\xi^c_{(i)})} \sum_{i=1}^{N_B} \int_V D(\varepsilon^{pc}_{(i)}, \xi^c_{(i)}, \omega)\,\mathrm{d}V \tag{52a}$$

subject to :

$$\left.\begin{array}{l} \displaystyle\sum_{i=1}^{N_B} \int_V \sigma^E_{\omega\,(i)} : \varepsilon^{pc}_{(i)}\,\mathrm{d}V = 1 \\ \text{— Constraints (41) and (42)} \end{array}\right\} \tag{52b}$$

where $\beta_s[\omega]$ is the shakedown safety factor of the chosen damaged structure.

A lower bound to the true shakedown safety factor (β^*_s) can be obtained by writing:

$$\beta^-_s = \min_{(\omega)} \beta_s[\omega] \quad \text{s.t. } \omega \in \mathcal{D}. \tag{53}$$

It was proven in Polizzotto *et al.* (2001) that $\beta^-_s = \beta_s[\omega = \omega^a]$ and that $\beta^+_s = \beta_s[\omega = 0]$ is a upper bound to β^*_s such that

$$\beta^-_s \leq \beta^*_s \leq \beta^+_s. \tag{54}$$

Note that a direct evaluation of β^*_s cannot be achieved since no shakedown theorem being simultaneously necessary and suffcient is available for this weak-form shakedown; also note that the bounds of (54) can hardly be made more stringent as long as the damage evolutive law is left unspecified.

3.5 An Upper Bound to Damage

We have to check whether the structure subjected to loads $\beta\,\boldsymbol{P}$ with $\beta < \beta^*_s$ may suffer a damage exceeding ω^a or not. This task can be achieved using some sort of simplified analysis, e.g. a bounding technique.

In case of ductile damage, an upper bound to damage can be derived in the shape of upper bound to the driving plastic strain. This kind of exercise is particularly simple in case of ductile isotropic damage, which —according to a model of Lemaitre and Chaboche (1985)— is proportional to the accumulated plastic strain, that is

$$\dot{\omega} = R\dot{p} \tag{55}$$

where R depends on the stress tri-axiality ratio and other material constants and $\dot{p} = \left(\frac{2}{3}\,\dot{\varepsilon}^p : \dot{\varepsilon}^p\right)^{1/2}$. For a Mises yield function, it is also $\dot{p} = \dot{\lambda}$ and $D = (1-\omega)k\,\dot{\lambda}$, with $k =$ yield stress, and thus

$$\dot{\omega} = \frac{R}{k}\,\frac{D(\dot{\varepsilon}^p, \dot{\boldsymbol{\xi}}, \omega)}{(1-\omega)}. \tag{56}$$

Assuming R constant for the present purpose, it is thus sufficient to construct an upper bound to D. As known, this task can be pursued by applying a suitable perturbation to the yield condition (32) on which the shakedown theorem is based, that is writing

$$f_p(g\,\widehat{\boldsymbol{\sigma}}, g\,\widehat{\boldsymbol{\chi}}, \widehat{\omega}) \le 0 \quad \text{in } V,\ \forall\, \boldsymbol{P} \in \Pi_B \tag{57}$$

where $g = 1 + \alpha\,\gamma(\boldsymbol{x})$, $\gamma(\boldsymbol{x}) \ge 0$ in V and α positive scalar. The above inequality must hold for $0 \le \widehat{\omega} \le \omega^a$. As (57), hence also (56), is valid also for $\forall \boldsymbol{P} \in \Pi$, and thus for any ALH with $\widehat{\omega}$ coincident with the actual damage, i.e. $\widehat{\omega}(t) \equiv \omega(t)$, we can write the Druckerian inequality

$$(\boldsymbol{\sigma} - g\,\widehat{\boldsymbol{\sigma}}) : \dot{\varepsilon}^p - (\boldsymbol{\chi} - g\,\widehat{\boldsymbol{\chi}}) \cdot \dot{\boldsymbol{\xi}} \ge 0 \quad \text{in } V,\ \forall t \ge 0 \tag{58}$$

which can be easily shown to be equivalent to

$$\alpha\,\gamma\,D - g\left[(\boldsymbol{\sigma} - \widehat{\boldsymbol{\sigma}}) : \dot{\varepsilon}^p - (\boldsymbol{\chi} - \widehat{\boldsymbol{\chi}}) \cdot \dot{\boldsymbol{\xi}}\right] \ge 0 \quad \text{in } V,\ \forall t \ge 0 \tag{59}$$

Considering that the following inequality holds, i.e.

$$g\,\frac{1}{2}\,(\boldsymbol{\sigma} - \widehat{\boldsymbol{\sigma}}) : \dot{\boldsymbol{C}} : (\boldsymbol{\sigma} - \widehat{\boldsymbol{\sigma}}) \ge 0 \quad \text{in } V,\ \forall t \ge 0, \tag{60}$$

summing (59) and (60) with each other, integrating over V and using the identity (35) —herein still valid—, we obtain, by (56) with R considered constant:

$$\alpha \int_V \gamma D\,\mathrm{d}V = \frac{\alpha k}{R} \int_V \gamma(1-\omega)\dot{\omega}\,\mathrm{d}V \le -g\,\frac{\mathrm{d}}{\mathrm{d}t} W(t) \tag{61}$$

where $W(t)$ is given by (37). Note that, denoting by $\bar{\gamma}$ the maximum value of γ in V and posing $\bar{g} = 1 + \alpha\bar{\gamma}$, we have $\bar{g}/g \ge 1$ and thus (61) can be enforced by writing

$$\int_V \gamma(1-\omega)\dot{\omega}\,\mathrm{d}V \le -\frac{\bar{g}R}{\alpha k}\,\frac{\mathrm{d}W}{\mathrm{d}t}. \tag{62}$$

Then, integrating in time in the interval $(0, t_1)$ gives

$$\frac{1}{2} \int_V \gamma(2\omega - \omega^2)\,\mathrm{d}V \bigg|_{t_1} \le \frac{\bar{g}k}{\alpha R}\,[W(0) - W(t_1)], \tag{63}$$

that is, dropping the positive subtractive term (so enforcing the inequality),

$$\int_V \gamma(2\omega - \omega^2)\,\mathrm{d}V\bigg|_{t_1} \leq \frac{2\bar{g}R}{\alpha k} W_0 \tag{64}$$

where

$$W_0 = W(0) = \int_V \left[\frac{1}{2}\widehat{\boldsymbol{\sigma}}^r : \boldsymbol{C}_0 : \widehat{\boldsymbol{\sigma}}^r + \Omega(\widehat{\boldsymbol{\chi}})\right] \mathrm{d}V \tag{65}$$

and $\boldsymbol{C}_0 = \boldsymbol{C}(\boldsymbol{0})$ and $\widehat{\boldsymbol{\chi}} = \boldsymbol{\chi}(\widehat{\boldsymbol{\xi}})$.

Assuming $\gamma = 1$ in a small region $\Delta V \subset V$ and $\gamma = 0$ elsewhere, such that $\bar{g} = 1$, and denoting by $\widetilde{\omega}$ the mean value of $\omega(\boldsymbol{x}, t_1)$ in ΔV, inequality (64) reads

$$\widetilde{\omega}^2 - 2\widetilde{\omega} + K \geq 0. \tag{66}$$

Therefore, provided that $K \leq 1$, the following inequality holds

$$\widetilde{\omega} \leq 1 - \sqrt{1-K} \tag{67}$$

which is a bound on $\widetilde{\omega}$. This result was given by Polizzotto *et al.* (2001).

Upper bounds on isotropic damage through upper bounds on plastic deformation parameters were given by Feng and Yu (1994, 1995) and by Hachemi and Weichert (1997, 1998).

4 Conclusions

Shakedown theory for damaging materials has still to be improved before reaching a development level as satisfactory as for other nondamaging material models. Two types of shakedown have been here considered, i) the *enlarged shakedown* in which plastic deformation and damage are considered as independent dissipative mechanisms, both of which cease eventually, after a transient initial period during which plastic strain and damage occur without producing failure; and ii) *weak-form shakedown* in which ductile damage is considered as induced by plastic deformation (with an evolutive law left unspecified) and plastic deformation eventually ceases together with ductile damage.

The notion of enlarged shakedown is most appropriate for elastic-plastic-damage materials, but still remain some difficulties for a complete satisfactory proof of a related static-type theorem. It has been shown that the shakedown problem is equivalent to a stability problem in Drucker's sense (D–stability) and that such a theorem holds for a class of D–stable structures, but there remain the need for establishing some pre-requisites to recognize structures in this class.

The notion of weak-form shakedown may be of importance considering that the majority of metals and alloys exhibit ductile damage. However, the proposed shakedown theorems are perhaps rather restrictive as they substantially require that the weakest structure obtained with damage fixed at the maximum allowable value (below the critical value) is still capable to shake down to the elastic regime under the given loading programme.

It is to point out that, in any case, shakedown does not guarantee that plastic deformations, as well as damage, do not reach dangerous levels during the transient phase. Like plastic deformations, also damage must be assessed by the use of bounding techniques. These techniques are well developed for plastic deformations, not for damage. In case of ductile damage, bounding techniques of plastic deformations can be usefully employed, as it has been shown by Feng and Yu (1994, 1995), Hachemi and Weichert (1997, 1998), Polizzotto *et al.* (2001). However, bounding techniques for damage of greater generality need to be developed in the future research work.

References

Borino, G., Fuschi, P., and Polizzotto, C. (1996). Elastic-plastic-damage constitutive models with coupling internal variables. *Mechanics Research Communications* 23:19–28.

Drucker, D. C. (1960). Plasticity. In Goodier, J. N., and Hoff, J. H., eds., *Structural Mechanics.* London: Pergamon Press. 407–455.

Druyanov, B., and Roman, I. (1995). Shakedown theorem extended to elastic nonperfectly plastic bodies. *Mechanics Research Communications* 22:571–576.

Druyanov, B., and Roman, I. (1998). On adaptation (shakedown) of a class of damaged elastic bodies to cyclic loading. *European Journal of Mechanics A/Solids* 17:71–78.

Feng, X.-Q., and Yu, S.-W. (1994). An upper bound on damage of elastic-plastic structures at shakedown. *Int. Journal of Damage Mechanics* 3:277–289.

Feng, X.-Q., and Yu, S.-W. (1995). Damage and shakedown analysis of structures with strain-hardening. *Int. Journal of Plasticity* 11:237–249.

Hachemi, A., and Weichert, D. (1992). An extension of the static shakedown theorem to a certain class of inelastic materials with damage. *Archives of Mechanics* 44:491–498.

Hachemi, A., and Weichert, D. (1997). Application of shakedown theory to damaging inelastic material under mechanical and thermal loads. *Int. J. of Mechanical Sciences* 39:1067–1076.

Hachemi, A., and Weichert, D. (1998). Numerical shakedown analysis of damaged structures. *Computer Methods in Applied Mechanics and Engineering* 160:57–70.

Halphen, B. (1979). Accomodation et adaptation des structures elasto-visco-plastiques et plastiques. In Ecole Polytechnique Palaiseau, ed., *Matériaux et Structures Sous Chargement Cyclique.* Paris: Association Amicale des Ingénieurs Anciens Elèves, Ecóle National Ponts et Chaussees. 203–229.

Ju, J. W. (1989). On energy-based coupled elastoplastic damage theories: constitutive modelling and computational aspects. *Int. Journal of Solids and Structures* 25:803–833.

Lemaitre, J., and Chaboche, J.-L. (1985). *Mécanique des Materiaux Solides.* Paris: Dunod.

Maier, G. (1987). A generalization to nonlinear hardening of the first shakedown theorem for discrete elastic-plastic models. *Atti Accademia dei Lincei, Rendiconti Fisica* 8, LXXXI:161–174.

Perzyna, P. (1984). Constitutive modelling of dissipative solids for postcritical behaviour and failure. *Journal of Engineering Materials and Technologies, ASME* 106:410–419.

Polizzotto, C., Borino, G., Caddemi, S., and Fuschi, P. (1991). Shakedown problems for material models with internal variables. *European Journal of Mechanics A/Solids* 10:621–639.

Polizzotto, C., Borino, G., and Fuschi, P. (1996). An extended shakedown theory for elastic-plastic-damage material models. *European Journal of Mechanics A/Solids* 15:825–858.

Polizzotto, C., Borino, G., and Fuschi, P. (2001). Weak forms of shakedown for elastic-plastic structures exhibiting ductile damage. *Meccanica* (to appear)

Siemaszko, A. (1993). Inadaptation analysis with hardening and damage. *European Journal of Mechanics A/Solids* 12:237–248.

Advanced Material Modelling in Shakedown Theory

Dieter Weichert and Abdelkader Hachemi

Aachen University of Technology, Aachen, Germany

Abstract. The aim of this lecture is to present possibilities how to extend the validity of shakedown theory to other than linear elastic-ideal plastic or linear elastic- unlimited linear hardening material behaviour in conjunction with the validity of the normality rule. More precisely, the following items will be discussed: application of the General Standard Material Model, introduction of material damage in shakedown theory, use of no-associated flow rules and the notion of the Sanctuary of Elasticity.

1.1 Introduction

In the pioneering works on shakedown theory by Melan (1936, 1938) and Koiter (1956, 1960) the plastic effects covered by the theoretical development were restricted to linear-elastic, ideal plastic and linear-elastic, unlimited linear kinematic hardening material behaviour, associated with the validity of the normality rule. However, real material behaviour is much more complicated and therefore, much efforts have been undertaken to enlarge the class of material laws, to which shakedown analysis can be applied. In this lecture some recent extensions of this kind will be presented. In particular the issues of material hardening, material damage, non-associated flow-rules and a general formulation of the static shakedown theorem covering a large class of materials will be addressed.

1.2 Formulation of the Problem and Basic Assumptions

We consider the behaviour of a three-dimensional elastic-plastic body $\mathscr{B}$ under the action of quasistatically varying external agencies **a*** consisting of surface tractions **p*** and surface displacements **u*** acting on the disjoint parts S_p and S_u of the surface S of $\mathscr{B}$, respectively, and volume forces $\mathbf{f}^*$. In the initial configuration C_i at the time $t = 0$, $\mathscr{B}$ occupies the volume V. The motion of $\mathscr{B}$ is described by the use of Cartesian coordinates, where the positions of the particles of $\mathscr{B}$ in the undeformed and deformed state are given by the coordinates $\mathbf{X} = [X_1, X_2, X_3]$ and $\mathbf{x} = [x_1, x_2, x_3]$, respectively. The actual configuration C_t of $\mathscr{B}$ is then defined by the displacement function **u** (Figure 1)

$$\mathbf{u}(\mathbf{X}, t) = \mathbf{x}(\mathbf{X}, t) - \mathbf{X}. \tag{1.1}$$

Then, the static and kinematic equations of the geometrically non-linear boundary value problem referred to the initial configuration are defined by

(a) *Statical equations*

$$\mathrm{Div}(\mathbf{F}.\boldsymbol{\sigma}) = -\mathbf{f}^* \qquad \text{in } V \tag{1.2}$$

$$\mathbf{n}.(\mathbf{F}.\boldsymbol{\sigma}) = \mathbf{p}^* \qquad \text{on } S_p \tag{1.3}$$

(b) *Kinematical equations*

$$\mathbf{F} = \partial\mathbf{x}/\partial\mathbf{X} \qquad \text{in V} \tag{1.4}$$

$$\mathbf{u}(\mathbf{X}) = \mathbf{u}^*(\mathbf{X}) \qquad \text{on } S_u \tag{1.5}$$

$$\boldsymbol{\varepsilon} = 1/2(\mathbf{F}^T.\mathbf{F} - \mathbf{I}) \qquad \text{in } V \tag{1.6}$$

with Div as divergence operator, **F** as deformation gradient, $\boldsymbol{\sigma}$ as second Piola-Kirchhoff stress tensor and $\boldsymbol{\varepsilon}$ as Green-Lagrange strain tensor. **I** denote the metric tensor of second rank and **n** is the outer normal vector to S in C_i.

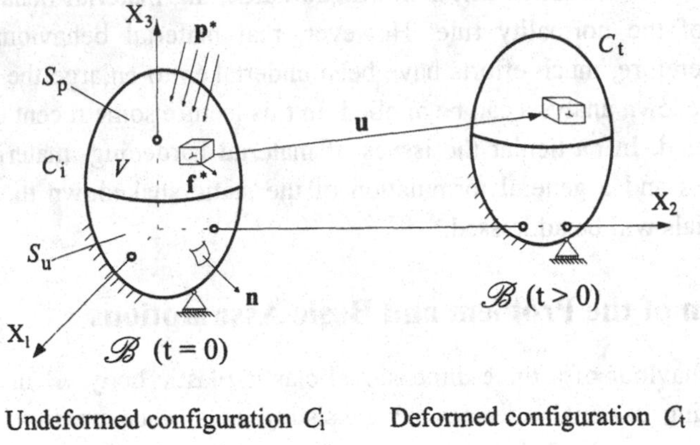

Figure 1. Structure or body $\mathscr{B}$.

Elastic-plastic deformations are described by means of a fictitious intermediate configuration C_K, derived from the multiplicative decomposition of the deformation gradient **F** into an elastic part $\mathbf{F}^e$ and a plastic part $\mathbf{F}^p$ (Figure 2) (Lee, 1969)

$$\mathbf{F} = \mathbf{F}^e\,\mathbf{F}^p \tag{1.7}$$

where $\mathbf{F}^e$ is obtained by unloading all infinitesimal neighbourhoods of the body $\mathscr{B}$. This decomposition provides the relation between elastic, plastic and total deformation valid for finites strains and leads to an additive decomposition of the Green-Lagrange strain tensor $\boldsymbol{\varepsilon}$ into a purely plastic part $\boldsymbol{\varepsilon}^p$ and an elastic part $\boldsymbol{\varepsilon}^e$ depending on the plastic deformation (Green and Naghdi, 1965)

$$\boldsymbol{\varepsilon} = \boldsymbol{\varepsilon}^e + \boldsymbol{\varepsilon}^p \tag{1.8}$$

with

$$\boldsymbol{\varepsilon}^e = \frac{1}{2}(\mathbf{F}^p)^T\,[(\mathbf{F}^e)^T(\mathbf{F}^e) - \mathbf{I}](\mathbf{F}^p) \quad \text{and} \quad \boldsymbol{\varepsilon}^p = \frac{1}{2}[(\mathbf{F}^p)^T(\mathbf{F}^p) - \mathbf{I}]. \tag{1.9}$$

In the geometrically linearised form we obtain

(a) *Statical equations*

$$\mathrm{Div}(\boldsymbol{\sigma}) = -\,\mathbf{f}^* \qquad \text{in } V \tag{1.10}$$

$$\mathbf{n}.(\boldsymbol{\sigma}) = \mathbf{p}^* \qquad \text{on } S_p \tag{1.11}$$

(b) *Kinematical equations*

$$\mathbf{F} = \partial\mathbf{x}/\partial\mathbf{X} \qquad \text{in V} \tag{1.12}$$

$$\mathbf{u}(\mathbf{X}) = \mathbf{u}^*(\mathbf{X}) \qquad \text{on } S_u \tag{1.13}$$

$$\boldsymbol{\varepsilon} = (\mathbf{F} - \mathbf{I})_{\,\mathrm{sym.}} \qquad \text{in } V \tag{1.14}$$

where no distinction is to be made between the different stress tensors (Kirchhoff, Cauchy, etc.). $\boldsymbol{\varepsilon}$ denotes the usual linearised strain tensor.

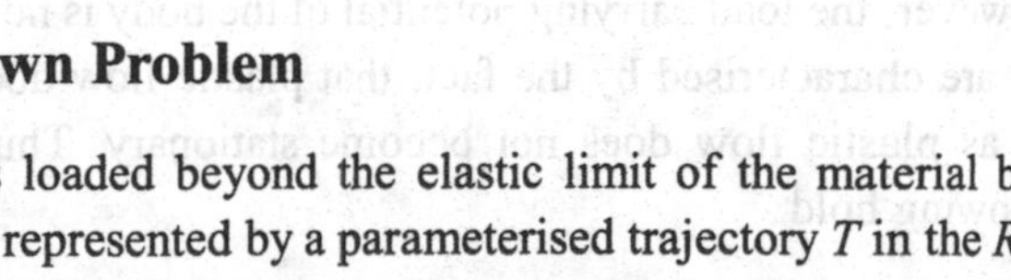

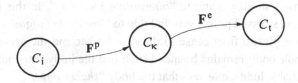

Figure 2. Kinematic of elastic-plastic deformation.

1.3 The Shakedown Problem

The solid body $\mathscr{B}$ is loaded beyond the elastic limit of the material by variable loads. The evolution of the loads represented by a parameterised trajectory T in the R-dimensional space of generalised loads (R denotes the number of independent loading systems) is not precisely given. However, the envelope of T for all times of the process is given (Figure 3).

Will there be *unlimited accumulation of inelastic deformations* in $\mathscr{B}$ or will the *evolution of inelastic deformations stop after some time* t?

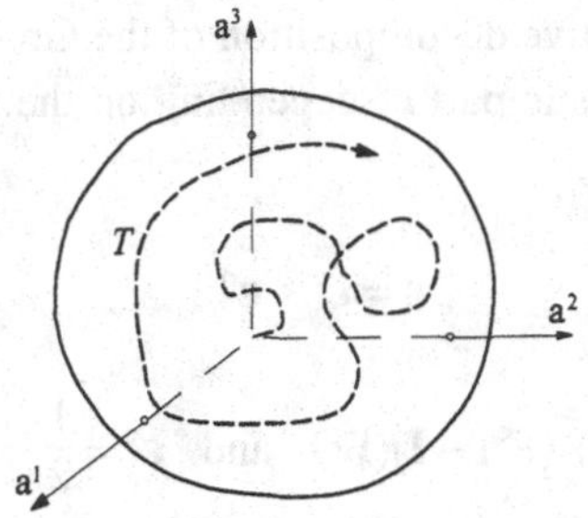

Figure 3. Trajectory in the loading domain.

The behaviour of the structure or body $\mathcal{B}$ subjected to variable loads $\mathbf{a}^*(t)$ (Figure 1), can be classified in the following ways (see Figure 4) (König, 1987):

i. If the load intensities remain sufficiently low, the response of the body is "*purely elastic*" (with the exception of stress singularities).

ii. If the load intensities become sufficiently high, the instantaneous load-carrying capacity of the body becomes exhausted and unconstrained plastic flow occurs. The body collapses in this case.

iii. If the plastic strain increments in each load cycle are of the same sign then, after a sufficient number of cycles, the total strains (and therefore displacements) become so large that the body departs from its original form and becomes unserviceable. This phenomenon is called "*incremental collapse*" or "*ratchetting*".

iv. If the strain increment change sign in every cycle, they tend to cancel each other and total deformation remains small leading to "*alternating plasticity*". In this case, however, the material at the most stressed points may fail due to "*low-cycle fatigue*".

v. If, after some time plastic flow cease to develop further and the accumulated dissipated energy in the whole body remains bounded such that the body responds purely elastically to the applied variable loads, one says that the body "*shakes down*".

The behaviour of the body according to the first point is not dangerous, since no plastic deformation occurs. However, the load carrying potential of the body is not fully exploited. The failure of types (*ii*)-(*iv*) are characterised by the fact, that plastic flow does not cease and that related quantities such as plastic flow does not become stationary. Thus, there exist points $\mathbf{x} \in V$ for which the following hold

$$\lim_{t \to \infty} \dot{\boldsymbol{\varepsilon}}^p(\mathbf{x}, t) \neq 0. \tag{1.15}$$

If the case (v) occurs, the body shakes down for the given history of loading $\mathbf{a}^*(t)$. It follows that

$$\lim_{t \to \infty} \dot{\boldsymbol{\varepsilon}}^p (\mathbf{x}, t) = 0. \tag{1.16}$$

If one accounts for plastic deformation in the structural design, it seems natural to require that, for any possible loading history, the plastic deformation in the considered body will stabilise, i.e. the body will shakes down. It is worthwhile mentioning that the phenomena of incremental collapse "*ratchetting*" and alternating plasticity "*low-cycle fatigue*" may appear simultaneously, e.g. if one component of the plastic strain tensor increases with each load cycle whereas another oscillates.

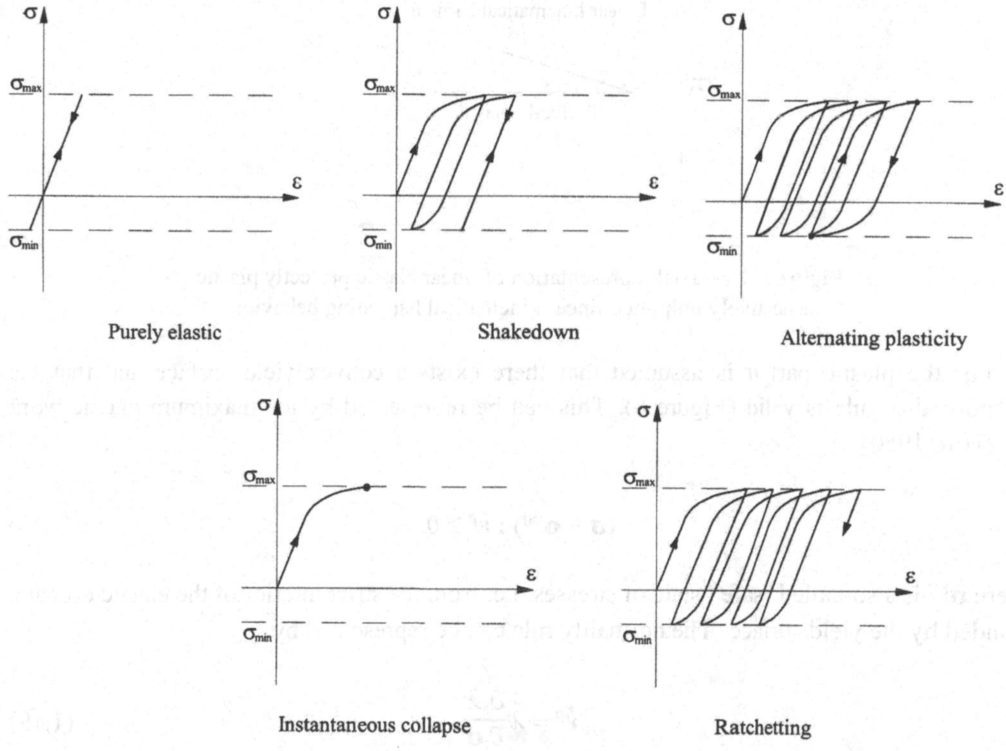

Figure 4. Possibilities of local response to cyclic loading.

1.4 Assumption on the Constitutive Relations

We first concentrate on the material effects in the framework of the geometrically linear theory of continuum mechanics. The assumptions on the constitutive relations in classical shakedown theory are as follows:

1. The additive decomposition of total strains into elastic and an inelastic part is justified

$$\boldsymbol{\varepsilon} = \boldsymbol{\varepsilon}^{e} + \boldsymbol{\varepsilon}^{p} \tag{1.17}$$

2. The material is either linear elastic-perfectly plastic or linear elastic-unlimited linear kinematic hardening (Figure 5).

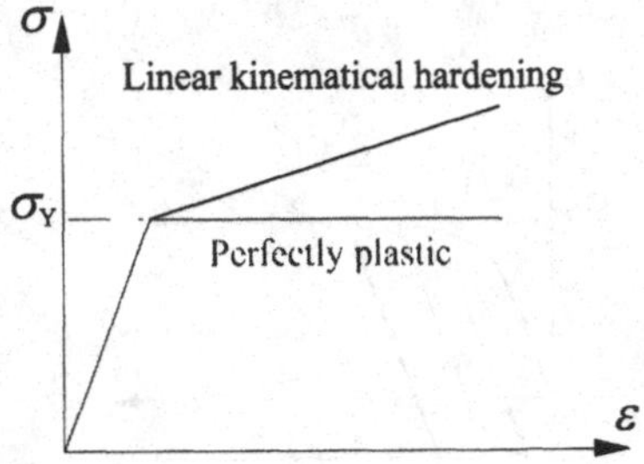

Figure 5. Uni-axial representation of linear elastic perfectly plastic respectively unlimited linear kinematical hardening behaviour.

3. For the plastic part it is assumed that there exists a convex yield surface and that the normality rule is valid (Figure 6). This can be represented by the maximum plastic work (Hill, 1950)

$$(\boldsymbol{\sigma} - \boldsymbol{\sigma}^{(s)}) : \dot{\boldsymbol{\varepsilon}}^{p} \geq 0 \tag{1.18}$$

where $\boldsymbol{\sigma}^{(s)}$ is a so-called "safe" state of stresses, i.e. from the strict interior of the elastic domain, bounded by the yield surface. The normality rule can be represented by

$$\dot{\boldsymbol{\varepsilon}}^{p} = \dot{\lambda} \frac{\partial \mathscr{F}}{\partial \boldsymbol{\sigma}} \tag{1.19}$$

with $\mathscr{F}(\boldsymbol{\sigma})$ as yield function and $\dot{\lambda}$ as plastic multiplier.

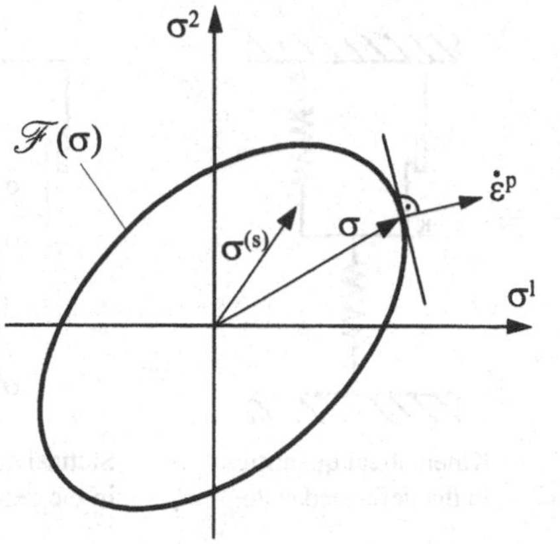

Figure 6. Convexity of the yield surface.

1.5 The Generalised Standard Material Model

By the introduction of internal parameters representing the change of the materials properties without affecting the equations of equilibrium and compatibility, this concept can be easily extended to a larger class of materials. One particularly attractive way to do so is to use the General Standard Material Model (GSMM) (Halphen and Nguyen, 1975): By introducing additional dimensions in the space of stresses and strains, limited kinematical hardening can be treated in the same manner as perfectly plastic material (Figures 7 and 8).

The generalised stresses and strains are then defined by

$$\mathbf{e} = [\boldsymbol{\varepsilon}, \mathbf{0}]^T;\ \mathbf{e}^e = [\boldsymbol{\varepsilon}^e, \boldsymbol{\omega}]^T;\quad \mathbf{e}^p = [\boldsymbol{\varepsilon}^p, \boldsymbol{\kappa}]^T;\quad \mathbf{s} = [\boldsymbol{\sigma}, \boldsymbol{\pi}]^T \tag{1.20}$$

where $\boldsymbol{\omega}$, $\boldsymbol{\kappa}$ and $\boldsymbol{\pi}$ denote the vectors of internal parameters. Then, the elastic behaviour of the material is represented by

$$\mathbf{s} = \mathbf{M} : \mathbf{e}^e \tag{1.21}$$

where **M** is a constant positive definite supermatrix defined by

$$\mathbf{M} = \begin{pmatrix} \mathbf{L} & \mathbf{0} \\ \mathbf{0} & \mathbf{Z} \end{pmatrix}. \tag{1.22}$$

The usual tensor of elastic moduli is represented by **L**, whereas the elastic relation between internal parameters of stresses and strains according to the illustration of the rheological model is represented by the tensor **Z**.

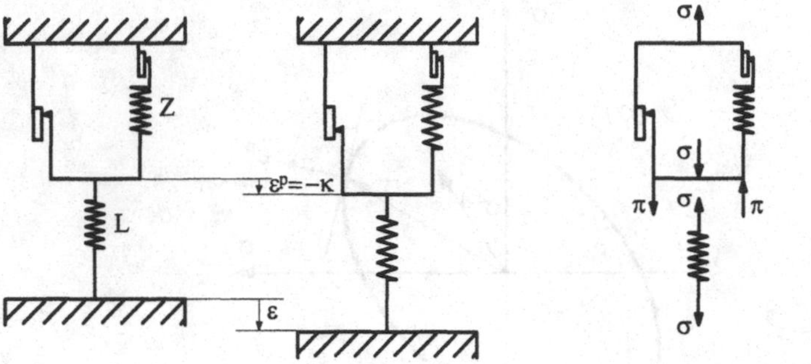

Figure 7. Rheological model of linear elastic-limited hardening plastic behaviour.

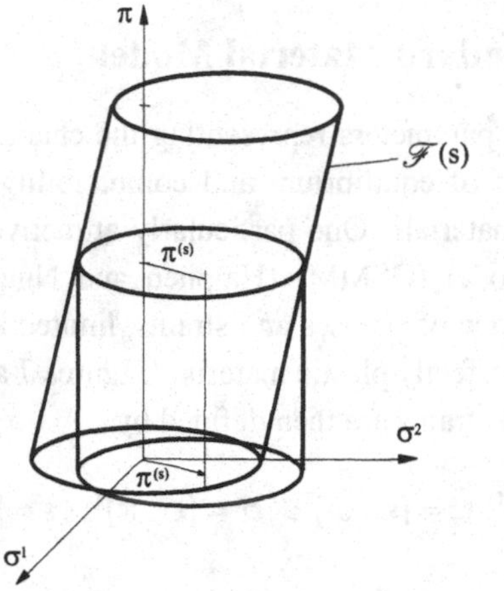

Figure 8. Representation of the yield surface in the space of generalised stresses.

1.6 Material Damage

With progressing plastic deformations, ductile materials, in particular metals, exhibit defects on a scale much larger than the scale of dislocations, which are responsible for plastic deformations, and much smaller than the characteristic length of the considered body. These defects will be called "damage" and can be represented in the framework of continuum mechanics by the reduction of the apparent stiffness. In order to find a convenient form for application in shakedown theory, the concept of effective stress is adopted in what follows (Figure 9) (Kachanov, 1958).

Here, $D = 0$ corresponds to the undamaged state, for a damaged state the damage variable D takes a value $D \in (0, D_c)$, where D_c corresponds to a critical value. Local rupture of the material is supposed to occur for $D = D_c$ with $D_c \in [0, 1]$.

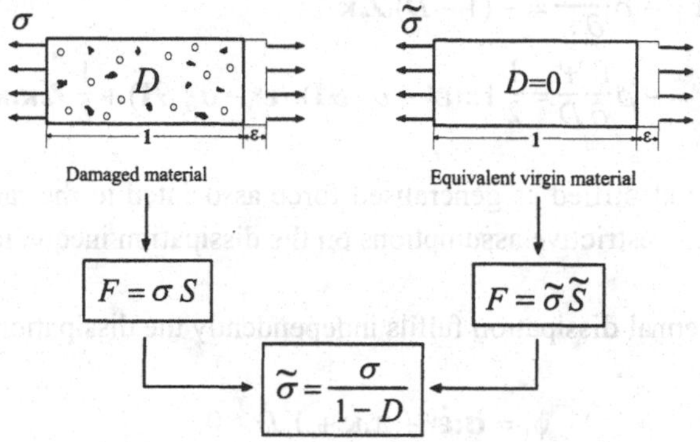

Figure 9. Illustration of the concept of effective stress.

1.7 Thermodynamic Framework

The second principle of thermodynamics applied to solid mechanics is used in the form of the dissipation inequality given by Clausius-Duhem

$$\boldsymbol{\sigma}:\dot{\boldsymbol{\varepsilon}} - \rho(\dot{\Psi} + s\dot{T}) + \mathbf{g}.\mathbf{q} \geq 0 \tag{1.23}$$

with the free energy Ψ composed of the parts Ψ_e and Ψ_p, related to the immediately recoverable and the stored elastic energy density, respectively and T denotes the absolute temperature

$$\Psi = \Psi_e(\boldsymbol{\varepsilon}^e, T, D) + \Psi_p(\boldsymbol{\kappa}, D) \tag{1.24}$$

where

$$\mathbf{g} = -\,\mathbf{grad}T/T. \tag{1.25}$$

Straightforward application to elastic-plastic hardening material with damage leads to (Lemaitre and Chaboche, 1985; Lemaitre, 1987; Ju, 1989)

$$\boldsymbol{\sigma}:\dot{\boldsymbol{\varepsilon}}^p + \boldsymbol{\pi}.\dot{\boldsymbol{\kappa}} + Y\dot{D} + \mathbf{g}.\mathbf{q} \geq 0 \tag{1.26}$$

with the definitions

$$\boldsymbol{\sigma} = \rho \frac{\partial \Psi}{\partial \boldsymbol{\varepsilon}^e} = (1 - D)\mathbf{L}:(\boldsymbol{\varepsilon}^e - \alpha_\vartheta\, \vartheta \mathbf{I}) \tag{1.27}$$

$$\boldsymbol{\pi} = -\rho \frac{\partial \Psi}{\partial \boldsymbol{\kappa}} = -(1 - D)\, \mathbf{Z}.\boldsymbol{\kappa} \tag{1.28}$$

$$Y = -\rho \frac{\partial \Psi}{\partial D} = \frac{1}{2}\mathbf{L}:(\boldsymbol{\varepsilon}^e - \alpha_\vartheta\, \vartheta \mathbf{I}):(\boldsymbol{\varepsilon}^e - \alpha_\vartheta\, \vartheta \mathbf{I}) + \frac{1}{2}\mathbf{Z}.\boldsymbol{\kappa}.\boldsymbol{\kappa}. \tag{1.29}$$

This way, Y can be identified as generalised force associated to the rate of damage $\dot{D}$. For the sequel, the following restrictive assumptions on the dissipation inequality are made:

- mechanical and thermal dissipation fulfils independently the dissipation inequality

$$\phi_1 = \boldsymbol{\sigma}:\dot{\boldsymbol{\varepsilon}}^p + \boldsymbol{\pi}.\dot{\boldsymbol{\kappa}} + Y\dot{D} \geq 0 \tag{1.30}$$

$$\phi_2 = \mathbf{g}.\mathbf{q} \geq 0 \tag{1.31}$$

- plastic and damage dissipation fulfils independently the dissipation inequality

$$\phi_p = \boldsymbol{\sigma}:\dot{\boldsymbol{\varepsilon}}^p + \boldsymbol{\pi}.\dot{\boldsymbol{\kappa}} \geq 0 \tag{1.32}$$

$$\phi_D = Y\dot{D} \geq 0 \tag{1.33}$$

To describe plasticity, the up mentioned GSMM is used. The normality rule for plastic flow is assumed valid, such that

$$\dot{\mathbf{e}}^p \in \delta\varphi(\mathbf{s}) \tag{1.34}$$

where $\delta\varphi(\mathbf{s})$ denotes the sub-gradients of the plastic potential $\varphi(\mathbf{s})$ (Halphen and Nguyen, 1975) which is the indicator function of a convex generalised elastic domain $\mathrm{P}(\mathbf{x})$ of all plastically admissible stress states

$$\mathbf{s}(\mathbf{x}) \in \mathrm{P}(\mathbf{x}), \quad \forall \mathbf{x} \in V. \tag{1.35}$$

$\mathrm{P}(\mathbf{x})$ is defined by means of a yield function $\mathscr{F}(\tilde{\mathbf{s}}, \mathbf{x})$

$$\mathrm{P}(\mathbf{x}) = \{\mathbf{s} \,/\, \mathscr{F}(\tilde{\mathbf{s}}, \mathbf{x}) \leq 0, \ \forall \mathbf{x} \in V\}. \tag{1.36}$$

Here, it is assumed that the yield function $\mathscr{F}(\tilde{\mathbf{s}}, \mathbf{x})$ is of von Mises type

$$\mathscr{F}(\tilde{\mathbf{s}}, \mathbf{x}) = \left\{ \frac{3}{2} \left(\frac{\boldsymbol{\sigma}^D}{1-D} - \frac{\boldsymbol{\pi}}{1-D} \right) : \left(\frac{\boldsymbol{\sigma}^D}{1-D} - \frac{\boldsymbol{\pi}}{1-D} \right) \right\}^{1/2} - \sigma_Y \leq 0 \tag{1.37}$$

where σ_Y denotes the yield stress and $\boldsymbol{\sigma}^D$ the deviatoric part of $\boldsymbol{\sigma}$. The convexity of $\mathscr{F}(\tilde{\mathbf{s}}, \mathbf{x})$ and the validity of the normality rule can be expressed by the generalised maximum plastic work inequality (Figure 10)

$$(\mathbf{s} - \mathbf{s}^{(s)}) : \dot{\mathbf{e}}^p \geq 0, \quad \forall \mathbf{s}^{(s)}(\mathbf{x}) \in \overline{\mathrm{P}}(\mathbf{x}) \tag{1.38}$$

in detail

$$(\boldsymbol{\sigma} - \boldsymbol{\sigma}^{(s)}) : \dot{\boldsymbol{\varepsilon}}^p + (\boldsymbol{\pi} - \boldsymbol{\pi}^{(s)}) . \dot{\boldsymbol{\kappa}} \geq 0 \tag{1.39}$$

where $\mathbf{s}^{(s)} = [\boldsymbol{\sigma}^{(s)}, \boldsymbol{\pi}^{(s)}]^T$ is any safe state of generalised stresses defined by

$$\overline{\mathrm{P}}(\mathbf{x}) = \{\mathbf{s}^{(s)} \,/\, \mathscr{F}(\tilde{\mathbf{s}}^{(s)}, \mathbf{x}) < 0, \ \forall \mathbf{x} \in V\}. \tag{1.40}$$

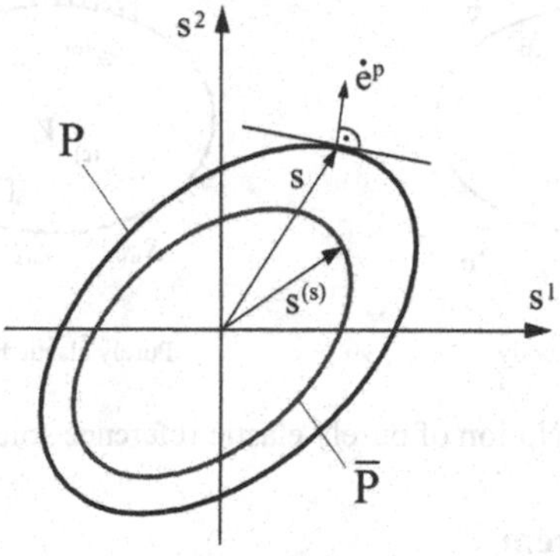

Figure 10. Illustration of the principle of maximum plastic work.

To complete this section, two simple models for the determination of the damage variable D are presented. These models will be used later to find local bounds for D (Hachemi, 1994; Hachemi and Weichert, 1997).

(a) *Model of Lemaitre (1985)*

$$D = \frac{D_c}{\varepsilon_R - \varepsilon_D} \left\langle \left\{ \frac{2}{3}(1+\nu) + 3(1-2\nu) \left(\frac{\sigma_H}{\sigma_{eq}} \right)^2 \right\} \varepsilon_{eq} - \varepsilon_D \right\rangle \tag{1.41}$$

(b) *Model of Shichun and Hua (1990)*

$$D = \frac{D_c}{\varepsilon_R - \varepsilon_D} \left\langle \exp\left\{\frac{3}{2}\left(\frac{\sigma_H}{\sigma_{eq}} - \frac{1}{3}\right)\right\} \varepsilon_{eq} - \varepsilon_D \right\rangle \tag{1.42}$$

with $\langle x \rangle = x$ if $x > 0$ and $\langle x \rangle = 0$ if $x \leq 0$. The scalars D_c, ε_R and ε_D are characteristic material coefficients and p is the equivalent plastic strain defined by $\varepsilon_{eq} = \int_0^t [2/3\dot{\boldsymbol{\varepsilon}}^p{:}\dot{\boldsymbol{\varepsilon}}^p]^{1/2}\, dt$.

1.8 The Extended Static Theorem of Shakedown

Of great importance in the theory of shakedown is the notion of the "*purely elastic reference solution*", usually supposed to be given (Figure 11). The associated boundary value problem differs from the actual elastic-plastic problem only by the fact, that the material behaviour is assumed to be linear elastic. Shape of the body, loads and kinematic boundary conditions are the same as in the original problem. All quantities related to this reference solution are indicated by superscript "$^{(c)}$".

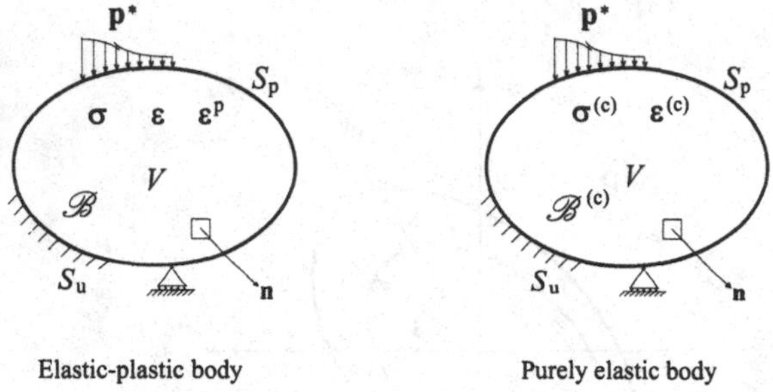

Figure 11. Notion of purely elastic reference solution.

1.8.1 Formulation of the Theorem

In this section, index notation will be used for easier understanding of calculus. Latin indices run form 1 to 3 and the summation convention is adopted. For simplicity we refer to Cartesian systems of co-ordinates so that no distinction is to be made between co- and contravariant co-ordinates. We then state: If there exists a field of time-independent residual generalised effective stresses $\tilde{\mathring{\mathbf{s}}}(\mathbf{x}) = [\tilde{\mathring{\boldsymbol{\rho}}}(\mathbf{x}), \tilde{\mathring{\boldsymbol{\pi}}}(\mathbf{x})]^T$ such that for $t > 0$, $\mathcal{F}(\tilde{\mathbf{s}}^{(c)}(\mathbf{x}, t) + \tilde{\mathring{\mathbf{s}}}(\mathbf{x}), \sigma_Y(\vartheta)) < 0$ then $\mathcal{B}$ shakes down.

The internal parameters to describe the state of hardening and damage in the material vanish naturally for the reference body $\mathscr{B}^{(c)}$, so that the generalised strains and stresses are given by

$$\mathbf{e}^{(c)} = \mathbf{e}^{e(c)} = [\boldsymbol{\varepsilon}^{(c)}, \mathbf{0}]^T, \ \mathbf{e}^{p(c)} = [\mathbf{0}, \mathbf{0}]^T, \ \mathbf{s}^{(c)} = [\boldsymbol{\sigma}^{(c)}, \mathbf{0}]^T. \tag{1.43}$$

The following system of equations is to be satisfied for the different fields

$$\sigma^{(c)}_{ij,j} = -f^*_i \quad \text{in } V \tag{1.44}$$

$$n_j\,\sigma^{(c)}_{ij} = p^*_i \quad \text{on } S_p \tag{1.45}$$

$$u^{(c)}_i = u^*_i \quad \text{on } S_u \tag{1.46}$$

$$\varepsilon^{(c)}_{ij} = \frac{1}{2}(u^{(c)}_{i,j} + u^{(c)}_{j,i}) \quad \text{in } V \tag{1.47}$$

$$\varepsilon^{(c)}_{ij} = L^{-1}_{ijkl}\,\sigma^{(c)}_{kl} + \alpha_\vartheta\,\vartheta\,\delta_{ij} \quad \text{in } V \tag{1.48}$$

The field of residual stresses $\mathring{\boldsymbol{\rho}}$ satisfies

$$\mathring{\rho}_{ij,j} = 0 \quad \text{in } V \tag{1.49}$$

$$n_j\,\mathring{\rho}_{ij} = 0 \quad \text{on } S_p \tag{1.50}$$

1.8.2 Proof

Following the idea of Melan (1936), a bounded, positive definite form is introduced. Here, the following form is chosen (see, Hachemi and Weichert, 1992; Feng and Yu, 1995; Polizzotto et al., 1996, 2001; Druyanov and Roman, 1998)

$$W = \frac{1}{2}\int_{(V)} \left\{(\rho_{ij} - \mathring{\rho}_{ij})\frac{L^{-1}_{ijkl}}{1-D}(\rho_{kl} - \mathring{\rho}_{kl}) + (\pi_m - \mathring{\pi}_m)\frac{Z^{-1}_{mn}}{1-D}(\pi_n - \mathring{\pi}_n)\right\} dV. \tag{1.51}$$

Using

$$\rho_{ij} = \sigma_{ij} - \sigma^{(c)}_{ij} \tag{1.52}$$

$$\mathring{\rho}_{ij} = \sigma^{(s)}_{ij} - \sigma^{(c)}_{ij} \tag{1.53}$$

$$\mathring{\pi}_m = \pi^{(s)}_m \tag{1.54}$$

and

$$\rho_{ij} = (1-D)\,\tilde{\rho}_{ij} \;\Rightarrow\; \dot{\rho}_{ij} = (1-D)\,\dot{\tilde{\rho}}_{ij} - \dot{D}\tilde{\rho}_{ij} \tag{1.55}$$

$$\pi_m = (1-D)\,\tilde{\pi}_m \;\Rightarrow\; \dot{\pi}_m = (1-D)\,\dot{\tilde{\pi}}_m - \dot{D}\tilde{\pi}_m \tag{1.56}$$

one gets the time-derivative of W

$$\begin{aligned} \dot{W} &= \int_{(V)} \left\{ (\sigma_{ij} - \sigma^{(s)}_{ij})\, L^{-1}_{ijkl}\, \dot{\tilde{\rho}}_{kl} + (\pi_m - \pi^{(s)}_m)\, Z^{-1}_{mn}\, \dot{\tilde{\pi}}_n \right\} \mathrm{d}V \\ &- \frac{1}{2}\int_{(V)} \left\{ (\tilde{\sigma}_{ij} - \tilde{\sigma}^{(s)}_{ij})\, L^{-1}_{ijkl}\, (\tilde{\sigma}_{kl} - \tilde{\sigma}^{(s)}_{kl}) + (\tilde{\pi}_m - \tilde{\pi}^{(s)}_m)\, Z^{-1}_{mn}\, (\tilde{\pi}_n - \tilde{\pi}^{(s)}_n) \right\} \dot{D}\, \mathrm{d}V. \end{aligned} \tag{1.57}$$

Using the generalised Hooke's law in rate form

$$\dot{\varepsilon}^{\rho}_{ij} = L^{-1}_{ijkl}\, \dot{\tilde{\rho}}_{kl}\,; \quad \dot{\omega}_m = Z^{-1}_{mn}\, \dot{\tilde{\pi}}_n \tag{1.58}$$

and the definitions

$$\dot{\varepsilon}^{\rho}_{ij} = \dot{\varepsilon}_{ij} - \dot{\varepsilon}^{p}_{ij} - \dot{\varepsilon}^{(c)}_{ij}\,; \quad \dot{\omega}_m = -\,\dot{\kappa}_m \tag{1.59}$$

one gets

$$\begin{aligned} \dot{W} &= \int_{(V)} \left\{ (\sigma_{ij} - \sigma^{(s)}_{ij})\, (\dot{\varepsilon}_{ij} - \dot{\varepsilon}^{p}_{ij} - \dot{\varepsilon}^{(c)}_{ij}) - (\pi_m - \pi^{(s)}_m)\, \dot{\kappa}_m \right\} \mathrm{d}V \\ &- \frac{1}{2}\int_{(V)} \left\{ (\tilde{\sigma}_{ij} - \tilde{\sigma}^{(s)}_{ij})\, L^{-1}_{ijkl}\, (\tilde{\sigma}_{kl} - \tilde{\sigma}^{(s)}_{kl}) + (\tilde{\pi}_m - \tilde{\pi}^{(s)}_m)\, Z^{-1}_{mn}\, (\tilde{\pi}_n - \tilde{\pi}^{(s)}_n) \right\} \dot{D}\, \mathrm{d}V. \end{aligned} \tag{1.60}$$

From the divergence theorem it follows

$$\int_{(V)} (\sigma_{ij} - \sigma^{(s)}_{ij})\, (\dot{\varepsilon}_{ij} - \dot{\varepsilon}^{(c)}_{ij})\, \mathrm{d}V = 0. \tag{1.61}$$

With

$$\Delta Y = \frac{1}{2} \left\{ (\tilde{\sigma}_{ij} - \tilde{\sigma}^{(s)}_{ij})\, L^{-1}_{ijkl}\, (\tilde{\sigma}_{kl} - \tilde{\sigma}^{(s)}_{kl}) + (\tilde{\pi}_m - \tilde{\pi}^{(s)}_m)\, Z^{-1}_{mn}\, (\tilde{\pi}_n - \tilde{\pi}^{(s)}_n) \right\} \tag{1.62}$$

one gets finally

$$\dot{W} = -\int_{(V)} \left\{(\sigma_{ij} - \sigma_{ij}^{(s)})\, \dot{\varepsilon}_{ij}^{p} + (\pi_m - \pi_m^{(s)})\, \dot{\kappa}_m\right\} dV - \int_{(V)} \Delta Y \dot{D}\, dV$$
$$= -\int_{(V)} (\mathbf{s} - \mathbf{s}^{(s)}) : \dot{\mathbf{e}}^{p}\, dV - \int_{(V)} \Delta Y \dot{D}\, dV. \quad (1.63)$$

One concludes that

$$W(0) \geq W(t) \text{ for } t > 0$$

and (1.64)

$$\dot{W} \to 0 \text{ and } W(t) \to \text{Const. for } t \to \infty.$$

1.8.3 Parameterisation of Loading

Be $\mathscr{D}$ the loading domain of external loads, defined by

$$\mathscr{D} = \left\{\mathscr{P} = \sum_{i=1}^{n} \mu_i \mathscr{P}_i^0;\quad \mu_i \in [\mu_i^-, \mu_i^+]\right\} \quad (1.65)$$

where $\mathscr{P}_i^0$ is the generalised vector of n independent (fixed) loads (Figure 12). In general, they are normalised and their values are defined by the multipliers μ_i. Two types of plastic failure under variable loads will be considered now in detail: Low-cycle plastic fatigue ("alternating plasticity") and incremental collapse ("ratchetting") (see, Figure 4). Here, the denomination "alternating plasticity" is favoured. The addressed type of structural behaviour is as follows: Although during cyclic loading no structural deformation is observed on a global level, locally alternating plastic deformations take place during the cycles of loading. This leads to the accumulation of local inelastic effects, material damage and finally local failure.

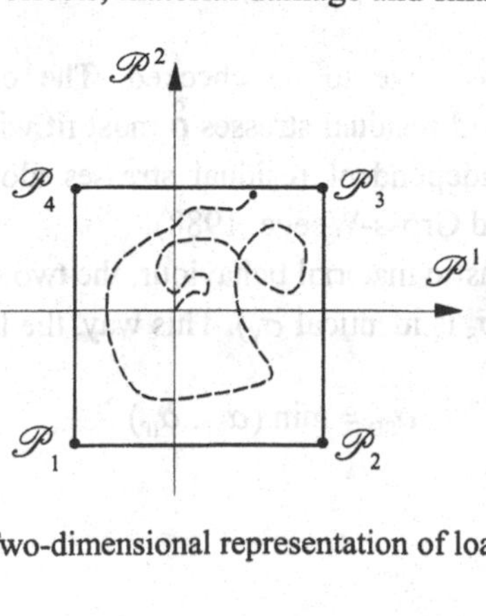

Figure 12. Two-dimensional representation of loading domain.

1.8.4 Alternating Plasticity

The problem writes (König, 1987; Pycko and Mróz, 1992)

Find

$$\alpha_{AP} = \max \alpha \tag{1.66}$$

under the constraint

$$\mathcal{F}_A\left(\sum_{i=1}^{n} \pm\frac{1}{2}[\mu_i^+ - \mu_i^-]\,\alpha\boldsymbol{\sigma}^{(c)}(\mathcal{P}_i),\ \sigma_Y(\mathcal{P}_\vartheta)\right) < 0 \quad \forall\mathcal{P}, \mathcal{P}_\vartheta \in \mathcal{D}. \tag{1.67}$$

It is in this case only to be checked if the extreme positions of the shifted elastic solutions fit within this domain. If yes, shakedown takes place.

1.8.5 Incremental Plasticity

In this case, the load multiplier is defined by (Hachemi, 1994)

$$\alpha_{IP} = \max \alpha \tag{1.68}$$

under the constraints (1.49-50) and

$$\mathcal{F}_I\left(\alpha\frac{\boldsymbol{\sigma}^{(c)}}{1-D}(\mathcal{P}) + \frac{\mathring{\boldsymbol{\rho}}}{1-D} - \frac{\mathring{\boldsymbol{\pi}}}{1-D},\ \sigma_Y(\mathcal{P}_\vartheta)\right) < 0 \quad \forall\mathcal{P}, \mathcal{P}_\vartheta \in \mathcal{D} \tag{1.69}$$

$$\mathcal{F}_L\left(\alpha\frac{\boldsymbol{\sigma}^{(c)}}{1-D}(\mathcal{P}) + \frac{\mathring{\boldsymbol{\rho}}}{1-D},\ \sigma_S(\mathcal{P}_\vartheta)\right) < 0 \quad \forall\mathcal{P}, \mathcal{P}_\vartheta \in \mathcal{D}. \tag{1.70}$$

In this case, both conditions have to be checked. The elastic states shifted by time independent internal stresses $\mathring{\boldsymbol{\pi}}$ and residual stresses $\mathring{\boldsymbol{\rho}}$ must fit within the inner domain and the elastic states shifted by time independent residual stresses alone must fit within the outer domain (Figure 13) (Weichert and Gross-Weege, 1988).

In case of elastic-perfectly plastic material behaviour, the two condition become one and the same ($\mathring{\boldsymbol{\pi}}$ is identical zero and/or σ_S is identical σ_Y). This way, the loading factor is given by

$$\alpha_{SD} = \min(\alpha_{AP}, \alpha_{IP}). \tag{1.71}$$

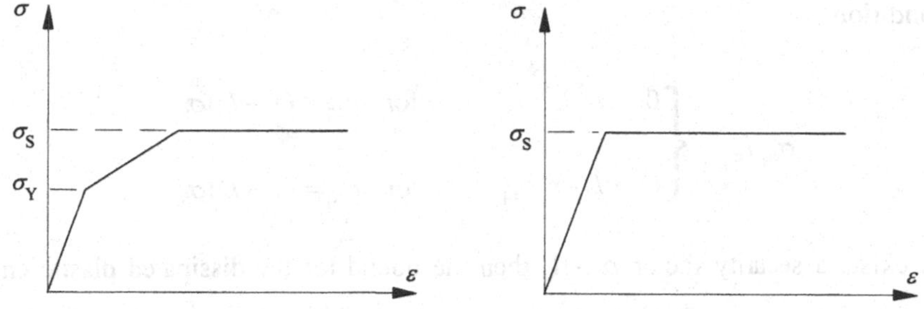

Figure 13. The concept of limited linear hardening material behaviour.

1.8.6 A Bound for the Damage Parameter

An important point to be considered if material damage is taken into account, is that in difference to plasticity, damage is by definition limited. If damage is represented by the scalar D, as here presented, value $D > 1$ is by definition excluded. Also, it is important to note that the quantity $1 - D$ appears as denominator in the definition of effective stresses. This means, that also from mathematical point of view, only value $D < 1$ make sense. Therefore, and in contrast to plasticity, damage must be controlled in shakedown analysis, if this mode of deterioration is taken into account. For the general case, this problem is still open. However, for certain cases, in particular if material damage is a function of plastic strains, one can find bounds for the damage parameter (Hachemi, 1994; Hachemi and Weichert, 1997).

We start with a von Mises-type yield function $\mathscr{F}_L$

$$\mathscr{F}_L = \left\{ \frac{3}{2} \left(\frac{\boldsymbol{\sigma}^D}{1-D} \right) : \left(\frac{\boldsymbol{\sigma}^D}{1-D} \right) \right\}^{1/2} - \sigma_S \leq 0. \tag{1.72}$$

Normality rule and the definition of equivalent plastic strain rate then give

$$\dot{\boldsymbol{\varepsilon}}^p = \dot{\lambda} \frac{\partial \mathscr{F}_L}{\partial \boldsymbol{\sigma}} = \frac{3}{2} \frac{\dot{\lambda}}{1-D} \frac{\tilde{\boldsymbol{\sigma}}^D}{\tilde{\sigma}_{eq}} \quad \text{and} \quad \dot{\varepsilon}_{eq} = \left(\frac{2}{3} \dot{\boldsymbol{\varepsilon}}^p : \dot{\boldsymbol{\varepsilon}}^p \right)^{1/2}$$

$$\Downarrow$$

$$\sigma_{eq} \, \dot{\varepsilon}_{eq} = \boldsymbol{\sigma} : \dot{\boldsymbol{\varepsilon}}^p \tag{1.73}$$

with the condition

$$\sigma_{eq}\,\dot{\varepsilon}_{eq}=\begin{cases}0 & \text{for } \sigma_{eq}<(1-D)\sigma_S\\ (1-D)\sigma_S\,\dot{\varepsilon}_{eq} & \text{for } \sigma_{eq}=(1-D)\sigma_S\end{cases}$$

If there exists a security factor $\alpha > 1$, then the bound for the dissipated plastic energy is defined by

$$\int_0^t\int_{(V)}\boldsymbol{\sigma}:\dot{\boldsymbol{\varepsilon}}^p\,dV dt \le \frac{\alpha}{\alpha-1}\frac{1}{2}\int_{(V)}\mathring{\boldsymbol{\rho}}:\mathbf{L}^{-1}:\mathring{\boldsymbol{\rho}}\,dV. \tag{1.74}$$

For initially virgin material it has been shown (Dorosz, 1976)

$$(\boldsymbol{\varepsilon}^p:\boldsymbol{\varepsilon}^p)^{1/2} \le \int_0^t(\dot{\boldsymbol{\varepsilon}}^p:\dot{\boldsymbol{\varepsilon}}^p)^{1/2}\,dt \tag{1.75}$$

or

$$\varepsilon_{eq} \le \int_0^t\dot{\varepsilon}_{eq}\,dt \tag{1.76}$$

equally

$$\zeta\,(\dot{\boldsymbol{\varepsilon}}^p:\dot{\boldsymbol{\varepsilon}}^p)^{1/2} \le \boldsymbol{\sigma}:\dot{\boldsymbol{\varepsilon}}^p \le \eta\,(\dot{\boldsymbol{\varepsilon}}^p:\dot{\boldsymbol{\varepsilon}}^p)^{1/2} \tag{1.77}$$

with

$$\zeta=\sqrt{2/3}\,(1-D_c)\sigma_S \quad\text{and}\quad \eta=\sqrt{2/3}\,\sigma_S.$$

A local bound for the accumulated plastic deformation is then obtained by

$$\varepsilon_{eq} \le \frac{1}{(1-D_c)\sigma_S}\int_0^t\boldsymbol{\sigma}:\dot{\boldsymbol{\varepsilon}}^p\,dt \tag{1.78}$$

$$\varepsilon_{eq} \le \frac{1}{V}\frac{1}{(1-D_c)\sigma_S}\frac{\alpha}{\alpha-1}\int_{(V)}\frac{1}{2}\mathring{\boldsymbol{\rho}}:\mathbf{L}^{-1}:\mathring{\boldsymbol{\rho}}\,dV \tag{1.79}$$

where V is an arbitrary (e.g. arbitrarily small) subdomain of the total volume of $\mathscr{B}$. The upper bound of the ductile damage parameter for any subdomain of $\mathscr{B}$ is then expressed by

$$D = \frac{D_c}{\varepsilon_R - \varepsilon_D} \left\langle \left\{ \frac{R}{V} \frac{1}{(1-D_c)\sigma_S} \frac{\alpha}{\alpha - 1} \int_{(V)} \frac{1}{2} \mathring{\boldsymbol{\rho}} : \mathbf{L}^{-1} : \mathring{\boldsymbol{\rho}} \, \mathrm{d}V \right\} - \varepsilon_D \right\rangle \tag{1.80}$$

with

$$R = \frac{2}{3}(1 + \nu) + 3(1 - 2\nu) \left(\frac{\sigma_H}{\sigma_{eq}}\right)^2 \tag{1.81}$$

for the model of Lemaitre (1985) and

$$R = \exp\left\{\frac{3}{2}\left(\frac{\sigma_H}{\sigma_{eq}} - \frac{1}{3}\right)\right\} \tag{1.82}$$

for the model of Shichun and Hua (1990).

1.8.7 Minimum Requirements

Independent from any specific kind of material behaviour, the minimum requirements from mathematical point of view for the occurrence of shakedown can be defined. For this, the notion of the "*Sanctuary of Elasticity*" is introduced (Nayroles and Weichert, 1993):

1. Be $\Gamma(\mathbf{x})$ an open domain in the space of stresses Σ such that if $\boldsymbol{\sigma}(\mathbf{x}) \in \Sigma$, no additional inelastic deformation occurs.
2. $\forall\, \boldsymbol{\sigma}^{(s)} \in \Gamma(\mathbf{x})$, holds $(\boldsymbol{\sigma} - \boldsymbol{\sigma}^{(s)}) : \dot{\boldsymbol{\varepsilon}}^p \geq 0$.
3. The Sanctuary of Elasticity C is defined by $C = \{\boldsymbol{\sigma}(\mathbf{x}) \in \Sigma \,/\, \forall \mathbf{x} \in V, \boldsymbol{\sigma} \in \Gamma(\mathbf{x})\}$

Theorem. If there exists a solution $\boldsymbol{\sigma}^{(c)}(\mathbf{x}, t)$ of a purely elastic reference problem, a real number $\alpha > 1$ and a field of time-independent residual stresses $\mathring{\boldsymbol{\rho}}(\mathbf{x})$ such that $\alpha\,(\boldsymbol{\sigma}^{(c)} + \mathring{\boldsymbol{\rho}}) \subset C$, then the inelastic work W is bounded by

$$W \leq \frac{1}{2} \frac{\alpha}{\alpha - 1} \langle \mathbf{L}^{-1} : (\boldsymbol{\rho}(0) - \mathring{\boldsymbol{\rho}}), (\boldsymbol{\rho}(0) - \mathring{\boldsymbol{\rho}}) \rangle \tag{1.83}$$

with

$$\langle \mathbf{a}, \mathbf{b} \rangle = \int_{(V)} (\mathbf{a} : \mathbf{b}) \, \mathrm{d}V. \tag{1.84}$$

One sees easily that for elastic-perfectly plastic material behaviour with an associated flow rule the "Sanctuary of Elasticity" (*SE*) corresponds to the usual elastic domain in the 6-dimensional space of stresses, bounded by the yield function (Figure 14). For the GSMM, the SE corresponds to the elastic domain in the space of generalised stresses of dimension 6+r, where r denotes the number of independent internal parameters describing the evolution of the internal structure of the material. However, due to the fact that the "*SE*" has not necessarily a specific physical meaning, other kinds of inelastic behaviour are covered by the theorem. The question arises, how such other laws can be put into practice.

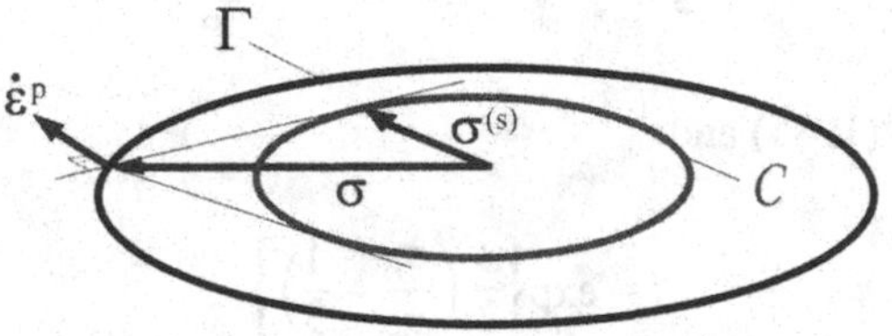

Figure 14. Illustration of the Sanctuary of Elasticity.

Another methodology to prove the boundedness of dissipation due to inelastic effects for specific materials, including or not geometrical effects, is now presented. It follows the pattern of the proof of the theorem by Melan (1936) by simply modifying the underlying quadratic form and the choice of generalised variables. The procedure consists of the following steps:

1. Choose an appropriate *positive and bounded* definite form $W = \langle \mathbf{M}^{-1}\mathbf{:v}, \mathbf{v}\rangle$, function of the problems generalised state variables and associated fictitious admissible fields, collected in the field $\mathbf{v}(\mathbf{x}, t)$.
2. Calculate the derivative of W with respect to time, $\dot{W} = \mathrm{d}W/\mathrm{d}t$, show that for *active failure mechanism* one part of $\dot{W}$, say $\dot{W}_{\mathrm{L}}$, is always *negative.*
3. Show that the time integral of the remaining part of $\dot{W}$, $\dot{W}_{\mathrm{NL}} = \dot{W} - \dot{W}_{\mathrm{L}}$ is bounded for the particular application.
4. Then the classical argument is that the considered failure mechanism must cease to be active beyond a certain instant, otherwise there is contradiction with the starting point.

This general procedure will be illustrated on the case of non-associated flow laws as used in soil mechanics (see, e.g., Maier, 1969). One can choose the quadratic form (Boulbibane, 1995; Boulbibane and Weichert, 1997)

$$W = \frac{1}{2}\langle \mathbf{A}^{\mathrm{ep}}\mathbf{:}(\boldsymbol{\rho} - \mathring{\boldsymbol{\rho}}), \boldsymbol{\rho} - \mathring{\boldsymbol{\rho}}\rangle \tag{1.85}$$

with

$$\mathbf{A}^{\mathrm{ep}} = \begin{cases} \mathbf{L}^{-1} + \dfrac{1}{H}\dfrac{\partial Q}{\partial \boldsymbol{\sigma}}\dfrac{\partial \mathscr{F}}{\partial \boldsymbol{\sigma}} & \text{for } \mathscr{F} = 0 \text{ and } \dfrac{\partial \mathscr{F}}{\partial \boldsymbol{\sigma}}\dot{\boldsymbol{\sigma}} \geq 0 \\ \mathbf{L}^{-1} & \text{for } \mathscr{F} < 0 \text{ or } \mathscr{F} = 0 \text{ and } \dfrac{\partial \mathscr{F}}{\partial \boldsymbol{\sigma}}\dot{\boldsymbol{\sigma}} < 0 \end{cases} \tag{1.86}$$

where $\mathscr{F}$ and Q denote the yield function and the plastic potential, respectively defined by

$$\mathscr{F}(\boldsymbol{\sigma}, R) \leq 0 \quad \text{and} \quad Q(\boldsymbol{\sigma}) = g(\boldsymbol{\sigma}) - \text{const.} = 0. \tag{1.87}$$

We remind that $\mathscr{F}$ equals Q in the case of associated flow rules. $\mathbf{L}$ is the tensor of elastic moduli and H a hardening parameter, defined in the sequel. For this model, plastic deformation rate is given by

$$\dot{\boldsymbol{\varepsilon}}^{\mathrm{p}} = \dot{\lambda}\frac{\partial Q}{\partial \boldsymbol{\sigma}}, \; \dot{\lambda} \geq 0 \tag{1.88}$$

and the application of the consistency condition

$$\left(\frac{\partial \mathscr{F}}{\partial \boldsymbol{\sigma}}\right)\dot{\boldsymbol{\sigma}} + \left(\frac{\partial \mathscr{F}}{\partial R}\right)\left(\frac{\partial R}{\partial \boldsymbol{\varepsilon}^{\mathrm{p}}}\right)\dot{\boldsymbol{\varepsilon}}^{\mathrm{p}} = 0 \tag{1.89}$$

gives

$$\left(\frac{\partial \mathscr{F}}{\partial \boldsymbol{\sigma}}\right)\dot{\boldsymbol{\sigma}} - H\dot{\lambda} = 0 \tag{1.90}$$

if the hardening parameter H is defined by

$$H = -\left(\frac{\partial \mathscr{F}}{\partial R}\right)\left(\frac{\partial R}{\partial \boldsymbol{\varepsilon}^{\mathrm{p}}}\right)\left(\frac{\partial Q}{\partial \boldsymbol{\sigma}}\right). \tag{1.91}$$

The safe state of stress $\boldsymbol{\sigma}^{(s)}$ is supposed to satisfy the condition $Q(\boldsymbol{\sigma}^{(s)}) < 0$ and is composed of $\boldsymbol{\sigma}^{(c)}$ and a fictitious, time independent field of residual stresses $\mathring{\boldsymbol{\rho}}$. It had been shown (Loret, 1987) that the pseudo-stiffness tensor $\mathbf{A}^{\mathrm{ep}}$ is in general strictly positive for $H > H_{cr}$, with

$$H_{cr} = \frac{1}{2}\left\{\left[\left(\frac{\partial \mathscr{F}}{\partial \boldsymbol{\sigma}}:\mathbf{L}^{-1}:\frac{\partial \mathscr{F}}{\partial \boldsymbol{\sigma}}\right)\left(\frac{\partial Q}{\partial \boldsymbol{\sigma}}:\mathbf{L}^{-1}:\frac{\partial Q}{\partial \boldsymbol{\sigma}}\right)\right]^{1/2} - \left(\frac{\partial \mathscr{F}}{\partial \boldsymbol{\sigma}}:\mathbf{L}^{-1}:\frac{\partial Q}{\partial \boldsymbol{\sigma}}\right)\right\} \geq 0. \tag{1.92}$$

As example, $\mathbf{L}^{-1}$ is defined by

$$\mathbf{L}^{-1} = \frac{(1+\nu)(1-\nu)}{E} \begin{pmatrix} 1 & \frac{-\nu}{1-\nu} & 0 \\ . & 1 & 0 \\ \text{Sym.} & . & \frac{2}{1-\nu} \end{pmatrix} \tag{1.93}$$

for isotropic materials in the state of plane strain, with E and ν are Young's modulus and Poisson's ratio, respectively.

In the special case of the Mohr-Coulomb yield criterion one can show that for constant angles ϕ and ψ and for constant elastic coefficients $\mathbf{L}$, the tensor $\mathbf{A}^{ep}$ is time-independent in the plastic range. In terms of principal stresses with $\sigma_1 \geq \sigma_2 \geq \sigma_3$ we get

$$\mathscr{F} = \left(\frac{1+\sin\phi}{2}\right)\sigma_1 - \left(\frac{1-\sin\phi}{2}\right)\sigma_3 - C\sin\phi = 0. \tag{1.94}$$

The plastic potential Q is in this case obtained by modifying the yield function

$$Q = \left(\frac{1+\sin\psi}{2}\right)\sigma_1 - \left(\frac{1-\sin\psi}{2}\right)\sigma_3 - \text{const.} = 0 \tag{1.95}$$

where ϕ and $\psi < \phi$ are angles of internal shearing resistance and dilatancy, respectively, and C is the coefficient of cohesion. So, the derivatives of $\mathscr{F}$ and Q are given by

$$\begin{aligned} \frac{\partial \mathscr{F}}{\partial \sigma_1} &= \frac{1+\sin\phi}{2}, & \frac{\partial \mathscr{F}}{\partial \sigma_3} &= -\frac{1-\sin\phi}{2} \\ \frac{\partial Q}{\partial \sigma_1} &= \frac{1+\sin\psi}{2}, & \frac{\partial Q}{\partial \sigma_3} &= -\frac{1-\sin\psi}{2} \end{aligned} \tag{1.96}$$

Then, the time-derivative of W is given by

$$\dot{W} = \langle \mathbf{L}^{-1}:(\boldsymbol{\rho} - \mathring{\boldsymbol{\rho}}), \dot{\boldsymbol{\rho}} \rangle + \langle \mathbf{L}^{-1}:(\boldsymbol{\rho} - \mathring{\boldsymbol{\rho}}) \left(\frac{1}{H}\frac{\partial \mathscr{F}}{\partial \boldsymbol{\sigma}}:\mathbf{L}^{-1}:\frac{\partial Q}{\partial \boldsymbol{\sigma}}\right), \dot{\boldsymbol{\rho}} \rangle \tag{1.97}$$

and

$$\dot{W} = \langle \boldsymbol{\sigma} - \boldsymbol{\sigma}^{(s)}, \dot{\boldsymbol{\varepsilon}}^{p} \rangle + \langle (\boldsymbol{\sigma} - \boldsymbol{\sigma}^{(s)}) \left(\frac{1}{H} \frac{\partial \mathcal{F}}{\partial \boldsymbol{\sigma}} : \mathbf{L}^{-1} : \frac{\partial Q}{\partial \boldsymbol{\sigma}} \right), \dot{\boldsymbol{\varepsilon}}^{p} \rangle \tag{1.98}$$

where

$$\dot{\boldsymbol{\varepsilon}} = \dot{\boldsymbol{\varepsilon}}^{(c)} + \dot{\boldsymbol{\varepsilon}}^{\rho} + \dot{\boldsymbol{\varepsilon}}^{p} \tag{1.99}$$

$$\dot{\boldsymbol{\varepsilon}}^{\rho} = \mathbf{L} : \dot{\boldsymbol{\rho}}. \tag{1.100}$$

We then get

$$\dot{W} = \langle \boldsymbol{\sigma} - \boldsymbol{\sigma}^{(s)}, \dot{\boldsymbol{\varepsilon}} - \dot{\boldsymbol{\varepsilon}}^{(c)} - \dot{\boldsymbol{\varepsilon}}^{\rho} \rangle + \langle (\boldsymbol{\sigma} - \boldsymbol{\sigma}^{(s)}) \left(\frac{1}{H} \frac{\partial \mathcal{F}}{\partial \boldsymbol{\sigma}} : \mathbf{L}^{-1} : \frac{\partial Q}{\partial \boldsymbol{\sigma}} \right), \dot{\boldsymbol{\varepsilon}} - \dot{\boldsymbol{\varepsilon}}^{(c)} - \dot{\boldsymbol{\varepsilon}}^{\rho} \rangle. \tag{1.101}$$

Here, the two first terms in the second bracket represent a kinematically admissible field of strain increments. From the application of Gauss's theorem follows therefore classically

$$\langle \boldsymbol{\sigma} - \boldsymbol{\sigma}^{(s)}, \dot{\boldsymbol{\varepsilon}} - \dot{\boldsymbol{\varepsilon}}^{(c)} \rangle = 0 \tag{1.102}$$

and we obtain

$$\dot{W} = - \langle \boldsymbol{\sigma} - \boldsymbol{\sigma}^{(s)}, \dot{\boldsymbol{\varepsilon}}^{\rho} \rangle - \langle (\boldsymbol{\sigma} - \boldsymbol{\sigma}^{(s)}) \left(\frac{1}{H} \frac{\partial \mathcal{F}}{\partial \boldsymbol{\sigma}} : \mathbf{L}^{-1} : \frac{\partial Q}{\partial \boldsymbol{\sigma}} \right), \dot{\boldsymbol{\varepsilon}}^{\rho} \rangle. \tag{1.103}$$

Eqn (1.103) can be written in the form

$$\dot{W} = \dot{W}_{\mathrm{L}} + \dot{W}_{\mathrm{NL}} \tag{1.104}$$

with

$$\dot{W}_{\mathrm{L}} = - \langle \boldsymbol{\sigma} - \boldsymbol{\sigma}^{(s)}, \dot{\boldsymbol{\varepsilon}}^{\rho} \rangle \tag{1.105}$$

and

$$\dot{W}_{\mathrm{NL}} = - \langle (\boldsymbol{\sigma} - \boldsymbol{\sigma}^{(s)}) \left(\frac{1}{H} \frac{\partial \mathcal{F}}{\partial \boldsymbol{\sigma}} : \mathbf{L}^{-1} : \frac{\partial Q}{\partial \boldsymbol{\sigma}} \right), \dot{\boldsymbol{\varepsilon}}^{\rho} \rangle. \tag{1.106}$$

From the assumption $Q(\boldsymbol{\sigma}^{(s)}) < 0$ follows that $(\boldsymbol{\sigma} - \boldsymbol{\sigma}^{(s)}):\dot{\boldsymbol{\varepsilon}}^p$ is always positive. Therefore, $\dot{W}_L \leq 0$ in any case. For $\dot{W}_{NL}$ we suppose that H is positive, which is in general the case for hardening materials. Using eqns (1.91) and (1.96), the value of the square bracket in eqn (1.106) is given as

$$\frac{1}{H}\frac{\partial \mathscr{F}}{\partial \boldsymbol{\sigma}}:\mathbf{L}^{-1}:\frac{\partial Q}{\partial \boldsymbol{\sigma}} = \frac{1}{H}\frac{E}{(1+\nu)(1-2\nu)}\left(\frac{1}{4}(1+\sin\phi)(1+\sin\psi) + \frac{1-2\nu}{4}(1-\sin\phi)(1-\sin\psi)\right) \quad (1.107)$$

which is positive for all possible friction and dilatancy angles. So, $\dot{W}_{NL}$ is negative. Consequently, $\dot{W}$ is equal to zero only for $\dot{\boldsymbol{\varepsilon}}^p = 0$ and otherwise negative. This proves that the evolution of plastic strains must cease beyond a certain time and the quadratic form (eqn (1.85)) is bounded. We note that even in the case of negative values of H, shakedown may occur if it can be shown that $|\dot{W}_L| > \dot{W}_{NL}$ always holds. For $H = 0$ (absence of any hardening), the term $\dot{W}_{NL}$ is identical to zero and the proof takes obviously a simplified form. We note that the plastic potential Q plays here the role of the "*ES*" as defined before.

References

Boulbibane, M. (1995). *Application de la Théorie d'Adaptation aux Milieux Elastoplastiques Non-Standards: Cas des Géomatériaux*. Ph. D. Dissertation, University of Lille.

Boulbibane M., and Weichert D. (1997). Application of shakedown theory to soils with non-associated flow-rules. *Mech. Res. Comm.* 24:513–519.

Dorosz, S. (1976). An upper bound to maximum residual deflections of elastic-plastic structures at shakedown. *Bull. Acad. Polon. Sci. Ser. Sci. Tech.* 24:167–174.

Druyanov, B. and Roman, I. (1998). On adaptation (shakedown) of a class of damaged elastic plastic bodies to cyclic loading. *Eur. J. Mech. A/Solids* 17:71–78.

Feng, X.Q. and Yu, S.W. (1995). Damage and shakedown analysis of structures with strain-hardening. *Int. J. Plasticity* 11:237–249.

Green, A.E. and Naghdi, P.M. (1965). A general theory of elastic-Plastic continuum. *Arch. Rat. Mech. Analys.* 18:251–281.

Hachemi, A. and Weichert, D. (1992). An extension of the static shakedown theorem to a certain class of inelastic materials with damage. *Arch. Mech.* 44:491–498.

Hachemi, A. (1994). *Contribution à l'Analyse de l'Adaptation des Structures Inélastiques avec Prise en Compte de l'Endommagement*. Ph. D. Dissertation, University of Lille.

Hachemi, A. and Weichert, D. (1997). Application of shakedown theory to damaging inelastic material under mechanical and thermal loads. *Int. J. Mech. Sci.* 39:1067–1076.

Halphen, B. and Nguyen, Q.S. (1975). Sur les matériaux standards généralisés. *J. Mécanique* 14: 39–63.

Hill, R. (1950). *The Mathematical Theory of Plasticity*. Oxford.

Ju, J.W. (1989). On energy-based coupled elastoplastic damage theories: Constitutive modeling and computational aspects. *Int. J. Solids Struct.* 25:803–833.

Kachanov, L.M. (1958). Time of the rupture process under creep conditions. *Izv. Akad. Nauk, S.S.R., Otd. Tech. Nauk* 8:26–31.

König, J.A. (1987). *Shakedown of Elastic-Plastic Structures*. Amsterdam: Elsevier.

Koiter, W.T. (1956). A new general theorem on shake-down of elastic-plastic structures. *Proc. Koninkl. Ned. Akad. Wet.* B59:24–34.

Koiter, W.T. (1960). General theorems for elastic-plastic solids. In Sneddon, I.N. and Hill, R., eds. *Progress in Solid Mechanics*. Amsterdam: North-Holland. 165–221.

Lee, E.H. (1969). Elasto-plastic deformation at finite strains. *J. Appl. Mech.* 36:1–6.

Lemaitre, J. (1985) A continuous damage mechanics model for ductile fracture. *J. Engng. Mat. Tech.* 107: 83–89.

Lemaitre, J. and Chaboche, J.L. (1985). *Mécanique des matériaux solides*. Paris: Dunod.

Lemaitre, J. (1987). Formulation unifiée des lois d'évolution d'endommagement. *C. R. Acad. Sci.* 305:1125–1130.

Loret, B. (1987). *Elastoplasticité à Simple Potentiel*. Paris: Presse E.N des Ponts et Chaussées.

Maier, G. (1969). Shakedown theory in perfect elastoplasticity with associated and non-associated flow-laws. Meccanica 6:250–260.

Melan, E. (1936). Theorie statisch unbestimmter Systeme aus ideal-plastischem Baustoff. *Sitber. Akad. Wiss., Wien, Abt. IIA* 145:195–218.

Melan, E. (1938). Zur Plastizität des räumlichen Kontinuums. *Ing. Arch.* 9:116–126.

Nayroles, B. and Weichert, D. (1993). La notion de sanctuaire d'élasticité et l'adaptation des structures. *C.R. Acad. Sci.* 316:1493–1498.

Polizzotto, C., Borino, G. and Fuschi, P. (1996). An extended shakedown theory for elastic-plastic-damage material models. *Eur. J. Mech. A/Solids* 15:825–858.

Polizzotto, C., Borino, G. and Fuschi, P. (2001). Weak forms of shakedown for elastic-plastic structures exhibiting ductile damage. *Meccanica* 36:49–66.

Pycko, S. and Mróz, Z. (1992). Alternative approach to shakedown as a solution of a min-max problem. *Acta Mechanica* 93:205–222.

Shichun, W. and Hua, L. (1990). A kinetic equation for ductile damage at large plastique strain. *J. Mat. Proc. Tech.* 21:295–302.

Weichert, D. and Gross-Weege, J. (1988). The numerical assessment of elastic-plastic sheets under variable mechanical and thermal loads using a simplified two-surface yield condition. *Int. J. Mech. Sci.* 30:757–767.

Shakedown of Thin-Walled Structures with Geometrical Non-Linear Effects

Dieter Weichert and Abdelkader Hachemi
Aachen University of Technology, Aachen, Germany

Abstract. The aim of this lecture is to show, how the general theory, as presented in the first lecture, can be applied to structural elements. Here, we concentrate on thin-walled shells and plates, taking certain geometrical effects into account that may occur during the loading history.

2.1 Special Kind of Loading

A practical method will be presented here allowing the use of large displacement in shakedown theory (Tritsch, 1993; Weichert and Hachemi; 1998). For this, we assume that the loads $\mathbf{a}*$ are of a special type: Up to an instant t^R the body $\mathscr{B}$ undergoes finite and given displacement $\mathbf{u}^R$ with respect to the initial configuration Ω_i at time $t = 0$ in such a way that $\mathscr{B}$ is in the known configuration Ω_R in equilibrium under time-independent loads $\mathbf{a}^{R*}$. For times $t > t^R$ the body $\mathscr{B}$ is submitted to additional variable loads $\mathbf{a}^{r*}$ such that

$$\mathbf{a}^* (\mathbf{X}, t) = \mathbf{a}^{R*} (\mathbf{X}) + \mathbf{a}^{r*}(\mathbf{X}, t) \tag{2.1}$$

and occupies the actual configuration Ω_t (see, e.g., Weichert, 1986; Gross-Weege, 1990; Saczuk and Stumpf, 1990; Pycko and König, 1991). Since the actual configuration should also be an equilibrium configuration and the following equations hold

(a) *Statical equations*

$$\text{Div}\,(\boldsymbol{\tau}^R + \boldsymbol{\tau}^r) = -\,\mathbf{f}^{R*} - \mathbf{f}^{r*} \quad \text{in } V \tag{2.2}$$

$$\mathbf{n}.(\boldsymbol{\tau}^R + \boldsymbol{\tau}^r) = \mathbf{p}^{R*} + \mathbf{p}^{r*} \quad \text{on } S_p \tag{2.3}$$

with

$$\boldsymbol{\tau}^R + \boldsymbol{\tau}^r = (\mathbf{F}^r.\mathbf{F}^R)(\boldsymbol{\sigma}^R + \boldsymbol{\sigma}^r) \tag{2.4}$$

(b) *Kinematical equations*

$$\mathbf{u} = \mathbf{u}^R + \mathbf{u}^r \qquad \text{in } V \qquad (2.5)$$
$$\mathbf{F} = \mathbf{F}^r.\mathbf{F}^R = \mathbf{I} + \text{grad}(\mathbf{u}^R) + \text{grad}(\mathbf{u}^r) \qquad \text{in } V \qquad (2.6)$$
$$\boldsymbol{\varepsilon} = \boldsymbol{\varepsilon}^R + \boldsymbol{\varepsilon}^r = \frac{1}{2}(\mathbf{C} - \mathbf{I}) \qquad \text{in } V \qquad (2.7)$$
$$\mathbf{u} = \mathbf{u}^{R*} + \mathbf{u}^{r*} \qquad \text{on } S_u \qquad (2.8)$$

with

$$\mathbf{C} = (\mathbf{F}^r.\mathbf{F}^R)^T(\mathbf{F}^r.\mathbf{F}^R) \qquad (2.9)$$

where all quantities caused by the time-independent loads $\mathbf{a}^{R*}$ are marked by a superscript "R", whereas the additional field quantities caused by the time-dependent loads $\mathbf{a}^{r*}$ are marked by superscript "r". The additional field quantities caused by $\mathbf{a}^{r*}$ have to satisfy the following equations

(a) *Statical equations*

$$\text{Div}\,(\boldsymbol{\tau}^r) = -\,\mathbf{f}^{r*} \qquad \text{in } V \qquad (2.10)$$
$$\mathbf{n}.\boldsymbol{\tau}^r = \mathbf{p}^{r*} \qquad \text{on } S_p \qquad (2.11)$$

with

$$\boldsymbol{\tau}^r = \mathbf{H}^r.\mathbf{F}^R.\boldsymbol{\sigma}^R + \mathbf{F}^R.\boldsymbol{\sigma}^r + \mathbf{H}^r.\mathbf{F}^R.\boldsymbol{\sigma}^r \qquad (2.12)$$

(b) *Kinematical equations*

$$\mathbf{F}^r = \mathbf{I} + \mathbf{H}^r \qquad \text{in } V \qquad (2.13)$$
$$\boldsymbol{\varepsilon}^r = \frac{1}{2}(\mathbf{F}^R)^T\,[(\mathbf{H}^r)^T + \mathbf{H}^r + (\mathbf{H}^r)^T\mathbf{H}^r](\mathbf{F}^R) \qquad \text{in } V \qquad (2.14)$$
$$\mathbf{u}^r = \mathbf{u}^{r*} \qquad \text{on } S_u \qquad (2.15)$$

with

$$\mathbf{H}^r = \text{grad}_R(\mathbf{u}^r) \qquad (2.16)$$

In the sequel, we restrict our considerations to loading histories characterised by the motion of a fictitious comparison body $\mathscr{B}^{(c)}$, having at time t^R the same field quantities as $\mathscr{B}$ but reacting, in contrast to $\mathscr{B}$, purely elastically to the additional time-dependent loads $\mathbf{a}^{r*}$, superimposed on $\mathbf{a}^{R*}$ for $t > t^R$ (Figure 15) (cf. Weichert, 1986; Gross-Weege, 1990; Saczuk and Stumpf, 1990). Henceforth, all quantities related to this comparison problem are indicated by superscript "$^{(c)}$". Obviously, eqns (2.10-16) also hold for the comparison problem with the only exception, that in the comparison body $\mathscr{B}^{(c)}$ no additional plastic strains and no damage can occur. The differences between the states in $\mathscr{B}$ and $\mathscr{B}^{(c)}$ are then described by the difference fields

$$\Delta \mathbf{u} = \mathbf{u}^r - \mathbf{u}^{r(c)}\,; \quad \Delta \mathbf{F} = \mathbf{F}^r - \mathbf{F}^{r(c)}\,; \quad \Delta \boldsymbol{\varepsilon} = \boldsymbol{\varepsilon}^r - \boldsymbol{\varepsilon}^{r(c)} \tag{2.17}$$

$$\Delta \boldsymbol{\tau} = \boldsymbol{\tau}^r - \boldsymbol{\tau}^{r(c)}\,; \quad \Delta \boldsymbol{\sigma} = \boldsymbol{\sigma}^r - \boldsymbol{\sigma}^{r(c)} \tag{2.18}$$

and have to fulfil the following equations

$$\mathrm{Div}\,(\Delta \boldsymbol{\tau}) = 0 \qquad \text{in } V \tag{2.19}$$

$$\mathbf{n}.(\Delta \boldsymbol{\tau}) = 0 \qquad \text{on } S_p \tag{2.20}$$

and

$$\Delta \mathbf{F} = \mathbf{H}^r - \mathbf{H}^{r(c)} \qquad \text{in V} \tag{2.21}$$

$$\begin{aligned} \Delta \boldsymbol{\varepsilon} &= \frac{1}{2}(\mathbf{F}^R)^T[(\Delta \mathbf{F})^T + (\Delta \mathbf{F})](\mathbf{F}^R) \\ &\quad + \frac{1}{2}(\mathbf{F}^R)^T[(\mathbf{H}^r)^T(\mathbf{H}^r) - (\mathbf{H}^{r(c)})^T(\mathbf{H}^{r(c)})](\mathbf{F}^R) \qquad \text{in } V \end{aligned} \tag{2.22}$$

$$\Delta \mathbf{u} = 0 \qquad \text{on } S_u \tag{2.23}$$

with

$$\begin{aligned} \Delta \boldsymbol{\tau} &= (\Delta \mathbf{F}).\mathbf{F}^R.\boldsymbol{\sigma}^R + \mathbf{F}^R.(\Delta \boldsymbol{\sigma}) \\ &\quad + \mathbf{H}^r.\mathbf{F}^R.\boldsymbol{\sigma}^r - \mathbf{H}^{r(c)}.\mathbf{F}^R.\boldsymbol{\sigma}^{r(c)} \end{aligned} \tag{2.24}$$

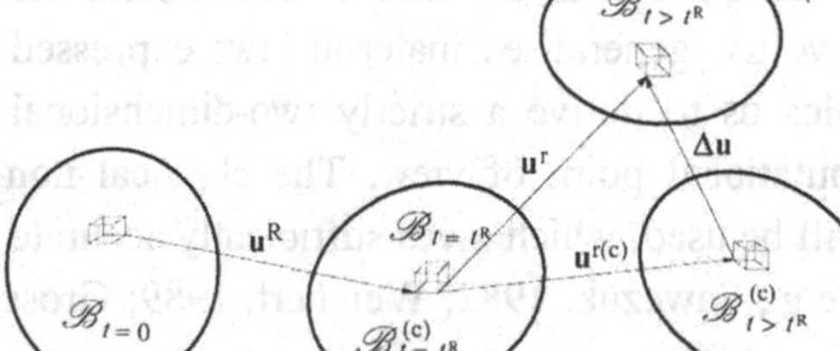

Figure 15. Evolution of real body $\mathscr{B}$ and comparison body $\mathscr{B}^{(c)}$.

In the following, we restrict our considerations to situations where the state of deformation and the state of stress in $\mathscr{B}^{(c)}$ are subjected to small variations in time (Weichert, 1986; Gross-Weege, 1990). Consequently, we neglect in the governing eqns (2.19-24) all terms, which are nonlinear in the time-dependent additional field quantities marked by a superscript "r". This excludes to study buckling effects induced by the additional time-dependent loads. Then the following extension of Melan's theorem holds (Weichert and Hachemi, 1998):

If there exists a time-independent field of effective residual stresses $\Delta\overset{\circ}{\tilde{\mathbf{s}}} = [\Delta\overset{\circ}{\tilde{\boldsymbol{\sigma}}}, \Delta\overset{\circ}{\tilde{\boldsymbol{\pi}}}]$ such that the following relations holds

$$\text{(i)} \qquad \text{Div}\,(\mathbf{F}^{R}.\boldsymbol{\sigma}^{R}) = -\,\mathbf{f}^{R*} \qquad \text{in } V \qquad (2.25)$$

$$\mathbf{n}.(\mathbf{F}^{R}.\boldsymbol{\sigma}^{R}) = \mathbf{p}^{R*} \qquad \text{on } S_p \qquad (2.26)$$

$$\mathbf{u} = \mathbf{u}^{R*} \qquad \text{on } S_u \qquad (2.27)$$

$$\text{(ii)} \qquad \text{Div}\,(\Delta\overset{\circ}{\boldsymbol{\tau}}) = 0 \qquad \text{in } V \qquad (2.28)$$

$$\mathbf{n}.(\Delta\overset{\circ}{\boldsymbol{\tau}}) = 0 \qquad \text{on } S_p \qquad (2.29)$$

$$\Delta\overset{\circ}{\mathbf{u}} = 0 \qquad \text{on } S_u \qquad (2.30)$$

$$\text{(iii)} \qquad \mathscr{F}\left(\frac{\mathbf{s}^{r(c)}}{1-D} + \frac{\mathbf{s}^{R}}{1-D} + \frac{\Delta\overset{\circ}{\mathbf{s}}}{1-D}, \sigma_Y\right) < 0 \qquad \text{in } V \qquad (2.31)$$

with

$$\Delta\overset{\circ}{\boldsymbol{\tau}} = (\Delta\overset{\circ}{\mathbf{F}})\mathbf{F}^{R}.\boldsymbol{\sigma}^{R} + \mathbf{F}^{R}.(\Delta\overset{\circ}{\boldsymbol{\sigma}})\,; \quad \tilde{\mathbf{s}}^{R} = [\tilde{\boldsymbol{\sigma}}^{R}, \tilde{\boldsymbol{\pi}}^{R}]^{T} \;\text{ and }\; D = D^{R} + D^{r}$$

for all time $t > t^R$, then the original body $\mathscr{B}$ will shakedown under given program of loading **a***.

2.2 Application to Shell-Like Bodies

The practical approach presented above in the case of geometrical non-linearity is applied to shell-like bodies. For this, we use generalised material laws expressed in terms of stress and strain resultants. This enables us to derive a strictly two-dimensional formulation, which is advantageous from a computational point of view. The classical non-linear shell theory of Donnell-Mushtari-Vlasov will be used, which gives sufficiently accurate results within imposed range of applicability (see, e.g., Sawczuk, 1982; Weichert, 1989; Gross-Weege and Weichert, 1995).

2.2.1 Basic Shell Relations

In the following, we restrict our consideration to shallow-shells undergoing moderate rotations about tangents and small rotations about normals to the mid-surface $\mathcal{M}$ of the shell. Let P the region of the three-dimensional Euclidean space E occupied by the shell in its undeformed configuration Ω_i, parameterised by the curvilinear coordinates $(\theta^1, \theta^2, \theta^3)$ such that $-h/2 \leq \theta^3 \leq h/2$; h is the thickness of the undeformed shell, assumed to be small compared with the smallest radius of curvature R of the mid-surface $\mathcal{M}$; i.e. $h/R << 1$ (Figure 16) (see, e.g. Basar and Krätzig, 1985). $\mathcal{M}$ is described by the position vector in the undeformed configuration by

$$\mathbf{r} = \mathbf{X}^k(\theta^\alpha)\,\mathbf{i}_k\ ;\ k = 1, 2, 3;\ \alpha = 1, 2 \tag{2.32}$$

where $\mathbf{i}_k$ is the orthonormal basis of a Cartesian system of coordinates with origin $O \in E$. Then, by usual definitions, we have

– base vectors:	$\mathbf{a}_\alpha = \mathbf{r}_{,\alpha}$	;	$\mathbf{a}^\alpha = a^{\alpha\beta}\,\mathbf{a}_\beta$	(2.33)
– unit normal vector:	$\mathbf{a}_3 = 1/2\ \varepsilon^{\alpha\beta}\,\mathbf{a}_\alpha \times \mathbf{a}_\beta$	;	$\mathbf{a}^3 = \mathbf{a}_3$	(2.34)
– metric tensor:	$a_{\alpha\beta} = \mathbf{a}_\alpha \cdot \mathbf{a}_\beta$	;	$a^{\alpha\beta} = \mathbf{a}^\alpha \cdot \mathbf{a}^\beta$	(2.35)
– curvature tensors:	$b_{\alpha\beta} = -\,\mathbf{a}_\alpha \cdot \mathbf{a}_{3,\beta}$	;	$b^\beta_\alpha = a_{\alpha\gamma}\,a^{\gamma\beta}$	(2.36)
– determinant:	$a = \lvert a_{\alpha\beta} \rvert = a_{11}\,a_{22} - (a_{12})^2$			(2.37)

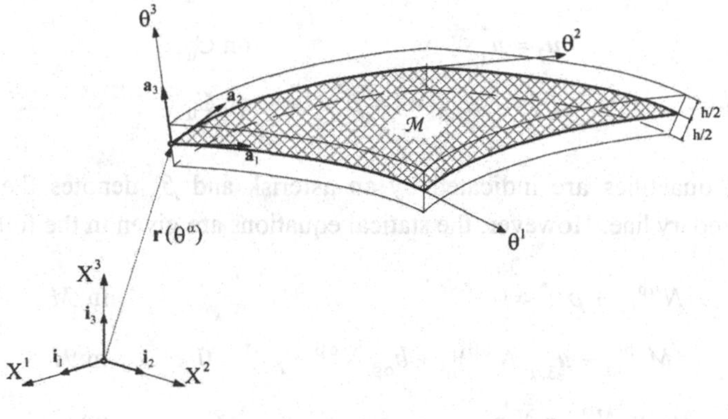

Figure 16. Shell geometry.

Here, $(\cdot)_{,\alpha}$ denotes partial derivatives with respect to θ^α. Greek and Latin indices take the values 1, 2 and 1, 2, 3, respectively. $\varepsilon^{\alpha\beta}$ denotes the contravariant components of the permutation tensor with $\varepsilon^{12} = -\,\varepsilon^{21} = a^{-1/2}$ and $\varepsilon^{11} = \varepsilon^{22} = 0$. The symbol $(\cdot)$ and $(\times)$ between two vectors represent scalar product and cross product, respectively. At line C, bounding the middle surface $\mathcal{M}$, we introduce an orthonormal vector system **v**, **t**, $\mathbf{a}_3$, where $\mathbf{t} = \partial\mathbf{r}/\partial s$ is the unit tangent vector, s being the arc length along C, and $\mathbf{v} = \mathbf{t} \times \mathbf{a}_3$ denotes the outward unit normal vector.

The deformation of the shell middle surface from the undeformed configuration $\mathcal{M}$ into the deformed configuration $\bar{\mathcal{M}}$ can be described by the displacement field

$$\mathbf{u} = \bar{\mathbf{r}} - \mathbf{r} = u^\alpha \, \mathbf{a}_\alpha + u^3 \, \mathbf{a}_3 \tag{2.38}$$

The deformation of the shell can be defined by the middle surface strain tensor $\boldsymbol{\gamma}$ and the change of curvature $\boldsymbol{\chi}$

$$\gamma_{\alpha\beta} = \frac{1}{2}\left(u_\alpha|_\beta + u_\beta|_\alpha - 2\, b_{\alpha\beta}\, u_3 + u_{3,\alpha}\, u_{3,\beta}\right) \qquad \text{in } \mathcal{M} \tag{2.39}$$

$$\chi_{\alpha\beta} = -\, u_3|_{\alpha\beta} \qquad \text{in } \mathcal{M} \tag{2.40}$$

where a vertical stroke preceding a subscript indicates covariant differentiation with respect to the corresponding direction.

To complete the set of kinematical equations, the geometrical boundary conditions on the part C_u of C are given by

$$u_v = u^\alpha \, v_\alpha = u_v^* \qquad \text{on } C_u \tag{2.41}$$

$$u_t = u^\alpha \, t_\alpha = u_t^* \qquad \text{on } C_u \tag{2.42}$$

$$u_3 = u_3^* \qquad \text{on } C_u \tag{2.43}$$

$$\beta_v = -\, u_{3,\alpha}\, v^\alpha = \beta_v^* \qquad \text{on } C_u \tag{2.44}$$

where prescribed quantities are indicated by an asterisk and β_v denotes the rotation about tangent to the boundary line. However, the statical equations are given in the following form

$$N^{\alpha\beta}|_\beta + p^{\alpha *} = 0 \qquad \text{in } \mathcal{M} \tag{2.45}$$

$$(M^{\alpha\beta}|_\alpha + u_{3,\alpha}\, N^{\alpha\beta})|_\beta + b_{\alpha\beta}\, N^{\alpha\beta} + p^{3*} = 0 \qquad \text{in } \mathcal{M} \tag{2.46}$$

$$v_\alpha \, v_\beta \, N^{\alpha\beta} = N_{vv}^* \qquad \text{on } C_p \tag{2.47}$$

$$t_\alpha \, v_\beta \, N^{\alpha\beta} = N_{tv}^* \qquad \text{on } C_p \tag{2.48}$$

$$v_\beta\,(M^{\alpha\beta}|_\alpha + u_{3,\alpha}\, N^{\alpha\beta}) - M_{tv,\alpha}\, t^\alpha = N_{nv}^* - M_{tv,t}^* \qquad \text{on } C_p \tag{2.49}$$

$$v_\alpha \, v_\beta \, M^{\alpha\beta} = M_{vv}^* \qquad \text{on } C_p \tag{2.50}$$

Here, C_p denotes the part of C, where static boundary conditions are prescribed and $p^{\alpha *}$ and p^{3*} are the components of the distributed surface loads per unit area of the undeformed middle surface $\mathcal{M}$. **N** and **M** denote the symmetric stress resultant tensor and the symmetric stress couple tensor, respectively.

2.2.2 Stress-Resultant Constitutive Relations

The strain measures $\boldsymbol{\gamma}$, $\boldsymbol{\chi}$ can be additively decomposed into elastic parts $\boldsymbol{\gamma}^e$, $\boldsymbol{\chi}^e$ and plastic parts $\boldsymbol{\gamma}^p$, $\boldsymbol{\chi}^p$, such that

$$\gamma_{\alpha\beta} = \gamma^e_{\alpha\beta} + \gamma^p_{\alpha\beta} \tag{2.51}$$

$$\chi_{\alpha\beta} = \chi^e_{\alpha\beta} + \chi^p_{\alpha\beta} \tag{2.52}$$

For isotropic linear elastic material behaviour, the strain measures $\boldsymbol{\gamma}^e$, $\boldsymbol{\chi}^e$ are given by the relations

$$\gamma^e_{\alpha\beta} = \frac{1}{A}\, G_{\alpha\beta\gamma\lambda}\, \tilde{N}^{\gamma\lambda} \tag{2.53}$$

$$\chi^e_{\alpha\beta} = \frac{1}{B}\, G_{\alpha\beta\gamma\lambda}\, \tilde{M}^{\gamma\lambda} \tag{2.54}$$

where **G** is the positive definite elasticity tensor and A and B are the extensional and bending stiffness, respectively, defined by

$$G_{\alpha\beta\gamma\lambda} = \frac{1}{2(1-\nu)}\left[a_{\alpha\gamma}\, a_{\beta\lambda} + a_{\alpha\lambda}\, a_{\beta\gamma} - \frac{2\,\nu}{1+\nu}\, a_{\alpha\beta}\, a_{\gamma\lambda}\right] \tag{2.55}$$

and

$$A = \frac{E\,h}{1-\nu^2}\,, \quad B = \frac{E\,h^3}{12(1-\nu^2)} \tag{2.56}$$

The plastic measures $\boldsymbol{\gamma}^p$, $\boldsymbol{\chi}^p$ are given by the normality rule

$$\dot{\gamma}^p_{\alpha\beta} = \dot{\lambda}_N \frac{\partial \mathcal{F}}{\partial N^{\alpha\beta}} \tag{2.57}$$

$$\dot{\chi}^p_{\alpha\beta} = \dot{\lambda}_M \frac{\partial \mathcal{F}}{\partial M^{\alpha\beta}} \tag{2.58}$$

where $\dot{\lambda}_N$ and $\dot{\lambda}_M$ are non-negative proportionality factors. In the following we use a so-called subsequent yield condition derived by Bieniek and Funaro (1976) and modified by several authors to take into account hardening parameter. As it has been suggested by Gross-Weege and Weichert (1992), we use a simplified two-surface yield condition, expressed exclusively in terms of the effective stress resultants

$$\mathcal{F}_I(\tilde{\mathbf{N}}, \tilde{\mathbf{M}}^*) = I_{NN} + \frac{9}{4} I^*_{MM} \pm 3 I^*_{NM} - 1 = 0 \tag{2.59}$$

$$\mathcal{F}_L(\tilde{\mathbf{N}}, \tilde{\mathbf{M}}) = I_{NN} + I_{MM} \pm \frac{1}{\sqrt{3}} I_{NM} - 1 = 0 \tag{2.60}$$

with

$$I_{NN} = I_{\alpha\beta\gamma\lambda} \frac{\tilde{N}^{\alpha\beta} \tilde{N}^{\gamma\lambda}}{N_p^2}\,; \quad I_{MM} = I_{\alpha\beta\gamma\lambda} \frac{\tilde{M}^{\alpha\beta} \tilde{M}^{\gamma\lambda}}{M_p^2}\,; \quad I_{NM} = I_{\alpha\beta\gamma\lambda} \frac{\tilde{N}^{\alpha\beta} \tilde{M}^{\gamma\lambda}}{N_p\, M_p} \tag{2.61}$$

and

$$I_{\alpha\beta\gamma\lambda} = \frac{3}{2} a_{\alpha\gamma}\, a_{\beta\lambda} - \frac{1}{2} a_{\alpha\beta}\, a_{\gamma\lambda} \tag{2.62}$$

$$N_p = \sigma_Y\, h; \quad M_p = \sigma_Y\, h^2/4 \tag{2.63}$$

Here $\mathcal{F}_I$ and $\mathcal{F}_L$ denote the initial yield surface (Bieniek and Funaro, 1976) and the limit yield surface (Ilyushin, 1956), respectively. N_p and M_p denote the uniaxial yield force and the uniaxial yield moment, respectively, and σ_Y is the uniaxial yield stress. I^*_{MM} and I^*_{NM} are also given by eqns (2.61) if we replace **M** by the differences **M** – **M***, where **M*** plays the role of a "*hardening*" parameter whose evolution is defined by

$$\dot{M}^{*\alpha\beta} = \begin{cases} C^{\alpha\beta\gamma\lambda}\, \dot{\chi}^p_{\gamma\lambda} & \text{for } \mathcal{F}_I = 0 \text{ and } \dot{\mathcal{F}}_I > 0 \\ 0 & \text{for } \mathcal{F}_I < 0 \text{ or } \mathcal{F}_I = 0 \text{ and } \dot{\mathcal{F}}_I < 0 \end{cases} \tag{2.64}$$

where

$$C^{\alpha\beta\gamma\lambda} = a^{\alpha\gamma}\, a^{\beta\lambda} \frac{Eh^3}{12} \tag{2.65}$$

2.2.3 A Stress-Resultant Formulation of Shakedown Theorem

The problem. An elastic-plastic shell P is at time $t = t^R$ in equilibrium in the configuration Ω^R under the time-independent external loading $\mathbf{a}^R$. Will the shell shakedown under the action of additional variable loads $\mathbf{a}^r$?

Proposition. The shell P shakes down if there exists time-independent fields of effective bending moments and effective membrane forces $\tilde{\mathring{\mathbf{M}}}^*$, $\Delta\tilde{\mathring{\mathbf{N}}}$ and $\Delta\mathring{\mathbf{M}}$, such that for $t > t^R$ the following conditions are valid

$$\Delta\mathring{N}^{\alpha\beta}|_\beta = 0 \qquad \text{in } \mathcal{M} \qquad (2.66)$$

$$(\Delta\mathring{M}^{\alpha\beta}|_\alpha + \overset{R}{u}_{3,\alpha}\Delta\mathring{N}^{\alpha\beta} + \Delta\mathring{u}_{3,\alpha}N^{R\alpha\beta})|_\beta + b_{\alpha\beta}\Delta\mathring{N}^{\alpha\beta} = 0 \qquad \text{in } \mathcal{M} \qquad (2.67)$$

$$v_\alpha\, v_\beta\, \Delta\mathring{N}^{\alpha\beta} = 0 \qquad \text{on } C_p \qquad (2.68)$$

$$t_\alpha\, v_\beta\, \Delta\mathring{N}^{\alpha\beta} = 0 \qquad \text{on } C_p \qquad (2.69)$$

$$v_\beta\,(\Delta\mathring{M}^{\alpha\beta}|_\alpha + \overset{R}{u}_{3,\alpha}\Delta\mathring{N}^{\alpha\beta} + \Delta\mathring{u}_{3,\alpha}N^{R\alpha\beta}) - \Delta\mathring{M}_{tv,\alpha}\, t^\alpha = 0 \qquad \text{on } C_p \qquad (2.70)$$

$$v_\alpha\, v_\beta\, \Delta\mathring{M}^{\alpha\beta} = 0 \qquad \text{on } C_p \qquad (2.71)$$

$$\Delta\mathring{u}_3 = 0 \qquad \text{on } C_u \qquad (2.72)$$

$$\beta_v = -\Delta\mathring{u}_{3,\alpha}\, v^\alpha = 0 \qquad \text{on } C_u \qquad (2.73)$$

$$\mathcal{F}_I\left(\frac{\mathbf{N}^{r(c)}}{1-D} + \frac{\mathbf{N}^R}{1-D} + \frac{\Delta\mathring{\mathbf{N}}}{1-D}; \frac{\mathbf{M}^{r(c)}}{1-D} + \frac{\mathbf{M}^R - \mathbf{M}^{*R}}{1-D} + \frac{\Delta\mathring{\mathbf{M}} - \mathring{\mathbf{M}}^*}{1-D}, \sigma_Y(\vartheta)\right) < 0 \qquad \text{in } \mathcal{M} \qquad (2.74)$$

$$\mathcal{F}_L\left(\frac{\mathbf{N}^{r(c)}}{1-D} + \frac{\mathbf{N}^R}{1-D} + \frac{\Delta\mathring{\mathbf{N}}}{1-D}; \frac{\mathbf{M}^{r(c)}}{1-D} + \frac{\mathbf{M}^R}{1-D} + \frac{\Delta\mathring{\mathbf{M}}}{1-D}, \sigma_S(\vartheta)\right) < 0 \qquad \text{in } \mathcal{M} \qquad (2.75)$$

with

$$D = D^R + D^r \qquad (2.76)$$

In eqns (2.66-76), the quantities fields induced by the time-independent loads $\mathbf{a}^R$ are indicated by superscript "R", however the quantities resulting from time-dependent additional loads $\mathbf{a}^r$ are indicated by "r" (see, e.g., Gross-Weege and Weichert, 1995; Tritsch and Weichert, 1995).

References

Basar, Y. and Krätzig, W.B. (1985). *Mechanik der Flächentragwerke*. Braunschweig: Vieweg.

Bieniek, M.P. and Funaro, J.R. (1976). *Elastic-plastic Theory of Shells and Plates*. Report No. DNA 3954 T, New York: Weidlinger Associates.

Gross-Weege, J. (1990). A unified formulation of statical shakedown criteria for geometrically nonlinear problems. *Int. J. Plasticity* 6:433–447.

Gross-Weege, J. and Weichert, D. (1992). Elastic-plastic shells under variable mechanical and thermal loads. *Int. J. Mech. Sci.* 34:863–880.

Gross-Weege, J. and Weichert, D. (1995). Shakedown of shells undergoing moderate rotations. In Mróz, Z., Weichert, D. and Dorosz, S., eds. *Inelastic Behaviour of Structures under Variable Loads*. Amsterdam: Kluwer Academic Publishers. 263–277.

Ilyuschin, A. A. (1956). *Plasticity*. Paris: Eyrolles.

Pycko, S. and König, J.A. (1991). Steady plastic cycles on reference configuration in the presence of second-order geometric effects. *Eur. J. Mech. A/Solids* 10:563–574.

Saczuk, J. and Stumpf, H. (1990). *On Statical Shakedown Theorems for Non-linear Problems*. IfM-Report, No 74, Ruhr-Universität, Bochum.

Sawczuk, A. (1982). On plastic shell theories at large strains and displacements. *Int. J. Mech. Sci.* 24:231–244.

Tritsch, J. B. (1993). *Analyse d'Adaptation des Structures Elasto-plastiques avec Prise en Compte des Effets Géométriques*. Ph. D. Dissertation, University of Lille.

Tritsch, J. B. and Weichert, D. (1995). Case studies on the influence of geometric effects on the shakedown of structures. In Mróz, Z., Weichert, D. and Dorosz, S., eds. *Inelastic Behaviour of Structures under Variable Loads*. Amsterdam: Kluwer Academic Publishers. 309–320.

Weichert, D. (1986). On the influence of geometrical nonlinearities on the shakedown of elastic-plastic structures. *Int. J. Plasticity* 2:135–148.

Weichert, D. (1989). Shakedown of shell-like structures allowing for certain geometrical nonlinearities. *Arch. Mech.* 41:61–71.

Weichert, D. and Hachemi, A. (1998). Influence of geometrical nonlinearities on the shakedown of damaged structures. *Int. J. Plasticity* 14:891–907.

Application of Shakedown Theory and Numerical Methods

Dieter Weichert and Abdelkader Hachemi

Aachen University of Technology, Aachen, Germany

Abstract. In this lecture, the methodological framework for the numerical application of shakedown analysis is developed. Examples of applications such as the analysis and optimisation of composite materials and the analysis of various plates and shells problems are presented.

3.1 Application to Composite Materials

The advantage of composite materials compared to conventional materials stemming just from the possibility to design materials for specific technological purposes where in some sense controversial material properties are required turns out to complicate the assessment of the material's performance. The development of such materials demands an optimal design of the microstructure and a fundamental understanding of the role of the microstructure on the overall properties. The microstructural parameters controlling the macroscopic properties are on one hand the morphology of the microstructure and on the other hand the constitutive behaviour of each individual component. The correlation between the microstructure and the macroscopic properties is addressed by homogenisation technique (Suquet, 1983).

3.1.1 Considered Composite

We consider a material which is composed of inclusions, embedded according to a regular pattern as shown in Figure 17 in a homogeneous elastic-plastic metal matrix. In a first step, it is assumed that the material can be regarded as two-dimensional. In this case, these inclusions represent fibres with constant cross sections in the case of plane strains and disk-shaped inclusions in the case of plane stresses. The macroscopic behaviour of this heterogeneous material is observed on the scale $\mathbf{x}$ and the microscopic behaviour on the scale $\mathbf{y}$. For reasons of simplicity, only infinitesimal geometrical transformations are considered.

The generalised stresses $\mathbf{\Sigma}(\mathbf{x})$ and strains $\mathbf{E}(\mathbf{x})$ on the macro-level are linked to generalised stresses $\mathbf{s}(\mathbf{y})$ and strains $\mathbf{e}(\mathbf{y})$ on the micro-level through (Hill, 1963)

$$\mathbf{\Sigma}(\mathbf{x}) = \langle \mathbf{s}(\mathbf{y}) \rangle = \frac{1}{V} \int_{(V)} \mathbf{s}(\mathbf{y})\, dV \tag{3.1}$$

$$\mathbf{E}(\mathbf{x}) = \langle \mathbf{e}(\mathbf{y}) \rangle = \frac{1}{V} \int_{(V)} \mathbf{e}(\mathbf{y})\, dV \tag{3.2}$$

where V denotes the volume of the (periodic) representative volume element (RVE) as shown in Figure 17.

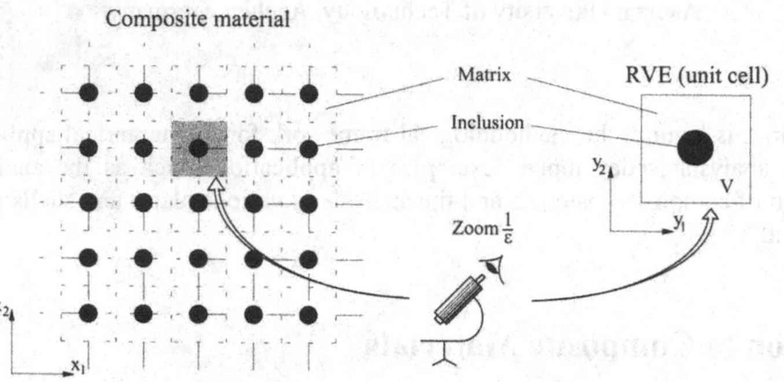

Figure 17. A periodic composite.

3.1.2 Formulation of Shakedown Theorem

The macroscopic admissible domain $P^m(\mathbf{x})$ of composite material may be conveniently defined as the set of macroscopic states of generalised stress $\boldsymbol{\Sigma}(\mathbf{x})$ for all microscopic states of plastically admissible generalised stress $\mathbf{s}(\mathbf{y})$

$$P^m(\mathbf{x}) = \{\boldsymbol{\Sigma}(\mathbf{x}) \,/\, \boldsymbol{\Sigma}(\mathbf{x}) = \langle \mathbf{s}(\mathbf{y}) \rangle \,,\, \mathbf{s}(\mathbf{y}) \in P(\mathbf{y}),\ \forall \mathbf{y} \in V\}. \tag{3.3}$$

The computation of the macroscopic admissible domain will give a reliable prediction of the failure of composite materials. To determine $P^m(\mathbf{x})$, the shakedown analysis is carried out on the micro-level. For this, we introduce the notion of a *reference representative volume element* (RVE$^{(c)}$) differing from the actual one only by the fact that the material is supposed to behave purely elastically.

The statical shakedown theorem states the following: if there exists a safety factor $\alpha > 1$, a time-independent field of generalised stresses $\mathring{\mathbf{s}}(\mathbf{y}) = [\mathring{\boldsymbol{\rho}}(\mathbf{y}), \mathring{\boldsymbol{\pi}}(\mathbf{y})]^T$ and a *Sanctuary of Elasticity* (see §1.8.7)

$$\overline{P}^m(\mathbf{x}) \subset P^m(\mathbf{x}) \tag{3.4}$$

with

$$\overline{P}^m(\mathbf{x}) = \{\boldsymbol{\Sigma}^{(s)}(\mathbf{x}) \,/\, \boldsymbol{\Sigma}^{(s)}(\mathbf{x}) = \langle \mathbf{s}^{(s)}(\mathbf{y}) \rangle \,,\, \mathbf{s}^{(s)}(\mathbf{y}) \in \overline{P}(\mathbf{y}),\ \forall \mathbf{y} \in V\} \tag{3.5}$$

then the periodic composite material shakes down (Weichert et al., 1999a, 1999b). Here, the safe state of generalised stresses $\mathbf{s}^{(s)}$ is defined as usual (see, e.g., Mandel, 1976)

$$\mathbf{s}^{(s)} = \alpha\, \mathbf{s}^{(c)} + \mathring{\mathbf{s}} \tag{3.6}$$

where $\mathbf{s}^{(c)} = [\boldsymbol{\sigma}^{(c)}, \mathbf{0}]$ is the generalised stress field which would occur in the RVE$^{(c)}$ under the same boundary conditions as the actual RVE such that the following relations hold for given macroscopic strains $\mathbf{E}$

$$\mathrm{Div}\, \boldsymbol{\sigma}^{(c)} = \mathbf{0} \quad \text{in } V \tag{3.7}$$

$$\mathbf{u}^{(c)} = \mathbf{E}.\mathbf{y} \quad \text{on } \partial V \tag{3.8}$$

$$\boldsymbol{\varepsilon}^{(c)} = \mathbf{grad}_{\mathrm{sym}}(\mathbf{u}^{(c)}) \quad \text{in } V \tag{3.9}$$

$$\boldsymbol{\sigma}^{(c)} = \mathbf{L}:\boldsymbol{\varepsilon}^{(c)} \quad \text{in } V \tag{3.10}$$

The field of the residual stresses $\mathring{\boldsymbol{\rho}}$ satisfies

$$\mathrm{Div}\, \mathring{\boldsymbol{\rho}} = \mathbf{0} \quad \text{in } V \tag{3.11}$$

$$\mathbf{n}.\mathring{\boldsymbol{\rho}} = \mathbf{0} \quad \text{on } \partial V \tag{3.12}$$

$$\langle \mathring{\boldsymbol{\rho}} \rangle = \mathbf{0} \quad \text{in } V \tag{3.13}$$

3.1.3 Optimum Design

The preceding formulation allows for the assessment of a composite through shakedown analysis according to the presented scenario of failure: more precisely, for given material properties of the individual components, their volume fraction, their individual geometrical form and their pattern of periodicity, one can determine the safe domains of macroscopic stresses applicable to the composite. This, however, does not always correspond to the technical problem to be solved. One can imagine that for given service conditions for a composite one may be confronted to the problem of optimum design of such material with respect to the size of the safe domain of macroscopic stresses for expected external fluctuating loads. One may ask e.g. for that volume fraction of reinforcing fibres for which the safe domain of macroscopic stresses is maximum according to some chosen norm. Another practical problem may be to look for the optimum pattern of periodicity of a reinforcing second phase or the best choice of material properties of the different phases. Be the vector of all design parameters $\boldsymbol{\beta}$. Then, the problem of optimum design with respect to the maximisation of the safe domain of macroscopic stresses can be formulated by

$$\alpha_{\mathrm{OPT}} = \sup_{\boldsymbol{\beta}} \alpha_{\mathrm{SD}} = \sup_{\boldsymbol{\beta}} \max_{\mathring{\boldsymbol{\rho}}, \mathring{\boldsymbol{\pi}}, D} \alpha \tag{3.14}$$

with the subsidiary conditions (3.11-13) and

$$D < D_c \tag{3.15}$$

$$\mathcal{F}_I\left(\alpha\frac{\boldsymbol{\sigma}^{(c)}}{1-D}(\mathcal{P})+\frac{\overset{\circ}{\boldsymbol{\rho}}}{1-D}-\frac{\overset{\circ}{\boldsymbol{\pi}}}{1-D},\sigma_Y\right)<0 \tag{3.16}$$

$$\mathcal{F}_L\left(\alpha\frac{\boldsymbol{\sigma}^{(c)}}{1-D}(\mathcal{P})+\frac{\overset{\circ}{\boldsymbol{\rho}}}{1-D},\sigma_S\right)<0 \tag{3.17}$$

such that

$$\boldsymbol{\Sigma}^{(s)} = \alpha <\boldsymbol{\sigma}^{(c)}> + <\overset{\circ}{\boldsymbol{\rho}}> \tag{3.18}$$

where α_{OPT} is the optimal shakedown loading factor. Attention has to be paid to the fact, that certain design parameters have limitations to be respected by geometrical and/or physical constraints. Furthermore it should be noted, that in many practical cases the properties of the different components of a composite are discontinuous and have to be chosen from catalogues, which then requires the use of discontinuous optimisation methods. In contrast to the optimisation problem underlying the shakedown analysis, the optimum design problem is not necessarily convex with the well-known consequence of loss of uniqueness of the solution.

3.2 Discrete Formulation of Shakedown Problem

Any discretised version of the statical formulation of the shakedown theorem presented above for a continuum medium in the case of geometrical linear theory preserves the relevant bounding properties if the following conditions are satisfied simultaneously:

1. The solution of the purely elastic response $\boldsymbol{\sigma}^{(c)}$ is exact;
2. The residual stress field $\overset{\circ}{\boldsymbol{\rho}}$ satisfies pointwise the homogenous equilibrium equations;
3. The damage parameter D does not reach the critical value in any point of the body;
4. The yield condition $\mathcal{F}$ is satisfied in each point of the body.

Passing to the discretised problem, these conditions are not all satisfied in the general case. Therefore, special case has to be taken as to the solution of the elastic reference problem and the choice of points where to check conditions (3) and (4).

3.2.1 Finite Element Discretisation of Plates and Shells

(a) *Discretisation of the purely elastic response*

To find the response of the purely elastic reference body, we use statically admissible finite elements following the formulation proposed by Morelle and Nguyen Dang Hung (1983). We assume that the equilibrated stress field $\boldsymbol{\sigma}^{(c)}$ is represented within the finite element by a polynomial form depending on two vectors of parameters **b** and **c** as follows

$$\{\boldsymbol{\sigma}^{(c)}\} = [\mathbf{S}]\{\mathbf{b}\} + [\mathbf{T}]\{\mathbf{c}\} \tag{3.19}$$

where [S] and [T] are the matrix of stress shape functions such as [S]{b} fulfils the homogenous equilibrium equations and [T]{c} is in equilibrium with the imposed forces.

The internal complementary energy is then defined by

$$\begin{aligned} U_{\text{int}} &= \frac{1}{2}\int_{(V)} \left[\{\boldsymbol{\sigma}^{(c)}\}^{T}[\mathbf{L}^{-1}]\{\boldsymbol{\sigma}^{(c)}\} + \{\boldsymbol{\sigma}^{(c)}\}^{T}\{\boldsymbol{\varepsilon}^{\vartheta}\}\right] dV \\ &= \frac{1}{2}\{\mathbf{b}\}^{T}[\mathbf{F}_{bb}]\{\mathbf{b}\} + \frac{1}{2}\{\mathbf{c}\}^{T}[\mathbf{F}_{cc}]\{\mathbf{c}\} + \{\mathbf{b}\}^{T}[\mathbf{F}_{bc}]\{\mathbf{c}\} + \{\mathbf{b}\}^{T}\{\mathbf{G}_{b}\} + \{\mathbf{c}\}^{T}\{\mathbf{G}_{c}\} \end{aligned} \tag{3.20}$$

and the external energy is defined by

$$U_{\text{ext}} = -\int_{(S_p)} \{\mathbf{u}\}^{T}\{\mathbf{p}^{*}\}\, dS = -\{\mathbf{g}\}^{T}\{\mathbf{q}\} \tag{3.21}$$

where,

$$[\mathbf{F}_{bb}] = \int_{(V)} [\mathbf{S}]^{T}[\mathbf{L}^{-1}][\mathbf{S}]\, dV; \quad [\mathbf{F}_{bc}] = \int_{(V)} [\mathbf{S}]^{T}[\mathbf{L}^{-1}][\mathbf{T}]\, dV; \quad [\mathbf{F}_{cc}] = \int_{(V)} [\mathbf{T}]^{T}[\mathbf{L}^{-1}][\mathbf{T}]\, dV$$

and

$$\{\mathbf{G}_{b}\} = \int_{(V)} [\mathbf{S}]^{T}\{\boldsymbol{\varepsilon}^{\vartheta}\}\, dV; \quad \{\mathbf{G}_{c}\} = \int_{(V)} [\mathbf{T}]^{T}\{\boldsymbol{\varepsilon}^{\vartheta}\}\, dV.$$

Here, {q} is the vector of generalised displacements corresponding to the vector of generalised forces {g} and where "T" stands for transposed. To ensure the exact transmission of the traction surface {p*}, generalised forces {g} can be defined in analogy to eqn (3.19)

$$\{\mathbf{g}\} = [\mathbf{C}_{b}]\{\mathbf{b}\} + [\mathbf{C}_{c}]\{\mathbf{c}\} \tag{3.22}$$

where $[\mathbf{C}_b]$ and $[\mathbf{C}_c]$ are the statical connection matrix.

We use here the principle of minimum complementary potential energy $\delta(U_{\text{int}} + U_{\text{ext}}) = 0$, which leads to

$$\{\mathbf{g}\} + \{\mathbf{g}_F\} + \{\mathbf{g}_\vartheta\} = [\mathbf{K}]\{\mathbf{q}\} \tag{3.23}$$

with

$$[\mathbf{K}] = [\mathbf{C}_b][\mathbf{F}_{bb}^{-1}][\mathbf{C}_b]^T \tag{3.24}$$

as the stiffness matrix,

$$\{\mathbf{g}_F\} = ([\mathbf{C}_b][\mathbf{F}_{bb}^{-1}][\mathbf{F}_{bc}] - [\mathbf{C}_c])\{\mathbf{c}\} \tag{3.25}$$

as the surface loads vector and

$$\{\mathbf{g}_\vartheta\} = [\mathbf{C}_b][\mathbf{F}_{bb}^{-1}]\{\mathbf{G}_b\} \tag{3.26}$$

the temperature vector. One then obtains

$$\{\mathbf{b}\} = [\mathbf{F}_{bb}^{-1}]\{[\mathbf{C}_b]^T\{\mathbf{q}\} - [\mathbf{F}_{bc}]\{\mathbf{c}\} - \{\mathbf{G}_b\}\}. \tag{3.27}$$

The displacements vector $\{\mathbf{q}\}$ given by eqn (3.23) is substituted into eqn (3.27). The resulting vector $\{\mathbf{b}\}$ yields after substitution into eqn (3.19) the elastic stresses $\{\boldsymbol{\sigma}^{(c)}\}$ in each Gaussian point. For details we refer to Morelle and Nguyen Dang Hung (1983), Gross-Weege (1988), Hachemi and Weichert (1998).

(b) *Discretisation of the residual stresses*

Following the formulation proposed by Nguyen Dang Hung and König (1976), we discretise the residual stresses field in the same way as the field of elastic stresses (eqn (3.19)) (see also Morelle and Nguyen Dang Hung (1983), Gross-Weege (1988), Hachemi and Weichert (1998)). This field satisfies the homogenous equilibrium eqns (1.49-50)

$$\{\mathring{\boldsymbol{\rho}}\} = [\mathbf{S}]\{\mathbf{b}\} \tag{3.28}$$

To take into account the equilibrium with the surface forces (eqn (3.22)), we divide the vector $\{\mathbf{g}\}$ into two parts

$$\{\mathbf{g}\} = \begin{pmatrix} \{\mathbf{g}\}_{(S_p)} \\ \{\mathbf{g}\}_{(S_u)} \end{pmatrix} \quad \text{and} \quad [\mathbf{C}_b]_{(S)} = \begin{pmatrix} [\mathbf{C}_b]_{(S_p)} \\ [\mathbf{C}_b]_{(S_u)} \end{pmatrix} \tag{3.29}$$

with

$$[\mathbf{C}_b]_{(S_p)}\{\mathbf{b}\} = 0. \tag{3.30}$$

where $[\mathbf{C}_b]_{(S_p)}$ is the global statical connection matrix in which the elements corresponding to the displacement degrees of freedom are eliminated.

It turns out that the elements of **b** are not independent of each other and so a Gauss-Jordan elimination procedure (Stiefel, 1960) is applied to the equality constraints (eqns (1.49-50)). Then, eqn (3.30) leads to

$$\{\mathbf{b}\} = [\mathbf{J}]\{\mathbf{X}\} \tag{3.31}$$

where $[\mathbf{J}]$ is the matrix obtained by the Gauss-Jordan elimination procedure. The vector $\{\mathbf{b}\}$ is completely determined by the linear independent parameters $\{\mathbf{X}\}$, calculated by the optimisation process. The resulting vector $\{\mathbf{b}\}$ yields after substitution into eqn (3.28) the residual stresses $\{\mathring{\boldsymbol{\rho}}\}$ in each Gaussian points

$$\{\mathring{\boldsymbol{\rho}}\} = [\mathbf{S}][\mathbf{J}]\{\mathbf{X}\} = [\mathbf{B}]\{\mathbf{X}\}. \tag{3.32}$$

3.2.2 Finite Element Discretisation for Soils

We use here, the finite element force method based on the principle of minimum complementary energy (see, e.g., Gallagher and Dhalla, 1975; Weichert and Gross-Weege, 1988). This approach use stress functions for the construction of the complementary energy function and represents an algebraic dual to the finite element displacement method. The method allows to compute the stresses under the assumption of plane stress as in disk problems, with the normal- and shear-stresses in *z*-direction set equal to zero, or alternatively of plane strain, with $\sigma_{zz} = f(\sigma_{xx}, \sigma_{yy})$, which gives a better modeling of the physical problem in our case. In both cases the stresses are function of two coordinates x and y.

(a) *Discretisation of the purely elastic response*
Neglecting body-forces, the purely elastic stress vector is expressed in terms of Airy's stress function $F(x_1, x_2)$ by

$$\{\boldsymbol{\sigma}^{(c)}\}^T = \{\sigma_{11}^{(c)}; \sigma_{22}^{(c)}; \sigma_{12}^{(c)}\} = \{F_{,22}; F_{,11}; -F_{,12}\} \tag{3.33}$$

The equilibrium in the volume is of the form

$$F_{,221} - F_{,122} = 0 \tag{3.34}$$
$$-F_{,121} + F_{,112} = 0 \tag{3.35}$$

The kinematic equation holds, if the strains fulfill the following condition, known as compatibility criterion (two-dimensional)

$$\varepsilon_{11,22}^{(c)} + \varepsilon_{22,11}^{(c)} - 2\,\varepsilon_{12,12}^{(c)} = 0. \tag{3.36}$$

Using Hook's law and the definition of the Airy function, we get

$$\Delta\Delta F = 0 \text{ or } F_{,1111} + F_{,2222} + 2\,F_{,1212} = 0. \tag{3.37}$$

Any stress field derived from a stress function F, which is a solution of the differential eqn (3.37), fulfills the static equations of equilibrium and the compatibility criterion. A solution composed of functions solving the differential equation has to fulfill the boundary conditions. The functions, which constitute the approximate solution for F, must not fulfill the differential eqn (3.37), so the compatibility criterion may be violated in the structure, but the stresses fulfill the equations of equilibrium exactly, because they are derived from Airy function. The approximate solution has to satisfy the static boundary conditions. The stresses, which fulfill the equations of equilibrium and the static boundary conditions are statically admissible stresses.

In the finite element model, approximate solutions for the stress function are constructed from a set of functions, Hermite polynomes in our case (Gallagher and Dhalla, 1975). The criterion for a best approximation through these functions is, that the approximation for the stress function fulfills the static boundary conditions and minimises the complementary energy.

The stress field is dependent of x and y, so the integration in V can be transformed into an integration in the x-y-plane over the area A. With vanishing body forces $\mathbf{f}^*$ in V and imposed displacements $\mathbf{u}^* = 0$ on S_u, the expression for the complementary energy of the elastic comparison body $\mathscr{B}^{(c)}$ in the presence of thermal strains $\boldsymbol{\varepsilon}^{\vartheta}$ simplifies to the equation

$$U_{\text{int}} = \frac{1}{2}\int_{(A)} \left[\{\boldsymbol{\sigma}^{(c)}\}^{T}[\mathbf{L}^{-1}]\{\boldsymbol{\sigma}^{(c)}\}\right] h\,\mathrm{d}A + \int_{(A)} \left[\{\boldsymbol{\sigma}^{(c)}\}^{T}\{\boldsymbol{\varepsilon}^{\vartheta}\}\right] h\,\mathrm{d}A \tag{3.38}$$

with thickness h ($\mathrm{d}V = h\,\mathrm{d}A$).

If we replace the stresses $\boldsymbol{\sigma}^{(c)}$ by the derivatives of Airy's stress function F (eqn (3.33)) we, get

$$U_{\text{int}} = \frac{1}{2}\int_{(A)} \left[\{\Delta F\}^{T}[\mathbf{L}^{-1}]\{\Delta F\}\right] h\,\mathrm{d}A \tag{3.39}$$

We approximate Airy's stress function by

$$F(x, y) = \{\mathbf{N}\}^{T}\{\boldsymbol{\Phi}^{e}\} \tag{3.40}$$

$$\Delta F(x, y) = \{\mathbf{N}''\}^{T}\{\boldsymbol{\Phi}^{e}\} \tag{3.41}$$

with the vectors $\{\mathbf{N}\}$ and $\{\boldsymbol{\Phi}^{e}\}$ contain the element shape functions and the nodal stress function parameters of the elastic solution, respectively. Superscript "e" refers to quantities defined in the individual elements and $\{\mathbf{N}''\}$ stands for the vector of second derivatives of $\{\mathbf{N}\}$ as defined in the square brackets.

Substituting (3.41) into (3.33) and (3.39), we obtain

$$\{\boldsymbol{\sigma}^{(c)}\}^{T} = \{\mathbf{N}^{T}_{,22};\, \mathbf{N}^{T}_{,11};\, \mathbf{N}^{T}_{,12}\}\{\boldsymbol{\Phi}^{e}\} = \{\mathbf{N}''^{T}\}\{\boldsymbol{\Phi}^{e}\} \tag{3.42}$$

and

$$U_{\text{int}} = \frac{1}{2}\{\boldsymbol{\Phi}^{e}\}^{T}[\mathbf{f}^{e}]\{\boldsymbol{\Phi}^{e}\} + \{\boldsymbol{\Phi}^{e}\}^{T}[\mathbf{e}^{e}] \tag{3.43}$$

with

$$[\mathbf{f}^{e}] = \int_{(A)} \left[\{\mathbf{N}''\}[\mathbf{L}^{-1}]\{\mathbf{N}''\}^{T}\right] h\,\mathrm{d}A \quad \text{and} \quad [\mathbf{e}^{e}] = \int_{(A)} \left[\{\mathbf{N}''\}\{\boldsymbol{\varepsilon}^{\vartheta}\}\right] h\,\mathrm{d}A$$

where $[\mathbf{f}^{e}]$ denotes the element flexibility matrix.

The implementation of the statical boundary conditions (eqn (1.45)) requires additional considerations. If S_p^e denotes that part of an element boundary where statical conditions are prescribed, we get, with eqn (3.33),

$$\begin{aligned} F_{,22}\, n_1 - F_{,12}\, n_2 &= \overset{*}{p_1} \\ -F_{,12}\, n_1 + F_{,11}\, n_2 &= \overset{*}{p_2} \end{aligned} \qquad \text{on } S_p^e \tag{3.44}$$

Substitution of eqn (3.42) into (3.44) then delivers

$$\left\{ \begin{pmatrix} \mathbf{N}_{,22}^{\mathrm{T}} \\ -\mathbf{N}_{,12}^{\mathrm{T}} \end{pmatrix} n_1 + \begin{pmatrix} \mathbf{N}_{,12}^{\mathrm{T}} \\ -\mathbf{N}_{,11}^{\mathrm{T}} \end{pmatrix} n_2 \right\} \{\mathbf{\Phi}^e\} = \{\mathbf{p}^*\} \qquad \text{on } S_p^e \tag{3.45}$$

The eqns (3.45) from the boundary conditions constitute a set of constraint equations for the system, which can be written as a linear problem

$$[\mathbf{C}]\{\mathbf{\Phi}\} = \{\mathbf{p}\} \tag{3.46}$$

where $[\mathbf{C}]$ is the global system constraint matrix. The vector of external loads $\{\mathbf{p}\}$ results from integration of surface tractions $\{\mathbf{p}^*\}$ over the individual element boundaries S_p^e. Adding the constraint eqns (3.46) with Lagrange multipliers $\{\lambda\}$ to the functional of the complementary energy, we obtain the augmented function of the complementary energy

$$U_{tot} = U_{int} + ([\mathbf{C}]\{\mathbf{\Phi}\} - \{\mathbf{p}\})^{\mathrm{T}}\{\lambda\} \tag{3.47}$$

The variation of (3.47) leads to the set of linear equations

$$\begin{pmatrix} [\mathbf{f}] & [\mathbf{C}]^{\mathrm{T}} \\ [\mathbf{C}] & [\mathbf{0}] \end{pmatrix} \begin{pmatrix} \{\mathbf{\Phi}\} \\ \{\lambda\} \end{pmatrix} = \begin{pmatrix} -\{\mathbf{e}\} \\ \{\mathbf{p}\} \end{pmatrix} \tag{3.48}$$

where the flexibility matrix $[\mathbf{f}]$ and the vector $\{\mathbf{\Phi}\}$ are obtained by the superposition of all element flexibility matrices $[\mathbf{f}^e]$ and of all element vectors $\{\mathbf{\Phi}^e\}$. The resulting vector $\{\mathbf{\Phi}^{(c)}\}$ of the system of linear eqns (3.48) yields after substitution into eqn (3.42) the elastic element stresses in each element

$$\{\boldsymbol{\sigma}^{(c)}\}^{\mathrm{T}} = \{\mathbf{N}''\}^{\mathrm{T}}\{\mathbf{\Phi}^{(c)}\} \tag{3.49}$$

(b) *Discretisation of the residual stresses*
For the solution of the shakedown problem we now represent the residual stresses $\mathring{\boldsymbol{\rho}}$, defined by eqns (1.49-50), in terms of $\{\boldsymbol{\Phi}\}$ and $[\mathbf{C}]$. As $\mathring{\boldsymbol{\rho}}$ is subjected only to homogenous boundary conditions on S_p, we get

$$[\mathbf{C}]\{\bar{\boldsymbol{\Phi}}^{\rho}\} = \{\mathbf{0}\} \quad \text{on } S_p \tag{3.50}$$

Turn out that element of $\{\bar{\boldsymbol{\Phi}}^{\rho}\}$ are not independent from each other and so a Gauss-Jordan elimination procedure is applied to find the vector containing only independent elements. Eqn (3.50) is then equivalent to

$$\big[[\mathbf{C}_1];[\mathbf{C}_2]\big]\begin{pmatrix}\{\bar{\boldsymbol{\Phi}}^{\rho}_1\}\\ \{\bar{\boldsymbol{\Phi}}^{\rho}_2\}\end{pmatrix} = \{0\} \tag{3.51}$$

where $\mathbf{C}_1^{-1}$ is supposed to exist, and finally we have

$$\{\bar{\boldsymbol{\Phi}}^{\rho}_1\} = -[\mathbf{C}_1^{-1}]\,[\mathbf{C}_2]\{\bar{\boldsymbol{\Phi}}^{\rho}_2\} \;\Rightarrow\; \{\boldsymbol{\Phi}^{\rho}\} = [\mathbf{J}]\{\bar{\boldsymbol{\Phi}}^{\rho}_2\} = \begin{pmatrix}-[\mathbf{C}_1^{-1}][\mathbf{C}_2]\\ \mathbf{I}\end{pmatrix}\{\bar{\boldsymbol{\Phi}}^{\rho}_2\} \tag{3.52}$$

The resulting vector $\{\boldsymbol{\Phi}^{\rho}\}$ yields after substitution into eqn (3.42) the residual stresses $\{\mathring{\boldsymbol{\rho}}\}$ in each element

$$\{\mathring{\boldsymbol{\rho}}\}^{\mathrm{T}} = \{\mathbf{N}''\}^{\mathrm{T}}\{\boldsymbol{\Phi}^{\rho}\} \tag{3.53}$$

3.2.3 Finite Element Discretisation of Composites

For the discretisation of the shakedown problem, in the case of composite materials, the displacement-based finite element method is employed (Dhatt et al., 1984; Zienkiewicz, 1984).

(a) *Discretisation of the purely elastic response*
To calculate the purely elastic stresses $\boldsymbol{\sigma}^{(c)}$, we use the virtual work principle

$$\delta U_{\text{int}} = \delta U_{\text{ext}} \Rightarrow \frac{1}{V}\int_{(V)} \{\delta\boldsymbol{\varepsilon}^{(c)}(\mathbf{y})\}^{\mathrm{T}}\,\{\boldsymbol{\sigma}^{(c)}(\mathbf{y})\}\; dV = \{\boldsymbol{\Sigma}\}\{\delta\mathbf{E}\} \tag{3.54}$$

By using the well-known Gauss-Legendre technique, this integral can be calculated for each finite element. The integration has to be carried out over all Gaussian points *NGE* with their weighting factors w_i in the considered element "e"

$$\int_{(V)} \{\boldsymbol{\delta\varepsilon}^{(c)}(\mathbf{y})\}^T \{\boldsymbol{\sigma}^{(c)}(\mathbf{y})\}\, dV = \sum_{j=1}^{NK} \{\delta\mathbf{u}_j^e\}^T \left\{ \int_{(V)} [\mathbf{B}(\mathbf{y})]^T [\mathbf{L}]\, [\mathbf{B}(\mathbf{y})]\, dV \right\} \{\mathbf{u}_j^e\}$$

$$= \sum_{j=1}^{NK} \{\delta\mathbf{u}_j^e\}^T \left\{ \sum_{i=1}^{NGE} w_i\, |\mathbf{J}_i|\, [\mathbf{B}_j(\mathbf{y}_i)]^T [\mathbf{L}][\mathbf{B}_j(\mathbf{y}_i)] \right\} \{\mathbf{u}_j^e\}$$

$$= \sum_{j=1}^{NK} \{\delta\mathbf{u}_j^e\}^T [\mathbf{K}]\, \{\mathbf{u}_j^e\}$$

$$= \sum_{j=1}^{NK} \{\delta\mathbf{u}_j^e\}^T \{\mathbf{F}_j\} \tag{3.55}$$

which leads for the *i*-th Gaussian point to

$$\{\boldsymbol{\sigma}^{(c)}(\mathbf{y}_i)\} = \sum_{j=1}^{NK} [\mathbf{L}][\mathbf{B}_j(\mathbf{y}_i)]\{\mathbf{u}_j^e\} \tag{3.56}$$

Here, $[\mathbf{B}_j]$ and $\{\delta\mathbf{u}_j^e\}$ denote the matrix of the derivatives of the shape functions and the vector of virtual displacement of the *j*-th node of the element "e", respectively. $[\mathbf{F}_j]$ and $\{\mathbf{u}_j^e\}$ are the vectors of nodal forces and displacements of the *k*-th node, respectively. *NK* denotes the total number of nodes of each element, $|\mathbf{J}_i|$ the determinant of the Jacobian matrix and $[\mathbf{K}]$ the stiffness matrix.

(b) *Discretisation of the residual stresses*
Analogously to (3.55), the field of residual stresses $\mathring{\boldsymbol{\rho}}$ is defined by

$$\int_{(V)} \{\boldsymbol{\delta\varepsilon}(\mathbf{y})\}^T \{\mathring{\boldsymbol{\rho}}(\mathbf{y})\}\, dV = 0 \tag{3.57}$$

The integration over all Gaussian points *NGE* in the considered element "e", gives

$$\int_{(V)} \{\delta\boldsymbol{\varepsilon}(\mathbf{y})\}^{\mathrm{T}}\{\mathring{\boldsymbol{\rho}}(\mathbf{y})\}\, dV = \sum_{j=1}^{NK} \{\delta\mathbf{u}_j^e\}^{\mathrm{T}} \int_{(V)} [\mathbf{B}(\mathbf{y})]^{\mathrm{T}}\{\mathring{\boldsymbol{\rho}}\}\, dV$$

$$= \sum_{j=1}^{NK} \{\delta\mathbf{u}_j^e\}^{\mathrm{T}} \left\{ \sum_{i=1}^{NGE} w_i\, |\mathbf{J}_i|\, [\mathbf{B}_j(\mathbf{y}_i)]^{\mathrm{T}}\{\mathring{\boldsymbol{\rho}}(\mathbf{y}_i)\} \right\}$$

$$= \sum_{j=1}^{NK} \{\delta\mathbf{u}_j^e\}^{\mathrm{T}}[\mathbf{C}_i]\{\mathring{\boldsymbol{\rho}}_i\} \tag{3.58}$$

By summation of the contributions of all elements and by variation of the virtual node-displacements $\{\delta\mathbf{u}_j^e\}$ with regard to the boundary conditions, one finally gets the linear system of equations (see, e.g., Stein et al., 1993; Gross-Weege, 1997)

$$\sum_{i=1}^{NG} [\mathbf{C}_i]\{\mathring{\boldsymbol{\rho}}_i\} = [\mathbf{C}]\{\mathring{\boldsymbol{\rho}}\} = \{\mathbf{0}\}. \tag{3.59}$$

Here, *NG* denotes the total number of Gaussian points of the considered structure, [C] is a constant matrix, uniquely defined by the discretised system and the boundary conditions and $\{\mathring{\boldsymbol{\rho}}\}$ is the global residual stress vector of the discretised structure.

Then, the discrete formulation of shakedown problems for the determination of safety factor against failure due to inadmissible damage or accumulated plastic deformations is given by the solution of the following non-linear optimisation problems:

$$\alpha_{\mathrm{OPT}} = \sup_{\beta} \alpha_{\mathrm{SD}} = \sup_{\beta} \max_{\mathring{\rho}_i,\, \mathring{\pi}_i,\, D_i} \alpha \tag{3.60}$$

with the subsidiary conditions

$$[\mathbf{C}]\{\mathring{\boldsymbol{\rho}}\} = \{\mathbf{0}\} \tag{3.61}$$

$$D_i < D_c \tag{3.62}$$

$$\mathscr{F}_{\mathrm{I}}\left(\alpha \frac{\boldsymbol{\sigma}_i^{(c)}}{1-D_i}(\mathscr{P}_j) + \frac{\mathring{\boldsymbol{\rho}}_i}{1-D_i} - \frac{\mathring{\boldsymbol{\pi}}_i}{1-D_i}, \sigma_Y\right) < 0 \tag{3.63}$$

$$\mathscr{F}_{\mathrm{L}}\left(\alpha \frac{\boldsymbol{\sigma}_i^{(c)}}{1-D_i}(\mathscr{P}_j) + \frac{\mathring{\boldsymbol{\rho}}_i}{1-D_i}, \sigma_S\right) < 0 \tag{3.64}$$

$$\forall i \in [1, NG],\ \forall j \in [1, 2^n]$$

such that

$$\mathbf{\Sigma}^{(s)} = \alpha < \boldsymbol{\sigma}^{(c)} > + < \mathring{\boldsymbol{\rho}} >. \tag{3.65}$$

with α_{OPT} as objective function to be optimised with respect to $\boldsymbol{\beta}$, $\mathring{\boldsymbol{\rho}}_i$, $\mathring{\boldsymbol{\pi}}_i$ and D_i and with relations (3.61) and (3.62-64) as linear and non-linear constraints, respectively. The condition (3.62) assures structural safety against failure due to material damage and the condition (3.63) assures that safe states of stresses are never outside the loading surface $\mathscr{F}_L$ and so guarantees implicitly the boundedness of the back-stresses $\mathring{\boldsymbol{\pi}}$. The damage parameter and the yield criteria have to be fulfilled in *NG* points and for 2^n combinations of loads. The solution of this optimisation problem can be carried out by using any algorithm of non-linear mathematical programming (c.f. Pierre and Low, 1975; Conn et al., 1992).

3.3 Numerical Examples

On the basis of the concepts developed in the preceding sections, the specialised Shakedown And Limit-Analysis software SALIA has been used (Gross-Weege, 1988; Giese, 1988; Tritsch, 1993; Hachemi, 1994; Boulbibane, 1995; Schwabe, 2000). For illustration, some numerical examples of different structures will be considered.

3.3.1 Circular Plate

For the circular plate, two cases are investigated:

- Circular plate under the action of uniform pressure and a bending moment
- Circular plate under the action of uniform pressure and a variable temperature

(a) *Circular plate under the action of uniform pressure and a bending moment*
The shakedown analysis is carried out for a simply supported circular plate loaded as shown in Figure 18 by two independent variable loads, the uniform pressure $p = \mu_1 p_0$ and a bending moment $M = \mu_2 M_0$ $(0 \le \mu_1 \le \mu_1^+ ; \mu_2^- \le \mu_2 \le \mu_2^+)$. This example has also been treated by König (1969) and Morelle (1989) using Tresca yield condition for uniform and sandwich cross-section, respectively. The first author gives the lower bound via an analytical method and the second via a numerical method.

The following values for geometry, material properties and loads have been adopted in the numerical analysis:

$$h/R = 1/100;\ E = 1.6 \cdot 10^5 \text{ MPa};\ \sigma_Y = 3.6 \cdot 10^2 \text{ MPa};\ \nu = 1/3;$$
$$p_0 = 1.5\, \sigma_Y\, (h/R)^2;\ \ M_0 = \sigma_Y\, h^2/4.$$

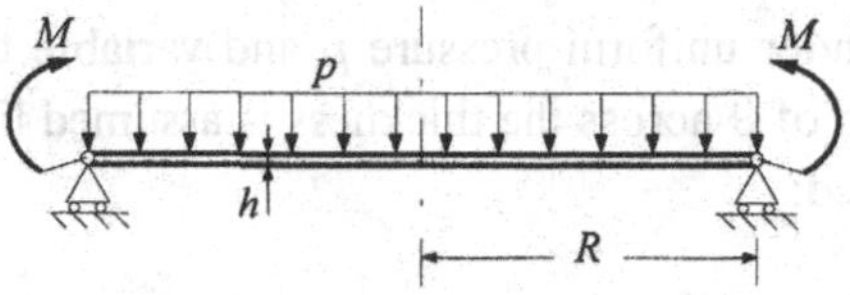

Figure 18. Circular plate under uniform pressure and a bending moment.

Two different classes of material behaviour are considered:

- Elastic perfectly-plastic material behaviour without damage (Figure 19). Both failure mechanisms are studied, namely failure due to alternating plasticity and failure due to accumulated plasticity. Curves (a) correspond to the failure due to alternating plasticity by considering uniform cross-section and curves (b) correspond to the failure due to accumulated plasticity by considering sandwich cross-section;
- Elastic perfectly-plastic material behaviour with ductile damage (Figure 20) using the model of Lemaitre (1985) and Shichun and Hua (1990). Only the failure mechanism due to accumulated plasticity is investigated because of the choice of the adopted damage models. We observe a stronger sensitivity of the shakedown load on the uniform pressure than on the moment variation. The range of the allowable safe moment variation is characterised by the failure mechanism of alternating plasticity.

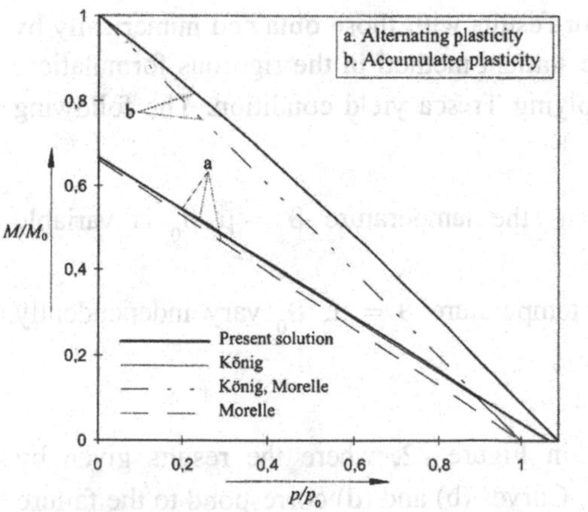

Figure 19. Loading domains without damage.

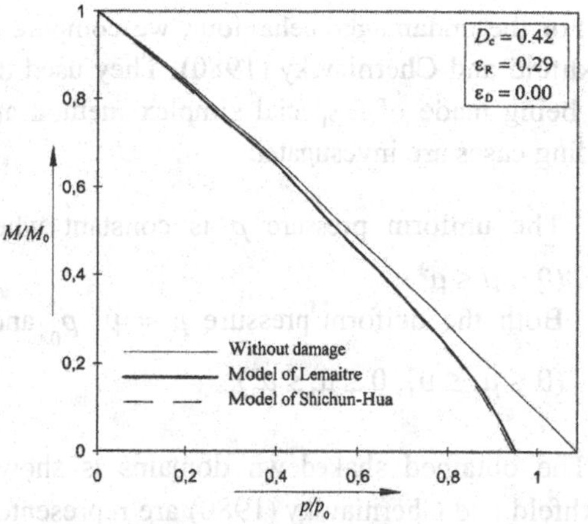

Figure 20. Influence of ductile damage.

(b) *Circular plate under the action of uniform pressure and variable temperature*
A clamped circular plate under uniform pressure p and variable temperature ϑ is considered (Figure 21). The distribution of ϑ across the thickness is assumed linear. Two different cases of material behaviour are studied:

- Perfectly plastic material behaviour without damage (Figure 22);
- Elastic perfectly plastic material behaviour with ductile damage (Figure 23), by using the model of Lemaitre (1985) and Shichun and Hua (1990).

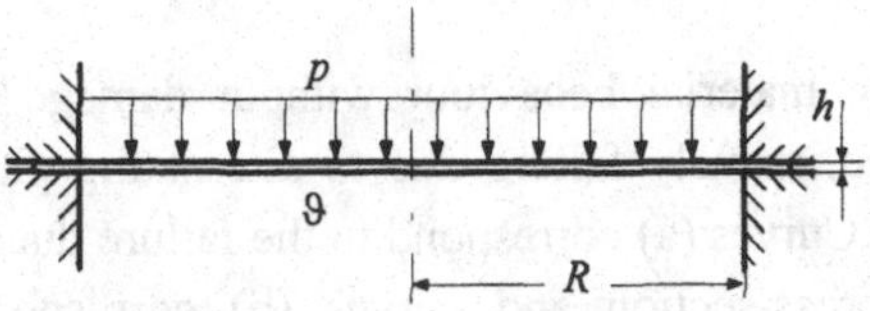

Figure 21. Clamped circular plate under uniform pressure and variable temperature.

The following values for geometry, material properties and loads have been adopted in the numerical analysis:

$$h/R = 1/25;\ E = 1.6\cdot 10^5\ \text{MPa};\ \sigma_Y = 3.6\cdot 10^2\ \text{MPa};\ \nu = 0.3;\ \alpha_\vartheta = 0.2\cdot 10^{-4}\ \text{K}^{-1};$$
$$p_0 = 4\,\sigma_Y\, h^2/((1+\nu)\,R^2);\ \vartheta_0 = 6(1-\nu)\,\sigma_Y/(E\,\alpha_\vartheta).$$

For the undamaged behaviour, we compare our results with those obtained numerically by Gokhfeld and Cherniavsky (1980). They used the statical method in the rigorous formulation, use being made of a special simplex method applying Tresca yield condition. The following loading cases are investigated:

- The uniform pressure p is constant whereas the temperature $\vartheta = \mu\,\vartheta_0$ is variable $(0 \le \mu \le \mu^+)$;
- Both the uniform pressure $p = \mu_1\, p_0$ and temperature $\vartheta = \mu_2\,\vartheta_0$ vary independently $(0 \le \mu_1 \le \mu_1^+;\ 0 \le \mu_2 \le \mu_2^+)$.

The obtained shakedown domains is shown in Figure 22, where the results given by Gokhfeld and Cherniavsky (1980) are represented. Curves (b) and (d) correspond to the failure due to alternating plasticity for uniform and sandwich plates, respectively, and curves (c) correspond to failure due to accumulated plastic deformation for uniform cross-section. When the loading program is defined by a variable thermal load at constant pressure, alternating plasticity (curves (a)) is relevant. In this case the drop in load-carrying capacity resulting from thermal loads is of the order of 33 per cent.

The shakedown domain in the case of combined plasticity and damage is shown in Figure 23 for uniform cross-section. The results show a considerable reduction of shakedown loads due to ductile damage compared to the undamaged behaviour, particularly when the thermal variable load is dominant.

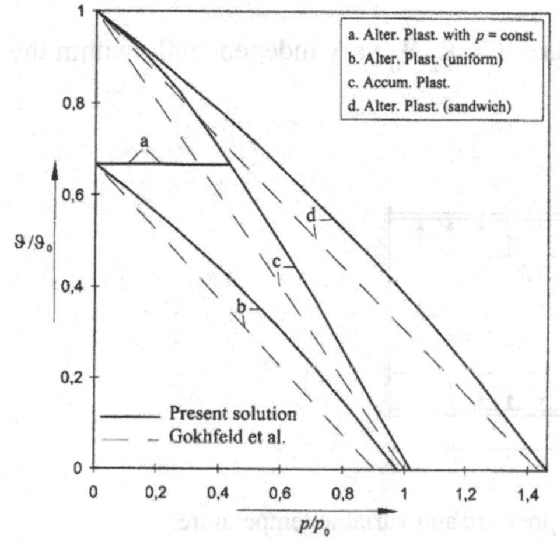

Figure 22. Loading domains without damage.

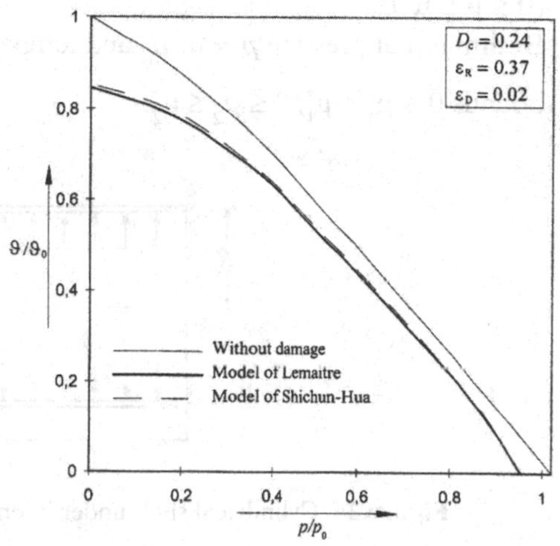

Figure 23. Influence of ductile damage.

3.3.2 Cylindrical and Conical Shells

We consider some examples of shells under different variable loads where the influence of damage, kinematical hardening, temperature dependent yield stress and large displacements are taken into account (see, e.g., Gross-Weege and Weichert, 1992; Tritsch and Weichert, 1995; Hachemi and Weichert, 1998; Weichert and Hachemi, 1998). The following cases are investigated:

- Cylindrical shell under internal pressure and variable temperature
- Cylindrical shell under internal pressure
- Cylindrical shell under internal pressure and a bending moment
- Conical shell under internal pressure and an axial ring load

(a) *Cylindrical shell under internal pressure and variable temperature*
The shell is subjected to internal pressure and a linear temperature distribution across the thickness (Figure 24). Temperature ϑ varies between ϑ_i and ϑ_a, where ϑ_a is the temperature on the outer wall of the shell and is assumed to be 0°C. For this case, the following loading conditions are investigated:

- The internal pressure p is constant whereas the temperature $\vartheta_i = \mu\, \vartheta_0$ is variable $(0 \leq \mu \leq \mu^+)$;
- Both internal pressure $p = \mu_1\, p_0$ and temperature $\vartheta_i = \mu_2\, \vartheta_0$ vary independently within the bounds $0 \leq \mu_1 \leq \mu_1^+$; $0 \leq \mu_2 \leq \mu_2^+$.

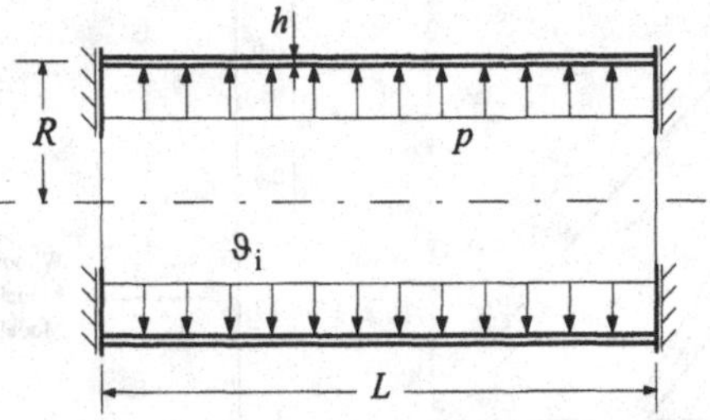

Figure 24. Cylindrical shell under internal pressure and variable temperature.

The following types of material behaviour are considered:

- Elastic perfectly-plastic material behaviour (Figure 25). The yield stress σ_Y is a function of the actual temperature with $\sigma_Y = \sigma_{Y_0}(1 + A\vartheta + B\vartheta^2)$, where σ_{Y_0}, A and B are given material parameters;
- Linear kinematical hardening material behaviour, the ratio of the tensile strength σ_S to the yield stress σ_Y is equal to 1.2 (Figure 26). The tensile strength σ_S and the yield stress σ_Y are temperature-dependent.

The following values for geometry, material properties and loads have been adopted in the numerical analysis:

$$h/R = 1/10;\ L/R = 1/2;\ E = 1.6\cdot 10^5\ \text{MPa};\ \sigma_{Y_0} = 2.05\cdot 10^2\ \text{MPa};\ \nu = 0.3;\ \alpha_\vartheta = 0.2\cdot 10^{-4}\ \text{K}^{-1};$$
$$A = -1.83\cdot 10^{-3}\ \text{K}^{-1};\ B = 1.83\ 10^{-6}\ \text{K}^{-2};\ p_0 = \sigma_{Y_0}\, h/R;\ \vartheta_0 = 2(1-\nu)\,\sigma_{Y_0}/(E\,\alpha_\vartheta).$$

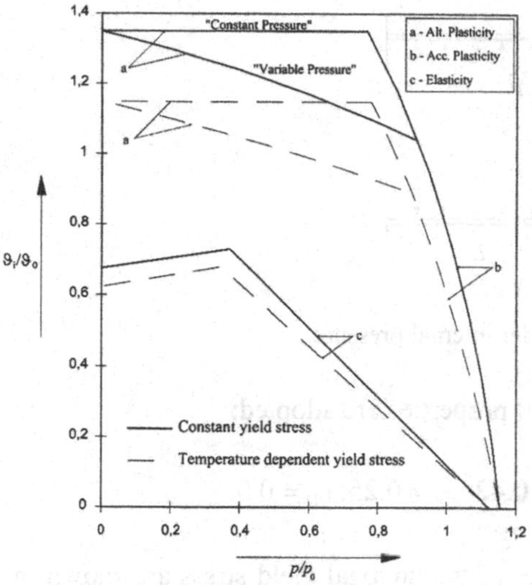

Figure 25. Loading domains.

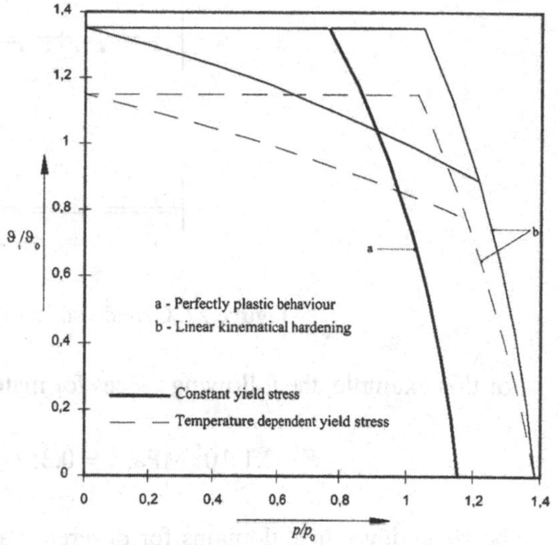

Figure 26. Influence of kinematical hardening.

(b) *Cylindrical shell under internal pressure*

To take into account the influence of large displacements, we consider a short clamped cylindrical shell studied by Gross-Weege (1988, 1990) with sandwich cross-section, length L, radius R and wall-thickness h (Figure 27), where L/R = 1/10 and h/R = 1/200. The shell is subjected to internal pressure p,

$$p = p^R + p^r$$

where p^R is a time-independent reference pressure and $p^r = \mu(t)\, p^R$ is a time-dependent additional pressure, where $\mu(t)$ varies between fixed bounds μ^- and μ^+, so that $\mu^- \le \mu \le \mu^+$. This is a typical problem in process engineering, when fluctuations of pressure in the neighborhood of the nominal pressure in pipelines and other pressure vessels may occur. To determine the state of stresses $\boldsymbol{\sigma}^R$, displacement $\mathbf{u}^R$ and damage parameter D^R in the reference configuration C_R under the time-independent reference pressure p^R, a modified version of the NONSAP step-by-step finite element program (Bathe et al., 1974) of Kreja et al. (1993) to take into account ductile damage model of Lemaitre (1985) has been used. The symmetry of shell and loads is assumed which allows to limit calculations to half of the shell.

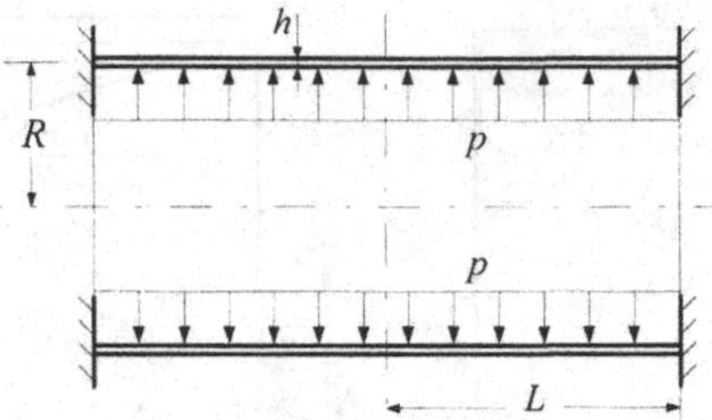

Figure 27. Cylindrical shell under internal pressure.

For this example, the following values for material properties are adopted:

$$E = 2.1\cdot 10^5 \text{ MPa};\ \nu = 0.3;\ D_c = 0.42;\ \varepsilon_R = 0.25;\ \varepsilon_D = 0.0.$$

The shakedown load domains for different values of the uniaxial yield stress are shown in Figure 28. For the sake of comparison, the results of undamaged shell are represented as well as geometrical linear solution using the von Mises sandwich yield condition.

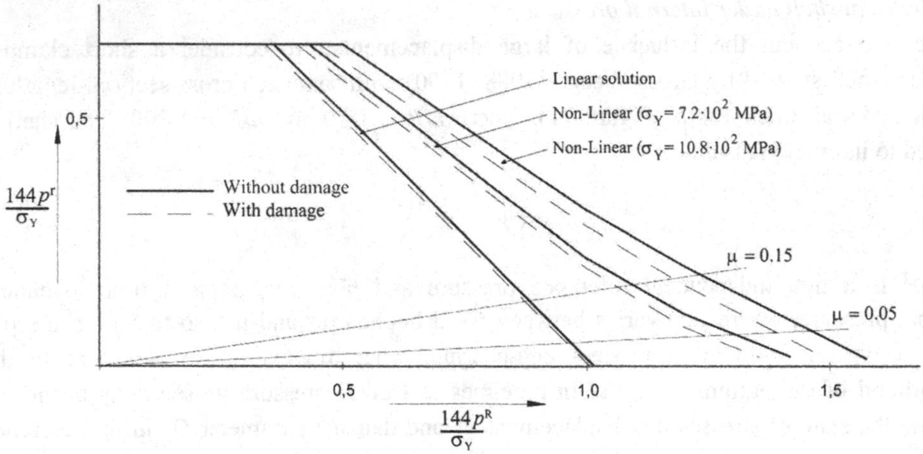

Figure 28. Shakedown load domain.

(c) *Cylindrical shell under internal pressure and a bending moment*

The cylindrical shell described above with the following dimensions $L/R = 1/10$ and $h/R = 1/100$ is loaded as shown in Figure 29 by two independent loads, the internal pressure $p = \mu_1\, p_0$ and a bending moment $M = \mu_2\, M_0$ $(0 \leq \mu_1 \leq \mu_1^+ ;\ 0 \leq \mu_2 \leq \mu_2^+)$.

The following values of material properties and loads have been adopted:

$$E = 1.6 \cdot 10^5 \text{ MPa}; \sigma_Y = 360 \text{ MPa}; \nu = 0.3; M_0 = 10.39 \cdot 10^3 \text{ N}; p_0 = 4.4 \text{ MPa}$$

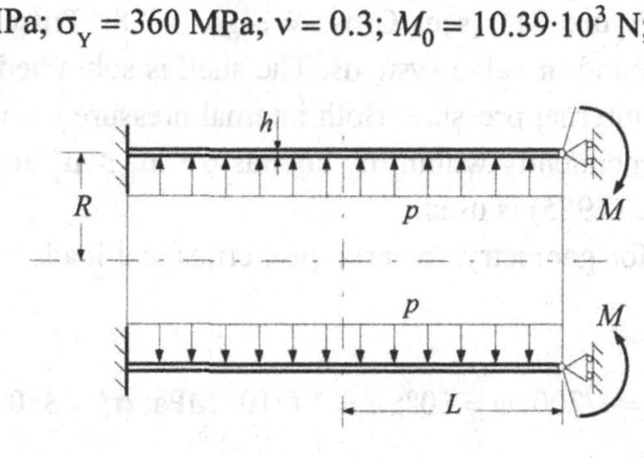

Figure 29. Cylindrical shell under internal pressure and a bending moment.

For this example, the Ilyushin yield condition (Ilyushin, 1956) and damage model of Lemaitre (1985) are used. The obtained shakedown domain is shown in Figure 30, where also the results of undamaged behaviour are presented. Curves (a) and (b) correspond to the shakedown limit load due to incremental collapse for uniform cross-section. We observe in the major part of the loading space a significant difference between geometrical linear, curves (a), and geometrical non-linear, curves (b), analysis and between damaged and undamaged behaviour.

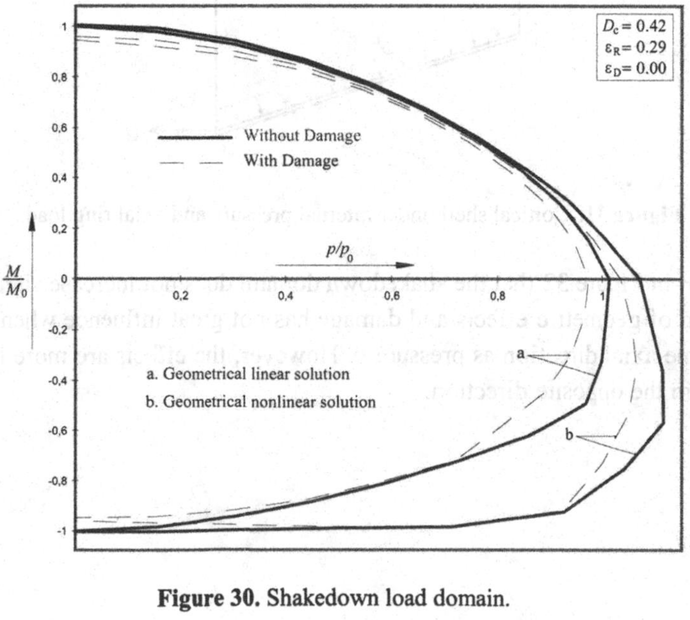

Figure 30. Shakedown load domain.

(d) *Conical shell under internal pressure and an axial ring load*
A simply supported conical shell with smaller radius R_i, external radius R_e, wall-thickness h and angle φ is considered (Figure 31) (see, Gross-Weege, 1988; Tritsch, 1993). This kind of mechanical structures is found in valve systems. The shell is submitted to an axial ring load at the large radius edge and internal pressure. Both internal pressure $p = \mu_1 p_0$ and axial ring load $Q = \mu_2 Q_0$ can vary independently within the bounds $0 \le \mu_1 \le \mu_1^+$ and $0 \le \mu_2 \le \mu_2^+$. Here the damage model of Lemaitre (1985) is used.

The following values for geometry, material properties and loads have been adopted in the numerical analysis:

$$R_i/R_e = 3/4;\ h/R = 1/200;\ \varphi = 60°;\ E = 1.6\cdot 10^5\ \text{MPa};\ \sigma_Y = 360\ \text{MPa};\ \nu = 0.3;$$

$$Q_0 = 1.2\cdot 10^3\ \text{N/mm};\ \ p_0 = 1.6\ \text{MPa}.$$

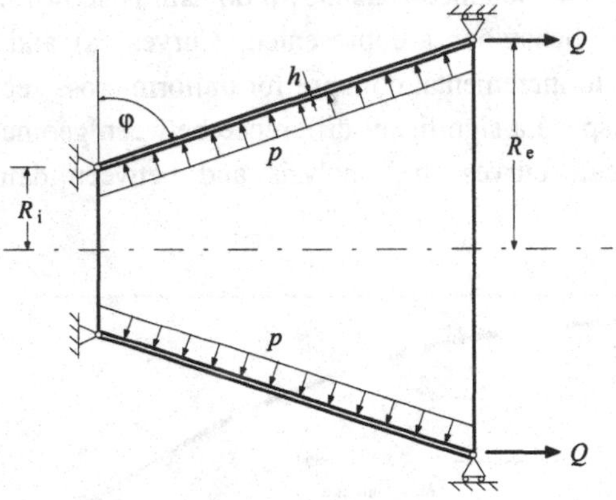

Figure 31. Conical shell under internal pressure and axial ring load.

It can be seen in Figure 32 that the shakedown domain does not increase significantly due to the consideration of geometric effects and damage has not great influence when the axial force Q acts in the same axial direction as pressure p. However, the effects are more important when these loads acts in the opposite direction.

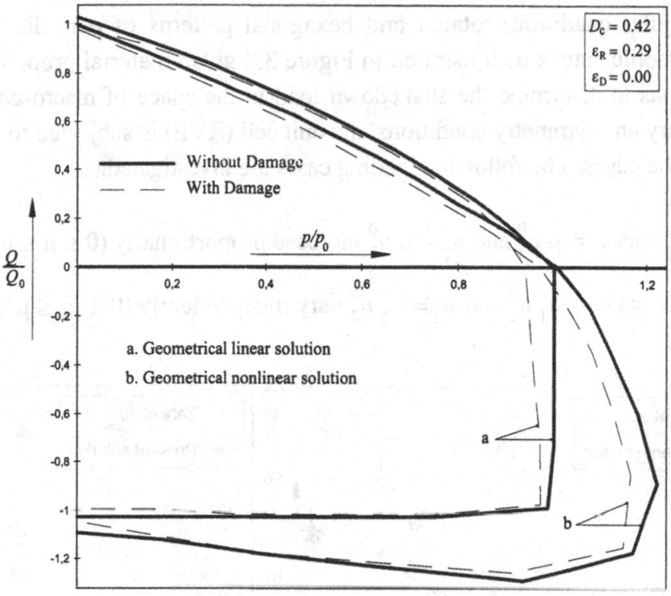

Figure 32. Shakedown load domain.

3.3.3 Composites

For the limit and shakedown analysis of composite materials, we consider a typical problem for an Al/Al_2O_3 Metal-Matrix-Composite, with the following properties for the matrix, $E = 70$ GPa, $\sigma_Y = 80$ MPa, $\nu = 0.3$ and for the fibres $E = 370$ GPa, $\nu = 0.3$.

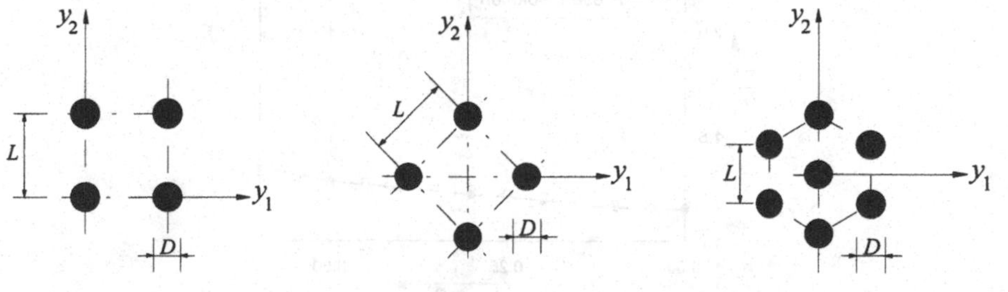

Figure 33. Arrangement of fibre-reinforced composite.

For given regular quadratic, rotated and hexagonal patterns of periodicity of reinforced elastic fibres in ductile matrix as illustrated in Figure 33, given material properties of the fibres and matrix, one has to determine the shakedown load in the space of macroscopic stresses. To satisfy the boundary and symmetry conditions, the unit cell (RVE) is subjected to bi-axial uniform displacements at the edges. The following loading cases are investigated:

- The displacements $u_1 = \mu\, u_1^0$ and $u_2 = \mu\, u_2^0$ increase proportionally $(0 \le \mu \le \mu^+)$;
- The displacements $u_1 = \mu_1\, u_1^0$ and $u_2 = \mu_2\, u_2^0$ vary independently $(0 \le \mu_1 \le \mu_1^+;\ 0 \le \mu_2 \le \mu_2^+)$.

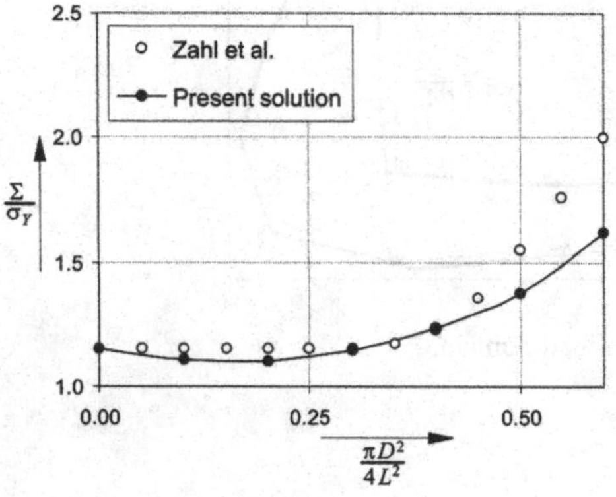

(a) quadratic pattern

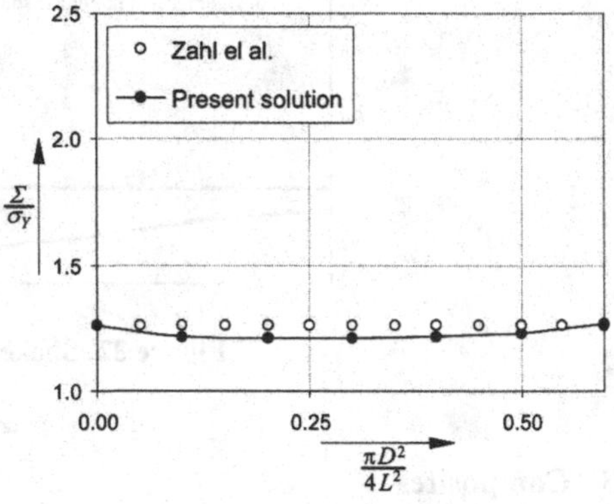

(b) rotated pattern

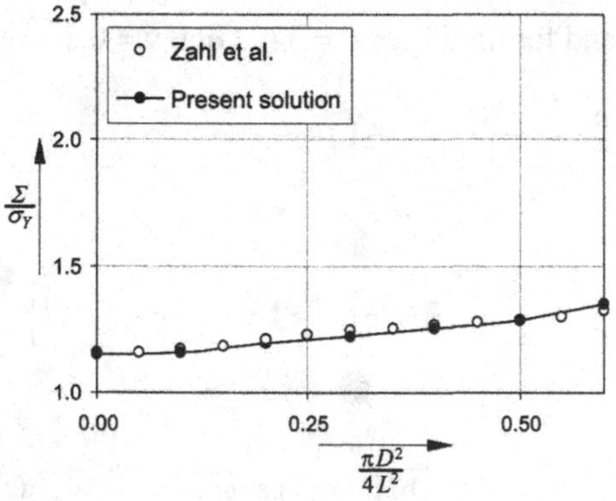

(c) hexagonal pattern

Figure 34. Variation of macroscopic stress with fibre volume fraction.

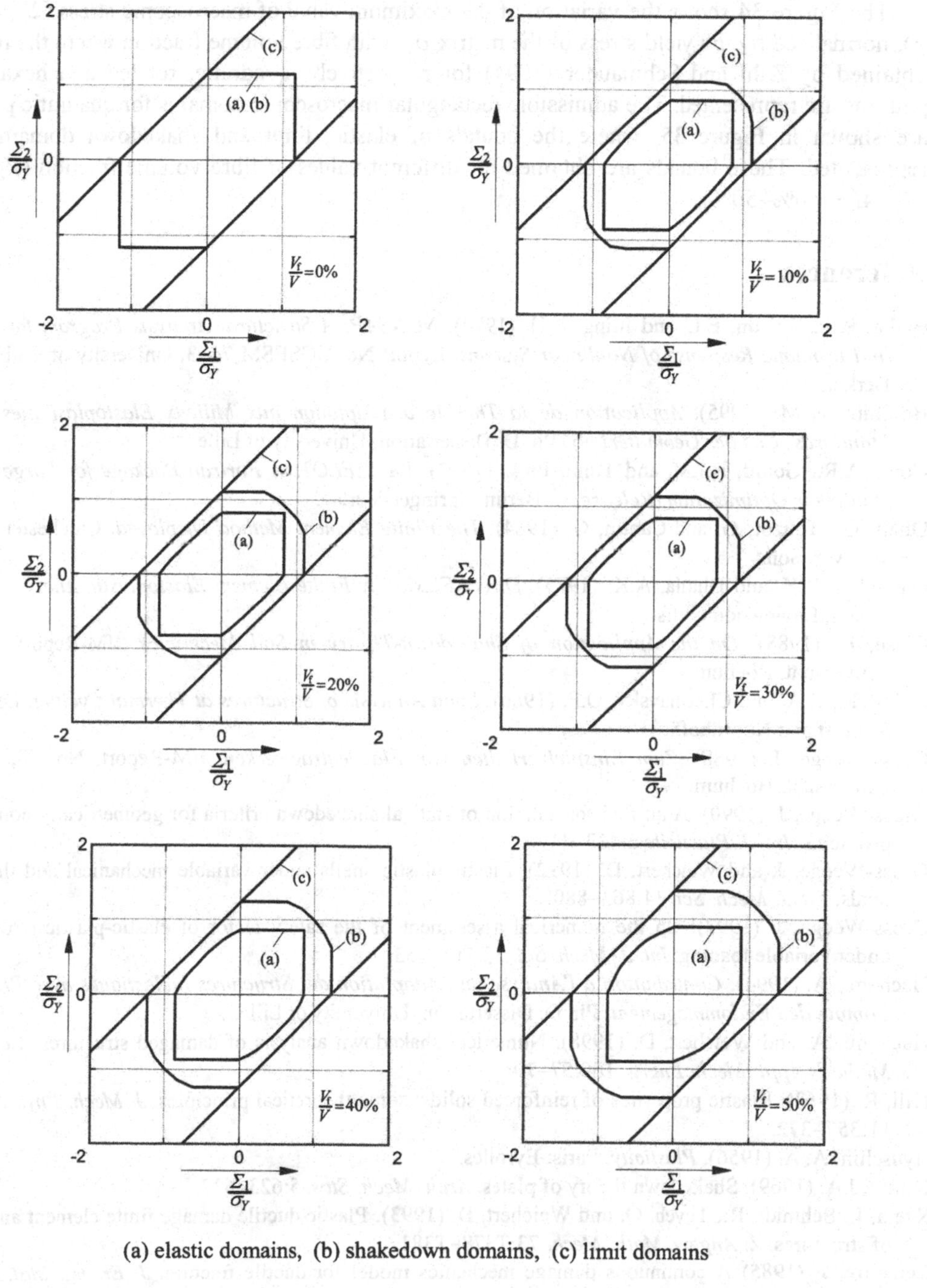

(a) elastic domains, (b) shakedown domains, (c) limit domains

Figure 35. Variation of safe rectangular macroscopic domains with fibre volume fraction (0%–50%).

The Figure 34 shows the variation of the maximum value of macroscopic stress ($\Sigma_1 = \Sigma_2 = \Sigma$), normalised by the yield stress of the matrix σ_Y, with fibre volume fraction where the results obtained by Zahl and Schmauder (1994) for respectively, quadratic, rotated and hexagonal patterns are represented. The admissible rectangular macroscopic domains for quadratic pattern are shown in Figure 35, where the bounds of elastic, limit and shakedown domains are represented. These bounds are obtained for different values of fibre volume fraction ($V_f / V = \pi D^2/4L^2$ = 0%–50%).

References

Bathe, K.J., Wilson, E.L. and Iding, R.H. (1974). *NONSAP: A Structural Analysis Program for Static and Dynamic Response of Nonlinear Systems.* Report No. UGSESM 74–3, University of California, Berkeley.

Boulbibane, M. (1995). *Application de la Théorie d'Adaptation aux Milieux Elastoplastiques Non-Standards: Cas des Géomatériaux.* Ph. D. Dissertation, University of Lille.

Conn, A.R., Gould, N.I.M. and Toint, Ph.L. (1992). *LANCELOT: A Fortran Package for Large-Scale Nonlinear Optimization (Release A).* Berlin: Springer-Verlag.

Dhatt, G., Touzot, G. and Cantin, G. (1984). *The Finite Element Method Displayed.* Chichester: John Wiley & Sons.

Gallagher, R.H. and Dhalla, A.K. (1975). *Direct Flexibility Finite Element Elastoplastic Analysis.* New Jersey: Englewood Cliffs.

Giese, H. (1988). *On the Application of Shakedown-Theory in Soil Mechanics.* IfM-Report, Ruhr-Universität, Bochum.

Gokhfeld, D.A. and Cherniavsky, O.F. (1980). *Limit Analysis of Structures at Thermal Cycling.* Leyden: Sijthoff and Noordhoff.

Gross-Weege, J. (1988). *Zum Einspielverhalten von Flächentragwerken.* IfM-Report, No. 58, Ruhr-Universität, Bochum.

Gross-Weege, J. (1990). A unified formulation of statical shakedown criteria for geometrically nonlinear problems. *Int. J. Plasticity* 6:433–447.

Gross-Weege, J. and Weichert, D. (1992). Elastic-plastic shells under variable mechanical and thermal loads. *Int. J. Mech. Sci.* 34:863–880.

Gross-Weege, J. (1997). On the numerical assessment of the safety factor of elastic-plastic structures under variable loading. *Int. J. Mech. Sci.* 39:417–433.

Hachemi, A. (1994). *Contribution à l'Analyse de l'Adaptation des Structures Inélastiques avec Prise en Compte de l'Endommagement.* Ph. D. Dissertation, University of Lille.

Hachemi, A. and Weichert, D. (1998). Numerical shakedown analysis of damaged structures. *Comput. Methods Appl. Mech. Engrg.* 160:57–70.

Hill, R. (1963). Elastic properties of reinforced solids: some theoretical principles. *J. Mech. Phys. Solids* 11:357–372.

Ilyuschin, A. A. (1956). *Plasticity.* Paris: Eyrolles.

König, J.A. (1969). Shakedown theory of plates. *Arch. Mech. Stos.* 5:623–637.

Kreja, I., Schmidt, R., Teyeb, O. and Weichert, D. (1993). Plastic ductile damage finite element analysis of structures. *Z. Angew. Math. Mech.* 73:T378–T381.

Lemaitre, J. (1985) A continuous damage mechanics model for ductile fracture. *J. Engng. Mat. Tech.* 107: 83–89.

Mandel, J. (1976). Adaptation d'une structure plastique écrouissable et approximations, *Mech. Res. Comm.* 3:483–488.

Morelle, P. and Nguyen Dang Hung (1983). Etude numérique de l'adaptation plastique des plaques et coques de révolution par les éléments finis d'équilibre. *J. Méc. Théo. Appl.* 2:567–599.

Morelle, P. (1989). *Analyse Duale de l'Adaptation Plastique des Structures par la Méthode des Eléments Finis et la Programmation Mathématique*. Ph. D. Dissertation, University of Liège.

Nguyen Dang Hung and König, J.A. (1976). A finite element formulation for shakedown problems using yield criterion of the mean. *Comput. Methods Appl. Mech. Engrg.* 8:179–192.

Pierre, D.A. and Lowe, M.J. (1975). *Mathematical Programming via Augmented Lagrangians*. London: Addison-Wesley.

Schwabe, F. (2000). *Einspieluntersuchungen von Verbundwerkstoffen mit periodischer Mikrostruktur*. Ph. D. Dissertation, RWTH Aachen.

Shichun, W. and Hua, L. (1990). A kinetic equation for ductile damage at large plastique strain. *J. Mat. Proc. Tech.* 21:295–302.

Stein, E., Zhang, G. and Huang, Y. (1993). Modeling and computation of shakedown problems for nonlinear hardening materials, *Comput. Methods Appl. Mech. Engrg.* 103:247–272.

Stiefel, E. (1960). Note on Jordan elimination, linear programming and Tchebycheff approximation. *Numer. Math.* 2:1–17.

Suquet, P. (1983). Analyse limite et homogénéisation. *C.R. Acad. Sci.* 296:1355–1358.

Tritsch, J. B. (1993). *Analyse d'Adaptation des Structures Elasto-plastiques avec Prise en Compte des Effets Géométriques*. Ph. D. Dissertation, University of Lille.

Tritsch, J. B. and Weichert, D. (1995). Case studies on the influence of geometric effects on the shakedown of structures. In Mróz, Z., Weichert, D. and Dorosz, S., eds. *Inelastic Behaviour of Structures under Variable Loads*. Amsterdam: Kluwer Academic Publishers. 309–320.

Weichert, D. and Gross-Weege, J. (1988). The numerical assessment of elastic-plastic sheets under variable mechanical and thermal loads using a simplified two-surface yield condition. *Int. J. Mech. Sci.* 30:757–767.

Weichert, D. and Hachemi, A. (1998). Influence of geometrical nonlinearities on the shakedown of damaged structures. *Int. J. Plasticity* 14:891–907.

Weichert, D., Hachemi, A. and Schwabe, F. (1999a). Shakedown analysis of composites. *Mech. Res. Comm.* 26:309–318.

Weichert, D., Hachemi, A. and Schwabe, F. (1999b). Application of shakedown theory to the plastic design of composites. *Arch. Appl. Mech.* 69:623–633.

Zienkiewicz, O.C. (1984). *Methode der finiten Elemente*. München: Carl Hanser Verlag.

Zahl, D.B. and Schmauder, S. (1994). Transverse strength of continuous fiber metal matrix composites. *Comput. Mat. Sci.* 3:293–299.

A Linear Matching Method for Shakedown Analysis

Alan R. S. Ponter *

University of Leicester, Leicester, United Kingdom

Abstract. This article describes the Linear Matching Method for the evaluation of shakedown limits for bodies subjected to cyclic load and temperature and composed of an elastic-perfectly plastic material. The method provides a development of the Elastic Compensation and related methods that have been used as a practical design tool in industry for some time. Such methods may be developed into general upper bound methods that are capable of providing the minimum upper bound associated with a class of possible displacement fields as described, for example, by a finite element mesh. At the same time a sequence of lower bounds are generated that converge to the least upper bound. The ability to implement these methods within a standard commercial finite element code makes them particularly attractive for engineering applications.

Introduction

Over the last thirty years there have been a considerable number of methods described in the literature for the direct evaluation of limit loads and shakedown limits. They general rely upon the application of linear and non-linear programming methods to the upper and lower bound shakedown theorems and the use of finite element approximations. But such methods have remained within the domain of specialist research workers and have generally not come into general use for engineering design. There are a number of reasons for this. Finite element simulation codes of great reliability are now available and these shakedown techniques require a wholely different set of computer algorithms. For large scale problems, programming techniques generally scale somewhat less attractively than linear elastic solutions. Some of the methods are not particularly robust and this is particularly so for methods based upon the lower bound theorem.

In parallel with such methods there has been a separate stream of developments where there has been a desire to make the most of standard linear elastic analysis. These methods developed into the Elastic Compensation method developed by Mackenzie and Boyle (1993). The idea behind the method is very simple when applied to limit analysis. The problem is first solved elastically for an assumed design load. Within the solution the stresses in regions of the body exceed the yield condition. The stresses in such locations are then reduced to a yield value by changing the elastic properties (usually Young's modulus), assuming the strain field remains unchanged. The new stress field does

* The author wishes to acknowledge the support of the Engineering and Physical Sciences Research Council and British Energy Ltd during the development of the methods described in this article. A number of colleagues have contributed towards the work, particularly Keith Carter, Markus Engelhardt, Sebastien Hentz and Francesco Parinello who carried out the computational work for the solutions described here.

not, of course, satisfy equilibrium. But, with this new distribution of elastic moduli the problem may be resolved. The process may then be repeated and the solution scaled to find the largest value of the applied load for which the stresses everywhere lie within yield. This provides a simple method of finding lower bound limit loads without the need for specialist software and has been used for some years in industry.

The interesting features of such methods are that they scale in exactly the same way as linear elastic solutions. But in the form described above the method is an essentially ad-hoc approach to limit analysis. The purpose of this article is to demonstrate that the essential idea of matching linear solutions to the non-linear problem may be developed into very robust upper bound methods where the objective is to compute, for a given finite element mesh, the minimum upper bound to either a limit load or a shakedown limit. We will see that the method may then be understood as a programming method where the linear solutions required for each iteration may be interpreted as a standard linear problem of the type for which finite element codes are designed. The core of this development is encapsulated within convergence theorems that show that the solution of a certain linear problem produces a strain rate field or history which reduces an upper bound. By repeated iteration the upper bound reduces to the minimum upper bound associated with the class of displacement field of the finite element mesh.

In the following section we review the upper and lower bound shakedown theorems (with limit analysis as a special case) and give a simple example of the application of the upper bound and the types of mechanisms that may be expected to occur. Readers familiar with this basic material may pass immediately to Section 2 where the linear matching method is described for limit analysis and the von Mises yield condition. A solution for a simple two bar structure demonstrates the relationship between the method and programming methods. In Section 3 the theory for shakedown analysis is given in detail for, again, a von Mises yield condition. Further examples and extensions of the method to arbitrary yield conditions is briefly discussed in Section 4.

In recent times the method has been extended to the evaluation of the ratchet limit by Ponter and Chen (2001a,b) and to a class of creep problems by Ponter (2001), demonstrating that the general methodology has the capacity to be used in a wide range of circumstances.

1 Shakedown and It's Applications

1.1 Introduction

Limit analysis and shakedown analysis have their origins in the plasticity schools of the 1940's and 50's when the lack of effective computational methods for complex problems of design provided a powerful incentive to discover theorems which gave fundamental insight into the ways structures behaved when subjected to realistic and complex loading. The design codes and life assessment methods for metallic structures which evolved from this period rely heavily on the insight so obtained. The development of finite element methods in the 1960's and the availability of fairly flexible and reliable commercial finite element code in the late 1980's caused a change in attitude. Provided accurate constitutive equations could be developed, direct simulation of inelastic behaviour of a structure became available at the nearest computer terminal. Indeed, in many industrial applications this has become a routine part of design practice, particularly where there is a high valued added element to detailed design. For example this is the case for the internal components of gas turbines where the geometries are few, the loading severe and the need for accurate predictions of component behaviour is extremely important.

However, interest in shakedown analysis failed to disappear, as important industrial problems still arose where the potential efficiency and the insight that applications of shakedown could provide allowed progress in design related issues which conventional finite element simulation methods had failed to provide. In all such cases the important aspect of the analysis has been that the evaluation of shakedown limits effectively inverts analysis. Hence a conventional finite element analysis will provide a very detailed description of the performance of a structure subjected to a complex history of load and temperature. If you require the behaviour for a neighbouring load history, the calculation must be repeated. The accuracy of the answer depends upon the accuracy of the constitutive relationship. A shakedown limit inverts the problem by giving the limits of a class of load histories when a significant event occurs, the beginning of either incremental strain growth or reverse plasticity. The advantage of this approach is the ability to understand the sensitivity of structural performance to the various parameters that govern the problem. If the intended loading history lies within the shakedown limit, the shakedown analysis gives a factor on that load before shakedown breaks down. Shakedown analysis of classes of problems can be represented as interaction diagrams which then join the literature of understanding of structural behaviour. The disadvantage is that there is only a limited class of constitutive equations where shakedown exists and there are many industrial problems which cannot be viewed in this way.

1.2 Upper Bound Shakedown Limit Theorem

Consider a body composed of an elastic-perfectly plastic solid which satisfies the yield condition;

$$f(\sigma_{ij}) \le 0 \tag{1.2.1}$$

The associated flow rule is given by,

$$\dot{\varepsilon}_{ij} = \dot{\alpha}\frac{\partial f}{\partial \sigma_{ij}} \tag{1.2.2}$$

where $\dot{\alpha}$ is a plastic multiplier. We assume that the yield function is convex and that the maximum work principle holds,

$$(\sigma_{ij}^{p} - \sigma_{ij}^{*})\dot{\varepsilon}_{ij}^{p} \ge 0 \tag{1.2.3}$$

where σ_{ij}^{p} is the stress associated with plastic strain rate $\dot{\varepsilon}_{ij}^{p}$ at yield and σ_{ij}^{*} is any state of stress which satisfies the yield condition (1.2.1).

Consider the following problem. A body of volume V and surface S is subjected to a cyclic history of loading with cycle time Δt. The history consists of mechanical loads

$\lambda P_i(x_j,t)$ over S_T, part of S, and a temperature history $\lambda\theta(x_j,t)$ within V, where λ is a positive load parameter. On the remainder of S, namely S_U, the displacement rate, $\dot{u}_i = 0$. The linear elastic solution to the problem, corresponding to $\dot{\varepsilon}_{ij}^p = 0$, is denoted by $\lambda\hat{\sigma}_{ij}$. The elastic material properties are assumed to be independent of temperature.

We define a class of kinematically admissible strain rate histories $\dot{\varepsilon}_{ij}^c$ with a corresponding displacement increment fields Δu_i^c and associated compatible strain increment,

$$\Delta\varepsilon_{ij}^c = \tfrac{1}{2}\left(\frac{\partial\Delta u_i^c}{\partial x_j} + \frac{\partial\Delta u_j^c}{\partial x_i}\right) \tag{1.2.4}$$

The strain rate history $\dot{\varepsilon}_{ij}^c$, which need not be compatible, satisfies the condition that,

$$\int_0^{\Delta t} \dot{\varepsilon}_{ij}^c dt = \Delta\varepsilon_{ij}^c \tag{1.2.5}$$

In terms of such a history of strain rate an upper bound on the shakedown limit is given by (see Koiter, 1960; Gokhfeld and Cherniavsky, 1980; Konig, 1987),

$$\int_V \int_0^{\Delta t} \sigma_{ij}^c \dot{\varepsilon}_{ij}^c dt dV = \lambda_{UB}^c \int_V \int_0^{\Delta t} \hat{\sigma}_{ij} \dot{\varepsilon}_{ij}^c dt dV \tag{1.2.6}$$

where $\lambda_{UB}^c \geq \lambda_s$, the exact shakedown limit and σ_{ij}^c denotes the stress at yield associated with $\dot{\varepsilon}_{ij}^c$. We wish to discover solution methods which describe a sequence of kinematically admissible strain increment fields which give a reducing sequence of upper bounds. The sequence converges to the shakedown limit λ_s or the least upper bound associated with the class of displacement fields and strain rate histories chosen.

1.3 The Lower Bound Theorem

For the same class of problems discussed above we seek a lower bound load parameter λ_{LB}. Consider a residual stress field ρ_{ij}^* in equilibrium within V and with zero surface tractions on S_T. If the resultant of this residual stress field and the elastic stress history;

$$\sigma_{ij}^* = \lambda_{LB}\hat{\sigma}_{ij} + \rho_{ij}^* \tag{1.3.1}$$

satisfies the yield condition $f(\sigma_{ij}^*) \leq 0$ everywhere within V, then λ_{LB} is a lower bound to the exact shakedown limit;

$$\lambda_{LB} \leq \lambda_s \tag{1.3.2}$$

The unique exact shakedown limit occurs when the upper and lower bounds coincide so that $\lambda_{UB}^c = \lambda_{LB}^* = \lambda_s$ and $\sigma_{ij}^c = \sigma_{ij}^*$ when $\dot{\varepsilon}_{ij}^c \neq 0$.

In the following sections the main emphasis will be on the evaluation of optimal upper bounds as the computational method employed, finite elements, relies upon classes of displacement fields and stresses are in equilibrium only in a mean sense. Designers, however, are often uncomfortable with the notion of upper bound techniques as they are regarded as inherently unsafe. This issue is discussed later as the linear matching method generates both upper and lower bounds.

1.4 Limit Analysis

Limit analysis may be considered as a special case of shakedown analysis when the load and temperature are constant in time. The strain rate field $\dot{\varepsilon}_{ij}^c = \Delta\varepsilon_{ij}^c / \Delta t$ is similarly constant and also compatible with a displacement rate field $\dot{u}_i^c$. The upper bound (1.2.6) becomes

$$\int_V \sigma_{ij}^c \dot{\varepsilon}_{ij}^c dV = \lambda_{UB}^c \int_V \hat{\sigma}_{ij} \dot{\varepsilon}_{ij}^c dV = \lambda_{UB}^c \int_S P_i \dot{u}_i^c dV \tag{1.4.1}$$

by the principle of virtual work. The lower bound remains unchanged on noting that the elastic solution is constant. Alternatively we may write

$$\sigma_{ij}^* = \lambda_{LB} \hat{\sigma}_{ij} + \rho_{ij}^* = \lambda_{LB} \tilde{\sigma}_{ij}$$

where $\tilde{\sigma}_{ij}$ is an arbitrary stress field in equilibrium with P_i.

1.5 Simplification of the Upper Bound for a von Mises Yield Condition

For most metals, the von Mises yield condition is a close approximation to the yield condition. Many of the examples discussed below will be for this condition.

Consider the case of the von Mises yield condition;

$$f(\sigma_{ij}) = \bar{\sigma} - \sigma_y = 0 \tag{1.5.1}$$

where $\bar{\sigma} = \sqrt{\tfrac{3}{2}\sigma'_{ij}\sigma'_{ij}}$ and $\sigma'_{ij} = \sigma_{ij} - \tfrac{1}{3}\sigma_{kk}\delta_{ij}$, the deviatoric stress. The associated flow rule becomes;

$$\dot{\varepsilon}'_{ij} = \dot{\alpha}\sigma'_{ij} \quad \text{and} \quad \dot{\varepsilon}_{kk} = 0 \tag{1.5.2}$$

or

$$\dot{\bar{\varepsilon}} = \tfrac{2}{3}\dot{\alpha}\bar{\sigma} \quad \text{where} \quad \dot{\bar{\varepsilon}} = \sqrt{\tfrac{2}{3}\dot{\varepsilon}'_{ij}\dot{\varepsilon}'_{ij}} \tag{1.5.3}$$

The upper bound (1.2.6) now simplifies to:

$$\int_V \int_0^{\Delta t} \sigma_y \bar{\varepsilon}(\dot{\varepsilon}_{ij}^c) dt dV = \lambda_{UB}^c \int_V \int_0^{\Delta t} \hat{\sigma}_{ij} \dot{\varepsilon}_{ij}^c dt dV \tag{1.5.4}$$

A further simplification occurs when the history of loading produces a history of elastic stress which describe polygons in stress space. In this case plastic strains only occur at r instances during the load cycle and we may replace the continuous strain rate history $\dot{\varepsilon}_{ij}^c$ by r discrete increments of plastic strain $\Delta\varepsilon_{ij}^1$, $\Delta\varepsilon_{ij}^2$, ... , $\Delta\varepsilon_{ij}^r$ and the upper bound (1.5.4) becomes:

$$\int_V \sum_{l=1}^{r} \sigma_y \bar{\varepsilon}(\Delta\varepsilon_{ij}^l) dV = \lambda_{UB}^c \int_V \sum_{l=1}^{r} \hat{\sigma}_{ij}(t_l) \Delta\varepsilon_{ij}^l dV \tag{1.5.5}$$

The times t_l correspond to the extremes of the elastic stress histories. In the following examples, solutions are found by using (1.5.5) for an assumed mechanism of deformation over the cycle and then choosing the incremental strains so that the volume integral on the right hand side of (1.5.5) is made as large as possible. The detailed calculations are not given here but I hope there is sufficient information for the reader to complete the calculations themselves.

1.6 Example of Shakedown Boundaries - The Bree Problem

This problem shown in Figure 1, involves a thin walled tube subjected to a constant axial stress σ_p and a cyclically changing through-thickness temperature difference $\Delta\theta$. The maximum thermoelastic stress induced by $\Delta\theta$ is given by

$$\sigma_t = E\alpha\Delta\theta / 2(1-\nu) \tag{1.6.1}$$

where E, α and ν are Young's modulus, the coefficient of linear thermal expansion and Poisson's ratio respectively. The problem was solved analytically by Bree (1967) who identified an elastic region E, a shakedown region S, a reverse plasticity region P where cyclic plastic strains occur but no cyclic strain growth, and ratchetting region R where rapid cyclic strain growth occurs. The position of the shakedown boundary can be calculated very easily from (1.5.5). If we treat the problem as a uniaxial problem and only consider the direct stresses along the tubes length, the x direction, the elastic solution is given by:

$$\hat{\sigma}_{xx}(t_1) = \sigma_p, \quad \hat{\sigma}_{xx}(t_2) = \sigma_p + \sigma_t(1 - 2z/h) \tag{1.6.2}$$

where the wall thickness is $0 \le z \le h$ and t_1 is the period of the cycle when $\Delta\theta = 0$. Note that the maximum positive elastic stress is given by:

$$\hat{\sigma}_{xx}^{\max} = \begin{cases} \hat{\sigma}_{xx}(t_2), 0 \le z \le h/2 \\ \hat{\sigma}_{xx}(t_1), h/2 \le z \le h \end{cases} \tag{1.6.3}$$

The mechanism of deformation is uniform through the thickness over the cycle and we may write;

$$\Delta\varepsilon_{xx}^{c}(z) = C = \Delta\varepsilon_{xx}^{1} + \Delta\varepsilon_{xx}^{2} \tag{1.6.4}$$

where C is a constant and $\Delta\varepsilon_{xx}^{1}$ and $\Delta\varepsilon_{xx}^{2}$ are the increments of plastic strain occurring at times t_1 and t_2. We may now maximise the right hand side of the upper bound (1.5.5) choosing;

$$\Delta\varepsilon_{xx}^{2} = \begin{Bmatrix} \Delta\varepsilon_{xx}^{c}, 0 \le z \le h/2 \\ 0, h/2 \le z \le h \end{Bmatrix}, \text{ and } \Delta\varepsilon_{xx}^{1} = \begin{Bmatrix} 0, 0 \le z \le h/2 \\ \Delta\varepsilon_{xx}^{c}, h/2 \le z \le h \end{Bmatrix} \tag{1.6.5}$$

and the upper bound becomes (absorbing the load parameter into the elastic stress field),

$$\sigma_y \Delta\varepsilon_{xx}^{c} = \tfrac{1}{h}\int_0^h (\hat{\sigma}_{xx}(t_1)\Delta\varepsilon_{xx}^{1} + \hat{\sigma}_{xx}(t_2)\Delta\varepsilon_{xx}^{2})dz \tag{1.6.6}$$

Substituting (1.6.3) to (1.6.5) into (1.6.6) produces the answer;

$$\sigma_y \Delta\varepsilon_{xx}^{c} = (\tfrac{1}{h}\int_0^{h/2}(\sigma_P + \sigma_t(1 - 2z/h)dz + \tfrac{1}{h}\int_{h/2}^{h}\sigma_p dz)\Delta\varepsilon_{xx}^{c}$$

i.e

$$\sigma_y = \sigma_p + \sigma_t/4 \tag{1.6.7}$$

and this is the exact answer for the section AB of the shakedown boundary of Figure 1. Equation (1.6.6) can be seen as the sum of the average mechanical stress plus the average maximum positive thermo-elastic stress equals the yield stress. The odd characteristic of thermo-elastic stresses is that the extremes occur at differing points of the structure at different times, unlike stresses due to mechanical load which tend to have the maximum stresses all at the same time, when the load has a maximum. We will discuss the consequence of this later.

The remaining section of the shakedown boundary, DB, can be equally easily understood from the upper bound theorem. The mechanism in this case involves a cycle of plastic strain at the surface which accumulates to zero and no plastic strain throughout the remainder of the cross section;

$$\Delta\varepsilon_{xx}^{1} = \begin{Bmatrix} \Delta\varepsilon_{xx}^{c}, 0 \le z \le \delta h \\ 0, \delta h \le z \le h \end{Bmatrix} \text{ and } \Delta\varepsilon_{xx}^{2} = \begin{Bmatrix} -\Delta\varepsilon_{xx}^{c}, 0 \le z \le \alpha h \\ 0, \alpha h \le z \le h \end{Bmatrix}$$

and

$$\Delta\varepsilon_{xx}^{c} = \Delta\varepsilon_{xx}^{1} + \Delta\varepsilon_{xx}^{2} = 0,\ 0 \le z \le h$$

where δ is small. Substituting this mechanism into the upper bound (1.5.5) (again absorbing the load parameter into σ_t), we obtain the upper bound,

$$\sigma_t = 2\sigma_y \tag{1.6.8}$$

As was the case of (1.6.7) this is again the correct answer.

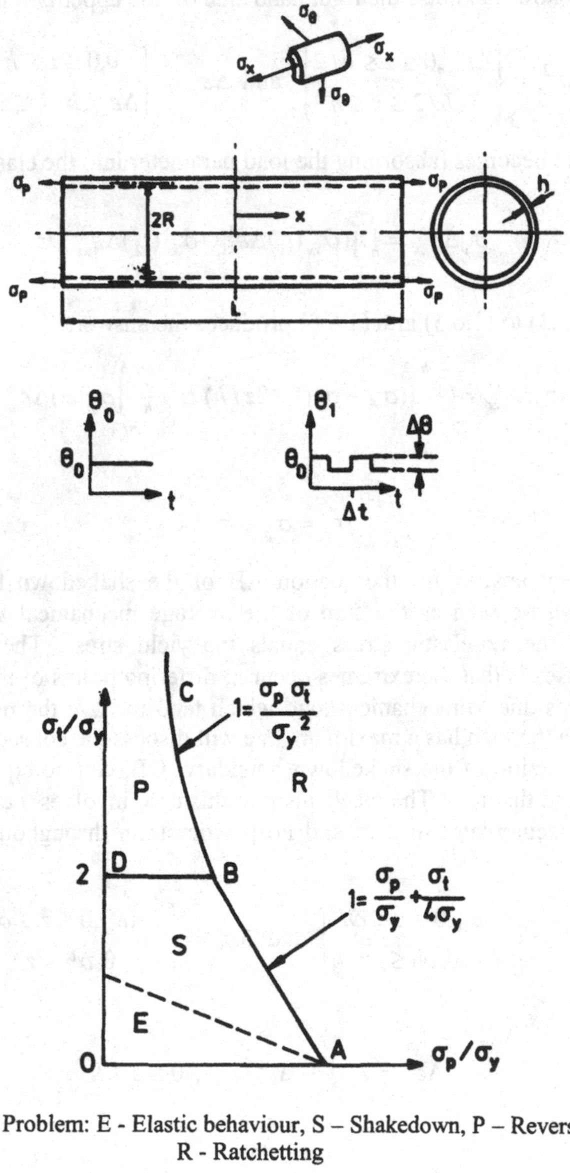

Figure 1. The Bree Problem: E - Elastic behaviour, S – Shakedown, P – Reverse Plasticity, R - Ratchetting

The purpose of going through this simple example is to demonstrate important general features of shakedown limits for histories of load and temperature which have distinct extremes. The shakedown limit mechanism tends to be dominated by either the mechanism associated with plastic collapse at the maximum mechanical load or a very localised reverse plasticity limit at the position of the maximum fluctuation of elastic stresses.

Before considering other examples it is worth considering the practical value that shakedown limits provide in design. The first, of course, is identifying the reduction of the load at the ratchet limit, AB in Figure 2, due to fluctuation of load and temperature. The contribution of the fluctuating temperature arises from the mean value of the thermo-elastic stresses in the direction of the dominant mechanism, the limit load mechanism. The second is the identification of the reverse plasticity limit where it is clear that $\sigma_t = 2\sigma_y$ is an upper bound where σ_t is the maximum change in the thermo-elastic effective stress. With this interpretation the shakedown boundary in an interaction diagram of the type shown in Figure 1 consists of two straight lines. Experience shows that this is not always the case but there are large groups of problems where this is a good approximation. As a piece of information for use in design, it excludes the possibility of identifying the line CB in Figure 1 in cases when, in the region P, the design limit is a fatigue endurance limit and the limit DB is unnecessarily restricting. In such cases we need to understand the location of CB to ensure that gross plastic strain growth does not occur. Methods of evaluating the position of CB are rather new and have recently been discussed by Ponter and Chen (2001) and Ponter and Chen (2001).

Consider the problem shown in Figure 2 where a thin walled tube of radius R and thickness h is subjected to discontinuity in temperature $\Delta\theta$ which moves cyclically along a length of tube Δx. Figure 2 shows the shakedown boundaries for a range of values of $\Delta\bar{x} = \Delta x / \sqrt{Rh}$ in terms of the axial load σ_p and the maximum thermo-elastic stress due to the temperature discontinuity;

$$\sigma_t = E\alpha\Delta\theta / 2 \tag{1.6.9}$$

There are two features of this diagram which are very striking. For $\Delta\bar{x} \geq 4.5$ the shakedown boundary and the elastic limit are virtually identical, i.e. the structural behaviour changes from purely elastic behaviour to ratchetting with only a very narrow range of loads within shakedown. The reverse plasticity line passes through $\sigma_t = \sigma_y$ rather than the expected $\sigma_t = 2\sigma_y$ of the Bree problem, and there is no P region. Hence the possibility of ratchetting in this case is very severe compared with Bree problem particularly for small values of σ_p / σ_y. Although superficially the problem may appear to be similar to the Bree problem it differs in an important respect. The thermo-elastic stress passes through the material element so that material in a whole section of the tube suffers identical elastic stress histories. Similar situations occur in rolling contact problems. The stress history is highly non-proportional and such problems tend to exhibit ratchetting problems where shakedown analysis can give very valuable insight.

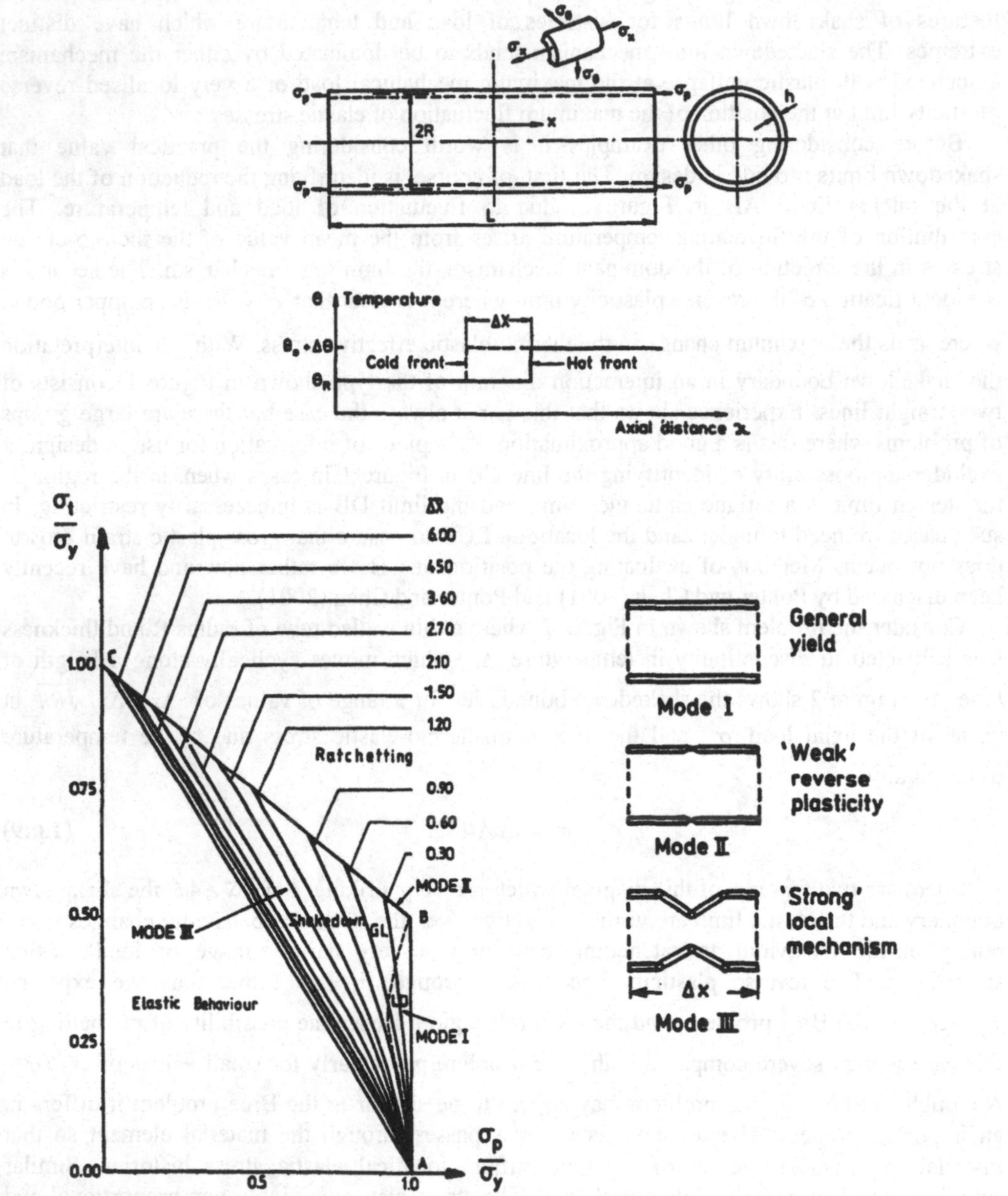

Figure 2. A cylinder with a cyclically moving temperature discontinuity, traversing a length Δx .

Note that $\Delta\bar{x} = \Delta x / \sqrt{Rh}$.

Hence, for thermal loading problems the position of the shakedown boundary and the existence of a P region is very sensitive to the details of the thermal loading histories. The realisation that this was the case posed a particular problem in certain problems in nuclear reactor design, particularly for fast breeder reactors where moving temperature fronts occur. The development of design codes for such cases has relied almost exclusively on the use of shakedown methods to look at a very large number of particular cases; my colleague, Keith Carter produced over 2000 interaction diagrams for thermally loaded tubes using a linear programming method . The linear programming methods used were similar to those described Franco and Ponter (1997a,b). It would have been impossible to produce the same information using convention finite element analysis. The solutions were then simplified into a few simple formulae which are now contained within a European design code, Riou et al. (1997). By the time the condensation was complete, much of the detail had been lost, and the rules are rather conservative in some cases. At that time (about ten years ago) we were encouraged to develop shakedown methods which could be used by industry and a prerequisite was that standard commercial finite element codes should be used. The method discussed here is the result of this study.

In summary, shakedown analysis can produce significant information for certain classes of problems in a way that conventional finite element methods are not able to do. Other examples are given later in this article.

2 Linear Matching Method for Limit Analysis

2.1 Simulation Technique for Limit Analysis

The method attempts to construct, as the limit of an iterative procedure, incompressible linear strain rate solution for the load P of the form,

$$\dot{\varepsilon}_{ij} = \frac{1}{\mu}\sigma'_{ij}, \quad \dot{\varepsilon}_{kk} = 0, \quad \text{i.e.,} \quad \frac{3}{2}\bar{\dot{\varepsilon}} = \frac{1}{\mu}\bar{\sigma} \tag{2.1.1}$$

where μ is a shear modulus. If, for some spatially varying $\mu(x_i)$ it is possible to find a solution so that

$$f(\sigma_{ij}) = \bar{\sigma}(\sigma_{ij}) - \sigma_Y = 0 \tag{2.1.2}$$

in the plastic zone V_p, where $\dot{\varepsilon}_{ij} \neq 0$ and

$$\dot{\varepsilon}_{ij} = 0, \quad f(\sigma_{ij}) \leq 0, \quad \mu \to \infty \tag{2.1.3}$$

in the rigid zone V_R then the solution would be identical to the limit state solution on making the identities

$$\sigma_{ij}^P = \sigma_{ij}, \dot{\varepsilon}_{ij}^P = k\dot{\varepsilon}_{ij} \tag{2.1.4}$$

where k is an arbitrary constant scaling factor. For any limit state solution there exists a distribution μ, unique except for an arbitrary scaling factor k, for which the linear solution σ_{ij} is identical to the limit state solution σ_{ij}^P. In a sense we are choosing a class of solutions that have the computational advantage of linearity and simulating the computationally less advantageous non linear problem.

2.2 Iterative Method

Commencing with an arbitrary uniform $\mu = \mu_o$ and load P_0, a linear solution $(\dot{\varepsilon}_{ij}^0, \sigma_{ij}^0)$ is generated. A new distribution $\mu_1(x_i)$ is now found so that, for fixed $\dot{\varepsilon}_{ij}^0$, the stress point would be brought to the yield surface:

$$\frac{3}{2}\dot{\bar{\varepsilon}}^0 = \frac{\bar{\sigma}(\sigma_{ij}^0)}{\mu_0} = \frac{\sigma_Y}{\mu_1}, \text{ i.e. } \mu_1 = \mu_0\left(\frac{\sigma_Y}{\bar{\sigma}(\sigma_{ij}^0)}\right) \tag{2.2.1}$$

This choice can be best understood from the construction shown in Figure 3 (where $E = \frac{3}{2}\mu$) for $\bar{\sigma} < \sigma_Y$. Essentially we are matching the linear and plasticity solutions at our current estimate of the strain rate distribution $\dot{\varepsilon}_{ij}^0$. A new linear solution is now constructed for $\mu = \mu_1(x_i)$ and the next iteration performed. In general

$$\frac{3}{2}\dot{\bar{\varepsilon}}^k = \frac{\bar{\sigma}(\sigma_{ij}^k)}{\mu_k} = \frac{\sigma_Y}{\mu_{k+1}}, \quad \mu_{k+1} = \mu_k\left(\frac{\sigma_Y}{\bar{\sigma}(\sigma_{ij}^k)}\right), \quad k = 1,2,.... \tag{2.2.2}$$

This process is continued until convergence occurs. In the following, as only a single load is involved we dispense with the use of the load parameter λ and discuss the magnitude of the load P directly. At each step both upper and lower bounds may be evaluated as follows,

Lower Bound

$$P_{LB}^k = \frac{P\sigma_Y}{Max(\bar{\sigma}(\sigma_{ij}^k))} \tag{2.2.3}$$

Upper Bound

If σ_{ij}^{ck} is the stress at yield associated with $\dot{\varepsilon}_{ij}^k$,

$$\sigma_{ij}'^k = \sqrt{\frac{2}{3}}\sigma_y \frac{\dot{\varepsilon}_{ij}^k}{\sqrt{\dot{\varepsilon}_{ij}^k \dot{\varepsilon}_{ij}^k}}, \quad \text{where} \quad \sqrt{\frac{3}{2}\sigma_{ij}'^{ck}\sigma_{ij}'^{ck}} = \sigma_Y \tag{2.2.4}$$

as required by the yield condition. Further

$$\sigma_{ij}^{\prime ck}\dot{\varepsilon}_{ij}^{k} = \sigma_Y \bar{\dot{\varepsilon}}^{k}, \quad \bar{\dot{\varepsilon}}^{k} = \sqrt{\frac{2}{3}\dot{\varepsilon}_{ij}^{k}\dot{\varepsilon}_{ij}^{k}} \tag{2.2.5}$$

Hence the upper bound (1.4.1) may be expressed explicitly in terms of $\dot{\varepsilon}_{ij}^{k}$,

$$P_{UB}^{k} = \sigma_Y \frac{\int_V \bar{\dot{\varepsilon}}^{k} dV}{\int_{S_T} \bar{p}_i \dot{u}_i^{k} dS} \tag{2.2.6}$$

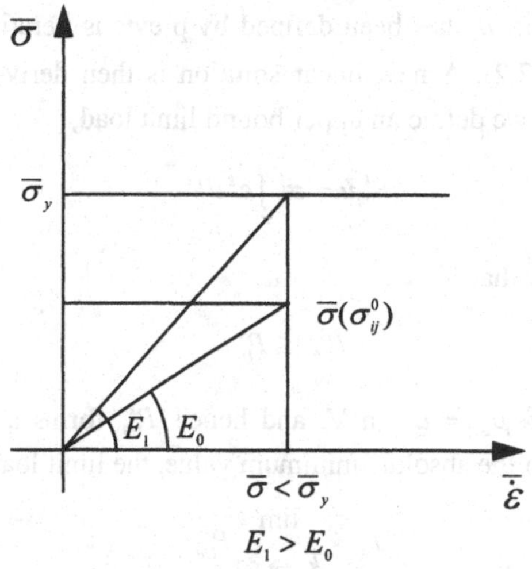

Figure 3. The Iterative Process

2.3 Convergence of the Method for Limit Analysis

Preliminaries

It is possible to prove convergence of the iterative process described above for either the case where loads are prescribed or, alternative, where the body is subjected to imposed boundary displacements, which is the case discussed below.

Consider the following class of problems. A total load P is applied to a body over a surface area S_T and direction d_i in such a way that the displacement rate $\dot{u}_i(x_i)$ is prescribed except for an arbitrary scalar multiplier, the positive scalar magnitude $\dot{u}$ i.e.

$$\dot{u}_i = \dot{u} d_i \tag{2.3.1}$$

The direction of the corresponding limit load P_L will be parallel to $\dot{u}_i$ as any component perpendicular to $\dot{u}_i$ will do no work. For this class of problems, the boundary condition may either be posed as a displacement rate $\dot{u}_i$ or as load P, and this applies equally to the limit load problem and the sequence of linear solutions that simulate the limit load solution. For simplicity we will prove a convergence theorem for the iterative process when the elastic problems are considered to result from the displacement boundary condition.

At the kth iteration, the iterative process provides a volume preserving strain field $\dot{\varepsilon}_{ij}^k$ with corresponding deviatoric stresses $\sigma_{ij}'^k$ given by

$$\sigma_{ij}'^k = \mu_k \dot{\varepsilon}_{ij}^k \tag{2.3.2}$$

where the spatial variation of μ_k has been defined by previous iterations. A new distribution μ_{k+1} is now defined by (2.2.2). A new linear solution is then derived from the relationship (2.2.1). For each strain field we define an upper bound limit load,

$$P_{UB}^k \dot{u} = \sigma_Y \int_V \bar{\dot{\varepsilon}}^k dV \tag{2.3.3}$$

We now proceed to show that

$$P_{UB}^{k+1} \leq P_{UB}^k \tag{2.3.4}$$

with equality if and only if $\mu_{k+1} = \mu_k$ in V, and hence P_{UB}^k forms a monotonically reducing sequence which converges to the absolute minimum value, the limit load

$$P_L = \lim_{k \to \infty} P_{UB}^k \tag{2.3.5}$$

Proof of Inequality (2.3.4)

We first define, for the linear material, the dissipation rate density U^k corresponding to an arbitrary strain rate field $\dot{\varepsilon}_{ij}$ as

$$U_k = \frac{1}{2}\mu_k \dot{\varepsilon}_{ij}\dot{\varepsilon}_{ij} = \frac{3}{4}\mu_k \bar{\dot{\varepsilon}}^2 \tag{2.3.6}$$

The corresponding deviatoric stress σ_{ij}' is given by

$$\sigma_{ij}' = \frac{\partial U_k}{\partial \dot{\varepsilon}_{ij}} = \mu_k \dot{\varepsilon}_{ij} \tag{2.3.7}$$

and, similarly

$$\bar{\sigma} = \frac{\partial U_k(\bar{\dot{\varepsilon}})}{\partial \bar{\dot{\varepsilon}}} = \frac{3}{2}\mu_k \bar{\dot{\varepsilon}} \tag{2.3.8}$$

Provided $\mu_k > 0$, a condition satisfied provided $\mu_0 > 0$, U is convex, both as a function of the components $\dot{\varepsilon}_{ij}$ and the effective strain rate $\bar{\dot{\varepsilon}}$. Hence

$$U_{k+1}(\bar{\dot{\varepsilon}}^1) - U_{k+1}(\bar{\dot{\varepsilon}}^2) - \frac{\partial U_{k+1}}{\partial \bar{\dot{\varepsilon}}^2}(\bar{\dot{\varepsilon}}^1 - \bar{\dot{\varepsilon}}^2) \geq 0 \tag{2.3.9}$$

where $\bar{\dot{\varepsilon}}^1$ and $\bar{\dot{\varepsilon}}^2$ are two arbitrary effective strain rate values.

By the rate version of the minimum potential energy theorem the strain rate field $\dot{\varepsilon}_{ij}^{k+1}$ provides the absolute minimum total elastic energy dissipation rate for any compatible strain rate field corresponding to prescribed displacement rate $\dot{u}_i$ and hence

$$\int_V U_{k+1}(\bar{\dot{\varepsilon}}^{k+1})dV \leq \int_V U_{k+1}(\bar{\dot{\varepsilon}}^{k})dV \tag{2.3.10}$$

where equality holds if and only if $\bar{\dot{\varepsilon}}^{k+1} \equiv \bar{\dot{\varepsilon}}^k$ and $\dot{\varepsilon}_{ij}^{k+1} \equiv \dot{\varepsilon}_{ij}^k$. Hence from (2.3.9) and (2.3.10), making the substitutions $\dot{\varepsilon}_{ij}^1 \equiv \dot{\varepsilon}_{ij}^{k+1}$ and $\dot{\varepsilon}_{ij}^2 \equiv \dot{\varepsilon}_{ij}^k$,

$$\int_V \frac{\partial U_{k+1}}{\partial \bar{\dot{\varepsilon}}^k}(\bar{\dot{\varepsilon}}^{k+1} - \bar{\dot{\varepsilon}}^k)dV \leq 0 \tag{2.3.11}$$

From (2.3.6), (2.3.2) and (2.2.2)

$$\frac{\partial U_{k+1}}{\partial \bar{\dot{\varepsilon}}^k} = \frac{3}{2}\mu_{k+1}\bar{\dot{\varepsilon}}^k = \frac{3}{2}\frac{\mu_{k+1}}{\mu_k}\bar{\sigma}^k = \sigma_Y \tag{2.3.12}$$

and hence from (2.3.3), (2.3.11) and (2.3.12)

$$P_{UB}^{k+1}\dot{u} = \int_V \sigma_Y \bar{\dot{\varepsilon}}^{k+1} dV \leq \int_V \sigma_Y \bar{\dot{\varepsilon}}^k dV = P_{UB}^k \dot{u} \tag{2.3.13}$$

i.e.

$$P_{UB}^{k+1} \leq P_{UB}^k \tag{2.3.14}$$

with equality if and only if $\mu_{k+1} \equiv \mu_k$.

2.4 Interpretation of the Method as a Non-linear Programming Technique

The proof above provides a fresh insight into the nature of the iterative process. The integrand of upper bound P_{UB}, (2.3.3) is a function which is locally differentiable in its argument $\dot{\varepsilon}_{ij}$ but suffers discontinuities in its derivatives. In function space (2.3.3) defines a surface composed of the intersection of surfaces with continuous derivatives. We may define the first variation of (2.3.3) as

$$\delta P_{UB}\dot{u} = \int_V \sigma_Y \delta\bar{\dot{\varepsilon}} dV \qquad (2.4.1)$$

where, locally, we assume the continuity of the right hand side. Effectively, we assume that a plastic strain rate vector pointing inward on the yield surface would produce negative plastic work.

The expression (2.4.1) may be compared with the first variation of the total dissipation rate corresponding to μ_{k+1} when $\bar{\dot{\varepsilon}} = \bar{\dot{\varepsilon}}^k$,

$$\delta \int_V U_{k+1}(\bar{\dot{\varepsilon}}^k)dV = \int_V \frac{\partial U_{k+1}}{\partial \bar{\dot{\varepsilon}}^k}\delta\bar{\dot{\varepsilon}} dV = \int_V \frac{3}{2}\mu_{k+1}\bar{\dot{\varepsilon}}^k \delta\bar{\dot{\varepsilon}}^k dV = \int_V \sigma_Y \delta\bar{\dot{\varepsilon}} dV = \delta P_{UB}\dot{u} \qquad (2.4.2)$$

i.e

$$\delta \int_V U_{k+1}(\bar{\dot{\varepsilon}}^k)dV = \delta P_{UB}\dot{u} \qquad (2.4.3)$$

from (2.4.1). Hence by the adoption of the linear modulus distribution (2.2.2), the first variation of the upper bound (2.2.6) and $\int_V U_{k+1} dV$ are equal, subject to the assumption of local smoothness. The solution of the linear problem for μ_{k+1} and the associated minimization of $\int_V U_{k+1} dV$ involves, in function space, a change from $\bar{\dot{\varepsilon}}^k$ to $\bar{\dot{\varepsilon}}^{k+1}$ along a path of steepest descent of $\int_V U_{k+1} dV$, and hence, along a path of reducing P_{UB} .The process, therefore, may be regarded as a programming method where the surface in displacement space is sequentially locally matched by a quadratic dissipation function with the same local slope.

This interpretation may be more easily understood through the one degree of freedom example in the next section.

2.5 Illustrative Example of Convergence Process

Consider the simple problem shown in Figure 4 where a uniaxial rod of length 21 consists of sections, of length 1, with cross sectional areas A and 2A. We treat the problem as a displacement boundary condition and assume the total deflection rate $\dot{\delta}$ is fixed. The problem has a single degree of freedom u, the mid-displacement of the rod, and the strain rates in the two sections are given by,

$$\dot{\varepsilon}_1 = \dot{u}/l, \quad \dot{\varepsilon}_2 = (\dot{\delta} - \dot{u})/l \qquad (2.5.1)$$

For arbitrary $\dot{u}$, the upper bound P_{UB} is given for $\dot{\delta} > 0$ by

$$P_{UB}\dot{\delta} = \begin{bmatrix} 2Al\sigma_y(\dot{u}/l) + Al\sigma_Y\left((\dot{\delta} - \dot{u})/l\right), & \dot{\delta} > \dot{u} \\ 2Al\sigma_y(\dot{u}/l) + Al\sigma_Y\left((\dot{u} - \dot{\delta})/l\right), & \dot{\delta} < \dot{u} \end{bmatrix} \tag{2.5.2}$$

or, in non-dimensional form, defining $\overline{u} = \dot{u}/\dot{\delta}$ and $\overline{P}_L = P_L/(A\sigma_y)$ then

$$\overline{P}_{UB} = \begin{bmatrix} \overline{u} + 1, & \overline{u} < 1 \\ 3\overline{u} - 1, & \overline{u} > 1 \end{bmatrix} \tag{2.5.3}$$

Figure 5 illustrates the variation of $\overline{P}_{UB}$ with $\overline{u}$. The absolute minimum is given by $\overline{u} = 0$ and $\overline{P}_{UB} = 1$ corresponding to section 1 rigid and plastic yielding in section 2.

If we denote by E_k^1 and E_k^2 the uniaxial moduli corresponding to the kth iteration in sections 1 and 2, we can write down the following equilibium and matching relationships from (2.2.2),

$$\sigma_2^k = 2\sigma_1^k, \quad E_{k+1}^1 = E_k^1 \frac{\sigma_y}{\sigma_1^k}, \quad E_{k+1}^2 = E_k^2 \frac{\sigma_y}{\sigma_2^k} \tag{2.5.4}$$

and hence

$$\frac{E_{k+1}^1}{E_{k+1}^2} = 2\frac{E_k^1}{E_k^2} \tag{2.5.5}$$

i.e. if $E_0^1/E_0^2 = 1$ then

$$E_k^1/E_k^2 = 2^k \tag{2.5.6}$$

The full linear solution at the kth iteration is given by

$$\begin{aligned} \dot{\varepsilon}_1^k + \dot{\varepsilon}_2^k &= \dot{\delta}/l \quad & \sigma_2^k &= 2\sigma_1^k = P/A \\ \dot{\varepsilon}_1^k &= \sigma_1^k / E_k^1 & \dot{\varepsilon}_2^k &= \sigma_2^k / E_k^2 \end{aligned} \tag{2.5.7}$$

which may be solved to give

$$\dot{\varepsilon}_1^k = \left(\frac{1}{1+2^{k+1}}\right)\frac{\dot{\delta}}{l}, \quad \dot{\varepsilon}_2^k = \left(\frac{2^{k+1}}{1+2^{k+1}}\right)\frac{\dot{\delta}}{l} \tag{2.5.8}$$

and the kth upper bound is given by

$$P_{UB}^{k} = A\sigma_{Y}\left(\frac{2+2^{k+1}}{1+2^{k+1}}\right) \tag{2.5.9}$$

The convergence of P_{UB}^{k} to $P_{UB} = 1$ in (2.5.9) is geometric, with the error reduced by 0.5 in each iteration for k large. The variation of the error is shown in Figure 6 and it is seen that a less than 1 % error is achieved after 6 iterations and 0.1 % after 9 iterations.

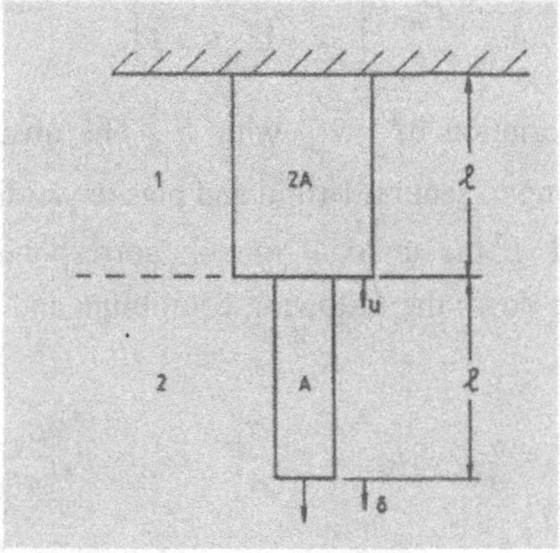

Figure 4. Geometry of problem discussed in Section 2.5

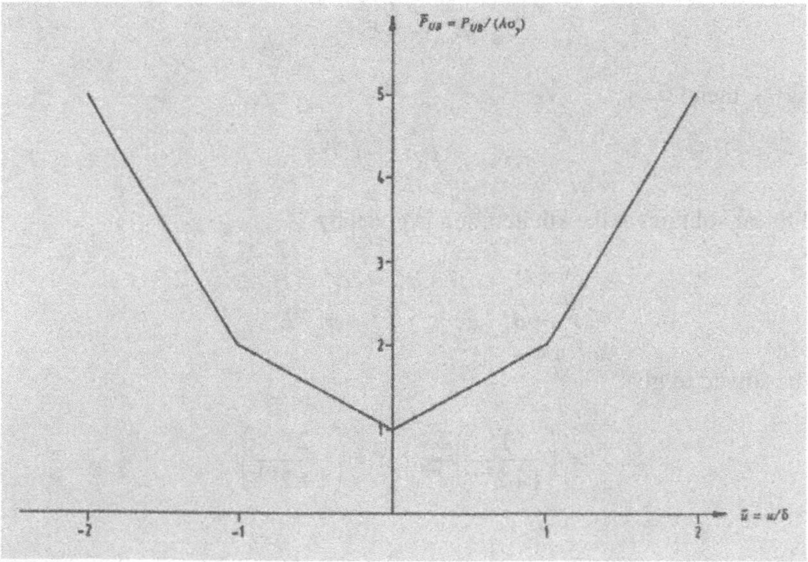

Figure 5. Variation of P_{UB} with u, in non-dimensional form, for structure shown in Figure 2

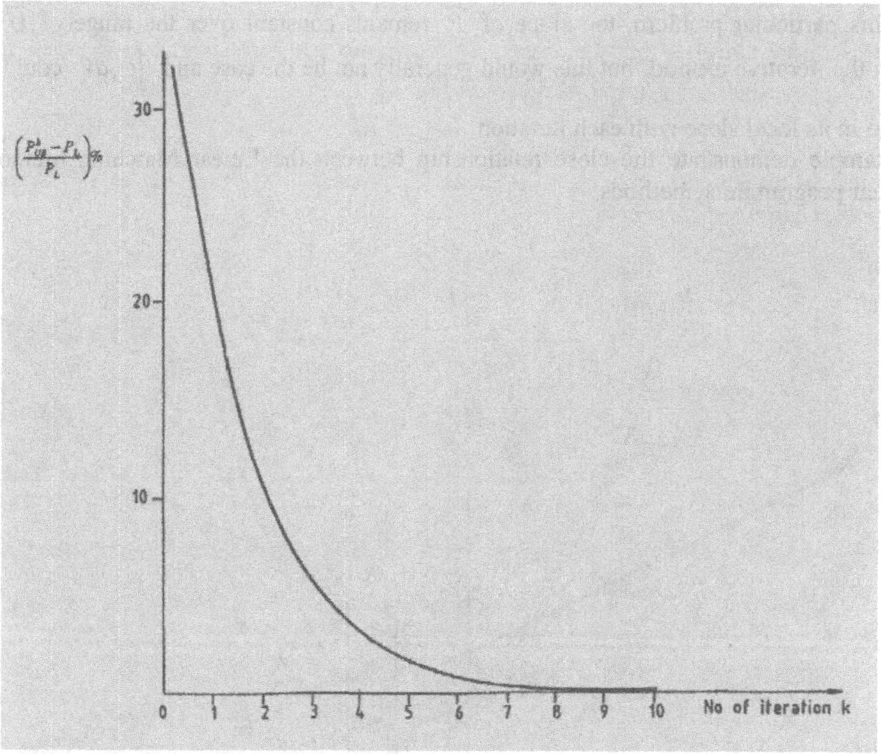

Figure 6. Variation of the percentage error in P_{UB}^{k} with number of iterations k.

The first two iterations are shown, in terms of the load $\overline{P}$ and

$$\overline{U}_k = \frac{\int_V U_k dV}{\dot{\delta} A \sigma_y} \tag{2.5.10}$$

in Figure 7.

For $E_0^1 = E_0^2$, the minimum of $\int_V U_0 dV$ is given by $\overline{u}_0 = 1/3$, and the upper bound $\overline{P}^0$ corresponds to point A in Figure 7. The construction of $\int_V U_1 dV$ provides a surface, labeled $\overline{U}_1$, where the slope at $\overline{u}_0 = 1/3$, point A', is identical to $\overline{P}$ at A. The minimum of $\overline{U}_1$, $\overline{u}_1 = 1/5$ gives a reduced upper bound $\overline{P}_1$, at B. The next linear solution corresponding to $\overline{U}_2$ gives a local slope at B' identical to that at B and so on. It can also be seen that

$$\overline{P}_{UB}(\overline{u}_k) = 2\overline{U}_{k+1}(\overline{u}_k)$$

In this particular problem, the slope of $\overline{P}$ remains constant over the range of $\overline{U}$ which occur in the iterative method, but this would generally not be the case and $\int_V U_k dV$ could suffer a change in its local slope with each iteration.
This example demonstrate the close relationship between the Linear Matching Method and non-linear programming methods.

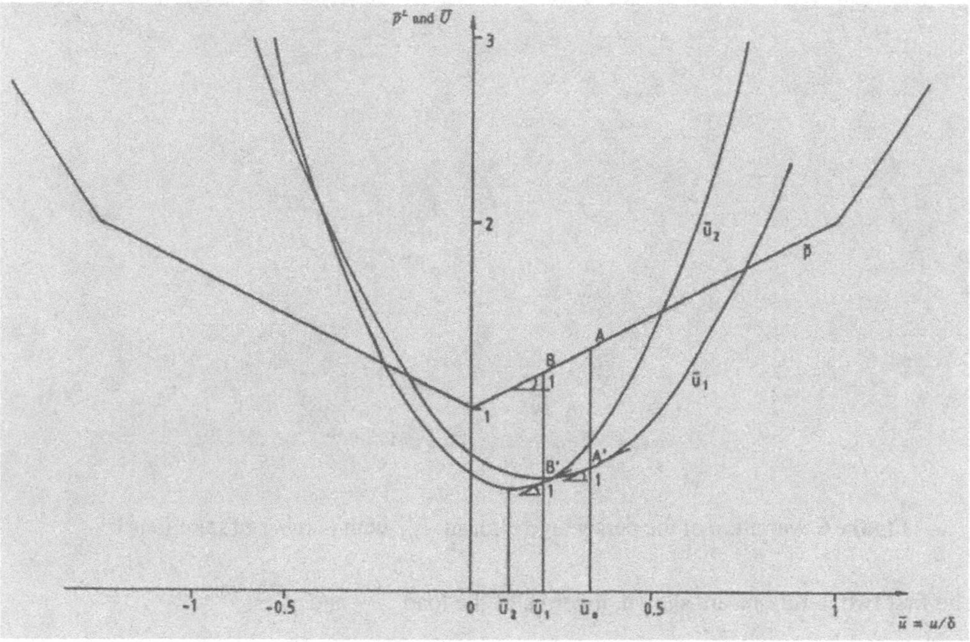

Figure 7. Geometric interpretation of the iterative process

3 Shakedown Limits for a von Mises Yield Condition

3.1 Introduction

In the last section it was demonstrated that Linear Matching Method may be developed that provides a decreasing sequence of upper bounds. In this section we generalise the method to shakedown, again for a von Mises yield condition. We then discuss the implementation of the method within a finite element code for both limit load and shakedown solutions. As the method calls upon procedures which form the basis of linear finite element analysis, it is possible to implement the method through the optional user procedures which are often included in commercial codes. For the solutions described here the general code ABAQUS was used.

The following consists of three main parts. Section 3.2 contains a summary of the method for shakedown. Section 3.3 contains a convergence proof. Section 3.4 is concerned with the implementation of the method within a finite element code. This is followed, in Section 4, by the solution of both limit load and shakedown problems involving variable load and variable temperature. Finally, in Section 4.3, the solution of an unconventional shakedown problem is discussed. The history of load is prescribed and the shakedown limit is required in terms of a minimum creep rupture stress for a maximum creep rupture life. This problems occurs in the life assessment method of British Energy, R5 (Goodall et al., 1991), and demonstrates the flexibility of the method.

The ease of implementation, efficiency and reliability of the method indicate that it has considerable potential for application in design and life assessment methods where efficient methods are required for generating indicators of structural performance of structures.

3.2 Shakedown Limit for a Von Mises Yield Condition

Consider, again, the problem described in section 1.2. A body of volume V and surface S is subjected to a cyclic history of load $\lambda P_i(x_j,t)$ over S_T, part of S, and temperature $\lambda\theta(x_j,t)$ within V, where λ is a positive load parameter. On the remainder of S, namely S_U, the displacement rate $\dot{u}_i = 0$. The linear elastic solution to the problem is denoted by $\lambda\hat{\sigma}_{ij}$.

We define a class of incompressible kinematically admissible strain rate histories $\dot{\varepsilon}_{ij}^c$ with a corresponding displacement increment fields Δu_i^c and associated compatible strain increment,

$$\Delta\varepsilon_{ij}^c = \tfrac{1}{2}\left(\frac{\partial \Delta u_i^c}{\partial x_j} + \frac{\partial \Delta u_j^c}{\partial x_i}\right) \tag{3.2.1}$$

The strain rate history $\dot{\varepsilon}_{ij}^c$, which need not be compatible, satisfies the condition that,

$$\int_0^{\Delta t} \dot{\varepsilon}_{ij}^c dt = \Delta\varepsilon_{ij}^c \tag{3.2.2}$$

In terms of such a history of strain rate an upper bound on the shakedown limit is given by (1.5.4),

$$\int_V\int_0^{\Delta t} \sigma_y \bar{\varepsilon}(\dot{\varepsilon}_{ij}^c) dt dV = \lambda_{UB}^c \int_V\int_0^{\Delta t} (\hat{\sigma}_{ij}\dot{\varepsilon}_{ij}^c) dt dV \tag{3.2.3}$$

where $\lambda_{UB}^c \geq \lambda_s$, with λ_s the exact shakedown limit. In the following we describe a convergent method where a sequence of kinematically admissible strain increment fields, with associated strain rate histories, corresponds to a reducing sequence of upper bounds. The sequence converges to the shakedown limit λ_s or the least upper bound associated with the class of displacement fields and strain rate histories chosen.

The iterative method relies upon the generation of a sequence of linear problems where the moduli of the linear problem are found by a matching process. For the von Mises yield condition the appropriate class of strain rates chosen are incompressible and the linear problem is defined by a single shear modulus $\mu(t)$ which varies both spatially and during the cycle. Corresponding to an initial estimate of the strain rate history $\dot{\varepsilon}^i_{ij}$, a history of a shear modulus $\mu(x_i, t)$ may be defined by the same matching condition as in the limit analysis case (2.2.1),

$$\tfrac{3}{2}\mu\bar{\dot{\varepsilon}}^i = \sigma_y \tag{3.2.4}$$

except that the matching condition occurs at each instant in the cycle. We now define a corresponding linear problem for a new kinematically admissible strain rate history, $\dot{\varepsilon}^f_{ij}$ and a time constant residual stress field $\bar{\rho}^f_{ij}$.

$$\dot{\varepsilon}'^f_{ij} = \frac{1}{\mu}\left(\bar{\rho}'^f_{ij} + \lambda\hat{\sigma}'_{ij}\right) \tag{3.2.5}$$

and, integrating over the cycle $0 \le t \le \Delta t$,

$$\Delta\dot{\varepsilon}'^f_{ij} = \frac{1}{\bar{\mu}}\left(\bar{\rho}'^f_{ij} + \sigma'^{in}_{ij}\right) \quad \text{and} \quad \Delta\dot{\varepsilon}'_{kk} = 0 \tag{3.2.6}$$

where we assume $\lambda = \lambda^i_{UB}$,

$$\frac{1}{\bar{\mu}} = \int_0^{\Delta t} \frac{1}{\mu(t)}dt \quad \text{and} \quad \sigma'^{in}_{ij} = \bar{\mu}\left\{\int_0^{\Delta t} \frac{1}{\mu(t)}\lambda\hat{\sigma}'_{ij}dt\right\} \tag{3.2.7}$$

The convergence proof, given below, then concludes that

$$\lambda^f_{UB} \le \lambda^i_{UB} \tag{3.2.8}$$

where equality occurs if and only if $\dot{\varepsilon}^i_{ij} \equiv \dot{\varepsilon}^f_{ij}$ and λ^f_{UB} is the upper bound corresponding to $\dot{\varepsilon}^c_{ij} = \dot{\varepsilon}^f_{ij}$. The repeated application of this procedure results in a monotonically reducing sequence of upper bounds which will converge to the least upper bound within a class of displacement fields and strain rate histories. A primary objective of the implementation of the technique is to ensure that the classes chosen results in a minimum is sufficiently close to the absolute minimum to be of practical use as a substitute for an analytic solution.

The choice of the linear problem (3.2.4) to (3.2.7) has a simple physical interpretation. For the initial strain rate history $\dot{\varepsilon}^i_{ij}$, the shear modulus is chosen so that the stress and the rate of energy dissipation in the linear material is matched to that of the perfectly plastic material for the same strain rate history. At the same time the load parameter is adjusted so that the value corresponds to a global balance in energy dissipation through equality (1.5.4). In other words, the linear problem is adjusted so that it satisfies as many of the conditions of the plasticity problem as is possible. The fact that the resulting solution, when equilibrium is reasserted, is closer to the shakedown limit solution and produces a reduced upper bound should be no surprise. However, we need not rely upon such intuitive arguments as a formal proof of convergence may be constructed and this is given in the next section.

3.3 Convergence Proof for a von Mises Yield Condition

The following convergence proof is based upon a more general proof by Ponter and Engelhardt (2000) for a general yield surface with a general class of linear problems. Here the form of proof is the same as that given in Ponter and Engelhardt (2000) but simplified and specialised to a von Mises yield condition. Essentially, we need to demonstrate that, beginning with a strain rate history $\dot{\varepsilon}_{ij}^{i}$, the linear problem, equations (3.2.4) to (3.2.7), for the residual stress field $\bar{\rho}_{ij}^{f}$ and strain rate history $\dot{\varepsilon}_{ij}^{f}$ produces an upper bound on the shakedown limit which satisfies inequality (3.2.8) with equality only when $\dot{\varepsilon}_{ij}^{i} \equiv \dot{\varepsilon}_{ij}^{f}$. Convergence then results from the repeated application of the argument producing a monotonically reducing sequence of upper bounds.

We begin with some preliminary observations. The linear incompressible material defined by (3.2.4) possesses a strain rate dissipation potential $U(\dot{\varepsilon}_{ij})$ so that

$$\sigma_{ij}^{L} = \frac{\partial U(\dot{\varepsilon}_{ij})}{\partial \dot{\varepsilon}_{ij}}, \qquad U = \tfrac{3}{4}\mu\bar{\dot{\varepsilon}}^{2} \tag{3.2.9}$$

where σ_{ij}^{L} is the stress generated by the linear material. The matching condition (3.2.4) may then be written in the more general form;

$$\sigma_{ij}^{pi} = \sigma_{ij}^{Li} = \frac{\partial U(\dot{\varepsilon}_{ij}^{i})}{\partial \dot{\varepsilon}_{ij}^{i}} \tag{3.2.10}$$

where σ_{ij}^{pi} is the stress at yield associated with $\dot{\varepsilon}_{ij}^{i}$.

The convergence proof requires a relationship between this linear material and the perfectly plastic material. This is provided by the following inequality which, in a more general context (Ponter and Engelhardt, 2000), may be regarded as a sufficient condition for convergence;

$$\sigma_{ij}^{pi}\dot{\varepsilon}_{ij}^{i} - \sigma_{ij}^{pf}\dot{\varepsilon}_{ij}^{f} \geq U(\dot{\varepsilon}_{ij}^{i}) - U(\dot{\varepsilon}_{ij}^{f}) \tag{3.2.11}$$

where $\dot{\varepsilon}_{ij}^{f}$ may be regarded as an arbitrary incompressible strain rate and where U is defined by the matching condition at $\dot{\varepsilon}_{ij}^{i}$. For a von Mises yield condition inequality (3.2.11) may be rewritten as;

$$\sigma_{y}(\bar{\dot{\varepsilon}}^{i} - \bar{\dot{\varepsilon}}^{f}) \geq U(\bar{\dot{\varepsilon}}^{i}) - U(\bar{\dot{\varepsilon}}^{f}) \tag{3.2.12}$$

This inequality is self evidently always satisfied, see Figure 8.

The convergence proof now commences by identifying $\dot{\varepsilon}_{ij}^{f}$ with the strain rate generated by the linear problem given by eqns (3.2.4) to (3.2.6). We first note that (3.2.10) and (3.2.5) may be written as;

$$\frac{\partial U(\dot{\varepsilon}_{ij}^{f})}{\partial \dot{\varepsilon}_{ij}^{f}} = \bar{\rho}_{ij}^{\prime f} + \lambda \hat{\sigma}_{ij}^{\prime} \tag{3.2.13}$$

On noting that U is a convex function of its argument and equation (3.2.13), the inequality (3.2.11) may be written in the extended form;

$$0 \le U(\dot{\varepsilon}_{ij}^{i}) - U(\dot{\varepsilon}_{ij}^{f}) - \frac{\partial U(\dot{\varepsilon}_{ij}^{f})}{\partial \dot{\varepsilon}_{ij}^{f}}(\dot{\varepsilon}_{ij}^{i} - \dot{\varepsilon}_{ij}^{f}) \le -(\sigma_{ij}^{pf} - \lambda\hat{\sigma}_{ij})\dot{\varepsilon}_{ij}^{f} + (\sigma_{ij}^{pi} - \lambda\hat{\sigma}_{ij})\dot{\varepsilon}_{ij}^{i} + \bar{\rho}_{ij}^{f}(\dot{\varepsilon}_{ij}^{f} - \dot{\varepsilon}_{ij}^{i}) \tag{3.2.14}$$

We now define $\lambda = \lambda_{UB}^{i}$ as the upper bound corresponding to $\dot{\varepsilon}_{ij}^{c} = \dot{\varepsilon}_{ij}^{i}$. Integrating the right hand side of inequality (3.2.14) over the volume V and cycle, we obtain the following;

$$0 \le -\int_{V}\int_{0}^{\Delta t}(\sigma_{ij}^{pf} - \lambda_{UB}^{i}\hat{\sigma}_{ij})\dot{\varepsilon}_{ij}^{f}\,dtdV + \int_{V}\int_{0}^{\Delta t}\bar{\rho}_{ij}^{f}(\dot{\varepsilon}_{ij}^{f} - \dot{\varepsilon}_{ij}^{i})dtdV \tag{3.2.15}$$

The contribution of the second term of the right hand side of (3.2.14) disappears due to the choice of $\lambda = \lambda_{UB}^{i}$. Note that the second term in (3.2.15) may be integrated to produce;

$$\int_{V}\int_{0}^{\Delta t}\bar{\rho}_{ij}^{f}(\dot{\varepsilon}_{ij}^{f} - \dot{\varepsilon}_{ij}^{i})dtdV = \int_{V}\bar{\rho}_{ij}^{f}(\Delta\varepsilon_{ij}^{f} - \Delta\varepsilon_{ij}^{i})dV \tag{3.2.16}$$

If the linear problem is solved exactly the right hand side of (3.2.16) is zero from the virtual work theorem. But equally, if the linear problem is solved by a Rayleigh Ritz method where the rate version of the potential energy is minimized for a class of displacement fields, the right hand side of (3.8) is also zero as the residual stress field so generated is orthogonal to all compatible strain fields derived from the class of displacement fields. If we now return to inequality (3.2.15) and denote by λ_{UB}^{f} the upper bound corresponding to $\dot{\varepsilon}_{ij}^{c} = \dot{\varepsilon}_{ij}^{f}$ in (1.5.4), we obtain;

$$0 \le (\lambda_{UB}^{i} - \lambda_{UB}^{f})\int_{V}\int_{0}^{\Delta t}\hat{\sigma}_{ij}\dot{\varepsilon}_{ij}^{f}\,dtdV \tag{3.2.17}$$

and hence $\lambda_{UB}^{f} \le \lambda_{UB}^{i}$ provided λ is positive. The integral is positive from the definition of the upper bound. Equality in (3.2.17) only occurs for $\dot{\varepsilon}_{ij}^{i} \equiv \dot{\varepsilon}_{ij}^{f}$ i.e. when equality occurs in the first inequality in the extended inequality (3.2.14).

The repeated application of the procedure will result in a monotonically reducing sequence of upper bounds which will converge to a minimum when the difference between successive strain rate histories has a negligible effect upon the upper bound. In common with all other programming methods for shakedown limits, the absolute minimum upper bound will be reached for the chosen class of displacement fields in a finite element method only if it is ensured that the process allows access to all possible strain rate histories which contain the absolute minimum. For example, assumptions can be made in defining the initial strain rate history concerning the instants when plastic strains occur during the cycle. Such assumptions will affect the nature of all subsequent strain rate histories and result in convergence to a minimum subject to contraints. In the method described in Ponter and Carter (1997b) consistency arguments were applied at each iteration. A simpler approach is possible and this is discussed below.

At each stage in the iterative process it is possible to evaluate a lower bound by scaling the current solution with respect to λ so that the stress history defined by the current value of the residual stress field lies within yield. When convergence has occurred the current value of λ used in the solution of the linear problem, the upper bound from the previous iteration, and the upper bound from the current iteration are identical. Hence at the points where the matching condition has been applied the maximum stress from the linear solution is at yield or less than yield. Hence the lower bound and upper bound are identical. The lower bound has converged to the least upper bound. We discuss this in more detain in Section 4. Generally the lower bound increases with iterations but this is by no means always the case and the lower bound can decrease as well as increase. This is particularly true for problems with cracks. This article places the principle emphasis on the method as an upper bound method, although practicing engineers prefer lower bounds. A particularly attractive feature of the method is, of course, that both upper and lower bounds are generated although it needs to be always kept in mind that the converged value is a least upper bound, hence the use of the term pseudo lower bound by Ponter and Carter (1997).

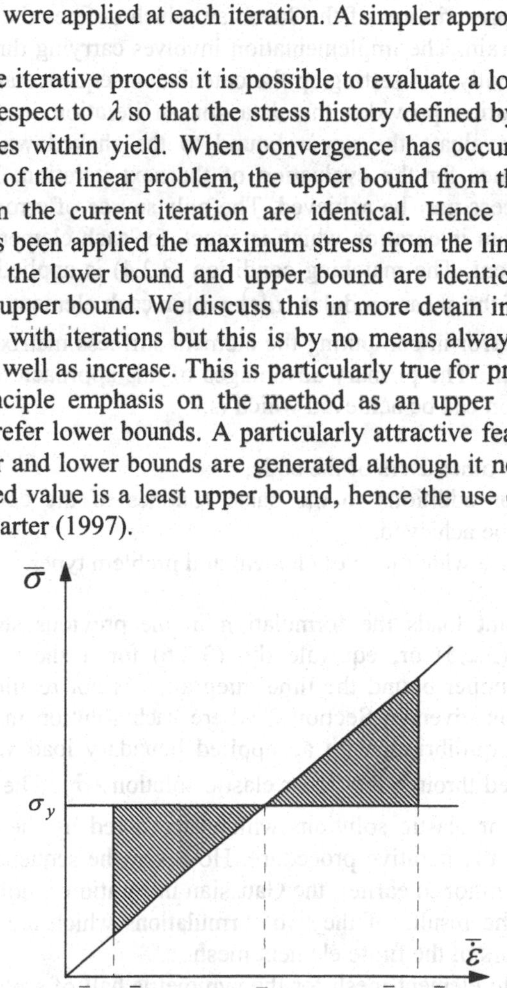

Figure 8. The inequality (3.2.12) for a von Mises yield condition requires that the shaded regions have a positive area. The linear material and the yield condition coincide at $\bar{\varepsilon}_i$.

4 Implementation of the Method

4.1 Limit Analysis

The method has been implemented in the commercial finite element code ABAQUS. The normal mode of operation of the code for material non-linear analysis involves the solution of a sequence of linearised problems for incremental changes in stress, strain and displacement in time intervals corresponding to a predefined history of loading. At each increment, user routines allow a dynamic prescription of the Jacobian which defines the relationship between increments of stress and strain. The implementation involves carrying through a standard load history calculation for the body, but setting up the calculation sequence and Jacobian values so that each incremental solution provides the data for an iteration in the iterative process. Volume integral options evaluate the upper bound to the shakedown limit which is then provided to the user routines for the evaluation of the next iteration. In this way an exact implementation of the process may be achieved. The only source of error arises from the fact that ABAQUS uses Gaussian integration which is exact, for each element type, for a constant Jacobian within each element. The matching condition (3.2.4) is applied at each Gauss point and results in variations of the shear modulus $\mu(t)$ within each element. There is, therefore, a corresponding integration error in computing the element stiffness matrix but the effect of this on the upper bound is small. The primary advantages of this approach to implementation are practical. An implementation can be achieved which is:

1) easily transferable to other users of the code,
2) requires fairly minor additions to the basic routines of the code so that a reliable implementation can be achieved,
3) can be introduced for a wide range of element and problem types.

For the case of constant loads the formulation in the previous section reduces to the solution of the equation (3.2.5) or, equivalently, (3.2.6) for a shear modulus distribution defined by (3.2.4). In the upper bound the time integration is not required. This formulation differs from the formulation given in Section 2 where each solution in the iterative process involves a stress field in equilibrium with an applied boundary load whereas in (3.2.6) the external loads are introduced through the linear elastic solution $\lambda\hat{\sigma}_{ij}$. The two formulations are entirely equivalent for linear elastic solutions which are solved by the same Galerkin finite element method as used in the iterative procedure. However, the sequence of calculations are not identical and, as was mentioned earlier, the Gaussian integration is not exact. In practice we find differences between the results of the two formulations which are negligible compared with the approximation errors of the finite element mesh.

Figure 9 shows the finite element mesh for the symmetric half of a plane strain indentation problem where a line load P is applied over a strip of width D. It may be recognised as one of the standard examples in the ABAQUS examples manual. The convergence of the upper bound is shown in Figure 10 for the mesh shown in Figure 9 and a refined mesh where each element has been subdivided into sixteen identical elements. The analytic solution (the Prandtl solution) for the half space is also shown. The observed behaviour illustrates two points which are common to all the following solutions. As the method is a strict upper bound the solution converges to the analytic solution from above. The accuracy of the converged solution depends entirely on the ability of the class of displacement fields defined by the mesh geometry to represent the displacement field in the exact solution. Generally there is the need for a sufficient density of elements in the regions where the deformation field varies most rapidly. In this case the refining of the mesh geometry has a very significant effect on the accuracy of the converged solution.

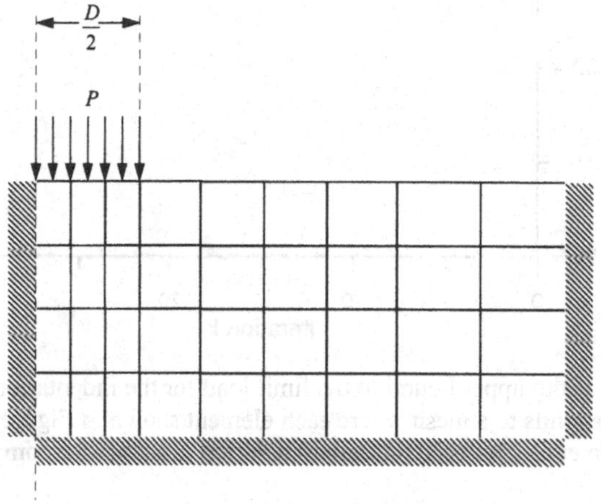

Figure 9. Finite element mesh for plane strain model of indentation of half space. Eight noded quadrilateral elements were used.

The method has been used to solve a large number of problems involving structural components with cracks. Accurate limit load solutions for such problems are required for the application of life assessment methods in the power industry (Goodall et al., 1991). Empirical investigations of the convergence of upper bound to available analytic solutions have been carried out for a range of geometries. For a fixed element type, the error diminishes near linearly with a length scale which characterises the mean size of the elements (see Section 5). From such data it is possible to find empirical rules for the production of limit load solutions with an error of less than, say, 1%.

It is worth commenting on the sensitivity of solutions to the assumptions within the method. In the examples considered above, near incompressibility was achieved by using hybrid elements with a Poissons ratio of 0.49999 . In the convergence proof discussed above it was assumed that the strain rates were incompressible. Practice has shown that this value of Poisson's ratio is sufficiently close to 0.5.

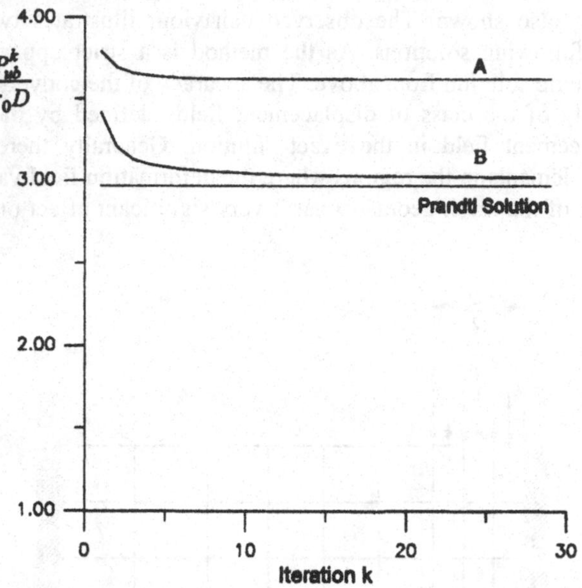

Figure 10. Convergence of the upper bound to the limit load for the indentation problem of Figure 9. The lower curve corresponds to a mesh where each element shown in Figure 9 has been divided into 16 elements. The analytic Prandtl solution is shown for comparison.

4.2 Shakedown Solutions

The procedure described by equations (3.2.4) to (3.2.7) requires the definition of a shear modulus at each instant during the cycle. There are problems where the distribution of the strain rate history in time during the cycle is unknown in advance but there is an important class of problems where we know *ab initio* that plastic strains may only occur at certain instants within the cycle. If the loading history consists of a sequence of proportional changes in loads between a set of extreme points, as shown schematically in Figure 11 for a problem involving two loads (P_1, P_2), then the linearly elastic stress history also describes a sequence of linear paths in stress space as shown. For a strictly convex yield condition, which includes the von Mises yield condition in deviatoric stress space, the only instants when plastic strains can occur are at the vertices of the stress history, $\hat{\sigma}_{ij}(t_l)$, $l = 1$ to r. The strain rate history then becomes the sum of increments of plastic strain;

$$\Delta\varepsilon_{ij} = \sum_{l=1}^{r} \Delta\varepsilon_{ij}^{l} \tag{4.2.1}$$

and equations (3.2.4) to (3.2.7) become

$$\Delta\varepsilon_{ij}'^{f} = \frac{1}{\bar{\mu}}\left(\bar{\rho}_{ij}'^{f} + \sigma_{ij}'^{in}\right) \tag{4.2.2}$$

$$\sigma_{ij}'^{in} = \bar{\mu}\left\{\sum_{m=1}^{r}\frac{1}{\mu_m}\lambda\hat{\sigma}_{ij}'(t_m)\right\} \tag{4.2.3}$$

where

$$\frac{1}{\bar{\mu}} = \sum_{l=1}^{r}\frac{1}{\mu_l} \quad \text{and} \quad \tfrac{3}{2}\mu_l\bar{\varepsilon}\left(\Delta\varepsilon_{ij}^{li}\right) = \sigma_y \tag{4.2.4}$$

The implementation of the method involves the following sequence of calculations: An initial solution assumes that plastic strains may occur at all r possible instants in the cycle. Initial, arbitary, values of the moduli $\mu_m = 1$ are chosen. As a result of this initial solution, the iterative method described in equations (3.2.4) to (3.2.7) is applied. The plastic strain components at instants where there is no strain in the converged solution then decline in relative magnitude until they make no contribution to the upper bound.

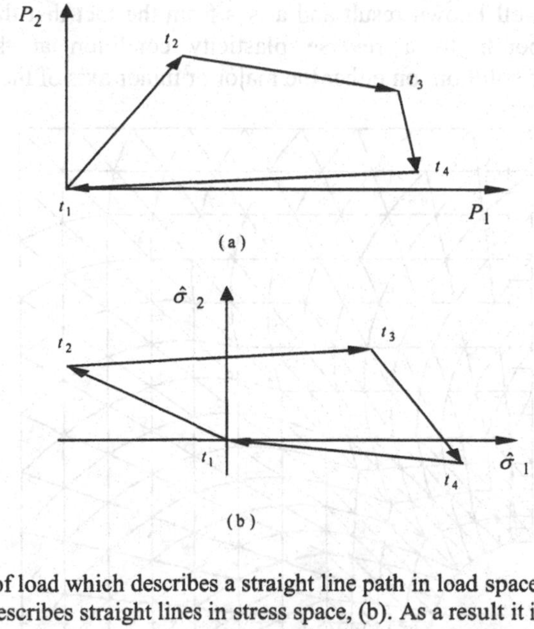

Figure 11. A history of load which describes a straight line path in load space (a) produces a history of elastic stresses which describes straight lines in stress space, (b). As a result it is know a priori that plastic strains only occur at the r vertices during the cycle. R=4 in the Figure.

In the following solutions the iterative method was continued until there was no changed in the fifth significant figure in the computed upper bound for five consecutive iterations. The number of iterations required depended upon the nature of the optimal mechanism. For reverse plasticity mechanisms the number of iterations required could be quite high, in excess of 100, whereas for mechanisms where all the plastic strains occurred at a single instant at each point in the body (although not necessarily the same instant) the number of iterations required was significantly less and 50 iterations was a typical number. For a less exacting convergence criteria a significantly smaller number of iterations are required and variation of the upper bound with iteration numbers shown in Figure 10 is typical of both limit load and shakedown solutions.

Figure 12 shows the symmetric section of a finite element discretisation for a plane stress plate, with a circular hole, subjected to biaxial tension. The shakedown limit has been evaluated for the two histories of $(P_1(t), P_2(t))$ shown in Figure 13. The interaction diagram of the shakedown limit evaluated by the method are shown in Figure 14 together with the limit load for monotonical increase in (P_1, P_2). The elastic limit is also shown, i.e, the highest load levels for which the elastic solutions just lie within the elastic domain for the prescribed yield stress and also for a yield stress of $2\sigma_y$. It may be observed that in all cases the shakedown limit is given either by the limit load for the loads at some point in the cycle or at the elastic limit for $2\sigma_y$. As the initial loading point is zero load, this later condition corresponds to the variation of the elastic stresses lying within the yield surface if superimposed upon an arbitrary residual stress. This is a well known result and arises from the fact that the mechanism at the shakedown limit corresponds to a reverse plasticity condition at the point of stress concentration in the elastic solution, on either the major or minor axis of the hole surface.

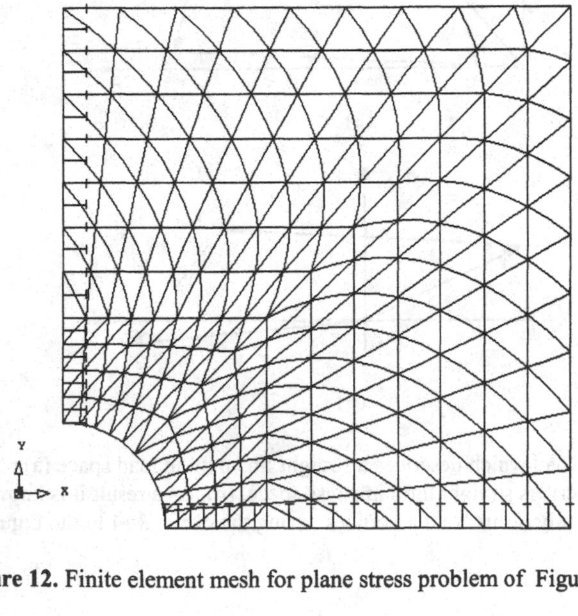

Figure 12. Finite element mesh for plane stress problem of Figure 13

Figure 15 shows the classic Bree problem where either a plate or a tube wall thickness is subjected to axial stress and a fluctuation temperature difference $\Delta\theta$ across the plate or through the wall thickness. The problem has been solved both as plane stress plate problem, where curvature of the plate due to thermal expansion is restrained, and as an axisymmetric cylinder. The finite element mesh consists of 6 eight-noded elements across the width and 30 elements in the direction of loading. The two converged solutions for a temperature independent yield stress are both shown in Figure 16 in terms of σ_t, the maximum principal thermoelastic stress due to $\Delta\theta$ for various ratios of load and thermal stress. The plate solution coincides with the classic Bree solution for a Tresca yield condition (the problem is essentially one dimensional) whereas the solution for the axisymetric problem lies outside the classic Bree solution to a maximum extent of 15%, the maximum difference between the Tresca and Von-Mises yield condition. The reverse plasticity solution, which corresponds to $\sigma_t = 2\sigma_y$ in both cases, is overestimated by the computed solutions. This is due to the way reverse plasticity limits are evaluated. The optimising strain rate history consists of increments of strain which result in a zero accumulation of strain over the cycle. The contribution of a single Gauss point (or in this problem a row of Gauss points) dominates over the contribution of all other Gauss points and the limit is governed by the variation of the elastic stress at that point. In Figure 16 we adopt for σ_t the value of thermal stress at the surface whereas the reverse plasticity shakedown limit corresponds to the slightly smaller thermal stress at a neighbouring Gauss points.

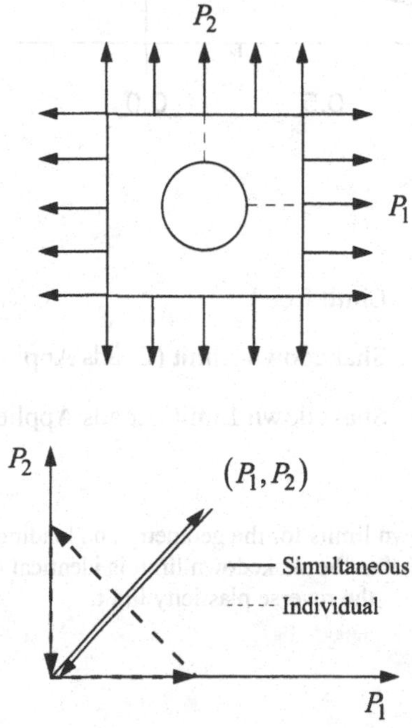

Figure 13. Loading histories for the shakedown solutions shown in Figure 14

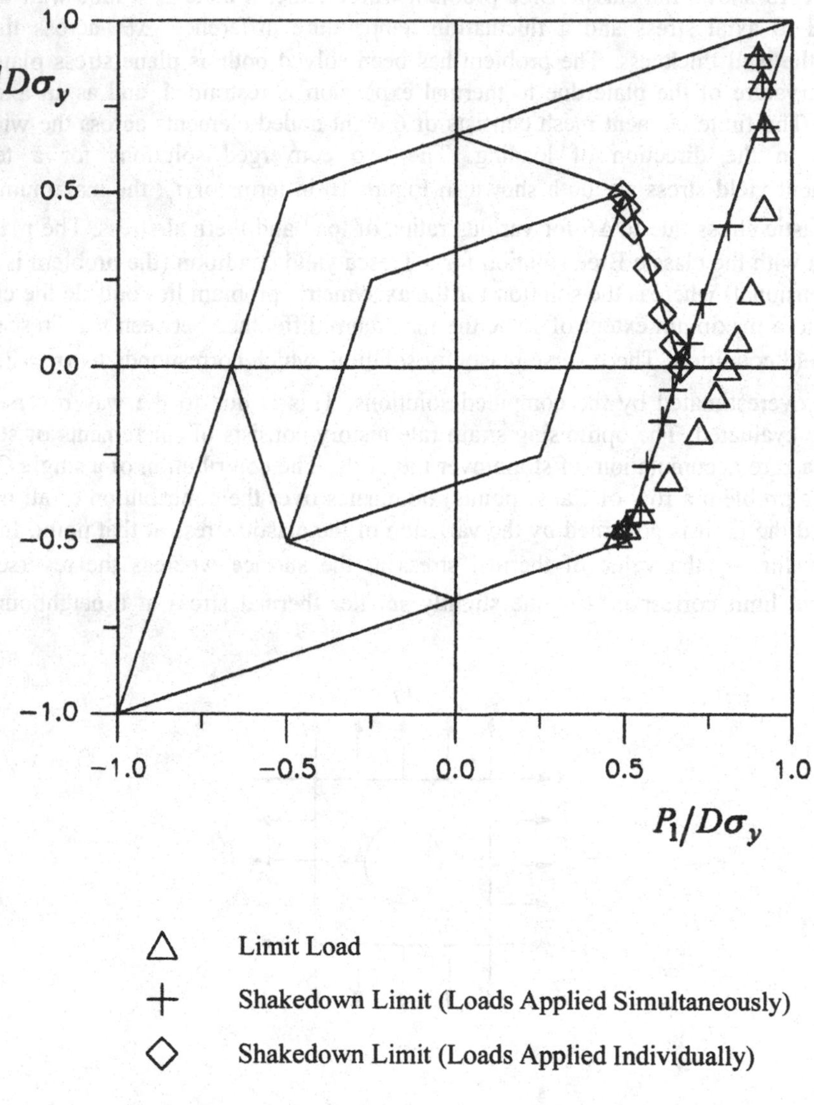

Figure 14. Limit load and shakedown limits for the geometry and loading histories shown in Figure 13 and mesh shown in Figure 12. Note that the shakedown limit is identical to the least of the limit load or the reverse plasticity limit.

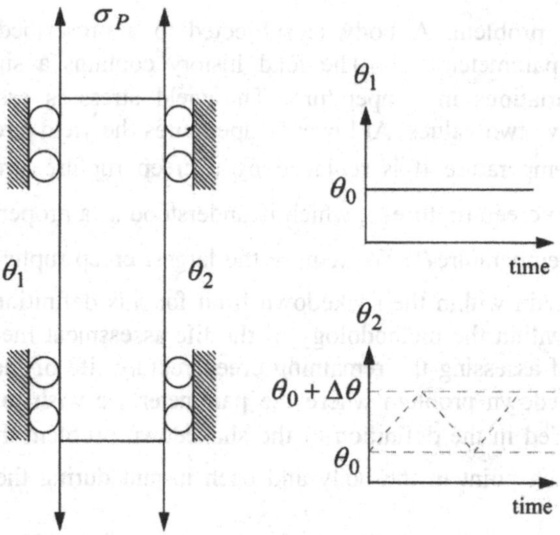

Figure 15. Bree Problem – a plate or axisymmetric tube is subjected to fluctuating temperature differences and axial stress.

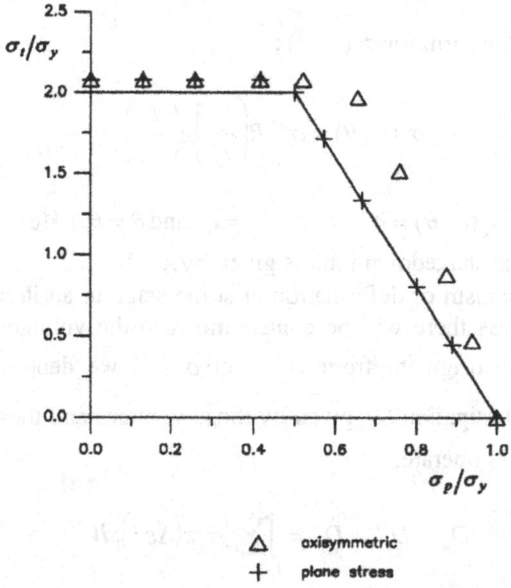

Figure 16. Shakedown limits for the Bree problem of Figure 15 modelled as both a plane stress plate and an axisymmetric thin walled tube. The solid line is Bree's solution for a Tresca yield condition.

4.3 An Extended Shakedown Limit

Consider the following problem. A body is subjected to a prescribed history of loading corresponding to load parameter $\lambda = 1$. The load history contains a significant component which derives from variations in temperature. The yield stress is assumed to vary with temperature and may have two values. At lower temperatures the yield stress equals a constant value σ_y^{LT}. At higher temperature it is replaced by a creep rupture stress $\sigma_c(t_f, \theta)$ which depends upon the time to creep rupture t_f, which is understood as a property of the structure as a whole, and the local temperature θ. We require the largest creep rupture time t_f for which the prescribed loads remain within the shakedown limit for this definition of the yield stress. This problem is posed within the methodology of the life assessment method R5 (Goodall et al., 1991) as a means of assessing the remaining creep rupture life of the structure. Here we treat it as a novel shakedown problem where the parameter we wish to optimise, the creep rupture time t_f, is included in the definition of the shakedown problem through the definition of the yield stress at each point in the body and each instant during the cycle when plastic strains can occur.

Hence the yield stress at each point of the body at time t_m is defined by;

$$\sigma_y(x_i, t_m) = \min\begin{Bmatrix} \sigma_y^{LT} \\ \sigma_c(t_f, \theta(x_i, t_m)) \end{Bmatrix} \tag{4.3.1}$$

We assume the following form for $\sigma_c(t_f, \theta)$;

$$\sigma_c(t_f, \theta) = \sigma_y^{LT} R\left(\frac{t_f}{t_0}\right) g\left(\frac{\theta}{\theta_0}\right) \tag{4.3.2}$$

and R(1)=g(1)=1 so that $\sigma_c(t_f, \theta) = \sigma_y^{LT}$ when $t_f = t_0$ and $\theta = \theta_0$. Hence we wish to compute the value of R for which the shakedown limit is given by $\lambda = 1$.

For a prescribed mechanism of deformation at some stage in an iterative process with this definition of the yield stress there will be contributions to the volume integral of the plastic energy dissipation which originate from σ_y^{LT} and σ_c. If we denote by $\dot{D}_p^{LT}$ and $\dot{D}_p^c$ the contributions to the total dissipation $\dot{D}_p$ given by those volumes and those times where the low temperature and creep stress operate;

$$\dot{D}_p = \dot{D}_p^{LT} + \dot{D}_p^C = \int_V \sum_{m=1}^{r} \left\{\sigma_y \dot{\varepsilon}\left(\Delta\varepsilon_{ij}^m\right)\right\} dV \tag{4.3.3}$$

then we can derive, from the upper bound (1.5.4) and (4.3.2), the following relationships between small changes in λ_{UB} and R for a particular mechanism of deformation;

$$\Delta\lambda\dot{D}_p^E = \frac{\Delta R}{R}\dot{D}_p^c \tag{4.3.4}$$

where

$$\dot{D}_p^E = \int_V \sum_{m=1}^{r} (\hat{\sigma}_{ij}(t_m)\Delta\varepsilon_{ij}^m) dtdV \tag{4.3.5}$$

This relationship forms the basis for an iterative process which converges to the value of R and hence the rupture time t_f corresponding to the shakedown for $\lambda = 1$.

We begin by choosing an initial value of $R = R_0$ and t_f so that the shakedown limit in the converged solution is expected to be $\lambda < 1$. For fixed R_0 the iterative process is allowed to converge until the k-th iteration yields the first upper bound value of the load parameter which satisfies $\lambda_{UB}^k < 1$. The value of R is then changed according to equation (4.3.4) at each iteration so that λ_{UB} returns to the preassigned value of R_0 i.e.,

$$\Delta R = R_o\left(1 - \lambda_{UB}^k\right)\left(\frac{\dot{D}_p^E}{\dot{D}_p^c}\right) > 0 \tag{4.3.6}$$

Hence

$$R_1 = R_0 + \Delta R > R_0 \tag{4.3.7}$$

and the process is repeated. At each iteration the value of R increases and converges, from below, to the value for which the shakedown limit is given by $\lambda = 1$.

In the following example we adopt the following simple form for the temperature dependence of σ_c;

$$g\left(\frac{\theta}{\theta_0}\right) = \frac{1}{\left(1 - \frac{\theta}{\theta_0}\right)} \tag{4.3.8}$$

In Figure 17 the solutions for the Bree problem are shown with $\theta_0 = 200^0 C$ and other material constants appropriate to a ferritic steel.

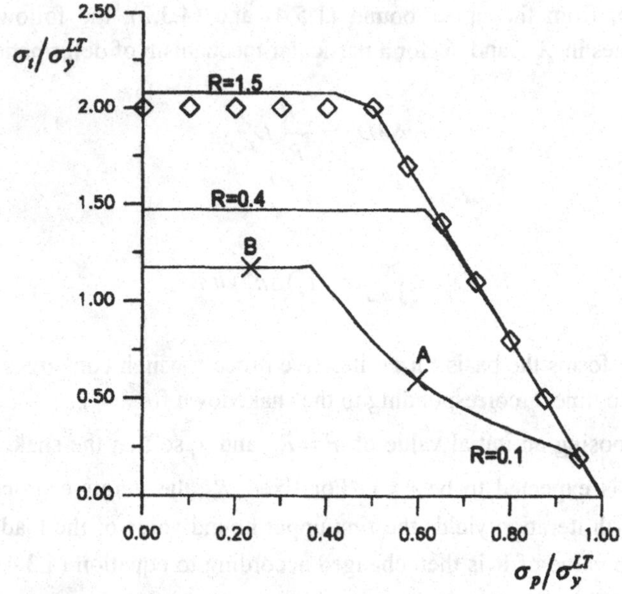

Figure 17. Shakedown limits for the problem discussed in section 4.3 for prescribed values of R. The diamonds correspond to the Bree solution which coincides with the computed solution for R=1.5. Points A and B refer to the solutions in Figure 18 for R=0.1.

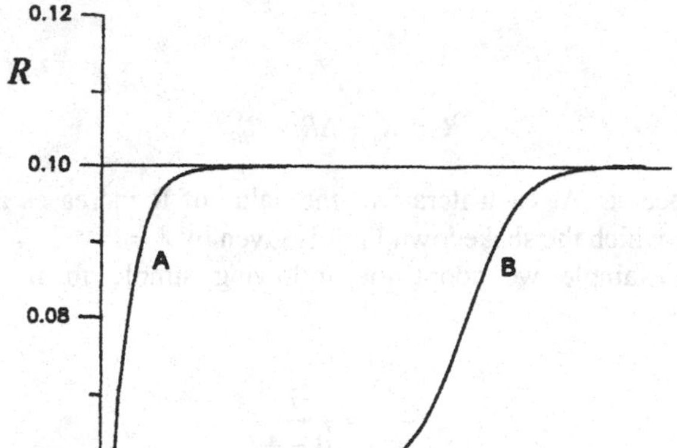

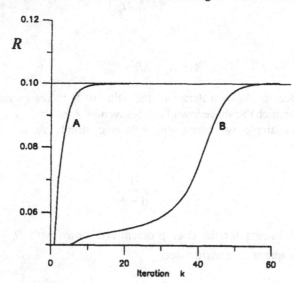

Figure 18. Convergence of R to R=0.1 for the extended shakedown method discussed in section 4.3. Curves labelled A and B correspond to the convergence to the corresponding points in Figure 17. The slow convergence of case B is due to the dominance of a reverse plasticity mechanism.

In Figure 17 the shakedown limit is shown for the three cases of R= 0.1, 0.4 and 1.5. The contours shown were evaluated by converging to the value of the load parameter corresponding to the prescribed yield conditions given by equation (4.3.1) although it is worth noting that the dependency of the yield stress on temperature causes a change in the yield stress at each iteration. For the case of R=1.5, $\sigma_y = \sigma_y^{LT}$ throughout the volume. In Figure 18 we show the inverse problem where the load parameter λ is prescribed and the value of R is evaluated corresponding to a shakedown state. The two solutions shown in Figure 18 corresponding to points A and B in Figure 17 for R=0.1. The two phases of the process can be seen where the initial value of $R_0 = 0.05$ is maintained constant for the first few iterations until $\lambda_{UB}^k < 1$, when R is allowed to increase according to the relationships (4.3.6) and (4.3.7) until convergence takes place. Convergence is slower in the case corresponding to point B in Figure 18, where, in the converged solution, a reverse plasticity mechanism operates.

This problem demonstrates the potential flexibility of the method. Traditionally, shakedown analysis has been seen as a method of defining a load parameter for a prescribed distribution of material properties and load history. It is clear from this example that the shakedown problem may be posed in other ways; in this particular problem the quantity optimised concerns a material property which enters the problem in only part of the volume and only during part of the load cycle. It is clearly possible, using the type of technique discussed in this section, to pose a variety of optimisation problems depending upon the needs of the problem. This introduces possibilities for shakedown analysis which have not previously been available.

5 Further Examples of the Implementation of the Linear Matching Method

The following examples of solutions are chosen to demonstrate the stability of the method and a range of applications that relate to problems in the life assessment of structures and geomechanics. All the solutions were generated using the commercial code ABAQUS (HKS Ltd), using a method devised by a research student Markus Engelhardt (1999), as described in Section 4.1. From experience the only source of computational error of any significance that can be identified arises from Gaussian integration of the stiffness matrix of the linear solutions. ABAQUS uses a level of Gaussian integration that gives an exact evaluation of the element stiffness matrices provided the element linear moduli are constant. As the linear moduli in the Linear Matching Method vary from Gauss point to Gauss point, there is a resultant error. However, there is an indirect check on the significance of this error. In the absence of this error; the upper and lower bound load parameters converge to a common value, the least upper bound associated with the class of displacement fields defined by the element structure. Hence differences between the converged bounds give an indirect indication of the significances of this error.

This phenomenon is shown in the first example, shown in Figure 19, where a plate with symmetrically place cracks is subjected to uni-axial tension. The elements are four noded quadrilateral elements. This is solved as a limit load problem for a von Mises yield condition. The convergence of the upper and lower bound load parameters is shown in Figure 20. The upper bound converges to the optimal upper bound for this mesh, a value that lies above the known analytic solution, as shown. The difference between the optimal upper bound and the analytic solution is primarily a function of the mesh geometry. The lower bound converges, more slowly, to a value close to but below the optimal upper bound and above the analytic solution. The difference between the two bounds is also a function of the mesh geometry but arises from the error in Gaussian integration. The effect is exaggerated here as the mesh is course and the element degrees of freedom are insufficient to capture the rapid changes of strain in the vicinity of the crack tip.

This example is included as a demonstration of the modes of behaviour of the method. Generally the mesh is refined until the converged upper bound does not significantly reduce for increasing mesh density. In Figure 21, from Hentz (1998), the relative error in the upper bound, defined as

Relative error = (Optimal upper bound - analytic value)/(analytic value)

is shown for uniform meshes for a range of crack problems. The characteristic dimension of a typical element h is defined as the diameter of the smallest circle that surrounds an element. The distance a is the size of the uncracked length, the crack ligament. It can be seen that convergence is near linear with h, implying that convergence is primarily concerned with the convergence of equilibrium within the element (rather between elements). Solutions with errors of less than 1% may be achieved without difficulty. This and other empirically derived information allows the generation of meshes that are likely to yield upper bounds with errors of less than 0.5%. With increasing mesh density the upper and lower bounds more closely approach at convergence and a difference of less than 1% is achieved. The lower bound does not, however, always converge monotonically and it can take a considerable number of cycles for the lower bound to converge. For a convergence criterion that there should be no change in the 5th significant figure in 5 iterations (the 5/5 criterion), problems involving the von Mises yield condition converge in about 50 iterations.

This convergence rate is reasonable independent of the complexity of the problem. The next example from Chen and Ponter (2001) is shown in Figure 22, with dimensions in Table 1, where a cylinder is subjected to internal pressure and axial load. The presence of various types of cut-outs turns the cylinder into a three dimensional problem and various examples of meshes and cut-outs are shown in Figure 23 where the elements are 20 noded brick elements. Such problems have been extensively investigated both as a limit load and a shakedown problem. In Tables 2 and 3 there is a comparison between the converged solutions and solutions obtained by an entirely different method (Chen et al., 1997) using conventional non-linear programming techniques with significantly less computational efficiency. The optimal upper bounds given by the Linear Matching Method lie slightly below these solutions but the agreement is very good. The convergence rate is shown for a typical case in Figure 24. This demonstrates an important feature of the method, there is no scaling effect with increase in either the complexity or the fineness of the mesh, convergence is achieved in about 50 cycles, i.e. as a result of 50 linear solutions. The convergence rate does, however, depend on the yield condition and this is discussed below.

The final set of solutions concern yield conditions appropriate for granular materials in terms of the von Mises effective stress and (compressive) hydrostatic pressure, denoted by s and p, respectively, in the following diagrams. The extension of the method to such cases has been discussed by Ponter, Fuschi and Engelhardt (2000) for limit analysis and by Parrinello (2000) and Ponter and Parrinello (2001) for shakedown. The following examples are taken from Parrinello and Ponter (2001). Figure 25 shows the general class of yield conditions under consideration. The yield surface is of the Drucker-Prager type with Mohr-Coulomb section terminated for sufficiently high values of hydrostatic pressure p by a cap yield surface as shown. Figure 26 demonstrates the additional conditions required for convergence for such cases. As before the linear material is matched to the yield condition at the current strain rate. In stress resultant space (p,s), the linear material may be characterised by the complementary energy $W(p,s)$, rather than the energy dissipation U. The corresponding strain rates are then given by the derivatives of W,

$$\dot{\bar{\varepsilon}} = \frac{\partial W}{\partial s}, \quad \dot{\varepsilon}_{kk} = -\frac{\partial W}{\partial p} \tag{5.1}$$

where the negative sign arises from the definition of p as the compressive hydrostatic pressure. Hence the matching process means that the contour of constant complementary energy, $W(p,s) = const$, for the linear matching material is tangential to the yield surface at the point on the yield surface where the current strain rate is associated. For convergence, however, an additional condition is required. The contour (surface) $W(p,s) = const$ must otherwise surround the yield surface as shown schematically in Figure 26. Hence the linear matching material must satisfy two conditions, the matching of the stress to the yield condition at the current strain rate, and this additional condition. From experience the most rapid convergence is obtained if the $W(p,s) = const$ surface most closely matches the yield surface. For the von Mises yield condition this additional condition is only satisfied if the linear material is incompressible when there is a perfect match between the yield surface and $W(p,s) = const$.

The first example of the application of this more general method is shown in Figure 27 and 28 where a half space is subjected to a lateral pressure p and a reversing tangential pressure t. Figure 29 shows convergence of the upper and lower bounds for a particular case and convergence is achieved in about 350 iterations. Both limit load and shakedown surfaces are shown in Figure 30. It is worth noting that the inner shakedown limit surface consists primary of regions where the maximum shear pressure t is constant, corresponding to a reverse plasticity mechanism.

The last example, shown in Figure 31, consists of a two dimensional model of a concrete dam subjected to water level varying between two limits. The yield condition for the concrete is shown in Figure 32. The converged shakedown parameter is shown in Figure 33 where it can be seen a converged upper bound is obtained in about 100 iterations but the lower bound converges more slowly. Figures 34 and 35 show contours of constant volumetric and effective strain in the converged solution that identifies a reverse plasticity solution at a point of the water side surface of dam at earth level. Optimising the design would consist in changes of geometry which relieve the stresses in this region. Note that the absolute values of strain in this case is arbitrary.

These example demonstrate that the method is a very stable practical method for computing optimal limit and shakedown limits with the following advantages:

1) Convergence criteria have been derived for an arbitrary yield condition. This allows the development of the method for any yield condition that satisfies convexity and normality.
2) The method may be understood, mathematically, as a programming method, but has the advantage that the linear solutions in each iteration have a simple interpretation, that the linear material and load are chosen so that the linear problem most closely matches the non-linear plasticity problem.
3) The linear solutions conform to the standard linearization already incorporated in commercial finite element codes and this allows implementations within such codes. As a result, the effort needed to produce an effective solution is vastly reduced compared with programming methods as all the facilities of the code are available to the user.
4) The implementations can be made near exactly, and the effect of the only error involved can be judged indirectly through the difference between the upper and lower bounds. This error has proved to be consistently small for sufficiently fine mesh densities.
5) There is no scaling effect in the method. The number of linear solutions for a converged solution is independent of the complexity of the geometry or the fineness of the mesh. For a given yield condition it is possible to judge the number of equivalent linear elastic solutions required. For the von Mises yield condition, this is about 50. For a Drucker-Prager yield condition it is greater and is likely to be reduced through optimisation of the implementation.

The Way Ahead

This article has been primarily concerned with the application of the Linear Matching Method to limit analysis and shakedown for the von Mises yield condition. The method has been extended to an arbitrary perfectly plastic yield condition by Ponter Fuschi and Engelhardt (2000), Ponter and Engelhardt (2000) and successfully applied to geotecnical problems, as described above, by Parrinello (2000) and Ponter and Parrinello (2001). There are, however, other problems of a similar nature that have the potential for the application of such a method. An extension to the evaluation of the ratchet limit for a class of thermal loading problems has been described by Ponter and Chen (2001) and Chen and Ponter (2001). Recently Ponter (2001) has discussed the theory behind the method for a class of creep problems. Without doubt there will be other examples in time.

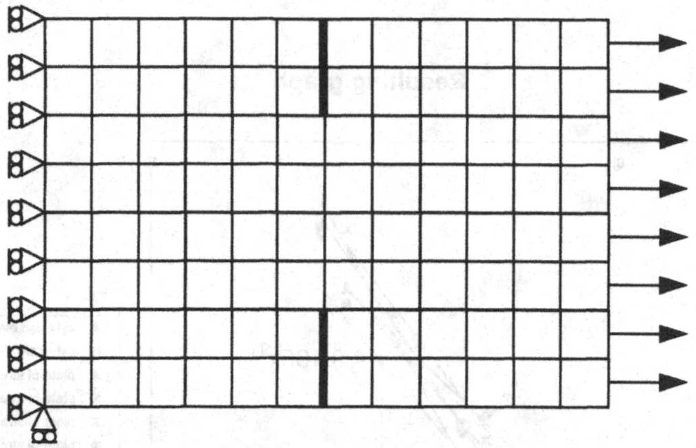

Figure 19. Double edge cracked plate subjected to uniaxial tension. The elements are four nodded quadrilateral plane stress elements.

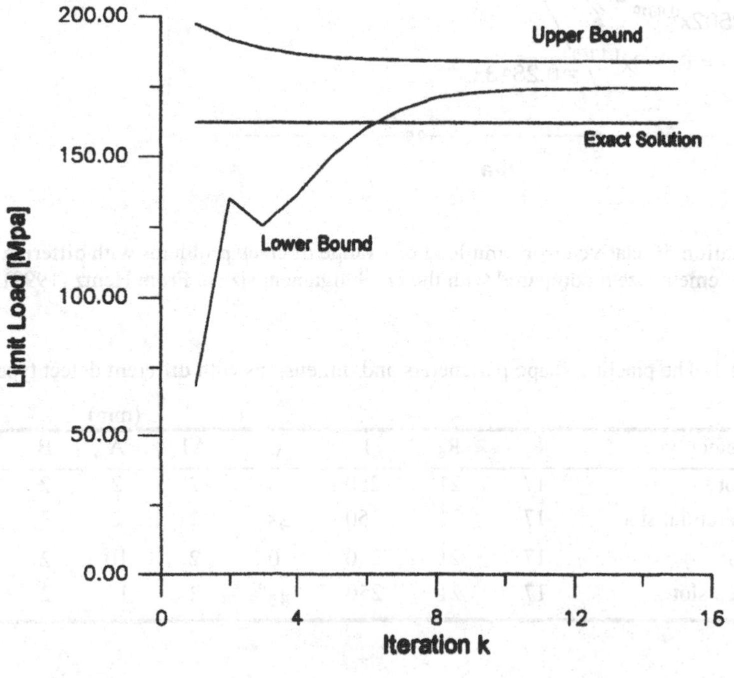

Figure 20. Convergence of upper and lower bounds fro the problem shown in Figure 19 for a yield stress of 200Mpa.

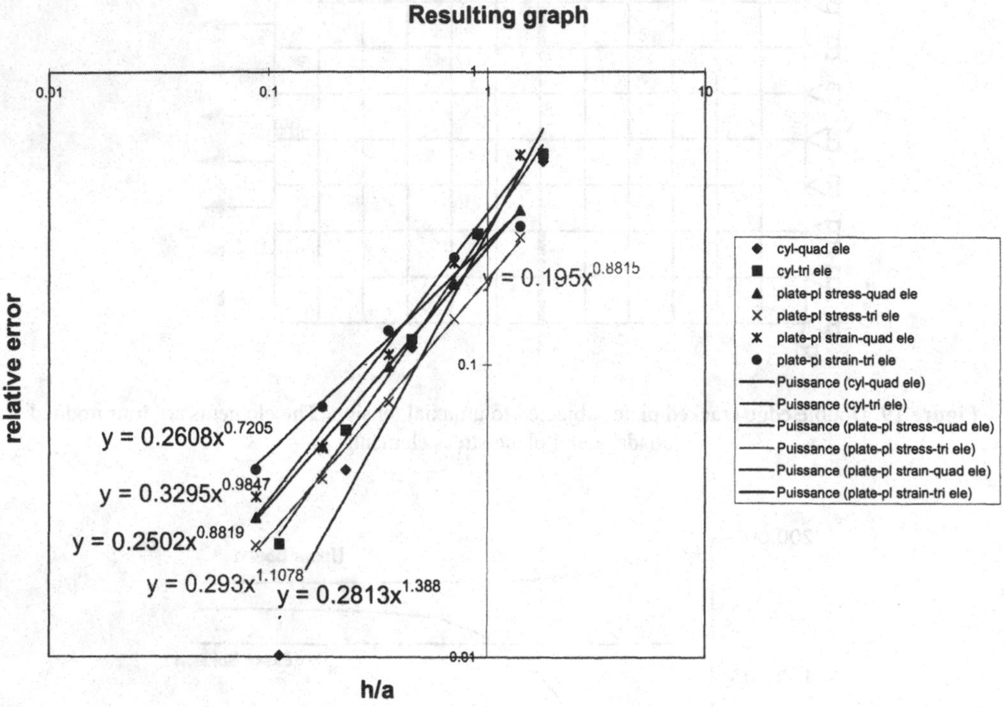

Figure 21. Variation of relative error limit load of a range of crack problems with differing mesh types and element size h compared with the crack ligament size a. From Hentz (1999).

Table 1. The pipeline shape parameters and dimensions with different defect types (mm)

Defect type	R_i	R_o	L	α	A1	A	B	C
Small slot	17	21	250	0	2	2	2	2
Circumferential slot	17	21	250	45°	2	2	2	2
Axial slot	17	21	250	0	2	10	2	2
Large area slot	17	21	250	45°	2	10	2	2

Table 2. Comparison of the limit loads for a defective pipeline subjected to axial tension (MPa)

Defect type	Results via present method	Lower bound Chen et al. (1997)	Upper bound Chen et al. (1997)
Small slot	244.7	240.0	244.8
Circumferential slot	199.2	185.3	203.0
Axial slot	244.4	239.1	244.4
Large area slot	184.1	168.3	185.1

Table 3. Comparison of the limit loads for a defective pipeline subjected to internal pressure (MPa)

Defect type	Results via present method	Lower bound Chen et al. (1997)	Upper bound Chen et al. (1997)
Small slot	59.65	56.9	59.68
Circumferential slot	55.67	55.2	57.7
Axial slot	49.52	46.4	50.8
Large area slot	39.70	37.2	39.9

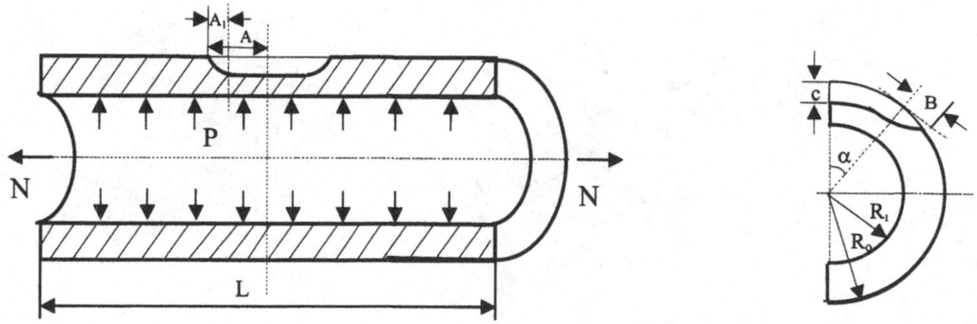

Figure 22. The geometry of pipeline with part-through slot subjected to internal pressure and axial tension.

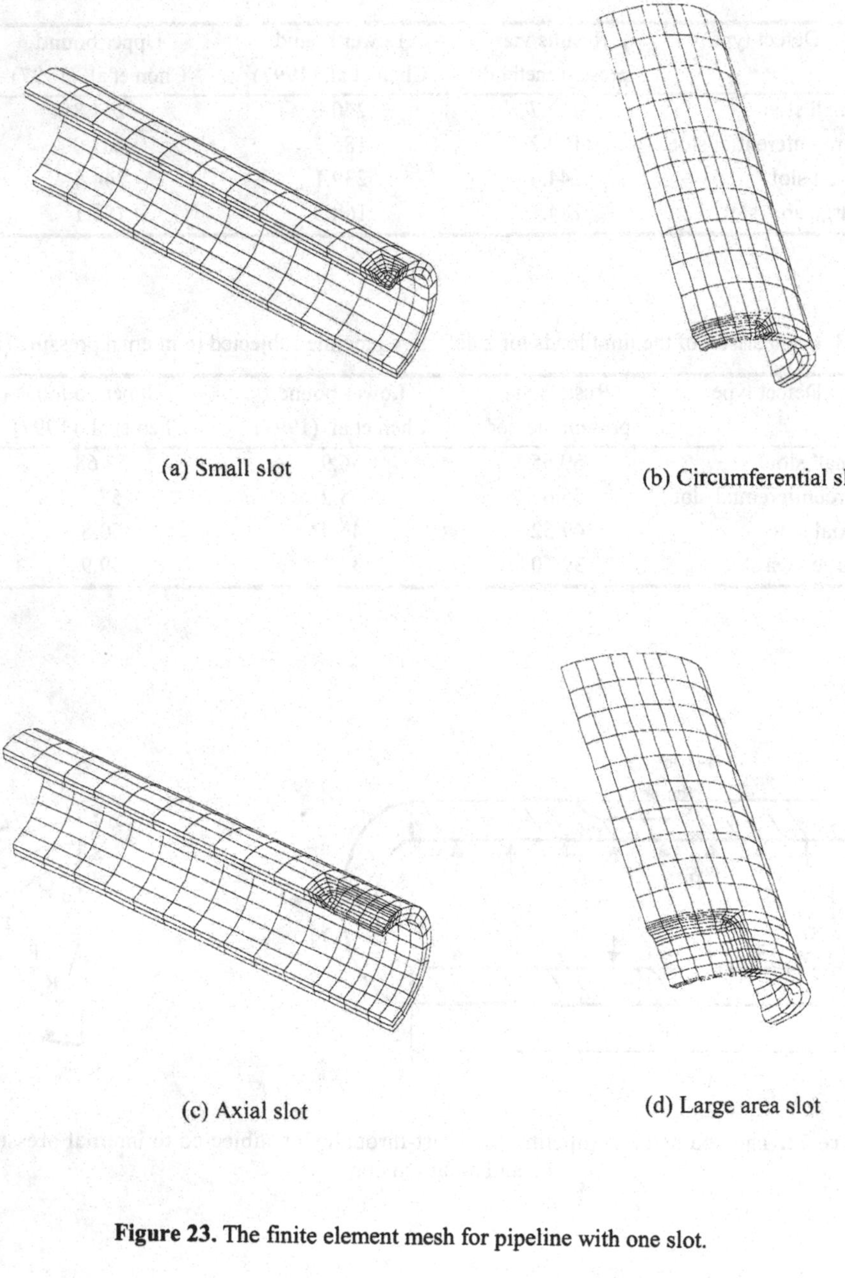

(a) Small slot

(b) Circumferential slot

(c) Axial slot

(d) Large area slot

Figure 23. The finite element mesh for pipeline with one slot.

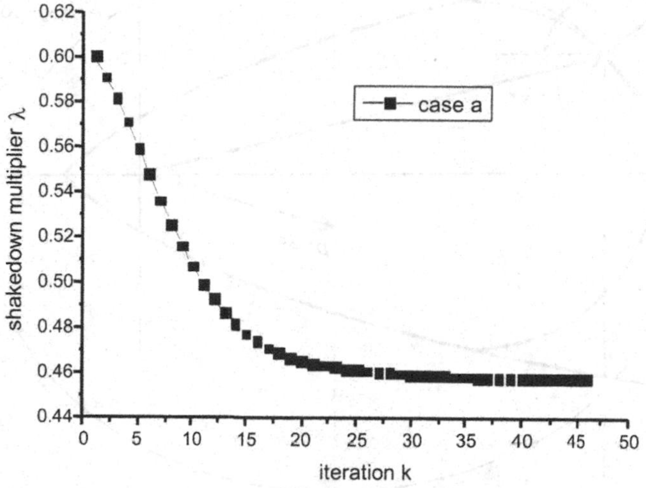

Figure 24. The convergence conditions of iterative processes for a pipeline with large area slot . This represents a typical convergence sequence for all cases.

$$p = \sigma_{kk}/3, s = \sqrt{\tfrac{1}{2}\sigma_{ij}\sigma_{ij}}$$

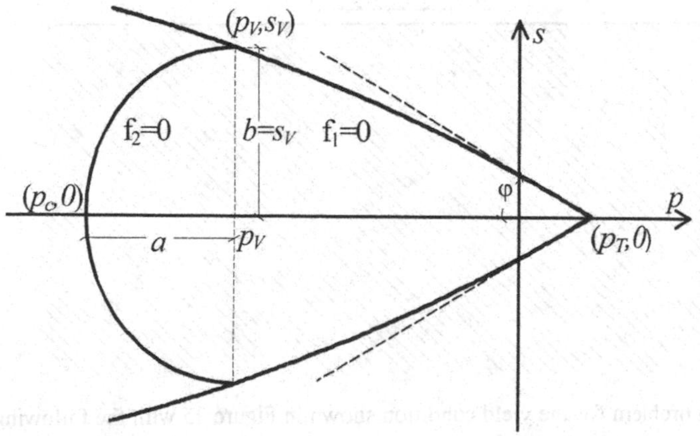

Figure 25. Modified Drucker-Prager with cap yield surface. $p = \sigma_{kk}/3$.

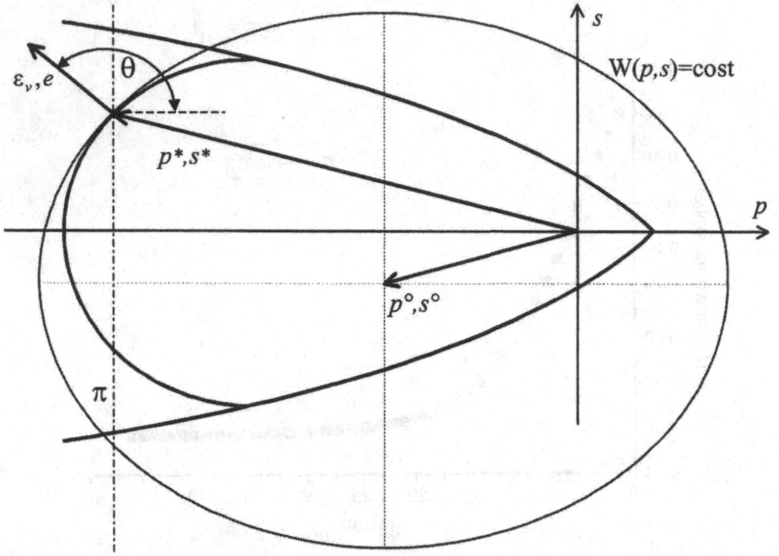

Figure 26. Linear Matching condition. The contour of constant complementary energy, $W(p,s) = \text{constant}$ for the linear matching material, touches the yield surface at the yield point (p^*, s^*) corresponding to the current plastic strain rate value and otherwise surrounds the yield surface.

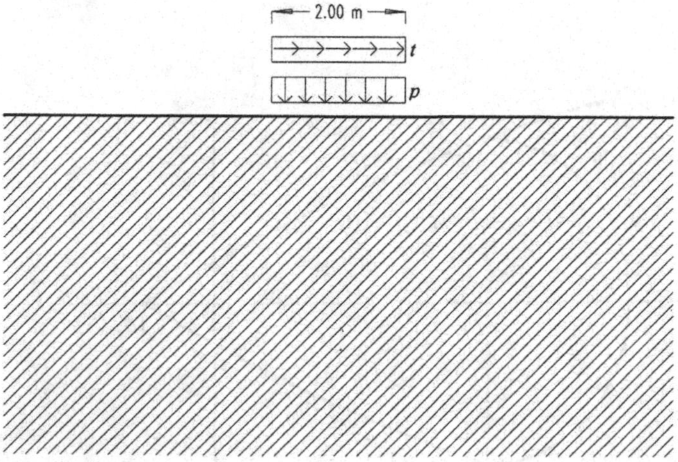

Figure 27. Shakedown problem for the yield condition shown in Figure 25 with the following properties; $\nu = 0.3$, $p_T = 5DaN/cm^2$, $p_V = -90DaN/cm^2, q = 0.077$, $\varphi = 30°$.

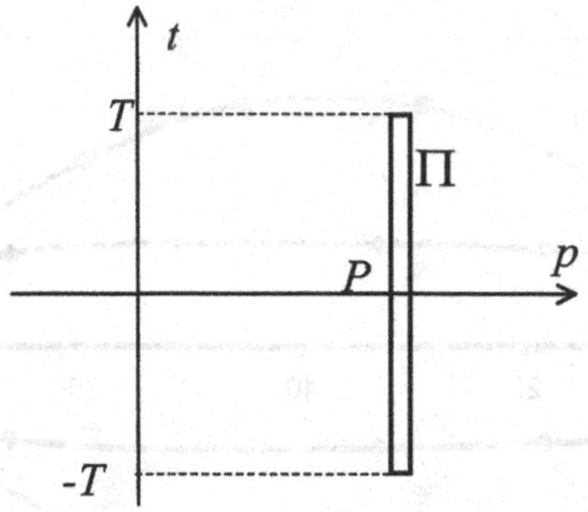

Figure 28. Load domain for the shakedown problem of Figure 29.

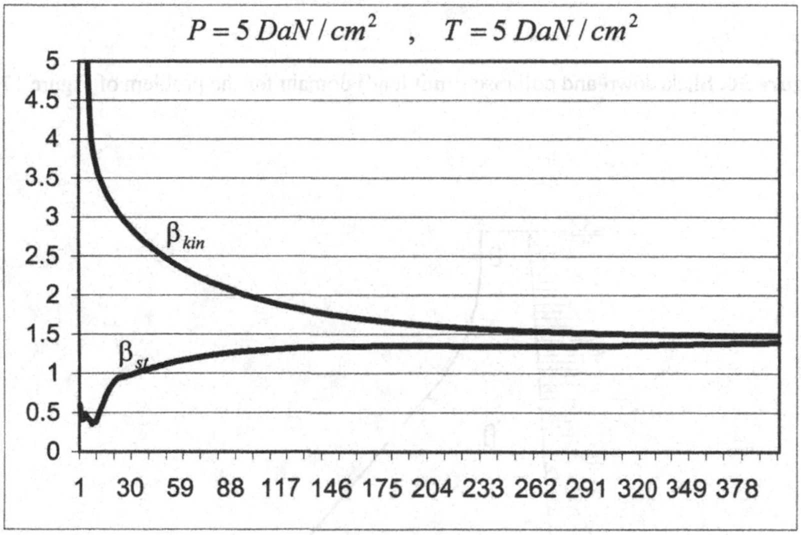

Figure 29. Upper and lower bound sequences for a particular case of the loading sequence of Figure 28.

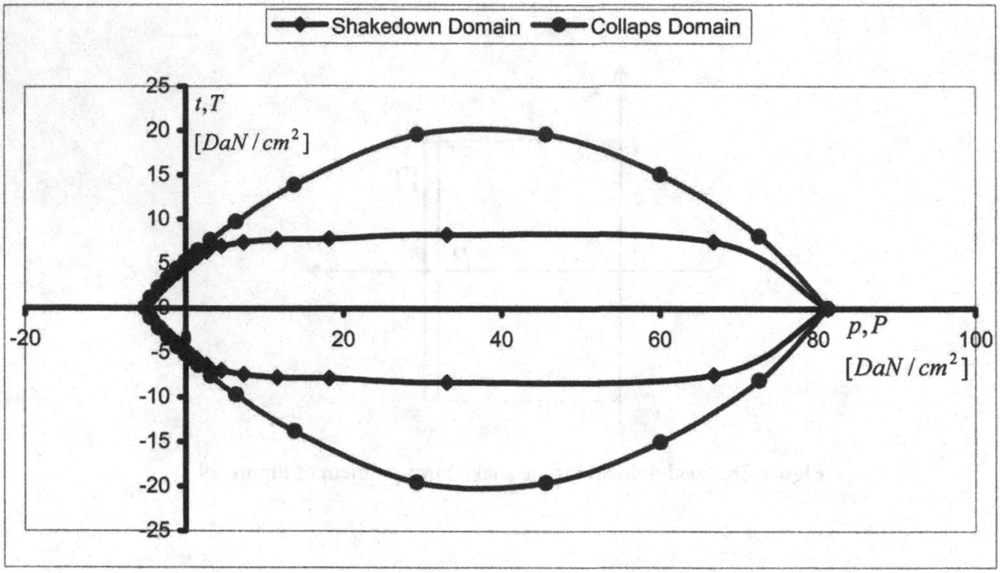

Figure 30. Shakedown and collapse (limit load) domain for the problem of Figure 17.

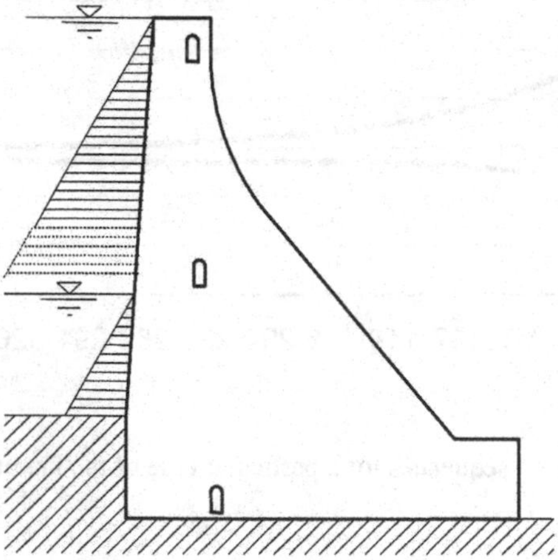

Figure 31. Shakedown problem of a concrete dam subjected to variable water level.

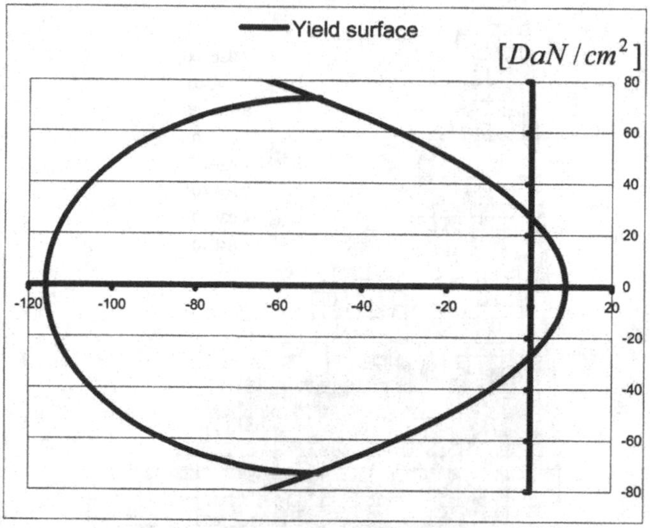

Figure 32. Concrete yield surface for the problem of Figure 20.

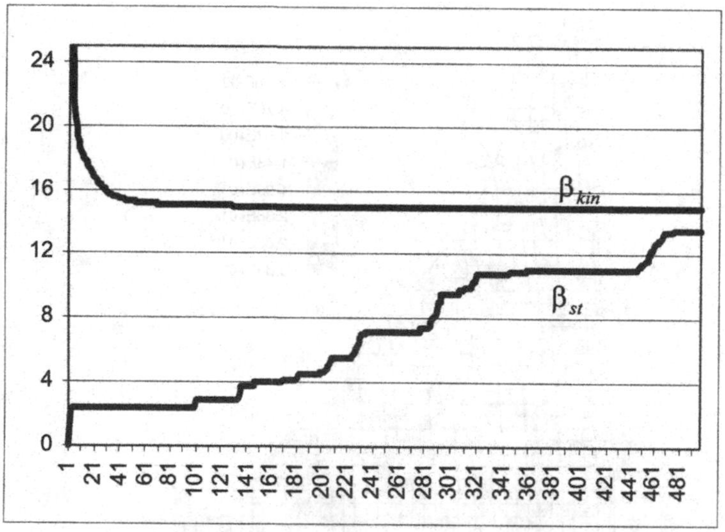

Figure 33. Upper and lower bound convergence sequences.

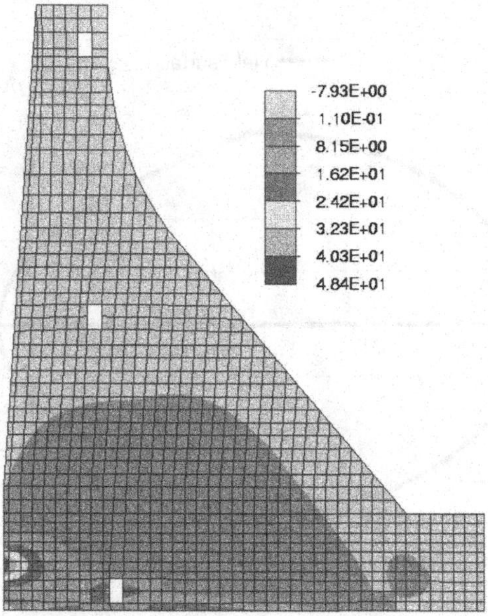

Figure 34. Volumetric strain distribution in the converged solution.

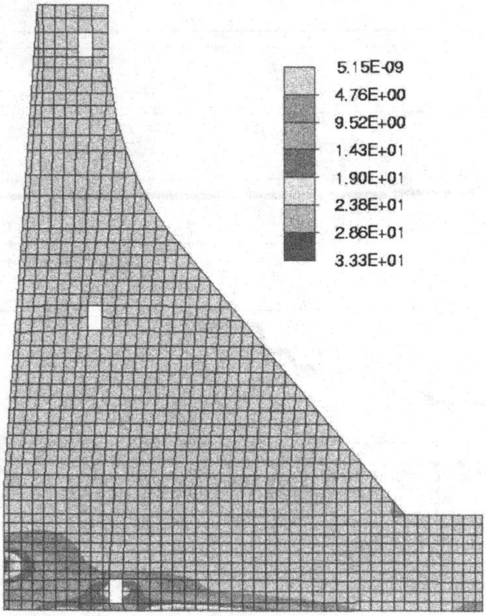

Figure 35. Effective strain distribution in the converged solution.

References

Bree, J. (1967). Elasto-plastic behaviour of thin walled tubes subjected to internal pressure and intermittent high-heat fluxes with application to fast reactor fuel elements. *J Strain Analysis* 2:226-238.

Chen, H. F., Cen, Z. Z., Xu, B. Y. and Zhan, S. G. (1997). A numerical method for reference stress in the evaluation of structure integrity," *Int Journal of Pressure Vessel & Piping* 71:47-53.

Chen, H. and Ponter, A. R. S. (2001). A method for the evaluation of a ratchet limit and the amplitude of plastic strain for bodies subjected to cyclic loading. *Euro. Jn. Mech., A/Solids* 20:555-571.

Chen, H. and Ponter, A. R. S. (2001). The 3-D shakedown and limit load using the elastic compensation method, *Int Journal of Pressure Vessel & Piping*, to appear.

Engelhardt, M. (1999), Computational Modelling of Shakedown, Ph.D thesis, University of Leicester.

Franco, J. R. Q. and Ponter, A. R. S. (1997). A general approximate technique for the finite element shakedown and limit load analysis of axisymmetric shells, Part 1: Theory and fundamental relationships. *Int. Jn. for Num. Methods in Engrg* 40:3495-3513

Franco, J. R. Q. and Ponter, A. R. S. (1997). A general approximate technique for the finite element shakedown and limit load analysis of axisymmetric shells, Part 2: Numerical applications. *Int. Jn. for Num. Methods in Engrg.* 40:3515-3536.

Gokhfeld, D. A. and Cherniavsky, D.F. (1980). *Limit Analysis of Structures at Thermal Cycling.* Sijthoff\& Noordhoff. Alphen an Der Rijn, The Netherlands.

Goodall, I. W., Goodman, A.M., Chell, G. C., Ainsworth, R. A., and Williams J. A. (1991). R5: An assessment procedure for the high temperature response of structures. Report, Nuclear Electric Ltd., Barnwood, Gloucester, UK.

Hentz, S.(1998). Finite element limit state solutions, accuracy of an iterative method, Internal Report, Department of Engineering, University of Leicester.

Koiter, W.T. (1960). General theorems of elastic-plastic solids. In Sneddon J.N. and Hill R., eds., *Progress in Solid Mechanics* 1:167-221.

König, J.A., (1987). *Shakedown of Elastic-Plastic Structures,* PWN-Polish Scientific Publishers, Warsaw and Elsevier, Amsterdam.

Mackenzie, D. and Boyle, J.T. (1993). A simple method of estimating shakedown loads for complex structures *Proc. ASME Pressure Vessel and Piping Conference*, Denver.

Parrinello, F. and Ponter, A. R. S. (2001). Shakedown limits based on linear solutions, for a hydrostatic pressure dependent material. *Proc. 2nd European Conference on Computational Mechanics*, Cracow, Poland, June 2001, Abstracts, ISBN83-85688-68-4, 718-719, Institute of Computer Methods in Civil Engineering, Cracow University of Technology, Poland.

Parrinello F, (2000). PhD thesis, University of Palermo, Italy.

Polizzotto, C., Borino, G., Caddemi, S and Fuschi, P. (1991). Shakedown problems for materials with internal variables. *Eur. J. Mech. A/ Solids* 10:621-639.

Ponter, A.R.S. and Carter, K.F. (1997). Limit state solutions, based upon linear elastic solutions with a statially varying elastic modulus. *Comput. Methods Appl. Mech. Engrg.* 140:237-258.

Ponter, A.R.S. and Carter, K.F., (1997). Shakedown state simulation techniques based on linear elastic solutions *Comput. Methods Appl. Mech. Engrg.* 140:259-279.

Ponter, A. R. S., Fuschi P. and Engelhardt, M. (2000). Limit analysis for a general class of yield conditions. *European Journal of Mechanics, A/Solids* 19:401-421.

Ponter, A. R. S. and Engelhardt, M. (2000). Shakedown limits for a general yield condition. *European Journal of Mechanics, A/Solids* 19:423-445

Ponter, A. R. S. and Chen H. (2001). A Minimum theorem for cyclic load in excess of shakedown, with applications to the evaluation of a ratchet limit. *Euro. Jn. Mech., A/Solids*, 20:539-553.

Ponter, A. R. S. (2001). Minimum theorems and iterative solution methods for creep cyclic loading problems, *Meccanica,* 296-8:1-11.

Riou, B., Ponter A.R.S., Carter K.F. and Guinovart J. (1997). Recent improvements in the ratchetting diagram method, Livolant M. ed. *Trans. 14th Int. Conf. on Structural Mechanics in Reactor Technology , International Ass. for Structural Mechanics in Reactor Technology*, 3:93-100.

Seshadri, R. and Fernando, C.P.D., (1991) Limit loads of mechanical components and structures based on linear elastic solutions using the GLOSS R-Node method", *Trans ASME Pressure Vessel and Piping Conference,* San Diego. 210-2:125-134.

Shakedown, Limit, Inadaptation and Post-Yield Analysis

Andrzej Siemaszko

Institute of Fundamental Technological Research, Warsaw, Poland

Abstract. An inelastic structure subjected to variable repeated loading may work either in elastic, shakedown (adaptation), inadaptation or limit regime. For the given load variation domain the safety factors against first yielding, inadaptation and limit state can be defined by the shakedown and limit analysis. The classical shakedown and limit analysis can be extended to account for nonlinear geometrical effects as well as material hardening. In this chapter the kinematic approach is described. An uniform formulation of shakedown, limit, inadaptation and post-yield analyses of discrete elastic-plastic structures is presented. Classical limit and shakedown problems are formulated as linear programs, whereas the post-yield and inadaptation as sequences of linear programs. Nonlinear geometric effects, nonlinear strain-hardening as well as material damage may be accounted for.

1 Introduction

Majority of engineering structures manifest some plastic properties. Their structural safety may be evaluated using the shakedown or limit analysis. For the given load variation domain it is possible to define a triad of safety factors against first yielding, ε, inadaptation, η, and limit state, ζ. The classical approaches are based upon assumption of geometrical linearity. Extension of the limit analysis to geometrically nonlinear one, called a post-yield analysis, were proposed among others by Duszek and Sawczuk (1976), Lodygowski (1980). Using the statical approach also the shakedown analysis has been extended towards nonlinear geometrical effects by Maier (1973), Weichert (1986), Gross-Weege (1990), Saczuk and Stumpf (1990). A different approach to nonlinear shakedown analysis, following the kinematic formulation, has been proposed by Siemaszko and König (1985, 1988, 1995). The method generalizes the post-yield analysis to the case of variable repeated loading. It is called an inadaptation analysis.

2 Working Regimes of Inelastic Structures Subjected to Variable Repeated Loading

Consider the load variation domain Δ^{ξ} defined in the n-dimensional space of applied loads. Depending on its size (volume) represented by the load domain multiplier ξ, the structure will work either in elastic, adaptation, inadaptation (ratcheting or alternating plasticity) or limit regime, see Figure 1. The shakedown and limit analysis allow to define proportional domains demarcating all these regimes: elastic domain Δ^{ε}, shakedown Δ^{η} and limit domain Δ^{ζ}. Their sizes may be represented by corresponding multipliers: elastic multiplier, ε, adaptation multiplier, η, and limit multiplier, ζ. It always holds that the adaptation multiplier is between elastic and limit multipliers (or coincides with one of them).

Comparing the given value ξ of the load domain multiplier with values of ε, η and ζ multipliers the working regime is immediately identified, see Figure 1. Normalizing the value $\xi = 1$ the multipliers ε, η and ζ will have the meaning of safety factors against first yielding, inadaptation and limit state, respectively.

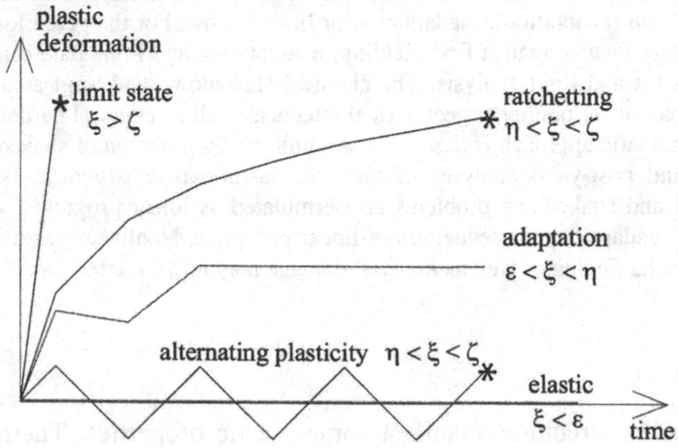

Figure 1. Working regimes of inelastic structures subjected to variable repeated loading.

Elastic multiplier. In the case of discrete structures the elastic multiplier ε can be evaluated from the first yielding problem written as

$$\varepsilon = \max\{\xi \mid \xi d \leq k\} \tag{1}$$

where k is the vector of plastic modulae defining the piece-wise linear yield condition

$$f = N^T s - k \leq 0 \tag{2}$$

and d is the vector of elastic stress envelope defined by

$$d = \max\{N^T EC(C^T EC)^{-1} p(t) = 0 \mid p(t) \in \Delta^{\xi}\} \tag{3}$$

where C is the compatibility matrix, s is the stress vector, E is the matrix of elasticity, N is the matrix of gradients of yield polyhedrons for all elements and $p(t)$ is the variable repeated loading.

Shakedown multiplier. The shakedown multiplier can be evaluated from the following linear program, (statical approach, Maier, 1970)

$$\eta = \max\left\{\xi \,\middle|\, \xi \boldsymbol{d} + \boldsymbol{N}^T \boldsymbol{s} \le \boldsymbol{k}, \quad \boldsymbol{C}^T \boldsymbol{s} = \boldsymbol{0}\right\} \tag{4}$$

where s is the residual stress.

Limit multiplier. The limit multiplier ζ_i for a certain proportionality of loads p_i^*, $i = 1, 2, \ldots, k$, k – number of critical proportionalities of loads, can be evaluated from the following linear program, (statical approach, Maier, 1970)

$$\zeta_i = \max\left\{\xi \,\middle|\, \boldsymbol{N}^T \boldsymbol{s} \le \boldsymbol{k}, \quad \boldsymbol{C}^T \boldsymbol{s} = \xi \boldsymbol{p}_i^*\right\} \tag{5}$$

where s is the total stress.

To safety considerations the minimum limit multiplier ζ should be taken

$$\zeta = \min\{\xi_i\} \tag{6}$$

Assuming the nominal load variation domain Δ^{ξ} as a reference domain $(\xi=1)$ the values of elastic multiplier ε, adaptation multiplier η, and limit multiplier ζ represent so called safety factors against first yielding, inadaptation and limit state, respectively, see Figure2.

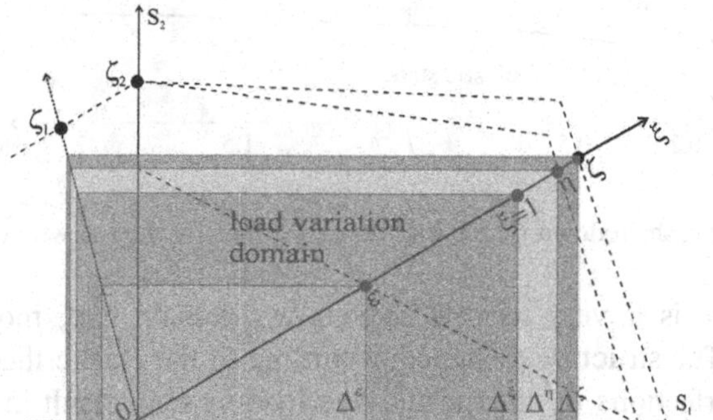

Figure 2. Load domain ξ, elastic ε, adaptation η, and limit ζ multipliers (ζ_1, ζ_2 other computed limit multipliers). The structure works in the shakedown regime $\varepsilon < \xi < \eta$.

The insight into post-critical (inelastic) behaviour can be crucial for the accurate evaluation of structural safety and reliability (cf. Bielawski et al., 1997). A mutual relation of elastic, shakedown, and limit carrying capacities can be an important indicative factor during structural design. Performing the design according to the elastic theory may lead to conservative, uneconomical designs since the shakedown reserves of strength are not exploited. However, the opposite situation is possible that the elastic approach leads to unsafe designs with a very narrow reserve of strength margins against the inadaptation or the limit state.

To discuss these problems an example of bar with a circular notch is considered, Figure 3. The radius of the notch is denoted by *r*, the radius of the bar is equal to *3a*, where *a* denotes the radius of the neck. The bar is subjected to variable tensile forces *p*. The elastic - perfectly plastic material model and the Mises yield condition are applied. The elastic, shakedown and limit analysis problems are solved.

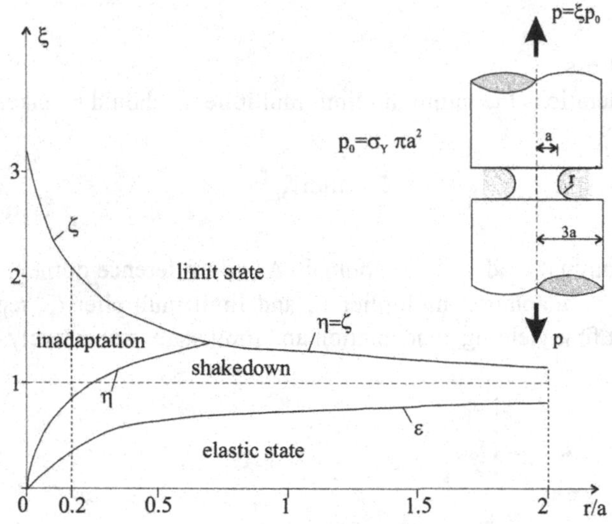

Figure 3. Elastic ε, shakedown η, and limit ζ multipliers for the rod with circular notch.

For *r/a* = *2*, there is a very narrow shakedown domain and, moreover, there is no inadaptation domain. The structure designed according to the elastic theory will have a low reliability. Some imperfections or unexpected overloading may result in a structural failure. Increasing loading by *38%* the ultimate state will be attained. For *r/a* = *0.2* the structure designed according to the elastic theory will be highly reliable, but on the other side uneconomical. To reach the ultimate state the loading must be increased *4.34* times.

For *r/a* = *0.2*, neglecting the variability of loads, the limit analysis or non-linear incremental analysis by a FE code will admit load levels exceeding *2.2* times the shakedown limit, Figure3. Even application of high safety factors, decreasing the admissible loads, will cause the inadaptation. The structure will fail after several load cycles. Therefore, the limit analysis and incremental FE analysis (with a single application of loading), providing the limit multiplier, may be inadequate for the evaluation of structural safety. This example shows that for structural assessment and design it would be desirable to develop a triad of analysis methods allowing to estimate the elastic (first yielding), shakedown, and limit multipliers.

3 Shakedown Analysis not Accounting for Geometric Effects

Let us consider an elastic-plastic discrete structure subjected to variable repeated loads. For such structures, shakedown is a widely recognized necessary condition of safe structural behaviour. It is assumed that loads $\boldsymbol{p(t)}$ may vary on arbitrary paths contained in a prescribed load domain Δ^{ξ}. The shakedown multiplier η is a maximal multiplier of the load domain which ensures shakedown. Stress, strain and displacement fields within the structure are assumed to be fully described by finite-dimensional vectors $\boldsymbol{s}$, $\boldsymbol{q}$ and $\boldsymbol{u}$, respectively.

Within the framework of geometrical linearity the fundamental kinematic shakedown theorem can be expressed as (cf. Maier, 1970)

$$\eta^0 = \min_{\dot{\lambda}^0, \dot{u}^0} \left\{ \boldsymbol{k}^T \cdot \dot{\boldsymbol{\lambda}}^0 \mid \boldsymbol{d}^T \cdot \dot{\boldsymbol{\lambda}}^0 = 1, \boldsymbol{N} \cdot \dot{\boldsymbol{\lambda}}^0 = \boldsymbol{C}(\boldsymbol{x}^0) \cdot \dot{\boldsymbol{u}}^0, \dot{\boldsymbol{\lambda}}^0 \geq \boldsymbol{0} \right\} \tag{7}$$

where $\boldsymbol{C}$ and $\boldsymbol{N}$ are rectangular matrices of compatibility and gradients of yield polyhedrons (a piece-wise linear yield condition is assumed), $\boldsymbol{k}$, $\boldsymbol{d}$ and λ are the vectors of yield moduli, elastic stress envelope and plastic multipliers, $\boldsymbol{x}^0$ is the vector of initial nodal coordinates, ξ and η are the load domain multiplier and shakedown multiplier, respectively. The initial shakedown multiplier, η^0, obtained from (7) is associated with the most stringent plastic mechanism described by $\dot{\boldsymbol{u}}^0$ and $\dot{\boldsymbol{\lambda}}^0$ which denote, respectively, the increments of plastic displacements and plastic multipliers.

For the given load variation domain Δ^{ξ}, the vector $\boldsymbol{d}$ of elastic stress envelope can be obtained as a solution of the problem

$$\boldsymbol{d} = \max\left\{ \boldsymbol{N}^T \boldsymbol{E}\boldsymbol{C}(\boldsymbol{C}^T \boldsymbol{E}\boldsymbol{C})^{-1} \boldsymbol{p}(t) = \boldsymbol{0} \,\middle|\, \boldsymbol{p}(t) \in \Delta^{\xi} \right\} \tag{8}$$

where $\boldsymbol{E}$ is the matrix of elasticity.

The problem (7) implies that the inadaptation will take place for a given load program ($\xi=1$) if we find any vector of the kinematically admissible plastic multipliers $\dot{\boldsymbol{\lambda}}^0$ for which the dissipation $\boldsymbol{k}^T \cdot \dot{\boldsymbol{\lambda}}^0$ is lower than the rate of internal work $\boldsymbol{d}^T \cdot \dot{\boldsymbol{\lambda}}^0$.

4 Limit Analysis not Accounting for Geometric Effects

In the limit analysis the rigid-perfectly plastic structures subjected to proportional loading are considered. The load $\boldsymbol{p}$ is described by a one-parametric function $\boldsymbol{p} = \xi\, \boldsymbol{p}^*$, where $\boldsymbol{p}^*$ is the reference load and ξ denotes the load multiplier. Within the kinematic approach the initial limit load multiplier ζ^0 is defined for discrete structures by the solution of the following linear program, cf. Maier (1970)

$$\zeta^0 = \min_{\dot{\lambda}^0, \dot{u}^0} \left\{ \boldsymbol{k}^T \cdot \dot{\boldsymbol{\lambda}}^0 \mid \boldsymbol{p}^{*T} \cdot \dot{\boldsymbol{u}}^0 = 1, \boldsymbol{N} \cdot \dot{\boldsymbol{\lambda}}^0 = \boldsymbol{C}(\boldsymbol{x}^0) \cdot \dot{\boldsymbol{u}}^0, \dot{\boldsymbol{\lambda}}^0 \geq \boldsymbol{0} \right\} \tag{9}$$

where $\boldsymbol{C}$ and $\boldsymbol{N}$ are, respectively, the matrices of compatibility and gradients of yield polyhedrons, $\boldsymbol{k}$ is the vector of plastic modulae, and $\boldsymbol{x}^0$ denotes the initial undeformed structural configuration. The initial limit load multiplier, ζ^0, obtained from (9) is associated with the most stringent plastic mechanism described by $\dot{\boldsymbol{u}}^0$ and $\dot{\boldsymbol{\lambda}}^0$ which denote, respectively, the increments of plastic displacements and plastic multipliers.

5 Inadaptation Analysis

The majority of engineering structures suffer from destabilization of the post-yield or shakedown processes due to geometrically nonlinear effects (geometrical softening) (cf. Siemaszko, 1988). On the other hand, most of structural materials manifest some hardening properties. If hardening effects prevail, we can obtain a stable post-critical structural behaviour. The influence of both these factors on structural behaviour can be determined by the inadaptation analysis proposed by Siemaszko and König (1985, 1988). It is a step-by-step procedure based upon a kinematic approach, which allows to determine the most stringent plastic deformation path of elastic-strain hardening structures.

The procedure may be applied under the following assumptions:

- Small displacements and strains due to a single load cycle
- Kinematically inadmissible plastic strains are small in comparison with kinematically admissible ones.

These assumptions seem to be realistic for the majority of sufficiently stiff engineering structures. The first assumption allows to use the linear formulation of shakedown problem on each step of the procedure. The second one means that the structural plastic deformation is localized in plastic hinges and, therefore, straight (undeformed) elements can be used for analysis.

The shakedown problem (7) is formulated with an assumption of geometric linearity. To account for the non-linear geometrical effects the most stringent mechanism, $\dot{\boldsymbol{u}}^0$, determined for the initial structural configuration, $\boldsymbol{x}^0$, is used to define a certain deformed configuration $\boldsymbol{x}^\delta$

$$\boldsymbol{x}^\delta = \boldsymbol{x}^0 + \mu\dot{\boldsymbol{u}}^0 \tag{10}$$

where μ is a scale factor, $\boldsymbol{x}^0$ and $\boldsymbol{x}^\delta$ are vectors of nodal coordinates, see Figure3.

Then, to account for the non-linear geometrical effects, the classical shakedown analysis problem is solved for the deformed configuration $\boldsymbol{x}^\delta$. As a result the reduced shakedown load multiplier ζ^δ can be obtained. It is defined by the following two-level linear program, (cf. Siemaszko, 1988):

$$\begin{cases} \eta^\delta = \min\limits_{\dot{\lambda}^\delta, \dot{u}^\delta} \left\{ \boldsymbol{k}^T \cdot \dot{\boldsymbol{\lambda}}^\delta \mid \boldsymbol{d}^T \cdot \dot{\boldsymbol{\lambda}}^\delta = 1, \quad \boldsymbol{N} \cdot \dot{\boldsymbol{\lambda}}^\delta = \boldsymbol{C}(\boldsymbol{x}^\delta) \cdot \dot{\boldsymbol{u}}^\delta, \quad \dot{\boldsymbol{\lambda}}^\delta \geq \boldsymbol{0} \right\} \\ \boldsymbol{x}^\delta = \boldsymbol{x}^0 + \mu \cdot \dot{\boldsymbol{u}}^0 \\ \eta^0 = \min\limits_{\dot{\lambda}^0, \dot{u}^0} \left\{ \boldsymbol{k}^T \cdot \dot{\boldsymbol{\lambda}}^0 \mid \boldsymbol{d}^T \cdot \dot{\boldsymbol{\lambda}}^0 = 1, \quad \boldsymbol{N} \cdot \dot{\boldsymbol{\lambda}}^0 = \boldsymbol{C}(\boldsymbol{x}^0) \cdot \dot{\boldsymbol{u}}^0, \quad \dot{\boldsymbol{\lambda}}^0 \geq \boldsymbol{0} \right\} \end{cases} \tag{11}$$

where η^δ and η^0 are the reduced and initial shakedown multipliers, respectively.

If $\eta^\delta < \eta^0$ then the geometric softening takes place. It means that due to the initial elastic deformation the inadaptation (non-shakedown) will occur for loads indicated as safe by the classical linear shakedown analysis. On the contrary, for $\eta^\delta > \eta^0$, the geometric hardening will occur. It means that the structure may work safely under loads exceeding the shakedown loads defined by the linear approach.

The value of the multiplier η^δ is dependent on the value of scale factor μ. To calibrate it Siemaszko (1988) has proposed to use some admissible displacements u^*

$$max\{\mu\dot{u}^0\} \leq u^* \tag{12}$$

The value of u^* should represent the elastic deformation of a structure at the limit state. For simplicity, this value can be specified according to admissible displacements recommended by design standards, for example, as a deflection equal to $L/350$ for a beam of the span L or a horizontal displacement equal to $H/1000$ for a frame of the height H.

A step-by-step repetition of the procedure described (with small plastic deformation increments) allows to define the most stringent plastic deformation path called an inadaptation path, see Figure 4. It is fully described by the displacement increments $\dot{\boldsymbol{u}}$ and increments of plastic multipliers $\dot{\boldsymbol{\lambda}}$. Since the plastic strain history along the inadaptation path is known, effects of hardening and damage can be considered. Updating appropriately plastic moduli an arbitrary strain-hardening law as well as an arbitrary strain-dependent evolution equation of the damage parameter can be accounted for.

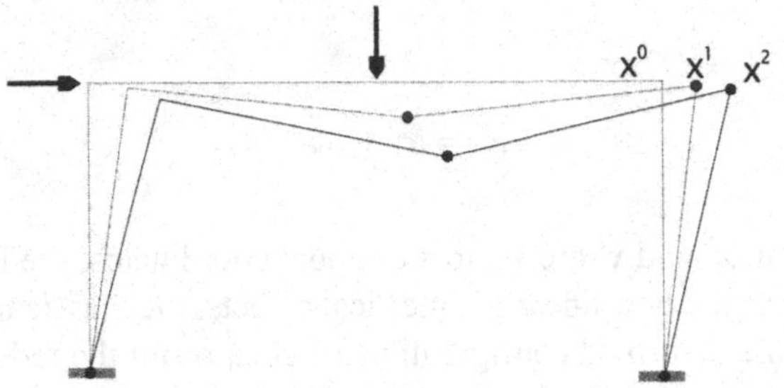

Figure 4. Subsequent deformed configurations.

For the description of elastic-plastic hardening-softening material we may use three internal state variables: backstress α, equivalent plastic strain $\in$ and damage parameter π. The inadaptation analysis accounting for nonlinear geometric effects, nonlinear hardening and damage may be formulated as the following sequence of linear programs (Siemaszko, 1993)

$$\begin{cases} \eta(\boldsymbol{x}^0) = \min\limits_{\dot{\boldsymbol{\lambda}}^0,\dot{\boldsymbol{u}}^0} \left\{ \boldsymbol{k}^T(\boldsymbol{\alpha}^0, \in^0, \boldsymbol{\pi}^0) \cdot \dot{\boldsymbol{\lambda}}^0 \mid \boldsymbol{d}^T(\boldsymbol{x}^0) \cdot \dot{\boldsymbol{\lambda}}^0 = 1, \quad \boldsymbol{N} \cdot \dot{\boldsymbol{\lambda}}^0 = \boldsymbol{C}(\boldsymbol{x}^0) \cdot \dot{\boldsymbol{u}}^0, \quad \dot{\boldsymbol{\lambda}}^0 \geq \boldsymbol{0} \right\} \\ \hline \boldsymbol{x}^\delta = \boldsymbol{x}^{\delta-1} + \mu^{\delta-1} \cdot \dot{\boldsymbol{u}}^{\delta-1} \qquad \delta = 1,2,\ldots \\ \boldsymbol{\alpha}^\delta = c_K \sum\limits_{i=0}^{\delta-1} \mu^i \boldsymbol{N}^T \boldsymbol{N} \dot{\boldsymbol{\lambda}}^i \\ \in^\delta = \sum\limits_{i=0}^{\delta-1} \mu^i \left| \boldsymbol{N}^T \boldsymbol{N} \right| \dot{\boldsymbol{\lambda}}^i \\ \boldsymbol{\pi}^\delta = \pi(\varepsilon^\delta) \\ \eta(\boldsymbol{x}^\delta) = \min\limits_{\dot{\boldsymbol{\lambda}}^\delta,\dot{\boldsymbol{u}}^\delta} \left\{ \boldsymbol{k}^T(\boldsymbol{\alpha}^\delta, \in^\delta, \boldsymbol{\pi}^\delta) \cdot \dot{\boldsymbol{\lambda}}^\delta \mid \boldsymbol{d}^T(\boldsymbol{x}^\delta) \cdot \dot{\boldsymbol{\lambda}}^\delta = 1, \quad \boldsymbol{N} \cdot \dot{\boldsymbol{\lambda}}^\delta = \boldsymbol{C}(\boldsymbol{x}^\delta) \cdot \dot{\boldsymbol{u}}^\delta, \quad \dot{\boldsymbol{\lambda}}^\delta \geq \boldsymbol{0} \right\} \end{cases} \tag{13}$$

where $\boldsymbol{x}^0$ is the vector of nodal coordinates for the initial undeformed structural configuration, $\boldsymbol{x}^\delta$ are vectors of the subsequent deformed configurations, μ is the small control parameter, c_K is the kinematical hardening modulus, α^δ, $\in^\delta$ and π^δ are vectors of internal state variables corresponding to generalized backstresses, equivalent plastic strains and damage parameters, respectively.

Vectors of internal state variables can be derived from the material function postulated by Perzyna (1984) for one-dimensional case as

$$\kappa(\alpha,\in,\pi) = [\kappa_0 + \alpha + \bar{\kappa}(\in)\in]\left[1 - \left(\frac{\pi(\in)}{\pi_F}\right)^{1/2}\right] \tag{14}$$

where κ_0 is an initial yield point, α is the backstress parameter defining kinematical hardening, $\bar{\kappa}(\in)\in$ is the nonlinear isotropic hardening function dependent on eqivalent plastic strain $\in$, $\pi(\in)$ is the porosity parameter function describing material damage and π_F is the critical porosity (fracture point).

From (14) it follows that loads tend to zero at the fracture point π_F . Hardening and damage parameters derived from (14), necessary for the solution of (13), are presented in (Siemaszko, 1995).

The function $\eta(x)$, obtained from (13), is called an inadaptation curve. Figure 5 presents the inadaptation curves for an example of two story frame subjected to variable repeated loading. When the curve is growing the inadaptation process is stabilized by geometrical and/or material hardening (or material hardening may prevail over the geometric softening). In this case it is possible that the structure will shake down for the load level higher than indicated by the classical shakedown analysis, see curve "b", Figure 5. The decreasing function $\eta(x)$ means the unstable inadaptation process. This is an effect of prevailing geometric softening, see curve "a", Figure 5, or material damage, cf. the final part of the curve "b". In the latter case the localization effects can be observed.

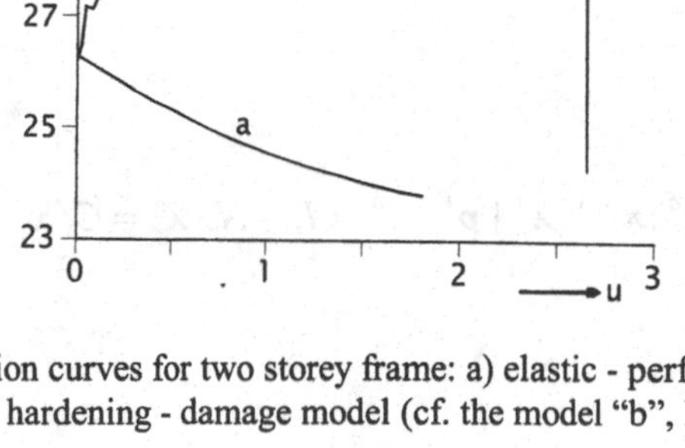

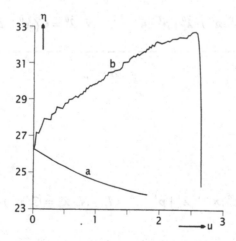

Figure 5. Inadaptation curves for two storey frame: a) elastic - perfectly plastic model; b) eastic - hardening - damage model (cf. the model "b", Figure 7).

6 Post-Yield Analysis

Similarly to the problem (11) the post-yield problem can be formulated. To account for the non-linear geometrical effects the classical limit analysis problem is solved for the deformed configuration x^{δ}. Therefore, the reduced limit load multiplier ζ^{δ} is defined by the following two-level linear program, cf. Siemaszko (1988)

$$\begin{cases} \zeta^{\delta} = \min\limits_{\dot{\lambda}^{\delta}, \dot{u}^{\delta}} \left\{ k^T \cdot \dot{\lambda}^{\delta} \mid p^{*T} \cdot \dot{u}^{\delta} = 1, \quad N \cdot \dot{\lambda}^{\delta} = C(x^{\delta}) \cdot \dot{u}^{\delta}, \quad \dot{\lambda}^{\delta} \geq 0 \right\} \\ x^{\delta} = x^{0} + \mu \cdot \dot{u}^{0} \\ \zeta^{0} = \min\limits_{\dot{\lambda}^{0}, \dot{u}^{0}} \left\{ k^T \cdot \dot{\lambda}^{0} \mid p^{*T} \cdot \dot{u}^{0} = 1, \quad N \cdot \dot{\lambda}^{0} = C(x^{0}) \cdot \dot{u}^{0}, \quad \dot{\lambda}^{0} \geq 0 \right\} \end{cases} \tag{15}$$

If $\zeta^{\delta} < \zeta^{0}$ then the geometric softening takes place. It means that due to the initial elastic deformation the limit state may occur for loads indicated as safe by the classical limit analysis. The load level corresponding to ζ^{0} will be never attained since the failure may take place at ζ^{δ}. On the contrary, for $\zeta^{\delta} > \zeta^{0}$, the geometric hardening occurs. It means that the structure may work under loads exceeding the limit loads defined by the linear approach.

The problem (15) can be generalized to account for material hardening and damage. Carrying out the step-by-step analysis the post-yield dependence of limit multiplier ζ on changes of geometry, strain-hardening and damage can be determined from the following sequence of linear programs (Siemaszko, 1995)

$$\begin{cases} \zeta(x^{0}) = \min\limits_{\dot{\lambda}^{0}, \dot{u}^{0}} \left\{ k^T(\alpha^{0}, \in^{0}, \pi^{0}) \cdot \dot{\lambda}^{0} \mid p^T \cdot \dot{u}^{0} = 1, \quad N \cdot \dot{\lambda}^{0} = C(x^{0}) \cdot \dot{u}^{0}, \quad \dot{\lambda}^{0} \geq 0 \right\} \\ \text{---} \\ x^{\delta} = x^{\delta-1} + \mu^{\delta-1} \cdot \dot{u}^{\delta-1} \qquad \delta = 1,2,\ldots \\ \alpha^{\delta} = c_K \sum\limits_{i=0}^{\delta-1} \mu^{i} N^T N \dot{\lambda}^{i} \\ \in^{\delta} = \sum\limits_{i=0}^{\delta-1} \mu^{i} \left| N^T N \right| \dot{\lambda}^{i} \\ \pi^{\delta} = \pi(\varepsilon^{\delta}) \\ \zeta(x^{\delta}) = \min\limits_{\dot{\lambda}^{\delta}, \dot{u}^{\delta}} \left\{ k^T(\alpha^{\delta}, \in^{\delta}, \pi^{\delta}) \cdot \dot{\lambda}^{\delta} \mid p^T \cdot \dot{u}^{\delta} = 1, \quad N \cdot \dot{\lambda}^{\delta} = C(x^{\delta}) \cdot \dot{u}^{\delta}, \quad \dot{\lambda}^{\delta} \geq 0 \right\} \end{cases} \tag{16}$$

If the obtained post-yield function $\zeta(x)$ is growing, it denotes a stable post-yield behaviour, i.e. the structure may still carry loads despite that the loading exceeds a level indicated as ultimate by the classical analysis. The decreasing function $\zeta(x)$ represents the unstable post-yield behaviour. It means that the structural collapse may take place for loads level lower than indicated by the classical analysis.

The post-yield and inadaptation analysis provide much more information necessary for the safe structural design than the classical limit and shakedown analysis. They define the most stringent post-yield and inadaptation deformation paths described by most dangerous plastic flow mechanisms as well as they indicate changes of these mechanisms and the way in which the structure will collapse. For stable behaviour it is possible to define the maximum load multiplier accounting for geometric and material hardening. For unstable behaviour it is reasonable to account for the sensitivity of the limit load or shakedown with respect to changes of geometry as an additional design parameter.

7 Example

Let us perform the post-yield and inadaptation analysis of the five story frame, Figure 6, subjected to six independent variable repeated load systems P_i, $i=1, 2, ..., 6$. For the inadaptation analysis it is assumed that each load may vary in the range of $P_i \in [0,1]$. For the post-yield analysis each load is equal to $P_i = 1$. Four different cross-sectional area F_i, $i=1, ..., 4$, are applied with the rhombic yield condition defined in the bending moment – axial force space. They have the following values $F = \{8 \cdot 10^{-3}, 6 \cdot 10^{-3}, 5 \cdot 10^{-3}, 4 \cdot 10^{-3}\}^T$.

Let us consider the elastic – linear isotropic and kinematic hardening material with hardening moduli $c_i = c_K = 10^6$, cf. line "a", Figure 7. This linear hardening model is combined with the piece-wise linear damage function and the resulting hardening – damaging material is presented as the curve "b", Figure 7.

The classical limit and shakedown analysis provide the initial limit multiplier $\zeta^0 = 23.3$, and the initial shakedown multiplier $\eta^0 = 20.7$, cf. Figure 8. These results do not inform on post-critical structural behaviour, in particular, on influence of the geometry changes and sensitivity to imperfections. These questions can be answered by the post-yield and inadapatation analysis. In both cases the structural plastic deformation proceeds according the same side-sway mechanism and, therefore, similar post-yield "ζ" and inadaptation "η" increasing curves are obtained, see Figure 8. It is seen that material hardening prevail over the geometric softening and a stable structural behaviour is observed. The maximum limit and shakedown multipliers, $\zeta_{max} = 36.2$, $\eta_{max} = 33.4$, are significantly greater than the initial multipliers. In this case, results of the classical analysis were on the safe side and as such recommendable for structural design.

By a detailed analysis of the plastic deformation process, Figures. 5 and 8, a phenomenon of changes of instantaneous side-sway mechanisms can be observed. Next, at the maximum values of load multipliers, the phenomenon of plastic strain localization at the support nodes shows up. The localization is observed immediately after the material of these nodes enters into the softening range corresponding to the decreasing part of material function "b", Figure 8. Despite significant reserves of strength at other nodes, only these two softening ones determine the unstable behaviour of the structure which eventually leads to the structural failure.

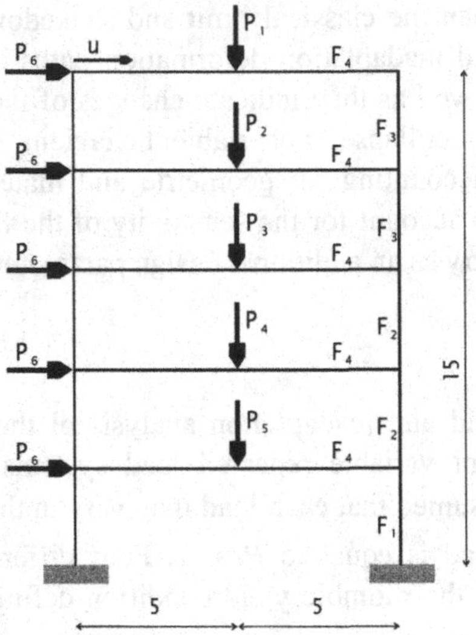

Figure 6. Five story frame subjected to six variable repeated load sets.

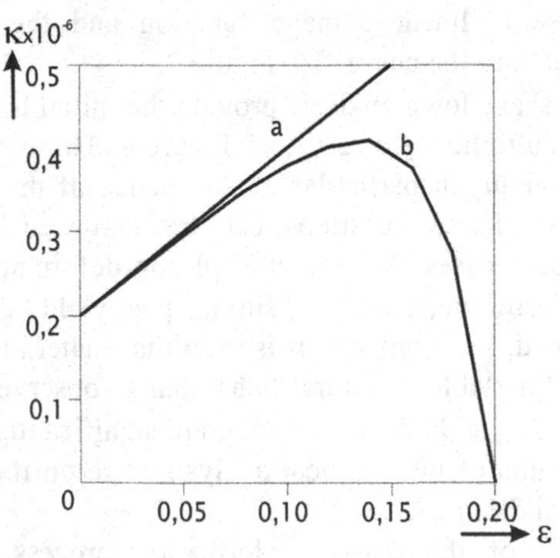

Figure 7. Material model: a) linear kinematic and isotropic hardening, b) linear hardening combined with nonlinear damage function.

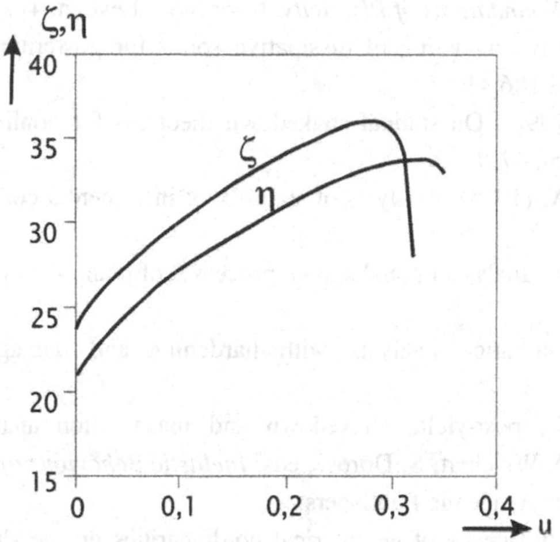

Figure 8. Post yield and inadaptation curvesfor the six story frame.

The post-yield and inadaptation analyses provide much more information necessary for the safe structural design than the classical limit and shakedown analyses. They define the most stringent plastic deformation paths, describe the most dangerous plastic mechanisms, changes of the instantaneous mechanisms and the kind of structural collapse. For the stable structural bahaviour it is possible to determine the maximum load multiplier which accounts for the nonlinear geometric effects, material hardening and damage. For the unstable behaviour, sensitivity of the load multiplier with respect to structural deformation can be defined. Higher (negative) sensitivities indicate a dangerous discrepancy between computed and real safety level. In this case the classical (initial) load multiplier should be reduced to avoid an overestimation of structural safety.

References

Bielawski, G., Siemaszko, A. and Zwolinski, J., (1997). An adaptation and limit analysis method in the design of inelastic structures. *Trans. SMiRT*, 14, 3, FW/4, 293–300.

Duszek, M. and Sawczuk, A. (1976). Stable and unstable states of rigid-plastic frames at the yield point load. *J.Struct.Mech.* 4:33–47.

Gross-Weege, J. (1990). A unified formulation of statical shakedown criteria for geometrically nonlinear problem. *Int.J.Plast.* 6:433.

Lodygowski, T. (1982). Geometrically nonlinear analysis of rigid plastic beams and plane frames. *IFTR Reports*, 9 (in Polish).

Maier, G. (1970). A matrix structural theory of piecewise linear elastoplasticity with interacting yield planes. *Meccanica* 5:54–66.

Maier, G. (1973). A shakedown matrix theory allowing for work-hardening and second-order geometric effects. A.Sawczuk, ed., *Foundations of Plasticity*. Noordhoff, Leyden, 417.

Perzyna, P., (1984). Constitutive modeling of dissipative solids for postcritical behaviour and fracture. *J.Eng.Mat.Techn.*, ASME 106:410.

Saczuk, J. and Stumpf, H. (1990). On statical shakedown theorems for nonlinear problems. Inst.Mech., Ruhr-Universität, Bochum, 47, 1.

Siemaszko, A. and König, J.A. (1985). Analysis of stability of incremental collapse of skeletal structures. *J.Struct.Mech.* 13:301.

Siemaszko, A. (1988). Stability analysis of shakedown processes of plane skeletal structures. *IFTR Reports* 12:1–176 (in Polish).

Siemaszko, A. (1993). Inadaptation analysis with hardening and damage. *Eur.J.Mech., A/Solids* 12:237–248.

Siemaszko, A. (1995). Limit, post-yield, shakedown and inadaptation analysis of inelastic discrete structures. In Z. Mróz, D. Weichert, S. Dorosz, eds. *Inelastic Behaviour of Structures under Variable Loads*. Dordrecht: Kluwer Academic Publishers.

Weichert, D., (1986). On the influence of geometrical nonlinearities on the shakedown of elastic-plastic structures. *Int.J.Plast.* 2:135.

Limit and Shakedown Reliability Analysis

Andrzej Siemaszko

Institute of Fundamental Technological Research, Warsaw, Poland

Abstract. In this chapter problems of load modelling for cyclic and variable repeated loading are discussed. Then, the general problem of time-invariant shakedown and limit reliability analysis in the load space is formulated. Also, the time-variant limit reliability analysis approach is shown. For load cases with low variance of maximum values, simplified formulations of the shakedown and limit reliability analysis methods are presented. Advantages of the presented methods are discussed in the conclusions.

1 Introduction

Inelastic structures such as for example pressure vessels and pipelines subjected to variable repeated/cyclic loading may work in four different regimes: elastic, shakedown (adaptation), inadaptation (non-shakedown), and limit (ultimate) state. Since for the elastic regime there are no plastic effects at all, whereas for the adaptation regime plastic effects are restricted to the initial loading cycles and then they are followed by exclusively elastic behavior, both regimes are considered as safe working regimes and they constitute a foundation for the structural design. They are incorporated explicitly to the design standards such as BS5500 and ASME VIII for pressure vessels. The inadaptation phenomena such as low cycle fatigue and/or ratcheting should be avoided since they lead to a structural failure. For the limit state the structure looses instantaneously its load bearing capacity.

For the global elastic-plastic analysis methods the loading history is insignificant (cf. König, 1987). These methods require only specification of limit values of variability of each loading. On this basis, some asymptotic structural behaviour can be evaluated falling into either elastic, shakedown, inadaptation or limit case. Corresponding to these states we have three important global structural failure criteria: first yielding, inadaptation and limit load. The evaluation of parameters of the working regime is crucial for the design as well as assessment of safety and reliability of existing structures. For a given load variation domain the safety factors (multipliers) against inadaptation and limit state can be computed.

Loading as well as material and geometrical parameters should be considered as random variables. Then, the structural limit and shakedown reliability problems may be formulated. Probabilities of failure (or corresponding reliability indices) with respect to inadaptation (non-shakedown) or limit state can be computed. It is worth to notice that they represent the system reliability and may be obtained in a single analysis without necessity of considering multiple componental reliability subproblems.

Since the loading is the most significant random factor influencing the structural reliability its proper description is of great importance. Usually, the loading is idealized as time-invariant, i.e. as random variables. In other words, the reliability so obtained corresponds to that under one load application though it may represent some extreme value of the load over the given period of time.

As most loads fluctuate in time and may act individually or in combination with other loads, the reliability problem may be significantly different from those with time-invariant loads. For example, the sequence and path of load application may be important; also, the failure (collapse) of a system may be a culmination of progressive failures of members over a long period of time than a sudden failure of all members at one time. Such considerations suggest that the time domain fluctuations of loadings and their interaction with the system be considered by the time-variant structural reliability analysis (Wen and Chen, 1989, Melchers, 1995).

However, there is an approach to structural analysis for which the time-invariant reliability approach is fully applicable, namely the shakedown analysis. For this analysis a sequence of loads is not important. The structure has a memory of all maximum stresses and for any repetition of loading it reacts elastically. The shakedown conditions are derived from analysis of cyclic loading or variable repeated loading contained in so called load variation domain.

The shakedown reliability corresponds to one application of a complex variable repeated loading path from the load variation domain. A single shakedown reliability analysis may be accompanied by a series of limit reliability analyses defined for the most critical load combinations resulting in dominant failure mechanisms. Each limit reliability value so obtained is conditioned on a certain proportionality of extreme loads.

2 Loading Modelling

There is usually a certain possibility of accurate modelling of history of cyclic loads which may be encountered in some technological installations such as pressure vessels/reactors. Unfortunately, for majority of structures, the exact modelling is rather impossible. Therefore, it is reasonable to consider some envelopes of loading.

Most of structural variable repeated/cyclic loadings $S(t)$ can be enveloped by a pulse loading $\boldsymbol{S_{max}}$ corresponding to maximum values. The probability density functions of the cyclic loading f_S and its maximum values $f_{S_{max}}$ are shown in Figure 1. It is worth to notice that for cyclic loading which appears in some technological processes the maximum value distribution $\boldsymbol{S_{max}}$ will be characterized by a very low variance.

In the multidimensional loading space $\boldsymbol{S}$ the exact loading path, cf. Figure 2a, can be enveloped by the load variation domain Δ^{ξ}, cf. Figure 2b.

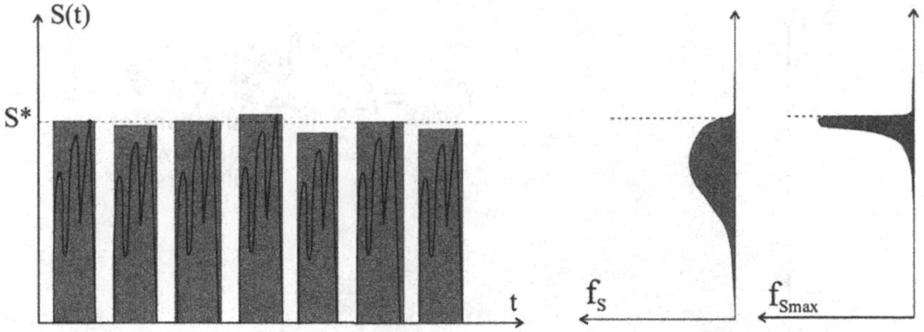

Figure1. Cyclic loading S and its envelope by the pulse loading S_{max}. Probability density functions of the loading f_S and of its maximum values f_{Smax}.

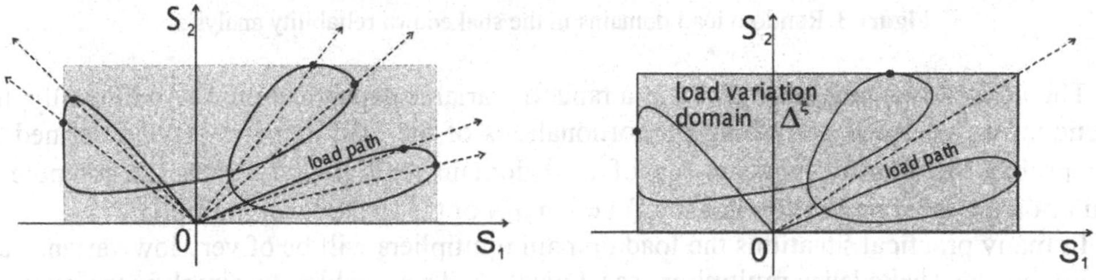

Figure 2. a) Load path with critical points; b) Corresponding load variation domain Δ^{ξ}.

3 Shakedown Reliability Analysis for Low Variance Maximum Loads

Consider cyclic loading with low variance of maximum loads, cf. Figure 1. As it has been already discussed, see Figure 2, each cycle of loads can be enveloped by the load variation domain Δ^{ξ}. For purpose of the shakedown analysis it is suitable to distinguish one of vertices of Δ^{ξ} as a control point, usually it should be the vertex corresponding to upper limits of variation of componental loads S_{max}. It is assumed that the joint probability density function f_{Smax} of all realizations of load variation domains is known, see Figure 3.

In the shakedown analysis the polar coordinates $(\xi, \boldsymbol{a})$ are used corresponding to the load domain multiplier Ξ (measured on the main diagonal of Δ^{ξ}) and the vector of cosines A of the main diagonal. By a suitable transformation from f_{Smax} it is easy to obtain a joint probability density function $f_{\Xi,A}$.

In the shakedown reliability analysis we have to account for randomness not only of loading but also of the shakedown multiplier. These problems (with respect to the limit state) were firstly considered by Augusti and Baratta (1972). Let us identify a number of variables by which the geometrical and material uncertainties related to the reliability of the structure can be described satisfactorily. They form a basic variable vector $\boldsymbol{X}$.

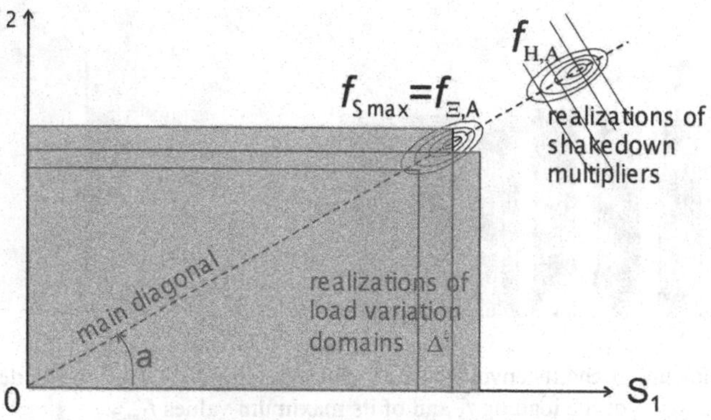

Figure 3. Random load domains in the shakedown reliability analysis.

The shakedown multiplier H(*X*) is a random variable dependent on ***X***. Additionally, it is dependent on vector ***A*** describing proportionalities of the load domain. Having defined the joint probability density function $f_{\Xi,A}$ of load domain multiplier it is easy to compute by simulation the joint probability density function $f_{H,A}$ of the shakedown multiplier.

In many practical situations the load domain multipliers will be of very low variance and so will be the shakedown multiplier, see Figure 3. This enables to simplify the analysis. Instead of full reliability analysis for all vectors ***A*** the analysis conditioned on a certain dominant direction $\bar{a}$ can be carried out. The representative radial (diagonal) direction may be selected as a median of the $f_{\Xi,A}$ function. Then, the one-dimensional probability density function $f_{\Xi|\bar{a}}$ will be used in analysis, see Figure4. Consequently, the one-dimensional conditional probability density function $f_{H|\bar{a}}$ of the shakedown multiplier must be used.

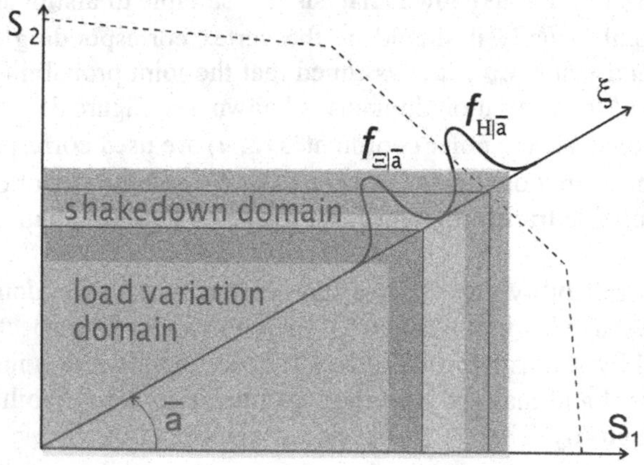

Figure 4. Shakedown reliability analysis problem conditioned on $\bar{a}$.

Consider the conditional shakedown reliability analysis defined for the dominant direction $\bar{a}$. Since in the framework of the shakedown (and limit analysis) the loading history has no influence on the structural resistance, the shakedown reliability problem may be presented as dependent only on two random variables: the random load domain multiplier Ξ and the shakedown multiplier H(*X*), see Figure 4. For brevity in the following expressions the condition $|\bar{a}$ is omitted. In the space of realizations $\boldsymbol{x}$ and ξ of random variables $\boldsymbol{X}$ and Ξ, respectively, the shakedown failure function g^{η} is written as follows

$$g^{\eta}(\boldsymbol{x},\xi)=\eta(\boldsymbol{x})-\xi \tag{1}$$

The failure function (1) divides the space of parameters into the safe domain, D_S^{η}, for which $g^{\eta}(\boldsymbol{x},\xi)\geq 0$, and the failure one D_F^{η}, where $g^{\eta}(\boldsymbol{x},\xi)<0$, see Figure 5.
Since the failure function (1) has a form of safety margin with only two random variables, the probability of failure with respect to inadaptation p_F^{η} can be determined by

$$p_F^{\eta}=\int_{-\infty}^{\infty}F_H(\xi)f_{\Xi}(\xi)d\xi \tag{2}$$

where f_{Ξ} is the probability density function of load domain multiplier, F_H is the cumulative probability function of shakedown multiplier.

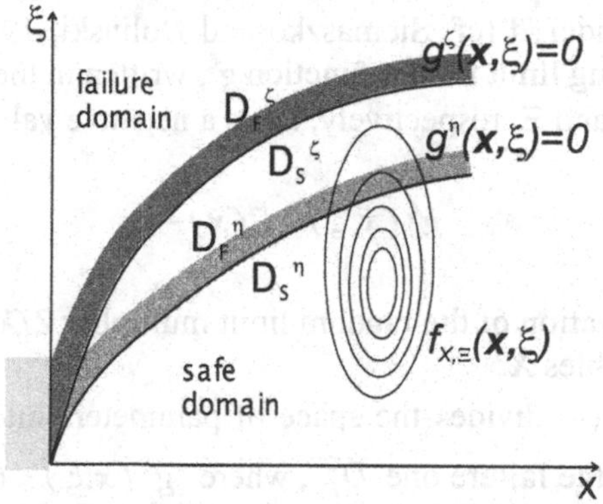

Figure 5. Safe and failure domains in the shakedown and limit reliability analysis.

If functions under integral (2) are normally distributed then probability can be exactly computed. Otherwise, standard numerical integration techniques or Monte Carlo simulation (MCS) may be used.

In another formulation, the probability of failure with respect to inadaptation can be evaluated from

$$p_F^{\eta} = \int_{D_F^{\eta}} f_{X,\Xi}(x,\xi)\,dx\,d\xi \tag{3}$$

where $f_{X,\Xi}(x,\xi)$ is the known joint probability density function of parameters X and Ξ, and D_F^{η} defines the failure domain.

To solve the problem (3) approximation techniques like Monte Carlo simulation (MCS) or the analytically based FORM method may be used. Recently, FORM approach was used by Heitzer and Staat (1999) and Siemaszko and Knabel (2000) in solving practical shakedown reliability problems.

4 Time-Invariant Limit Reliability Analysis for a Given Load Proportionality

The limit reliability analysis conditioned on a certain dominant direction $\bar{a}$ may be considered as a complementary to the shakedown reliability analysis. On the contrary to a single shakedown analysis, it may be necessary to carry out several limit reliability analyses conditioned on a number of critical load proportionalities, see Figure 2a. For brevity in the following expressions the condition $|\bar{a}$ is omitted.

In the limit reliability problem, randomness of both the load multiplier and limit multipliers must be considered (cf. Siemaszko and Dolinski, 1996). The limit state will be attained when the following limit failure function g^{ζ}, written in the space of realizations x and ξ of random variables X and Ξ, respectively, takes a negative value

$$g^{\zeta}(x,\xi) = \zeta(x) - \xi \tag{4}$$

where ζ is a certain realization of the random limit multiplier $Z(X)$ dependent of material and geometrical random variables X.

The failure function (4) divides the space of parameters into the safe domain, D_S^{ζ}, for which $g^{\zeta}(x,\xi) \geq 0$, and the failure one D_F^{ζ}, where $g^{\zeta}(x,\xi) < 0$, see Figure 5.

For a given load proportionalities, described by the angle $\bar{a}$, the probability of failure with respect to limit state p_F^{ζ} can be determined by, see Figure 6,

$$p_F^{\zeta} = \int_{-\infty}^{\infty} F_Z(\xi) f_{\Xi}(\xi)\,d\xi \tag{5}$$

where f_{Ξ} is the probability density function of load multiplier, F_Z is the cumulative probability function of limit multiplier. It can be computed exactly when both variables have normal distribution.

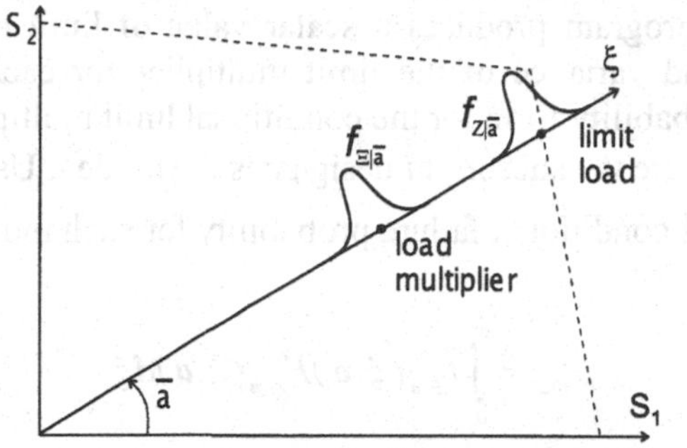

Figure 6. Limit reliability analysis problem conditioned on $\bar{a}$.

In another formulation, the probability of failure with respect to limit state can be evaluated from

$$p_F^{\zeta} = \int_{D_F^{\zeta}} f_{X,\Xi}(x,\xi)\,dx\,d\xi \tag{6}$$

where $f_{X,\Xi}(x,\xi)$ is the known joint probability density function of parameters X and Ξ, and D_F^{ζ} describes the failure domain defined by $g^{\zeta}(x,\xi)<0$, see Figure 5 .

5 Time-Invariant Limit and Shakedown Reliability Analysis

In some cases, instead of the conditional limit reliabilities defined for some critical load proportionalities it may be convenient to compute the total limit or shakedown reliability (probability of failure).

Consider a particular load vector, $\boldsymbol{S}$, defined by the load multiplier Ξ and the directional cosines $\boldsymbol{A}$. It will cross the limit state in a particular failure mode region, say mode i. The probability of failure with respect to the limit state for the assumed proportionalities is the probability that the load multiplier Ξ exceeds the limit multiplier Z

$$p_{f|a} = P[\Xi > Z] = \int_0^{\infty} f_{\Xi|a}(\xi,a)F_{Z|a}(\xi,a)\,d\xi \tag{7}$$

where the load multiplier and the limit multiplier have the conditional probability density functions $f_{\Xi|a}$, and $f_{Z|a}$, respectively, defined along the radial direction $\boldsymbol{A} = \boldsymbol{a}$.

The distribution of the limit multiplier Z is not known directly, but can be obtained from simulations using the linear programs of limit analysis (Corotis and Nafday, 1989). These simulations will lead to the dominant failure modes associated with ranges of load and limit multipliers. It is convenient to group the simulations according to the mode of failure. Solution of each linear program produces a scalar value of limit multiplier. Therefore, by simulations the mean and variance of the limit multiplier for each mode may computed. Selecting a particular probability form for the conditional limit multiplier provides a complete specification for $f_{Zi|\boldsymbol{a}}$, where the subscript i designates the mode i. Using this approach allows to compute an estimate of conditional failure probability for each mode

$$p_{fi|\boldsymbol{a}} = \int_0^\infty f_{\Xi|\boldsymbol{a}}(\xi,\boldsymbol{a})F_{Zi|\boldsymbol{a}}(\xi,\boldsymbol{a})d\xi \tag{8}$$

This may be converted from a conditional probability to a joint one by introducing the multivariate density function, $f_A(\boldsymbol{a})$

$$p_{fi,\boldsymbol{a}} = \int_0^\infty f_{\Xi|\boldsymbol{a}}(\xi,\boldsymbol{a})F_{Zi|\boldsymbol{a}}(\xi,\boldsymbol{a})f_A(\boldsymbol{a})d\xi = \int_0^\infty f_{\Xi,\boldsymbol{a}}(\xi,\boldsymbol{a})F_{Zi|\boldsymbol{a}}(\xi,\boldsymbol{a})d\xi \tag{9}$$

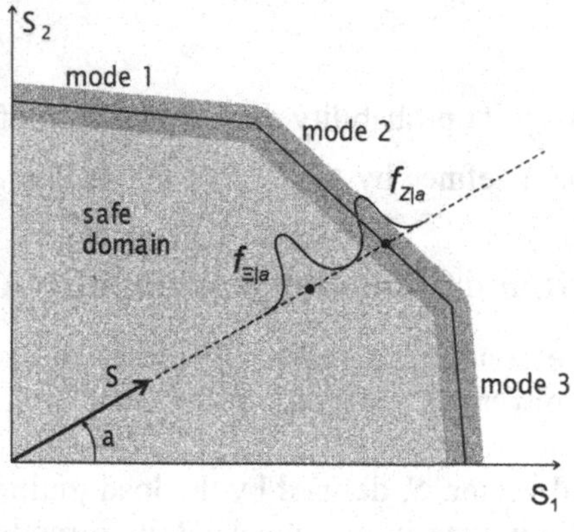

Figure 7. Limit reliability analysis.

For computational efficiency, Corotis and Nafday (1989) suggested that an approximation may be introduced by selecting a representative vector, $\boldsymbol{a}_i$, for each mode (the average from the simulations can be used). The problem then simplifies to

$$p_{fi,\boldsymbol{a}} = \int_0^\infty f_{\Xi,\boldsymbol{a}_i}(\xi,\boldsymbol{a}_i)F_{Zi|\boldsymbol{a}_i}(\xi,\boldsymbol{a}_i)d\xi \tag{10}$$

Integrating numerically (10) for all ranges Δa_i, associated with modes i, $i = 1,\ldots, N$ the total limit probability of failure can be obtained as

$$p_f = \sum_{i=1}^{N} \Delta a_i \int_0^{\infty} f_{\Xi,a_i}(\xi, a_i) F_{Z|a_i}(\xi, a_i) d\xi \tag{11}$$

In a similar way the shakedown reliability problem can be formulated when the maximum loads over cycles are of higher variance and the simplified approach (3) may not provide accurate results.

6 Time-Variant Limit Reliability Analysis

For majority of structures subjected to variable repeated/cyclic loading the structural material deteriorates in time and, therefore, the reliability will decrease in time. Then, the time-variant reliability analysis should be carried out. The history of variable loads is now of great importance and the loads must be modelled as load processes. Consider the space of n load processes $\boldsymbol{Q}(t)$. For a given load domain a conditioning scheme may be used to estimate the probability of failure at the time t_L for the failure function described as $g(\boldsymbol{q},\boldsymbol{r}) = \boldsymbol{r} - \boldsymbol{q}$

$$p_f(t_L) = \int_{r} p_f(t_L|\boldsymbol{r}) f_{\boldsymbol{R}}(\boldsymbol{r}) d\boldsymbol{r} \tag{12}$$

where $\boldsymbol{r}$ is the vector of resistances ($\boldsymbol{R}=\boldsymbol{R}(\boldsymbol{X})$), $\boldsymbol{q}$ is the load point, $p_f(t_L|\boldsymbol{r})$ is the probability of structural failure conditional on $\boldsymbol{R}$, $f_{\boldsymbol{R}}(\boldsymbol{r})$ is the joint probability density function of $\boldsymbol{R}$.

Introducing polar coordinates $\boldsymbol{R} = Z\boldsymbol{A}$ ($\boldsymbol{r} = \zeta\boldsymbol{a}$) and considering the total probability of failure with respect to the limit state it is possible to rewrite (12) as

$$p_f(t_L) = \int_{\Omega} f_A(\boldsymbol{a}) \left[\int_{h} p_f(\zeta|\boldsymbol{a}) f_{Z|A}(\zeta|\boldsymbol{a}) d\zeta \right] d\boldsymbol{a} \tag{13}$$

where Ω denotes an unit sphere, $p_f(\zeta|\boldsymbol{a})$ is the conditional limit probability of failure, $\boldsymbol{A}$ is the vector of direction cosines having the probability density function $f_A(\boldsymbol{a})$, Z is a random scalar representing the conditional limit multiplier, whereas $f_{Z|A}(\zeta|\boldsymbol{a})$ is the conditional probability density function of limit multiplier for a given direction $\boldsymbol{A} = \boldsymbol{a}$.

Expression (13) forms a basis for the directional simulation when each integral is replaced by an expectation operator (Ditlevsen et al. 1990, Melchers, 1990). The conditional failure probability $p_f(\zeta|\boldsymbol{a})$ due to the load $\boldsymbol{Q}(t)$ outcrossing out the safe domain at $Z = \zeta$ along the ray $\boldsymbol{A} = \boldsymbol{a}$, see Figure 8, can be evaluated from the following upper bound for high reliability systems on the Poisson outcrossings

$$p_f(\zeta \mid \boldsymbol{a}) \leq p_f(0,\zeta \mid \boldsymbol{a}) + \left\{1 - exp[-\nu_D^+(\zeta \mid \boldsymbol{a})t_L]\right\} \tag{14}$$

where $p_f(0,\zeta \mid \boldsymbol{a})$ is the failure probability at time $t = 0$ and ν_D^+ is the outcrossing rate of the vector process $\boldsymbol{Q}(t)$ out of the safe domain. Both quantities may be evaluated directly using directional simulation. For problems with rare outcrossings the expression (14) may be simplified to

$$p_f(\zeta \mid \boldsymbol{a}) \leq p_f(0,\zeta \mid \boldsymbol{a}) + \nu_D^+(\zeta \mid \boldsymbol{a})t_L \tag{15}$$

The probability density function $f_{Z|A}(\zeta|\boldsymbol{a})$ can be evaluated using simulations along $\boldsymbol{A} = \boldsymbol{a}$ (Melchers, 1994).

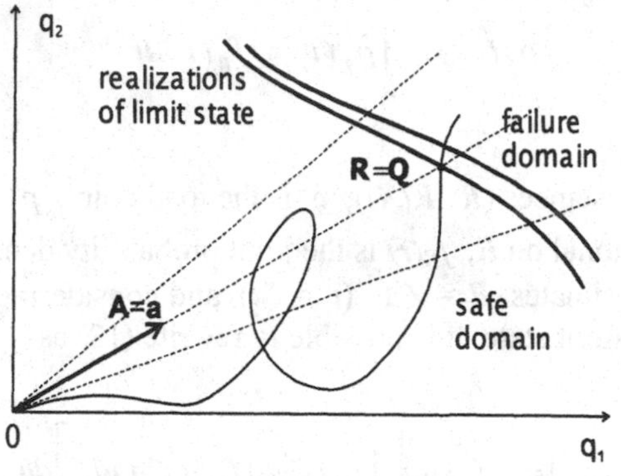

Figure 8. Conditioning scheme for evaluation of probability of failure. Outcrossing out of the safe domain

7 Safety Analysis

There are two approaches to the safety analysis for elastic-plastic structures subjected to variable repeated/cyclic loading, see Figure 2:

- limit analysis which must consider separately all critical point on the loading path (critical load proportionalities)

- shakedown analysis which considers in a single analysis all loading paths contained in the load variation domain Δ^{ξ}

Since the shakedown state corresponds to the safe working regime and the limit state corresponds to ultimate, critical one, it always holds that the shakedown multiplier (safety factor) is lower (or equal) to the limit load multiplier. This suggests the following strategy for assessing the reliability of the structure:

- carry out the shakedown reliability analysis considering all load variation domains Δ^{ξ}, see Figure 2b. The results should constitute a basis for structural design and assessment at the exploitation load level.
- carry out the conditional limit reliability analysis for certain critical load proportionalities, see Figure 2a, corresponding to some extreme (exceptional) loading or overloading.

For the structural design the following reliability requirements should be fulfilled

$$p_F^{\eta} \le p_F^{\eta}* \tag{16}$$

$$p_{Fi}^{\zeta} \le p_F^{\zeta}*$$

where p_F^{η} is the inadaptation probability of failure computed according to (2) or (3), p_{Fi}^{ζ} are limit probabilities of failure conditioned on certain critical proportionalities of loads computed according (5) or (6), k is the number of critical limit load proportionalities, $p_F^{\eta}*$ and $p_F^{\zeta}*$ are the admissible probabilities of failure for inadaptation and limit state, respectively.

Having in mind that shakedown corresponds to working conditions and limit load to the ultimate state it should be assumed

$$p_F^{\zeta}* \le p_F^{\eta}* \tag{17}$$

8 Conclusions

This paper presented methods of shakedown and limit reliability analysis for inelastic structures subjected to heavy cyclic/variable repeated loading. The problems may be efficiently solved by reliability simulation methods as well as the general shakedown/limit analysis computational codes combined with the response surface. The problems are formulated in the load space. This approach has several advantages:

- load domain approach allows for simple but accurate representation of cyclic/variable repeated loading of practical importance;

- shakedown and limit approach provides global failure criteria corresponding to the system reliability. Therefore, a single analysis may replace a number of cumbersome componental analyses of series/parallel subsystems;
- it permits a separate examination of load and strength parameters since the resistance limit state functions according to shakedown/limit analysis principles are assumed not to depend in any way on the load processes. If the sequence of loading applied to a structure may affect the capacity limit state the more refined time-variant reliability methods should be used;
- the dimensions of the load space may be kept small (depending only on a number of independent loadings) even for large structures;
- working in the X-space.

The application of a general shakedown analysis defined for the load variation domain Δ^{ξ}, combined with the limit analysis defined for certain critical load proportionalities, accompanied with definition of shakedown and limit reliability problems with different probabilities of failure corresponding to working and ultimate states makes an important basis for structural design and assessment. The methodology presented may be easily applied in the reliability design of realistic structures such as pressure systems.

References

Augusti, G. and Baratta A. (1972). Limit analysis of structures with stochastic strength variations. *J. Struct. Mech.* 1:43–62.

Corotis, R.B. and Nafday, A.M. (1989). Structural systems reliability using linear programming and simulation. *J.Struct.Eng., ASCE* 115:2435–2447.

Dietlevsen, O., Melchers, R.E. and Gluver, H, (1990). General multi-dimensional probability integration by directional simulation. *Comp. Struct.* 36:355–368.

Heitzer, M. and Staat M. (1999). Structural reliability analysis of elasto-plastic structures. In: Schueller G.I., Kafka P., eds., *Safety and Reliability*. A.A. Balkema, 513–518.

König, J.A. (1987). *Shakedown of elastic-plastic structures*. Warsaw-Amsterdam: PWN-Elsevier.

Melchers, R.E. (1990). Radial importance sampling for structural reliability. *J. Eng. Mech., ASCE*, 116:189–203.

Melchers, R.E. (1994). Structural system reliability assessment using directional simulation. *Struct. Safety* 16:23–39.

Melchers, R.E. (1995). Load space reliability formulation for Poisson pulse processes. *J. Eng. Mech. ASCE* 121:779–784.

Siemaszko, A. and Knabel, J. (2000). Reliability-based adaptation analysis. Proc. of 33-rd Solid Mechanics Conference, Zakopane, September 5–9.

Siemaszko, A. and Dolinski, K. (1996) Limit state reliability optimization accounting for geometric effects. *Struct. Opt.* 11:80–87.

Wen, Y.K. and Chen, H-C. (1989). System reliability under time varying loads. *J.Eng. Mech.* 115:808–823.

Computational Methods for Shakedown and Limit Reliability Analysis

Andrzej Siemaszko
Institute of Fundamental Technological Research, Warsaw, Poland

Abstract. In this chapter the computational system for the shakedown and limit analysis is presented. It is based upon an iterative min-max procedure proposed by Zwolinski and Bielawski. The system is called CYCLONE. Application to the realistic shakedown analysis problem of pressure valve is shown. Next, simulation methods of the reliability analysis are presented. A computational system is described which is composed of reliability analysis, response surface method, shakedown/limit analysis and FE analysis. It is able to solve the realistic shakedown and limit reliability analysis problems. The method is illustrated by an example of the shakedown and limit reliability analysis of high pressure chamber subjected to variable repeated pressure.

1 Introduction

Design codes of pressure vessels, e.g. BS 5500 and ASME VIII, divide stresses into several categories (e.g. primary, secondary stresses) and according to limits of elastic, shakedown and inadaptation regimes provide some safety factors (multipliers). The classical design-by-rules route is now being substituted by the design-by-analysis which requires the finite element analysis. The elastic multiplier is obtained by a linear elastic analysis, whereas the limit multiplier usually by a non-linear incremental analysis. The shakedown limit is approximated by a formula, which holds only for simple stress states

$$\eta = min\{\zeta, 2\varepsilon\} \tag{1}$$

where η, ζ, ε are the shakedown, limit, and elastic multipliers, respectively. In general, for combined loading the above formula overestimates the structural safety.

To assess the structural safety and reliability, for a given load variation domain, the safety factors (multipliers) against inadaptation and limit state should be computed, cf. König (1987). Knowing the variable repeated loading history the non-linear incremental procedures offered by modern finite element analysis codes may be used. However, since the variable loading path must be incrementally simulated, the computational effort may become prohibitively high (König and Kleiber, 1978). Moreover, the exact loading history is rather not available. The shakedown and limit analysis offer a powerful alternative or at least a complementary approach. They provide the accurate shakedown and limit multipliers without specification of exact loading history. It is enough to know only limits of load variation.

There are several attempts to develop general computational codes capable to solve the shakedown and limit analysis problems. Within the Brite-Euram project LISA there is a code developed which is integrated with the commercial FE code PERMAS, (cf. Heitzer and Staat, 1999). The shakedown problem is solved by a basis reduction technique in the residual stress space and by the sequential quadratic programming (cf. Stein et al, 1992, Staat and Heitzer, 1997). Ponter and Carter (1997) have developed a shakedown analysis method which is called the elastic compensation method. It can be used with any commercial FE code.

To compute the shakedown and limit multipliers, Zwolinski and Bielawski, (1987), Siemaszko et al. (2000) proposed an iterative min-max procedure, which minimises the maximum values of the elastic stresses by optimal selection of the residual stresses. The procedure is implemented for plane stress/strain and axisymmetric problems and it is actually extended to 3D problems. A computational system CYCLONE is developed which is fully integrated with commercial finite element codes.

2 Min-Max Procedure for the Reliability Analysis

Majority of the shakedown and limit analysis methods is based on mathematical programming procedures (cf. Maier, 1970, 1975) which for real-sized problems are not so efficient. To compute the shakedown and limit multipliers, Zwolinski and Bielawski (1987) and Pycko and Mroz (1992) proposed an iterative min-max procedure. The procedure is based upon the statical approach to shakedown and limit load problems. It aims at minimization of the maximal values of the reduced Mises stress by optimal selection of the residual stress

$$\max_{(x,t)} f\{\sigma(x,t)=\sigma_E(x,t)+\sigma_R[q(x)]\} \Rightarrow \min_{q} \tag{2}$$

where $f\{.\}$ is the reduced Mises stress, $\sigma(x,t)$ is the total variable stress, $\sigma_E(x,t)$ is the elastic stress corresponding to applied variable loads $\boldsymbol{p(t)}$, σ_R is the constant residual stress, $\boldsymbol{q}$ is the plastic strain, $\boldsymbol{x}$ are coordinates and t denotes time.

The limit multiplier ζ is given as a solution of the following min-max problem

$$\zeta^{-1} = \min_{(q)} \max_{(x)} \frac{f\{\sigma_E(x)+\sigma_R[q(x)]\}}{\sigma_Y[q(x)]} \tag{3}$$

whereas, the shakedown multiplier η results from the following min-max problem

$$\eta^{-1} = \min_{(q)} \max_{(x)} \frac{f\{\sigma_\varepsilon(x)+\sigma_R[q(x)]\}}{\sigma_Y[q(x)]} \tag{4}$$

where σ_ε is the envelope of elastic stresses, σ_Y is the yield stress which may be dependent on plastic strains $\boldsymbol{q}$ for hardening materials.

The procedure tries to select such residual stresses σ_R which reduce in a most efficient way all peaks (maxima) of elastic stresses σ_E (or σ_ε) within the whole structure. A highly efficient minimization algorithm (Zwolinski and Bielawski, 1987, 1994), specially tailored for residual stresses problems is implemented. The algorithm requires the elastic solution σ_E from the commercial FE code.

Consider a structure discretized by FE. Elastic analysis provides the vector of elastic stresses σ_E defined for all finite elements of the structure. Assuming that for the step $t\text{-}1$ of the computation process the plastic strain field q^{t-1} is known the vectors of residual and total stresses can be evaluated as

$$\sigma_R^{t-1} = A q^{t-1}$$

$$\sigma^{t-1} = \sigma_E + \sigma_R^{t-1} \tag{5}$$

where A is the influence matrix (cf. Maier, 1975).

The maximal function $f(\sigma_E + \sigma_R^{t-1})$ can be minimized by imposing some additional residual stress $\Delta\sigma_R^t$ such that

$$f\{\sigma_E + \sigma_R^{t-1} + \Delta\sigma_R^t\} < f\{\sigma_E + \sigma_R^{t-1}\} \tag{6}$$

The additional residual stress will result from a predicted plastic strain increment

$$\Delta q^t = \lambda^t \bar{q}^t \tag{7}$$

where the predicted field $\bar{q}$ may, for instance, follow the gradient $\partial f / \partial \sigma$, and where the scale (perturbation) coefficient λ is to be determined.

Therefore, plastic strains and total stresses at the step t are defined by

$$q^t = q^{t-1} + \Delta q^t$$

$$\sigma^t = \sigma^{t-1} + \Delta\sigma_R^t = \sigma^{t-1} + A\Delta q^t = \sigma^{t-1} + \lambda^t A \bar{q}^t \tag{8}$$

Substituting the above relations to the Mises stress functions, quadratic equations $f(\lambda^t)$ with respect to coefficients λ^t are obtained. For all elements the procedure identifies maximal parabola at $\lambda^t = 0$ and moves along it to the minimum or to the intersection point with another parabola, see Figure 1. This provides the perturbation value of λ^t to be used in scaling plastic strain increments. When λ^t is determined the actual plastic strains and stresses are computed and the next iteration is initiated.

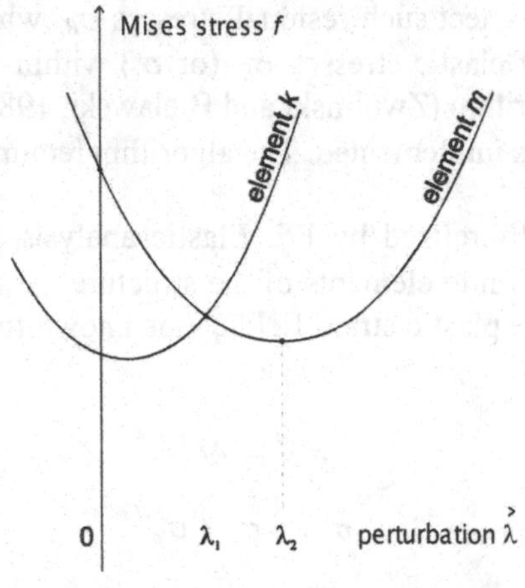

Figure 1. Perturbed reduced stress for element *k* and *m*.

When the procedure converges to the optimum, i.e. all peaks (maxima) of elastic stresses σ_E are reduced by residual stresses σ_R, there are more and more elements reaching the same value of yield function *f*. Some special rules are applied to ensure for these elements the equal descent of function *f*. The procedure is interrupted when some convergence criteria are fulfilled.

At this stage the final (asymptotic) stress and strain fields are fully defined. The procedure provides the plastic strains at the shakedown and limit state, $\boldsymbol{q}_\eta$ and $\boldsymbol{q}_\zeta$, respectively. Next, the residual stresses at the shakedown and limit state, $\boldsymbol{\sigma}_R^\eta$ and $\boldsymbol{\sigma}_R^\zeta$, respectively, are computed

$$\boldsymbol{\sigma}_R^\eta = \boldsymbol{A}\boldsymbol{q}_\eta$$

$$\boldsymbol{\sigma}_R^\zeta = \boldsymbol{A}\boldsymbol{q}_\zeta \tag{9}$$

Finally, the total stress at the shakedown and limit state, σ_η and σ_ζ, respectively, is defined

$$\boldsymbol{\sigma}_\eta = \boldsymbol{\sigma}_\varepsilon + \boldsymbol{\sigma}_R^\eta$$

$$\boldsymbol{\sigma}_\zeta = \boldsymbol{\sigma}_E + \boldsymbol{\sigma}_R^\zeta \tag{10}$$

The above fields of elastic, residual and total stress as well as plastic strain can be visualized by a commercial post-processor.

In a view of (3)(4) the shakedown multiplier η and the limit multiplier ζ are defined as

$$\eta = \sigma_Y / f_{max}(\sigma_\eta)$$

$$\zeta = \sigma_Y / f_{max}(\sigma_\zeta) \tag{11}$$

where σ_Y is the yield stress, $f_{max}(\sigma_\eta)$ is the maximum reduced (Mises) stress at the shakedown and $f_{max}(\sigma_\zeta)$ is the maximum reduced stress at the limit state.

Additionally, the elastic multiplier (corresponding to the first yielding) ε can be computed as

$$\varepsilon = \sigma_Y / f_{max}(\sigma_\varepsilon) \tag{12}$$

where $f_{max}(\sigma_\varepsilon)$ is the maximum reduced (Mises) stress from the envelope of elastic stresses. The method admits that effects of material hardening resulting from plastic strains could be accounted for as well as the influence of temperature on material properties.

The min-max method has been implemented into a computational system called CYCLONE. The computational code is written using the object oriented programming provided by C++ language. The system is installed on PC computers as well as on the SUN workstation. The shakedown and limit analysis module must be integrated with a FE code from which it retrieves the initial solution with the elastic stress σ_E. Actually, standard FE codes as ABAQUS or ANSYS are used as the background finite elements analysis code. The min-max procedure works for plane strains and stresses as well as axisymmetrical problems. Linear finite elements with the constant stress and strain assumption are implemented. Development of 3D capabilities is in progress.

3 Simulation Methods in Time-Invariant Reliability

3.1 Monte Carlo Simulation (MCS)

Reliability of failure for time-invariant reliability problem is given by

$$p_f = \int_{g(X)\le 0} f_X(x)dx \tag{13}$$

where X is the n-dimensional vector of random variables, $f_X(x)$ is the known multivariate probability density function of X.

Introducing the indicator function $I[g(X)\le 0]$ such that

$$I[g(X)\le 0] = \begin{cases} 1 & \text{if } g(X)\le 0 \\ 0 & \text{if } g(X)>0 \end{cases} \tag{14}$$

the probability of failure may be written as

$$p_f = \int_x I[g(X) \le 0] f_X(x) dx \tag{15}$$

and estimated by

$$p_f = \frac{1}{N} \sum_{j=1}^{N} I[g(\hat{x}_j) \le 0] \tag{16}$$

where $\hat{x}_j$ represents the j-th vector of random observations from $f_X(x)$.

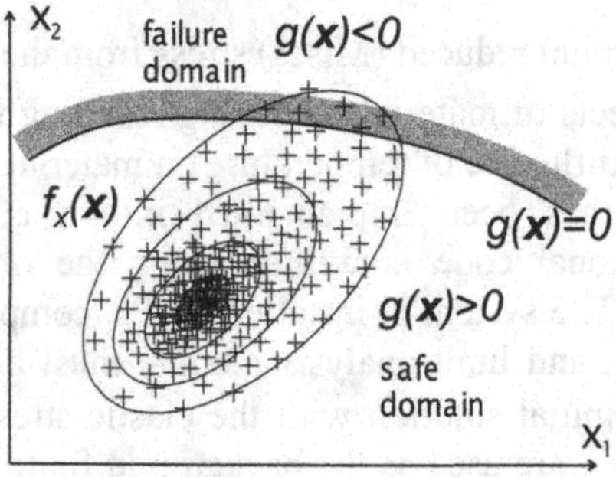

Figure 2. Monte Carlo Simulation.

3.2 Importance Sampling

To reduce the number of simulations necessary to estimate p_f, specially for low probabilities of failure, the importance sampling probability density function $h_X(x)$ is introduced. Then p_f is described by

$$p_f = \int_x I[g(X) \le 0] \frac{f_X(x)}{h_X(x)} h_X(x) dx \tag{17}$$

and can be estimated as

$$p_f = \frac{1}{N} \sum_{j=1}^{N} I[g(\hat{x}_j) \le 0] \frac{f_X(\hat{x}_j)}{h_X(\hat{x}_j)} \tag{18}$$

where $\hat{x}_j$ represents the j-th vector of random observations from $h_X(x)$.

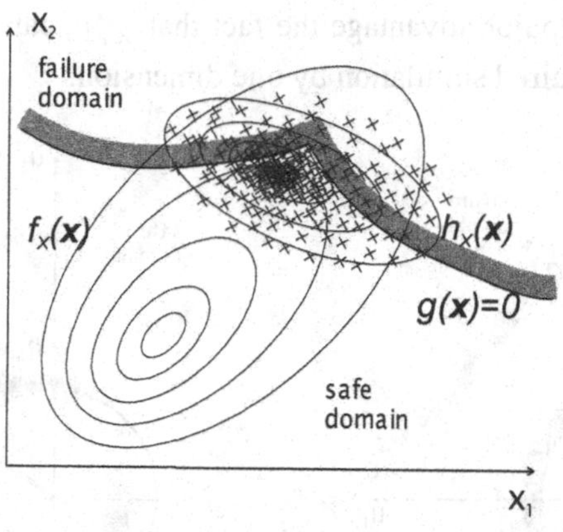

Figure 3. Importance sampling.

3.3 Directional Simulation

Consider the n-dimensional Gaussian vector $\boldsymbol{U}$, expressed as $\boldsymbol{U}=R\boldsymbol{A}$ $(R \geq 0)$, where $\boldsymbol{A}$ is an independent random unit vector, indicating direction in the space $\boldsymbol{U}$, see Figure 4. Assume that $\boldsymbol{A}$ is uniformly distributed $f_A(\boldsymbol{a})$ on the n-dimensional unit sphere. It is known that the radial distance R is such that R^2 is chi-square distributed with n degrees of freedom. Assuming that R is independent of $\boldsymbol{A}$ and by conditioning on $\boldsymbol{A}=\boldsymbol{a}$, the probability of failure can be written as

$$p_f = \int_{\Omega} P[g(R\boldsymbol{A}) \leq 0 | \boldsymbol{A} = \boldsymbol{a}] f_A(\boldsymbol{a}) d\boldsymbol{a} \qquad (19)$$

where $f_A(\boldsymbol{a})$ is the constant density of $\boldsymbol{A}$ on the unit sphere and $g()$ is the failure (limit state) function. Simulation using (22) proceeds by generating randomly a sample standard unit vector $\hat{\boldsymbol{a}}_j$ and finding the root for r in $g(r\hat{\boldsymbol{a}}_j) = 0, r \geq 0$. The root is the distance from the origin to the failure surface in the direction of $\boldsymbol{a}$. For this radius the probability of failure conditional on $\boldsymbol{A} = \hat{\boldsymbol{a}}_j$ is given by

$$\hat{p}_{fi} = P[g(R\hat{\boldsymbol{a}}_j) \leq 0] = 1 - \chi_n^2(r^2) \qquad (20)$$

where $\chi_n^2(.)$ denotes the chi-square distribution function with n degrees of freedom. Repeating this for N samples of $\boldsymbol{A}$ the probability of failure may be written as

$$p_f = \frac{1}{N} \sum_{i=1}^{N} \hat{p}_{fi} \qquad (21)$$

The method has as its major advantage the fact that $\chi_n^2(.)$ can be evaluated analytically, effectively reducing the required simulation by one dimension.

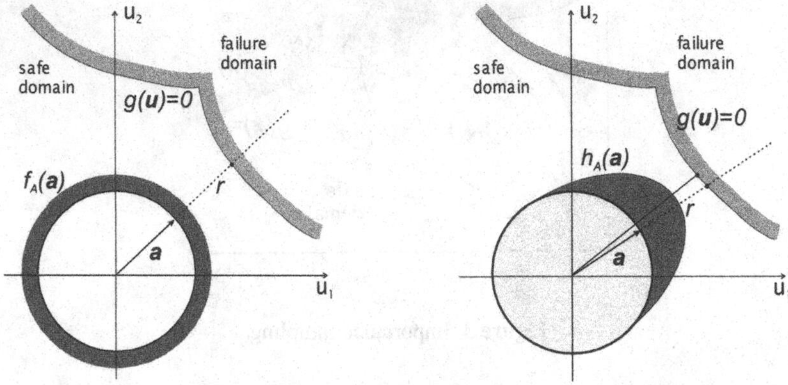

Figure 4. Directional Simulation.

The directional simulation method can be coupled with the importance sampling (cf. Bjerager, 1988). The uniform sampling density $f_A(a)$ can be modified by the directional sampling probability density function $h_A(a)$ on the unit sphere about the origin.

Assuming that R is independent of A and by conditioning on A=a, the probability of failure can be written as

$$p_f = \int_{\Omega} P[g(RA) \leq 0] \frac{f_A(a)}{h_A(a)} h_A(a) da \tag{22}$$

where $h_A(a)$ is the sampling probability function on the unit sphere. For any sample $\hat{a}_j$ generated from $h_A(a)$ the sample value of p_{fi} becomes

$$\hat{p}_{fi} = P[g(R\hat{a}_j) \leq 0] \frac{f_A(\hat{a}_j)}{h_A(\hat{a}_j)} \tag{23}$$

and then the p_f is estimated by

$$p_f = \frac{1}{N} \sum_{i=1}^{N} \hat{p}_{fi} \tag{24}$$

4 Computational Method for the Shakedown and Limit Reliability Analysis

There are efforts to develop efficient computational codes for solving shakedown problems. König and Kleiber (1978) proposed to formulate an equivalent cyclic loads problem and solve it by commercial FE codes. There were developed several codes for skeletal structures, for instance CEPAO (Nguyen D.H., 1983) and SDLA (Siemaszko, 1988). Recently, general FE-based shakedown analysis codes have been developed: CYCLONE (Siemaszko et al. 2000) and LISA (Heitzer and Staat, 1999). Ponter and Carter (1997) developed a shakedown method which may be carried out with any commercial FE code.

Solution of shakedown or limit reliability analysis problems for realistic structural problems requires integration of finite element analysis code (FEA), shakedown analysis code (e.g. CYCLONE), response surface module (RSM) and reliability analysis code (REL).
The first step is to define structural model (required by FEA) as well as random variables: load domain multiplier ξ as well as k-dimensional vector of material and geometrical parameters $\boldsymbol{X}$ (required by REL).

Reliability methods, like MCS, for computation of probabilities of failure (13)(16)(18)(22) will need values of shakedown multiplier $\eta(\boldsymbol{x})$ and limit multiplier $\zeta(\boldsymbol{x})$ for different realizations $\boldsymbol{x}$ of parameters $\boldsymbol{X}$. Number of samples may be very high, for instance in the case of MCS may reach 100 000. In this situation it would be desirable to have an explicit definition of shakedown $\eta(x)$ and limit $\zeta(x)$ functions. This would allow to save a significant number of experiments. It can be done by application of a Response Surface Method which applies a polynomial approximation (see Siemaszko and Knabel, 2000)

$$\eta(\boldsymbol{x}) = a^{\eta} + \sum_{i=1}^{k} b_i^{\eta} x_i + \sum_{i=1}^{k} c_i^{\eta} x_i^2 \tag{25}$$

$$\zeta(\boldsymbol{x}) = a^{\zeta} + \sum_{i=1}^{k} b_i^{\zeta} x_i + \sum_{i=1}^{k} c_i^{\zeta} x_i^2 \tag{26}$$

where a, b, c are unknown coefficients, k is the number of random size/material variables $\boldsymbol{X}$.

The coefficients of failure function approximations are computed by the regression analysis. In the simplest approach of linear approximation only $k + 1$ experiments, whereas for quadratic approximation 2^k experiments $\eta(\bar{\boldsymbol{x}})$ or $\zeta(\bar{\boldsymbol{x}})$ are required. The values $\eta(\bar{\boldsymbol{x}})$ or $\zeta(\bar{\boldsymbol{x}})$ are obtained directly from the shakedown analysis code CYCLONE (Zwolinski and Bielawski, 1987, 1994; Siemaszko et al. 2000). CYCLONE needs from FEA the elastic solution σ_E for each vector $\bar{\boldsymbol{x}}$ of experiments requested by RSM, see Figure 5.

Solving the regression analysis the coefficients in polynomial approximation of $\eta(\boldsymbol{x})$ or $\zeta(\boldsymbol{x})$ are fully defined. Then, RSM is ready to answer immediately any request for response values coming from REL. Therefore, despite of a very high number of required samples the computational time is very low. Finally, using the MCS method, REL computes the shakedown and limit probability of failure.

As a result two probabilities of failure are obtained p_F^η and p_F^ζ, shakedown and limit probabilities of failure, respectively. Their values should be lower than admissible probabilities of failure

$$p_F^\eta \leq p_F^\eta * \tag{27}$$

$$p_F^\zeta \leq p_F^\zeta * \tag{28}$$

where $p_F^\eta *$ and $p_F^\zeta *$ are the shakedown and limit admissible probabilities of failure, respectively. Instead of probabilities of failure, the corresponding reliability indices β may be used.

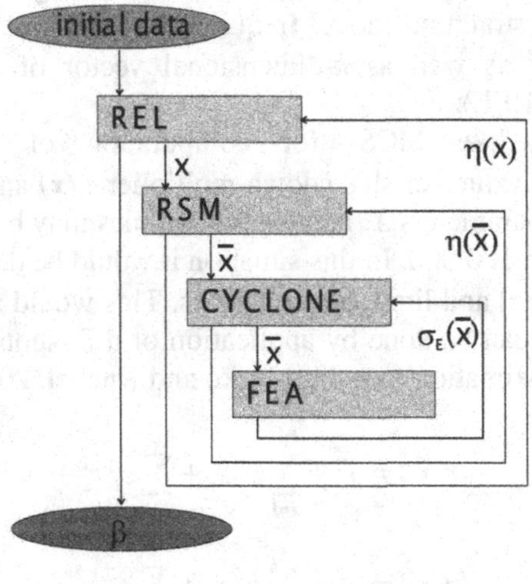

Figure 5. Flow chart for shakedown reliability evaluation using the shakedown analysis code CYCLONE.

5 Examples

The min-max procedure presented in this paper has been used to assess safety of the valve. The valve is subjected to variable repeated pressure p_0 = *265 MPa*. The geometry of the valve is shown in Figure 6. The discrete reaction of screws was replaced by equivalent axisymmetrical one. Three cases were considered: valve without the crack and valves with axisymmetrical cracks of lengths $l = 0.08r$ and $l = 0.46r$, (r being the external radius of the valve).

For the valve without any crack the maximum elastic Mises stress is equal to $f(\sigma_E) = 1484\ MPa$. Since the yield limit equals $\sigma_Y = 600\ MPa$, the elastic multiplier is equal to $\varepsilon = \sigma_Y / f(\sigma_E) = 0.40$. It means that the working pressure exceeds *2.5* times the elastic limit (first yielding). The maximum stress corresponding to the shakedown is equal to $f(\sigma_\eta) = 742\ MPa$. Therefore, the shakedown multiplier is equal to $\eta = \sigma_Y / f(\sigma_\eta) = 0.81$, see Figure 7. It means that the working pressure exceeds the shakedown limit and the inadaptation phenomena (alternating plasticity) should develop. Designing this valve the working pressure should be reduced at least to that corresponding to the shakedown limit $p_\eta = \eta\, p_0 = 214\ MPa$.

Finally, the Mises stress corresponding to the limit state was computed $f(\sigma_\zeta) = 165\ MPa$. This value means that the residual stress has allowed to reduce the maximum elastic stress more than *9* times (the reduction from *1484* to *165 MPa*). The corresponding limit multiplier is equal $\zeta = \sigma_Y / f(\sigma_\zeta) = 3.64$.

Since the actual pressure p_0 exceeds the shakedown pressure p_η the inadaptation phenomenon has been developed. Due to cyclic repetition of pressure the crack was formed. The analysis of the valve with crack shows that shakedown multiplier is always two times greater than elastic multiplier, see Figure 7. However, short cracks have no influence on ultimate multipliers. Only for large cracks the ultimate multiplier is affected. From the Figure 7, it seen that valve always works in the inadaptation regime. The crack progresses from the cycle to cycle. At the crack length $l = 0.5r$ the ultimate state will be reached and the valve will fail.

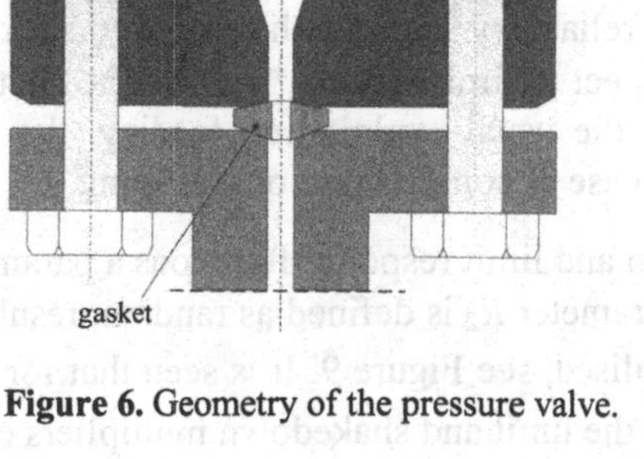

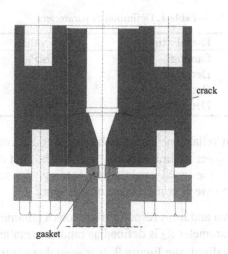

Figure 6. Geometry of the pressure valve.

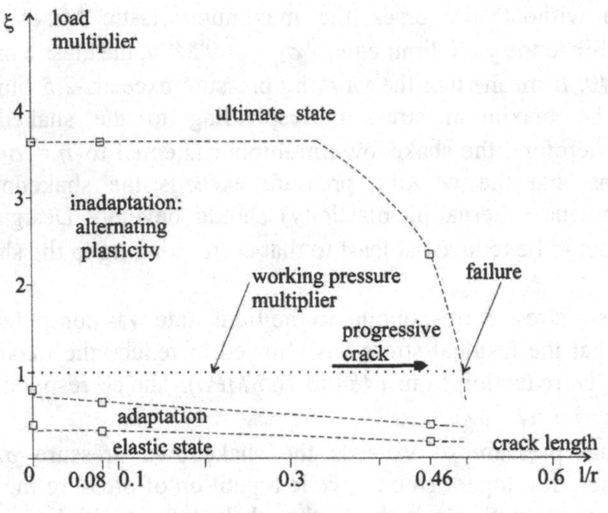

Figure 7. Working regimes for the valve with a progressive crack.

In the second example the shakedown and limit reliability-based design of a high pressure chamber was carried out. Geometry of the chamber is shown in Figure 8. It is subjected to cyclic pressure $p=130$ *MPa*. The internal and external radii are assumed as $R_1 = 12$ *mm* and $R_2 = 27$ *mm*, see Table 1 for the definition of parameters.

Table 1. Definition of parameters

No.	Parameter	Distribution	Mean value	Standard deviation
1	Pressure p	Gumbel	130 MPa	10 MPa
2	Radius R_1	Deterministic	12 mm	
3	Radius R_2	Beta	27 mm	0.1 mm
4	Yield stress σ_Y	Deterministic	200 MPa	

The required minimum reliability index with respect to shakedown $\beta^{\eta}* = 3.7$, whereas the reliability index with respect to limit state $\beta^{\zeta}* = 4.5$. The first reliability level is selected to ensure shakedown under the usual exploitation loading, the second one to prevent the global structural burst in the case of some exceptional loading.

To define the shakedown and limit response functions a parametric study was performed. Since only one geometric parameter R_2 is defined as random, resulted functions $\eta = \eta(x)$ and $\zeta = \zeta(x)$ can be easily visualised, see Figure 9. It is seen that for thin-walled chambers with $R_2 / R_1 \leq 2.1$ $(R_2 \leq 25mm)$ the limit and shakedown multipliers coincide.

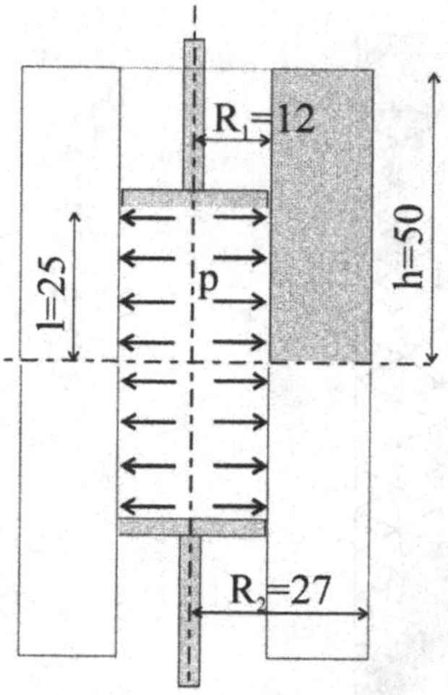

Figure 8. High pressure chamber.

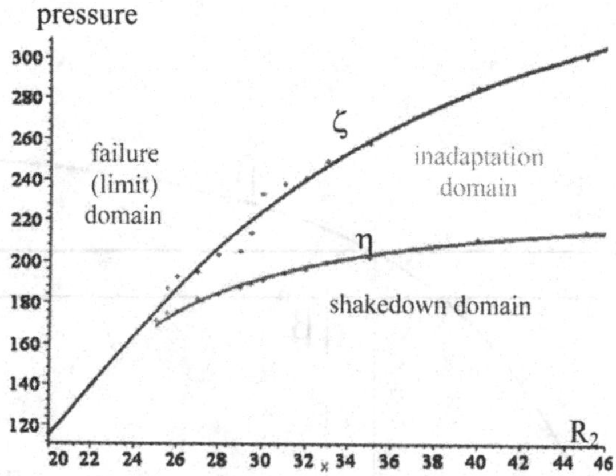

Figure 9. Limit ζ and shakedown η response functions.

For the analyzed example *(R_2 = 27 mm)* the computed values of the shakedown and limit multiplier are η = *182* and ζ = *194*. The total Mises stress within the structure in the shakedown and limit state is shown in Figure 10.

The admissible limit reliability index $\beta^{\zeta}{*} = 4.5$ is attained at $R_2 / R_1 = 2.46$. However, to fulfill the shakedown reliability requirements $\beta^{\eta}{*} = 3.7$ a thicker chamber must be used with $R_2 / R_1 = 2.61$.

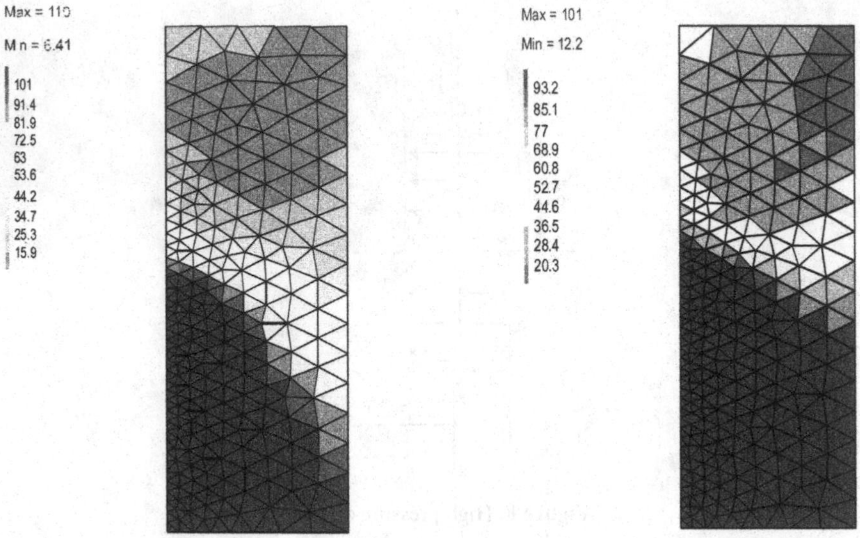

Figure 10. Total Mises stress for the high pressure chamber at the shakedown and limit state

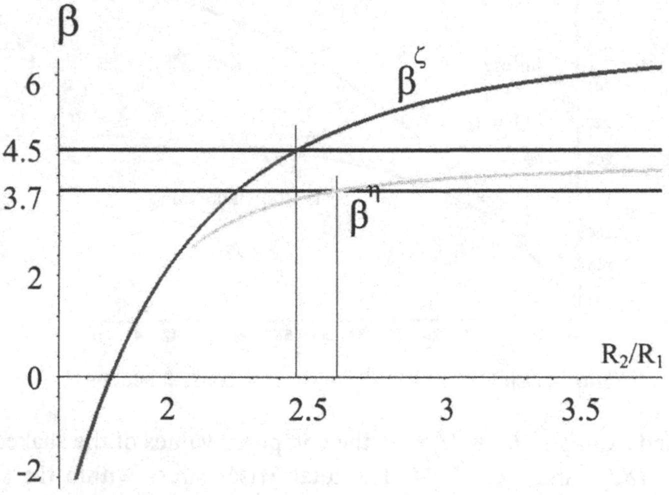

Figure 11. Limit and shakedown reliability indices with respect to the ratio R_2 / R_1.

References

Bielawski, G., Siemaszko A. and Zwolinski, J. (1997). An adaptation and limit analysis method in the design of inelastic structures. *Trans. SMiRT*, 14, 3, FW/4, 293–300.

Bjerager, P. (1988). Probability integration by directional simulation. *J.Eng.Mech., ASCE* 114:1285–1302.

Heitzer, M. and Staat, M. (1999). Structural reliability analysis of elasto-plastic structures. In: Schueller, G.I. and Kafka, P., eds., *Safety and Reliability*. A.A. Balkema. 513–518.

König, J.A. (1987). *Shakedown of elastic-plastic structures*. PWN-Elsevier: Warsaw-Amsterdam.

König, J.A. and Kleiber, M. (1978). On a new method of shakedown analysis. *Bull. Ac. Pol. Sci. Ser. Sci. Techn.* 26:165–171.

Maier, G. (1970). A matrix structural theory of piecewise-linear plasticity with interacting yield planes. *Meccanica* 5:55–66.

Maier, G. (1975). Mathematical programming method in analysis of elastic-plastic structures. *Arch.Inz.Lad.* 31: 387.

Nguyen, D. H. (1983). Aspects of analysis and optimization of structures under proportional and variable loadings. *Eng. Opt.*, 7, 35–57.

Ponter, A.R.S. and Carter, K.F. (1997). Limit state solutions, based upon linear solutions with a spatially varying elastic modulus. *Comp. Meth. Appl. Mech. Eng.* 140:237–258

Pycko, S. and Mróz, Z. (1992). Alternative approach to shakedown as a solution of min-max problem. *Acta Mechanica* 93:205–222.

Siemaszko, A. (1988). Stability analysis of shakedown processes of plane skeletal structures. IFTR Reports, 12, 1–176.

Siemaszko, A., Bielawski, G. and Zwolinski, J. (2000). CYCLONE-system for structural adaptation and limit analysis. In Weichert, D. and Maier, G., eds., *Inelastic Analysis of Structures under Variable Loads*. Dordrecht: Kluwer Academic Publishers. 135–146.

Siemaszko, A. and Knabel, J. (2000). Reliability-based adaptation analysis. Proc. of 33-rd Solid Mechanics Conference, Zakopane, September 5–9.

Staat, M. and Heitzer, M. (1997). Limit and shakedown analysis for plastic safety of complex structures. *Transactions of SMiRT*, 14, B02/2.

Stein, E., Zhang, G. and König, J.A. (1992). Shakedown with nonlinear strain-hardening including structural computation using finite element method. *Int. J. Plast.* 8:1–31.

Zwolinski, J., Bielawski G. (1987). An optimal selection of residual stress for shakedown and limit load analysis. *Proc.Conf. Comp. Meth.Struct. Mech.*, Jadwisin, 459–462.

Zwolinski, J. (1994). Min-max approach to shakedown and limit load analysis for elastic perfectly plastic and kinematic hardening materials. In Mróz, Z., Weichert, D. and Dorosz, S., eds. *Inelastic Behaviour of Structures under Variable Loads*. Dordrecht: Kluwer Academic Publishers. 363–380.

Limit and Shakedown Reliability Optimization Accounting for Nonlinear Geometric Effects

Andrzej Siemaszko

Institute of Fundamental Technological Research, Warsaw, Poland

Abstract. In this chapter, the relation between deterministic, stochastic and design parameters is explained. Differences between deterministic and reliability-based optimisation are pointed out. A formulation of the system reliability optimization with the limit and shakedown failure criteria is presented. For discrete structures it allows to account for nonlinear geometric effects. An example of limit reliability optimization is presented.

1 Introduction

Optimization methods have recently become one of fundamental tools in design of advanced structural systems. There are two basic formulations of optimization problem associated with the shakedown and limit state. In the first one the structural volume or weight is minimized under the shakedown or limit state constraint. In the second, the limit or shakedown multiplier is maximized under a constant volume/weight constraint, see e.g. Cohn et al. (1972), Cyras (1969), Nguyen (1983).

Results of the deterministic-based optimization provide designs which cannot often be considered as optimal since they are highly sensitive to variations of design variables. This may be caused by the non-linear geometrical effects (P-δ effect). Siemaszko & König (1990) and Siemaszko & Mróz (1991) proposed to use in the deterministic optimization a reduced limit (shakedown) multiplier obtained from the initial one by a reduction proportional to the sensitivity to non-linear geometric effects. As a general rule it may be concluded that the reliability of deterministically optimal structures may be significantly lower than of other designs. In this situation only the reliability-based optimization (cf. e.g. Thoft-Christensen, 1990) may provide an acceptable optimal design.

The reliability-based optimization using the limit state as the failure function and accounting for geometric effects was considered by Siemaszko and Dolinski (1996). Practically, the problem of the shakedown reliability optimization has not been deeply analyzed.

2 Limit and Shakedown Analysis

Let us consider an elastic-plastic discrete structure subjected to variable repeated loads. For such structures, shakedown is a widely recognized necessary condition of safe structural behaviour. It is assumed that loads ***p(t)*** may vary on arbitrary paths contained in a prescribed load domain Δ. The shakedown multiplier η is a maximum multiplier of the load domain which ensures shakedown. Stress, strain and displacement fields within the structure are assumed to be fully described by finite-dimensional vectors ***s***, ***q*** and ***u***, respectively.

Within the framework of geometrical linearity the fundamental kinematic shakedown and limit analysis theorems can be expressed as (cf. Maier, 1970)

$$\eta^0 = \min_{\dot{\lambda}^0, \dot{u}^0} \left\{ k^T \cdot \dot{\lambda}^0 \mid d^T \cdot \dot{\lambda}^0 = 1, \quad N \cdot \dot{\lambda}^0 = C^0 \cdot \dot{u}^0, \quad \dot{\lambda}^0 \geq 0 \right\} \tag{1}$$

$$\zeta^0 = \min_{\dot{\lambda}^0, \dot{u}^0} \left\{ k^T \cdot \dot{\lambda}^0 \mid p^{*T} \cdot \dot{u}^0 = 1, \quad N \cdot \dot{\lambda}^0 = C^0 \cdot \dot{u}^0, \quad \dot{\lambda}^0 \geq 0 \right\} \tag{2}$$

where N is the matrix of gradients of yield polyhedrons, C^0 is the initial matrix of compatibility, k, d and λ are the vectors of yield moduli, elastic stress envelope and plastic multipliers, ξ, η^0 and ζ^0 are the load domain multiplier, initial shakedown multiplier, and initial limit multiplier respectively.

Solving the two-level linear program, the reduced shakedown multiplier η^δ or the reduced limit multiplier ζ^δ is obtained (Siemaszko, 1995).

$$\begin{cases} \eta^\delta = \min_{\dot{\lambda}^\delta, \dot{u}^\delta} \left\{ k^T \cdot \dot{\lambda}^\delta \mid d^T \cdot \dot{\lambda}^\delta = 1, \quad N \cdot \dot{\lambda}^\delta = C^\delta(\mu \cdot \dot{u}^0) \cdot \dot{u}^\delta, \quad \dot{\lambda}^\delta \geq 0 \right\} \\ \eta^0 = \min_{\dot{\lambda}^0, \dot{u}^0} \left\{ k^T \cdot \dot{\lambda}^0 \mid d^T \cdot \dot{\lambda}^0 = 1, \quad N \cdot \dot{\lambda}^0 = C^0 \cdot \dot{u}^0, \quad \dot{\lambda}^0 \geq 0 \right\} \end{cases} \tag{3}$$

$$\begin{cases} \zeta^\delta = \min_{\dot{\lambda}^\delta, \dot{u}^\delta} \left\{ k^T \cdot \dot{\lambda}^\delta \mid p^{*T} \cdot \dot{u}^\delta = 1, \quad N \cdot \dot{\lambda}^\delta = C^\delta(\mu \cdot \dot{u}^0) \cdot \dot{u}^\delta, \quad \dot{\lambda}^\delta \geq 0 \right\} \\ \zeta^0 = \min_{\dot{\lambda}^0, \dot{u}^0} \left\{ k^T \cdot \dot{\lambda}^0 \mid p^{*T} \cdot \dot{u}^0 = 1, \quad N \cdot \dot{\lambda}^0 = C^0 \cdot \dot{u}^0, \quad \dot{\lambda}^0 \geq 0 \right\} \end{cases} \tag{4}$$

where the compatibility matrix C^δ, defined for the deformed configuration, is dependent on the solution $\dot{u}^0$ from the lower-level problem.

3 Definition of Parameters in the Reliability-Based Optimization

A formulation of the reliability-based optimization problem requires a clarification of nomenclature concerning different types of parameters. All size, shape, material and loading parameters should be divided into deterministic or stochastic parameters. The deterministic parameters x are defined by single characteristic values x^d. The stochastic parameters X can be described by moments (statistical descriptors): mean values x^μ and standard deviations x^σ. The stochastic parameter X will have realizations denoted by x. Design parameters, i.e. the parameters which are used in the optimization process, can be assigned to statistical descriptors of random variables such as mean value and standard deviation and they cannot be assigned to realizations of random variables, see Figure1.

In the problem considered it is assumed that size/material and loading parameters are random variables and are denoted by X and Ξ, respectively. Realizations of random variables, considered by reliability analysis, are denoted by x and ξ, respectively. We assume only the mean values of size parameters, x^u, as the variable design parameters. The mean value of the load multiplier, ξ^u, is not subjected to the optimization procedure and therefore is considered as a fixed design parameter. More complicated interrelations of stochastic as well as deterministic parameters and variable and fixed design parameters were described by Siemaszko and Santos (1993) and Santos et al. (1995).

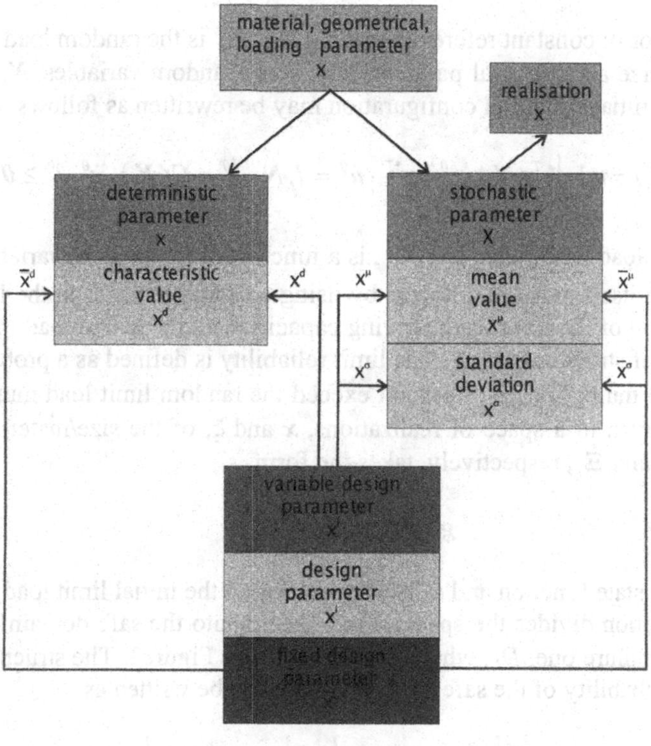

Figure1. Definition of deterministic and stochastic as well as design parameters.

4 Limit Reliability Analysis Accounting for Geometrical Effects

In the space of random structural parameters some failure functions are introduced to define a boundary between the safe and failure domain. A structure does not fail if a realization of random variables will belong to the safe domain. Due to randomness of structural parameters the state of being in the safe domain is a random event. Its probability is called a structural reliability R. The probability of the opposite event is a compliment to one of the reliability and defines the probability of structural failure p_F.

The failure functions may be local, i.e. associated with some particular points or elements of the structure such as excessive displacements or stresses. In this case the reliability analysis of the structures requires some multiple failure states to be considered. The failure function may be also global, i.e. associated with the entire structure behaviour such as, for example, admissible eigenfrequency, limit state or shakedown. Taking into account the global failure functions the reliability analysis can be considerably simplified.

Consider a limit reliability problem and define the random applied loading $\boldsymbol{p}$ as

$$\boldsymbol{p} = \Xi \cdot \boldsymbol{p}^* \tag{5}$$

where $\boldsymbol{p}^*$ is the vector of constant reference loading, and Ξ is the random load multiplier.

Assuming the size and material parameters as some random variables, $\boldsymbol{X}$, the limit state problem (2) for the initial structural configuration may be rewritten as follows

$$Z^0(\boldsymbol{X}) = \min_{\dot{\lambda}^0, \dot{u}^0} \left\{ \boldsymbol{k}^T(\boldsymbol{X}) \cdot \dot{\lambda}^0 \mid \boldsymbol{p}^{*T} \cdot \dot{\boldsymbol{u}}^0 = 1, \boldsymbol{N} \cdot \dot{\lambda}^0 = \boldsymbol{C}(\boldsymbol{X}) \cdot \dot{\boldsymbol{u}}^0, \dot{\lambda}^0 \geq \boldsymbol{0} \right\} \tag{6}$$

The initial limit load multiplier, $Z^0(\boldsymbol{X})$, is a function of the random variables $\boldsymbol{X}$. Thus, it is a random variable itself, what is reflected by using a capital letter Z. In the limit reliability analysis a comparison of the limit load carrying capacity with the actual load may be used for the global criterion of structural safety. The limit reliability is defined as a probability that the current random load multiplier, Ξ , does not exceed the random limit load multiplier, Z. The failure function g given in a space of realizations, $\boldsymbol{x}$ and ξ, of the size/material and loading random variables $\boldsymbol{X}$ and Ξ , respectively, takes the form

$$g^0(\boldsymbol{x},\xi) = \zeta^0(\boldsymbol{x}) - \xi \tag{7}$$

where g^0 is the limit state function and ζ^0 is a realization of the initial limit load multiplier Z^0.

The failure function divides the space of parameters into the safe domain, D_S, for which $g^0(\boldsymbol{x},\xi) \geq 0$, and the failure one, D_F, where $g^0(\boldsymbol{x},\xi) < 0$, see Figure2. The structural reliability, R, defined as the probability of the safe structural state can be written as

$$R = \mathsf{P}\left[g^0(\boldsymbol{X},\Xi) \geq 0\right] = \mathsf{P}\left[\Xi \leq Z^0(\boldsymbol{X})\right] \tag{8}$$

where $\mathsf{P}[A]$ denotes the probability of an event A.

Assuming the joint probability density function, $f_{X,\Xi}(\boldsymbol{x},\xi)$, of parameters $\boldsymbol{X}$ and Ξ, to be known, see Figure2, the reliability (8) has to be computed as an integral

$$R = \int_{D_S^0} f_{X,\Xi}(\boldsymbol{x},\xi)\, d\boldsymbol{x}\, d\xi \tag{9}$$

where $D_S^0 = \{(\boldsymbol{x},\xi): \xi \leq \zeta^0(\boldsymbol{x})\}$ denotes the initial safe domain.

The failure probability of the structure, p_F, is obtained as

$$p_F = 1 - R \tag{10}$$

The above formulation of the limit state reliability can be extended to problems accounting for geometric effects. The limit load criterion (4) should be generalized into the following two-level linear program:

$$\begin{cases} Z^{\delta}(X) = \min\limits_{\dot{\lambda}^{\delta},\dot{u}^{\delta}} \left\{ k^T(X) \cdot \dot{\lambda}^{\delta} \mid p^{*T} \cdot \dot{u}^{\delta} = 1, N \cdot \dot{\lambda}^{\delta} = C^{\delta}\left(X, \dot{u}^{0}(X)\right) \cdot \dot{u}^{\delta}, \dot{\lambda}^{\delta} \geq 0 \right\} \\ Z^{0}(X) = \min\limits_{\dot{\lambda}^{0},\dot{u}^{0}} \left\{ k^T(X) \cdot \dot{\lambda}^{0} \mid p^{*T} \cdot \dot{u}^{0} = 1, N \cdot \dot{\lambda}^{0} = C^{0}(X) \cdot \dot{u}^{0}, \dot{\lambda}^{0} \geq 0 \right\} \end{cases} \tag{11}$$

where $Z^{\delta}(X)$ is the random reduced limit load multiplier defined for the admissible deformed configuration. The limit state function is now defined as

$$g^{\delta}(x,\xi) = \zeta^{\delta}(x) - \xi \tag{12}$$

and the reliability R can be written as

$$R = \mathrm{P}\left[g^{\delta}(X,\Xi) \geq 0\right] = \mathrm{P}\left[\Xi \leq Z^{\delta}(X)\right] \tag{13}$$

The reliability is again defined by a joint probability density function as

$$R = \int_{D_S^{\delta}} f_{X,\Xi}(x,\xi)\, dx\, d\xi \tag{14}$$

where $D_S^{\delta} = \{(x,\xi): \xi \leq \zeta^{\delta}(x)\}$.

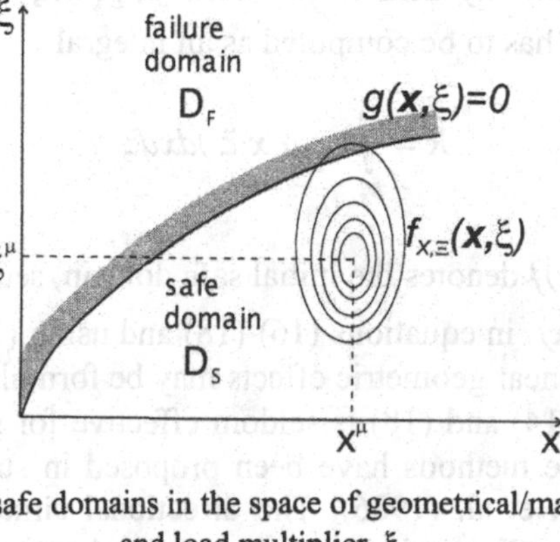

Figure 2. Failure and safe domains in the space of geometrical/material parameters, x, and load multiplier, ξ.

5 Shakedown Reliability Analysis Accounting for Geometrical Effects

Similarly, it is possible to consider a shakedown reliability problem accounting for geometrical effects. Now, Ξ denotes the random load domain multiplier. Assuming the size and material parameters as some random variables, ***X***, the shakedown problem (1) for the initial structural configuration may be rewritten as follows

$$H^0(X) = \min_{\dot{\lambda}^0, \dot{u}^0} \left\{ k^T(X) \cdot \dot{\lambda}^0 \mid d^T \cdot \dot{\lambda}^0 = 1, N \cdot \dot{\lambda}^0 = C(X) \cdot \dot{u}^0, \dot{\lambda}^0 \geq 0 \right\} \tag{15}$$

The initial shakedown load multiplier, $H^0(X)$, is a function of the random variables ***X***. Thus, it is a random variable itself. In the shakedown reliability analysis a comparison of the shakedown multiplier with the actual load domain multiplier may be used for the global criterion of structural safety. The shakedown reliability is defined as a probability that the current random load domain multiplier, Ξ, does not exceed the random shakedown multiplier, H.

The failure function given in a space of realizations, ***x*** and ξ, of the size/material and loading random variables ***X*** and Ξ, respectively, takes the form

$$g^0(x,\xi) = \eta^0(x) - \xi \tag{16}$$

where g^0 is the limit state function and η^0 is a realization of the initial shakedown multiplier H^0.

The failure function divides the space of parameters into the safe domain, D_S, for which $g^0(x,\xi) \geq 0$, and the failure one, D_F, where $g^0(x,\xi) < 0$, see Figure2. The structural reliability, *R*, defined as the probability of the safe structural state can be written as

$$R = \mathrm{P}\left[g^0(X,\Xi) \geq 0\right] = \mathrm{P}\left[\Xi \leq H^0(X)\right] \tag{17}$$

Assuming the joint probability density function, $f_{X,\Xi}(x,\xi)$, of parameters ***X*** and Ξ, to be known the reliability (17) has to be computed as an integral

$$R = \int_{D_S^0} f_{X,\Xi}(x,\xi)\, dx\, d\xi \tag{18}$$

where $D_S^0 = \{(x,\xi): \xi \leq \eta^0(x)\}$ denotes the initial safe domain, see Figure 2.

Replacing $\eta^0(x)$ by $\eta^\delta(x)$ in equations (16)-(18) and using (3) the shakedown reliability analysis accounting for nonlinear geometric effects may be formulated.

A direct integration in (14) and (18) is seldom effective for multidimensional problems. Therefore, some approximate methods have been proposed in structural reliability analysis, cf. Dolinski (1983), Madsen et al. (1986). The directional simulation technique (Bjerager, 1988), was proved to be very efficient in estimation of the above probabilities.

6 Reliability-Based Optimization

Deterministic optimization of structures is usually formulated as a problem of minimization of structural volume (weight, cost) subject to constraints reflecting safety or geometrical requirements. The vector of design parameters should collect all significant geometrical (size/shape), material and loading parameters. The optimization problem may be formulated as

$$\min_{x^d} V(x^d)$$

$$\text{s.t. } g(x^d) \geq 0 \tag{19}$$

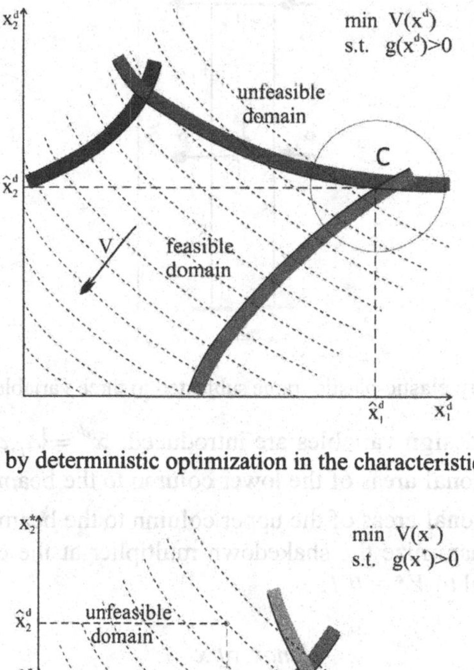

Figure 3. "Optimum" design by deterministic optimization in the characteristic (mean) values x^d space.

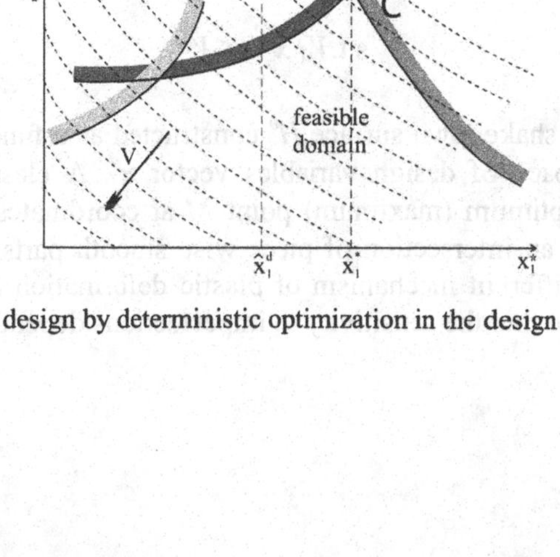

Figure 4. "Optimum" design by deterministic optimization in the design values x* space.

There is a question which values of parameters should be considered in the optimization: characteristic values x^d (related to mean values) or the design ones x^* (multiplied by safety factors). In both cases it may happen that due to imperfections or a wrong definition of design values the obtained optimum is not acceptable for practical applications. This is illustrated in Figures 3 and 4.

Let us consider the problem of imperfections in detail. We take an example of a two-story elastic-plastic frame subjected to three independent variable repeated load systems, cf. Figure 5. The assumed loading programme is described by $P_1 \in [0,1], P_2 \in [0,1], P_3 \in [-0.01, 0.02]$. The ideal sandwich cross-section is assumed with corresponding rhombic yield condition (axial force and bending moment interaction).

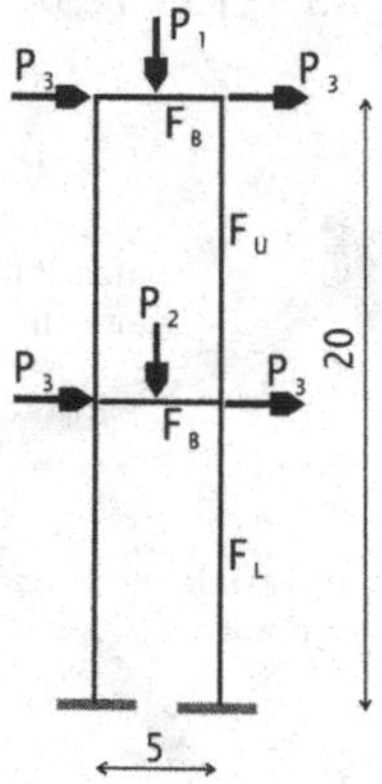

Figure 5. Two-story elastic-plastic frame subjected to three variable repeated load sets.

Two dimensionless design variables are introduced: $x^d = \{a_1, a_2\}^T$, $a_1 = F_L / F_B$ defined by the ratio of cross-sectional areas of the lower column to the beam and $a_2 = F_U / F_B$ defined by the ratio of cross-sectional areas of the upper column to the beam.

The problem is to maximize the shakedown multiplier at the condition of constant total volume of the frame equal to $V^* = 0.1$.

$$\max_{x^d} \eta(x^d) \tag{20}$$

$$\text{s.t.}\, V(x^d) \leq V^*$$

Figure 6 shows the shakedown surface H^0 constructed as a function $\eta^0(x^d)$ spanned over the two dimensional space of design variables vector x^d. A classical plastic optimization solution provides the optimum (maximum) point M at coordinates $x^d = (0.4123, 0.2214)$. This point is located at an intersection of piece-wise smooth parts of the H^0 surface. Each part corresponds to a different mechanism of plastic deformation and is characterized by a different gradient. Due to this the sensitivity to imperfections depends on the direction.

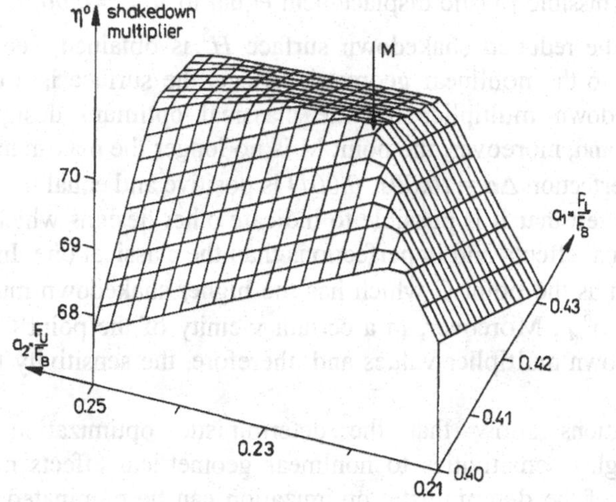

Figure 6. Initial shakedown surface H^0 constructed as the function $\eta^0(x^d)$.

For example for an imperfection $\Delta \boldsymbol{a}$ = *(0.001, 0)* the sensitivity of the shakedown multiplier $\phi_\eta = [\eta(a+\Delta a)-\eta(a)]/|\Delta a|$ is equal to ϕ_η = *-61*, whereas for the imperfection $\Delta \boldsymbol{a}$ = *(0, -0.001)* the sensitivity is equal to ϕ_η = *-217*.

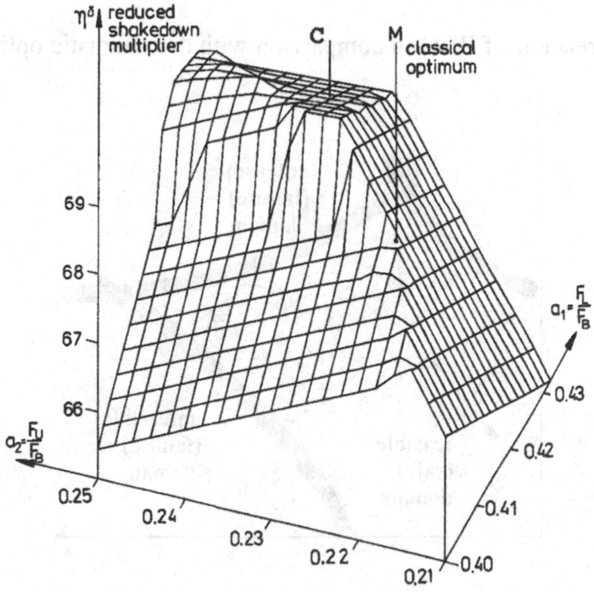

Figure 7. Reduced shakedown surface H^δ constructed as the function $\eta^\delta(x^d)$.

Assuming the admissible plastic displacement equal to $u_{adm} = 0.002L$ (L – characteristic length of elements) the reduced shakedown surface H^{δ} is obtained, see Figure 7. Due to different sensitivities to the nonlinear geometric effects the surface is a discontinuous one. The value of shakedown multiplier for the classical optimum design (point M) was significantly reduced and, moreover, the point M is no longer the maximum. For example the sensitivity to the imperfection $\Delta \boldsymbol{a} = (0.001, 0.001)$ is positive and equal to $\phi_{\eta} = 106$.

These results suggest that it is possible to indicate other designs which are less sensitive to nonlinear geometrical effects and imperfections than the classical one. Intuition suggests to consider designs such as the point C which has the higher shakedown multiplier, η^{δ}_{C}, than that of the point M, η^{δ}_{M}. Moreover, in a certain vicinity of the point C, there is no rapid change of the shakedown multiplier values and, therefore, the sensitivity to imperfections is rather low.

These considerations show that the deterministic optimization in the case of imperfections and higher sensitivities to nonlinear geometrical effects may lead to unsafe designs. Weaknesses of the deterministic optimization can be eliminated by formulation of the problem in the framework of reliability. On the contrary to (19) we arrive at the following reliability-based optimization (RBO) problem

$$\min_{x^{\mu}} V(\boldsymbol{x}^{\mu}) \tag{21}$$

$$\text{s.t.} \quad R[g(\boldsymbol{x}) \geq 0] \geq R^{*}$$

Simplified interpretation of RBO in comparison with deterministic optimization is shown in Figure 8.

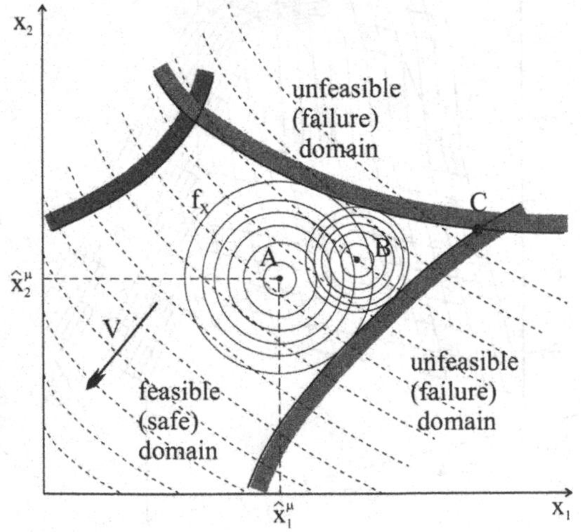

Figure 8. Deterministic- and reliability-based design.

We are looking for a point with minimum volume provided that the admissible reliability level R* is assured. In the deterministic case, without any imperfections under consideration, this leads to the optimum point C.

Accounting for imperfections we have to consider a bell-like probability density function f_x. We search for such position of the bell center (given by coordinates of the mean values, x^{μ}) which minimizing volume keeps the bell inside the safe domain, see Figure 8. For higher imperfections of parameters (higher coefficients of variation) the bell will assume a flatter shape with a larger radius, cf. the bell A, Figure 8. For smaller imperfections of design parameters the bell will take a pick shape with a small radius, cf. the bell B. In a limit case (no imperfections) it will be reduced to a rod, cf. point C.

The required reliability level R^* is a second factor shaping the bell. For higher reliabilities the tails must be accounted for and, therefore, the bell will be cut off with a larger radius. For smaller required reliability, the bell will be cut off with a smaller radius.

Hence, the reliability-based optimization will indicate the point A in the case of higher imperfections of design parameters (higher CoVs) and for higher required reliability R^*. For smaller imperfections and lower reliability it will provide the point B. This is in a full agreement with the intuitive approach presented in the case of two-story frame (Figure 7). Some steps of the optimization procedure for more general case (non-Gaussian variables) are shown in Figure 9 and 10.

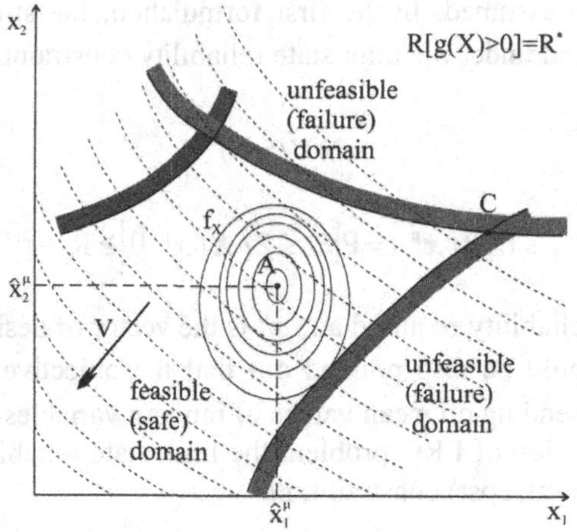

Figure 9. Reliability-based optimization. Feasible solution.

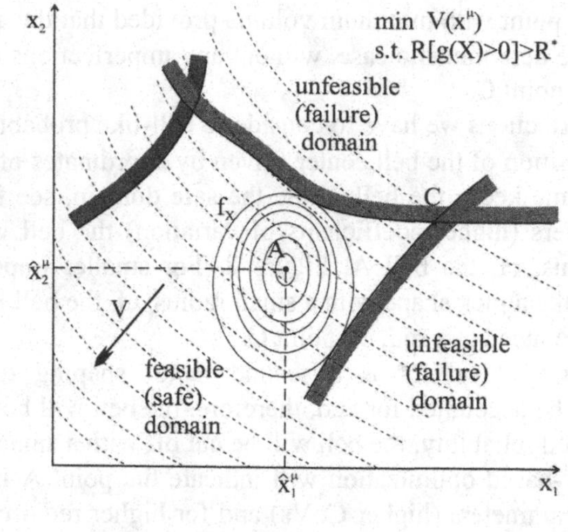

Figure 10. Reliability-based optimization. Optimum solution.

7 Limit Reliability Optimization

The limit reliability optimization (LRO) problem can be formulated in two ways depending on the objective function assumed. In the first formulation, the structural volume (weight, cost), $V(\boldsymbol{x}^{\mu})$, is minimized under the limit state reliability constraint, i.e.

$$\min_{\boldsymbol{x}^{\mu}} V(\boldsymbol{x}^{\mu}) \tag{22}$$

$$\text{s.t. } R(\boldsymbol{x}^{\mu}) = P[\Xi \le Z(\boldsymbol{X}(\boldsymbol{x}^{\mu}))] \ge R^{*}$$

where R^{*} is a minimum reliability required and $\boldsymbol{x}^{\mu}$ is the vector of design variables assigned to mean values of $\boldsymbol{X}$. It should be also pointed out that the objective function, $V(\boldsymbol{x}^{\mu})$, is a deterministic function depending on mean values of random variables $\boldsymbol{X}$.

In the second formulation of LRO problem the limit state reliability is maximized under the structural volume (weight, cost) constraint, i.e.

$$\max_{\boldsymbol{x}^{\mu}} R(\boldsymbol{x}^{\mu}) = \max_{\boldsymbol{x}^{\mu}} P[\Xi \le Z(\boldsymbol{X}(\boldsymbol{x}^{\mu}))] \tag{23}$$

$$\text{s.t. } V(\boldsymbol{x}^{\mu}) \le V^{*}$$

where V^{*} is a maximum allowable structural volume/weight/cost.

General formulations (22) and (23) allow to account for both linear and non-linear limit state formulations. The limit load multiplier Z may have the meaning of the reduced limit load multiplier Z^{δ} or the initial limit multiplier Z^{0}. In the case of structures insensitive to geometrical effects the failure function (7) may be used. However, for the sensitive structures, the extended formulation (12) should be used. Taking into account (10), the problems (22) and (23) may be formulated not in terms of the reliability *R* but of the failure probability p_F.

8 Shakedown Reliability Optimization

The shakedown reliability optimization (SDRO) problem may be formulated similarly to LRO. The structural volume (weight, cost), $V(x^{\mu})$, is minimized under the shakedown reliability constraint, i.e.

$$\min_{x^{\mu}} V(x^{\mu})$$

$$\text{s.t.}\quad R(x^{\mu}) = P[\Xi \le H(X(x^{\mu}))] \ge R^{*} \tag{24}$$

where R^{*} is a minimum reliability required and x^{μ} is the vector of design variables assigned to mean values of X.

In the second formulation of SDRO problem the shakedown reliability is maximized under the structural volume (weight, cost) constraint, i.e.

$$\max_{x^{\mu}} R(x^{\mu}) = \max_{x^{\mu}} P[\Xi \le H(X(x^{\mu}))] \tag{25}$$

$$\text{s.t.}\quad V(x^{\mu}) \le V^{*}$$

where V^{*} is a maximum allowable structural volume/weight/cost.

In both formulations (24) and (25), the shakedown multiplier *H* may have the meaning of the reduced shakedown multiplier H^{δ} to account for nonlinear geometric effects.

9 Example of Limit Reliability Optimization

Consider an example of a two-story elastic-plastic frame of Figure 5 subjected to proportional monotonic loads given by the reference loads $P_1 = 100, P_2 = 100, P_3 = 2.5$. The ideal sandwich cross-section is assumed with corresponding rhombic yield condition (axial force and bending moment interaction). Two dimensionless design variables are introduced: $x^{d} = \{a_1, a_2\}^{T}$, $a_1 = F_L / F_B$ defined by the ratio of cross-sectional areas of the lower column to the beam and $a_2 = F_U / F_B$ defined by the ratio of cross-sectional areas of the upper column to the beam.

The problem is to maximize the limit multiplier at the condition of constant total volume of the frame equal to $V^* = 0.119$.

$$\max_{x^d} \zeta(x^d) \tag{26}$$

$$\text{s.t.} \quad V(x^d) \leq V^*$$

The deterministic optimization provides the optimum design located at the edges of limit load surfaces at the point C, see Figure 11. The value of the initial limit multiplier is equal to $\zeta^0(0.286, 0.139) = 1.023$ and the reduced limit multiplier is not unique and has values $\zeta^\delta(0.286, 0.139) = 1.016$ as well as $\zeta^\delta(0.286, 0.139) = 0.993$, see Figure 12.

In order to rationalize the design of the two-story frame, the reliability-based optimization will be carried out according to the formulation

$$\min_{x^d} p_F(x^d) = 1 - \max_{x^d} R(x^d) \tag{27}$$

$$\text{s.t.} \quad V(x^d) \leq V^*$$

$$R = P[\xi \leq \zeta]$$

Figure 11 shows the contour plot of the initial limit load surface $Z^0(a_1, a_2)$ with the position of deterministic optimum C and the optimum designs L and D obtained for different coefficients of variance of geometric and loading parameters. As a rule it is observed that higher imperfections of geometric parameters (CoV = 5% and 10%) shift optimum designs towards a plateau, far away from the point C. Similarly, Figure 12 shows the contour plot of the reduced limit load surface $Z^\delta(a_1, a_2)$.

Table 1 shows probabilities of failure obtained for the points C and L obtained without considering the geometric effects (Z^0 surface) and with accounting for geometrical effects (Z^δ surface) for the coefficient of variance of geometric parameters CoV =5% and 10%.

Table 1. Probabilities of failure for load multiplier mean $\xi^\mu = 0.8$.

	CoV = 5%		CoV = 10%	
	Design C	Design $L^\alpha_{0.8;5\%}$	Design C	Design $L^\alpha_{0.8;10\%}$
Neglected geometric effects p_F^0	$1.2 \cdot 10^{-3}$	$8.8 \cdot 10^{-5}$	$9.2 \cdot 10^{-2}$	$5.3 \cdot 10^{-2}$
Geometric effects accounted for p_F^δ	$4.5 \cdot 10^{-3}$	$2.5 \cdot 10^{-4}$	$1.4 \cdot 10^{-1}$	$6.8 \cdot 10^{-2}$

It is easy to see that the deterministic optimisation provides a design which is much less reliable than the designs provided by limit reliability optimisation. The difference in failure probabilities may reach 2-3 orders. Also, neglecting geometric effects may cause that the failure probability will be significantly underestimated.

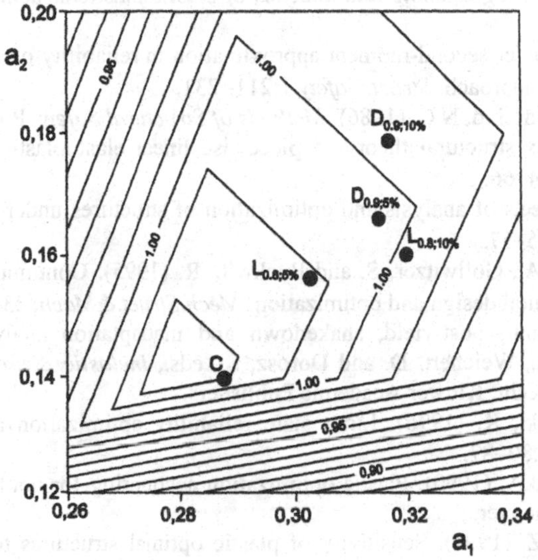

Figure 11. Contour plot of the initial limit load surface $Z^0(a_1, a_2)$.

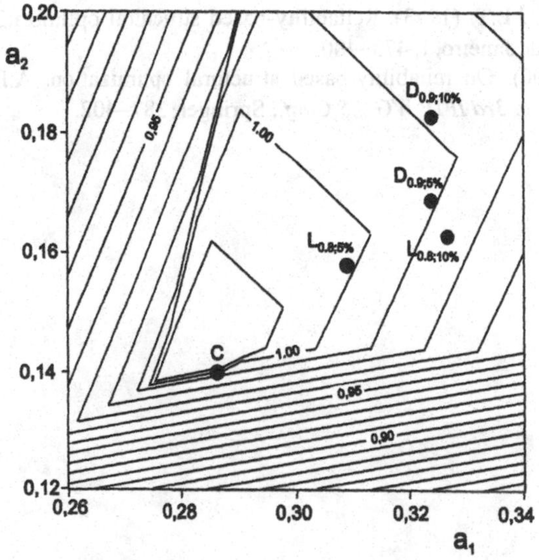

Figure 12. Contour plot of the reduced limit load surface $Z^\delta(a_1, a_2)$.

References

Bjerager, P. (1988). Probability integration by directional simulation, *J.Eng.Mech., ASCE* 114:1285-1302.

Cohn, M.Z., Ghosh, S.K. and Parimi, S.R. (1972). A unified approach to the theory of plastic structures. *J.Eng.Mech.Div., ASCE* 98, EM5, 1133–1158.

Cyras, A.A. (1969). *Linear programming and analysis of elastic-plastic structures.* Leningrad: Stroiizdat (in Russian).

Dolinski, K. (1983). First order second-moment approximation in reliability of structural systems: critical review and alternative approach. *Struct. Safety* 1:211–231.

Madsen, H.O., Krenk, S. and Lind, N.C. (1986). *Methods of Structural Safety*. Prentice-Hall.

Maier, G. (1970). A matrix structural theory of piecewise linear elastoplasticity with interacting yield planes. *Meccanica* 5:54–66.

Nguyen, D.H. (1983). Aspects of analysis and optimization of structures under proportional and variable loadings. *Eng.Opt.* 7:35–57.

Santos, J.L.T., Siemaszko A., Gollwitzer, S. and Rackwitz R. (1995). Continuum sensitivity method for reliability-based structural design and optimization. *Mech.Struct.&Mach.* 23:497–520.

Siemaszko, A. (1995). Limit, post-yield, shakedown and inadaptation analysis of inelastic discrete structures. In Mróz, Z., Weichert, D. and Dorosz, S., eds., *Inelastic Behaviour of Structures under Variable Loads*, Dordrecht: Kluwer Academic Publishers.

Siemaszko, A. and Dolinski, K. (1996). Limit state reliability optimization accounting for geometric effects. *Struct.Opt.*, 11:80–87.

Siemaszko, A. and König, J.A. (1990). Plastic optimization accounting for nonlinear geometrical effects. GAMM'90 Conf., Hannover.

Siemaszko, A. and Mróz, Z. (1991). Sensitivity of plastic optimal structures to imperfections and non-linear geometrical effects. *Struct.Opt.* 3:99–105.

Siemaszko, A. and Mróz, Z. (1991). Optimal plastic design of imperfect frame structures. Progress in Structural Engineering, D. Grierson , A. Franchi, P. Riva (eds.). Dordrecht: Kluwer Academic Publishers.

Siemaszko, A. and Santos, J.L.T. (1993). Reliability-based structural optimization. *Proc. Struct. Opt. 93 World Congress*, Rio de Janeiro, I, 473–480.

Thoft-Christensen, P. (1990). On reliability-based structural optimization. A.Der Kiureghian, P.Thoft-Christensen (eds.), *Proc. 3rd IFIP WG 7.5 Conf.*, Springer, 387–402.

Application of Shakedown Theory to Fatigue Analysis of Structures

Ky Dang Van

UMR 7649 - CNRS, Palaiseau, France

Abstract. Fatigue failure is final result of complex microscopic phenomena which occur under cyclic loading. Traditionally this phenomenon is studied by different ways depending on the fatigue regime and on the field of interest: fatigue limit analysis, life prediction in high or in low cycle fatigue, thermal fatigue... The diversity of proposed approaches is so great that design engineers meet many difficulties to have a clear idea of the fatigue calculations which have to be done. The purpose of this paper is to present an original unified approach to both high and low cycle fatigue based on shakedown theories and dissipated energy. The discussion starts with an explanation of fatigue phenomena at different scales (microscopic, mesoscopic, and macroscopic). Then some useful aspects of shakedown theory in relation with fatigue are presented. Applications to modeling of high cycle fatigue is then introduced: for instance, some multiaxial fatigue criteria (Dang Van, Papadopoulos) are essentially based on the hypothesis of elastic shakedown at the mesoscopic scale and therefore a bounded cumulated dissipated energy. In the low cycle fatigue regime, some recent results show that we can speak of a plastic shakedown at both mesoscopic and macroscopic scale and a cumulated energy bounded by the failure energy. These ideas are also justified by some infrared thermography test results permitting a direct determination of the fatigue limit.

1 Introduction to Fatigue Analysis of Structures

Current industrial design is highly concerned by fatigue, as this phenomenon became the main failure mode of mechanical structures under variable loads. During a long period, starting with the pioneering work of Wöhler (1860) and ending in the late fifties, high-cycle fatigue (HCF) was the most significant topic for engineers and researchers. Mechanical engineers were mainly interested in establishing S-N curves and in determining fatigue limits for metallic materials because they are concerned with the effect on lifetime of external loadings. During this period only simple methods like beam theory are available for engineering computations. The stress evaluated by that way are most of the time uniaxial and simple so that S-N curves approach is in adequation with these computational tools. However scientists are aware of the need of more general criteria and the first multiaxial criteria were proposed quite early in the beginning of the last century.

In the sixties two new approaches for studying fatigue were introduced.

First a special interest occurred in studying low-cycle fatigue (LCF). Instead of developing stress approaches, Coffin and Manson proposed fatigue models based on the strain amplitude or the plastic strain amplitude.

Nearly at the same time Paul Paris proposed to use linear fracture mechanic approach for studying fatigue crack propagation. He has first to overcome the doubts of fatigue reviewers who ignore at that time this new subject. This new way for studying this old science has rapidly a great audience. With the help of new experimental devices the development of research on fatigue crack propagation became "à la mode" and increases exponentially. Nowadays some authors consider that, in fact, fatigue is only a crack propagation phenomenon and fatigue limit corresponds to a fatigue crack propagation threshold.

These two new trends for studying fatigue are favoured by new experimental apparatus and more and more precise observation devices. The two cited approaches showed their effectiveness in some cases for structures in the aeronautical and the nuclear industry. However these successes were obtained when the stress or the strain cycle were uniaxial and simple. As stresses and strain are often multiaxial and present a complex path during a loading cycle, application of these models is then difficult and the predicted results are often not in reasonable agreement with test results. Nevertheless these new fatigue approaches has so much success that the more classical domain of high cycle was more or less neglected: few new ideas concerning this field of application was proposed despite the great number of existing fatigue criteria. Surprisingly most of the time, the only theoretical tools used by the mechanical industries are always those developed during the 19th or at the beginning of the 20th century. Nowadays S-N curve, Goodman diagram approach are still used as design tools to predict fatigue resistance of mechanical structures, even if they are only applicable in uniaxial situations. The computational methods like FEM or integral equations methods are much more efficient in comparison with beam theory so that the stress and the strain states on a modern structures are defined not only by nominal stress or strain value but by fields of tensorial components. The S-N curve method is clearly not sufficient and more sophisticated methods are needed. This is particularly true when the fatigue resistance of mechanical structures which have to resist to complex cyclic loadings is evaluated, which necessitates the use of multiaxial fatigue criteria. In the high cycle regime, many proposals exist. They are most of the time based on stress or stress amplitude parameters, corresponding for instance to adaptation of plasticity criteria to fatigue applications. For instance, many fatigue criteria are obtained by replacing Tresca or Mises equivalent stress by similar quantities formulated by using amplitude of stress. Predictions are often poor and not in reasonable agreement with experiments so that the design of structures which have to resist to fatigue and particularly to high cycle fatigue is still a problem: engineers have to perform difficult and time consuming experiments to find fatigue limit of mechanical structures. This is done generally on the structure itself, which necessitates iterative experiments on one to one scale test models.

The need of more efficient design methods, the generalization of finite elements method calculations favour the development of researches. Therefore a sustained effort has been deployed for deriving reliable fatigue computational methods applicable to the previous situations. Beside the fatigue crack propagation methods (which are not studied hereafter), the actual prediction techniques are generally based on multiaxial fatigue criteria using a stress approach in the HCF regime and on strain or inelastic strain approaches in the LCF domain. The materials fail in both domains in similar ways, however nowadays the approaches for modelling these two types of fatigue are different. In the following, a tentative explanation will be given.

In the present paper a unified approach is proposed in order to overcome this difficulty. It is derived from a HCF multiaxial fatigue theory based on a multiscale approach (Dang Van, 1999), from some recent results in LCF and shakedown theories (Koiter, 1960; Mandel et al., 1977; Nguyen, 2000).

2 Recall of the Multiscale Approach in Fatigue

In discussing fatigue phenomena we shall distinguish three scales:

- The microscopic scale of dislocations
- The mesoscopic scale of grains
- The macroscopic scale representing phenomena at the scale of the engineering

In a simplified analysis we could say that fatigue phenomena start generally with appearance of slip bands in grains which broaden progressively during the cycles. The proportion of grains in which slip bands develop increases with the applied load.

In the *high-cycle fatigue regime* (HCF), in general no irreversible deformation, (i.e. plastic or viscous), is detected at the macroscopic level. The material behaviour seems to be purely elastic and as a consequence, the use of stress or strain at this engineering scale are equivalent. In practice stress is often preferred to strain. However at a *mesoscopic* level, plasticity occurs in certain number of grains and generates a *heterogenous* plastic strain. Only misoriented crystals undergo plastic slip corresponding to a heterogeneous distribution of micro cracks. The initiation of the first visible crack, at the macroscopic scale represents a large part of the fatigue life.

The *low-cycle fatigue regime* (LCF) implies significant macroscopic deformations conducting to irreversible deformations at this level. At the mesoscopic level, the metal grains are subjected to plastic deformation in a more homogeneous manner than in HCF regime. The first micro cracks in the persistent slip bands appear quite early in the life of the structure. The strain and the plastic strain are no more related to the stress through a simple relation since, as it is very well known, it depends also on the loading path.

In both LCF and HCF, damage phenomena occur in the grains, and therefore the use of mesoscopic fields seem to be relevant for studying fatigue phenomena. Let us recall, that the macroscopic fields (stress Σ, strain E, plastic strain P...) are in a certain sense approximately the mean value of the mesoscopic fields (stress σ, strain ε, plastic strain p...). The macroscopic fields are therefore supposed constant in a small volume, surrounding the point under consideration. In the theories of polycrystalline aggregates, this volume is called "representative volume element" or RVE. For instance the mesoscopic stress σ and the macroscopic stress Σ are related by the following formula:

$$\sigma = A.\Sigma + \rho \qquad (1)$$

where ρ is the local residual stress and A is the elastic stress localization tensor. A is the identity tensor if local and macroscopic elastic moduli are similar. This relation shows that it is incorrect to use Σ for characterizing phenomena which occur at the grain scale since the *local stress σ is not proportional to Σ* and does not include information about ρ.

As fatigue is caused by irreversible phenomena let us compare the dissipated energy at both mesoscopic and macroscopic scales. It is well known (see for instance Dang Van, 1999) that the total macroscopic work rate is the mean value in the RVE of the local total work rate. However, the equality between the mesoscopic and macroscopic energy does not hold for plastic dissipation as proven by H.D. Bui and recalled also in Dang Van (1999). The difference between macroscopic plastic dissipation and mean value of mesoscopic plastic dissipation decreases with increasing plastic strain, as the plastic heterogeneity from grain to grain decreases. This also justifies why macroscopic plastic deformation is a reasonable approximation in LCF.

The evaluation of the local mesoscopic fields from the macroscopic ones is in general a difficult task since *the material is locally heterogeneous and has to be considered as a structure when submitted to complex loading histories*. Depending on the loading characteristics one can accept different simplifying assumptions which will permit a solution to the problem. The multiscale approaches in fatigue which we proposed are precisely based on the use of mesoscopic parameters instead of engineering macroscopic quantities. In order to derive a unified theory of fatigue, we shall suppose that the elastic shakedown occurs at the level of the microstructure as well as at the macroscopic one.

Before studying the application of the shakedown theory to the microstructure, let us examine some physical models which were at the origin of the proposed approach.

3 On Some Physical Modellings of Fatigue

The first multiscale fatigue model was proposed by Orowan (1939). In fact, this model is only qualitative. Orowan want to give a plausible explanation of fatigue fracture near the fatigue limit. In this regime, the stress could be well below the macroscopic yield limit. To derive his model, this author remarks that fatigue is generally due to stress concentrations, heterogeneity... and that the first fatigue phenomena are microscopic and local i.e. they appear in some grains which have undergone plastic deformations. These deformations are localized in intra-crystalline shear bands, the rest of the matrix behaves elastically because the macroscopic plastic strain is negligible. The first cracks initiates precisely in these shear bands.

Using this image, Orowan proposed the following model depicted in Figure 1.

A weak plastic element is embedded between two elastic springs, which impose their deformation to the whole system. The weak element undergoes plastic strain and hardens. If the plastic behaviour of this element is governed by pure isotropic hardening, as shown on the Figure 1, then its response tends towards an elastic shakedown state precluding fatigue. The limit state in that case oscillates between A (the corresponding shear is τ) and B (the corresponding shear is $-\tau$). The local shear loading is symmetric even if the prescribed external loading is not symmetric. Then, if the limit range is less than some definite value, there is no fatigue. On the contrary, fatigue occurs if this condition is not verified.

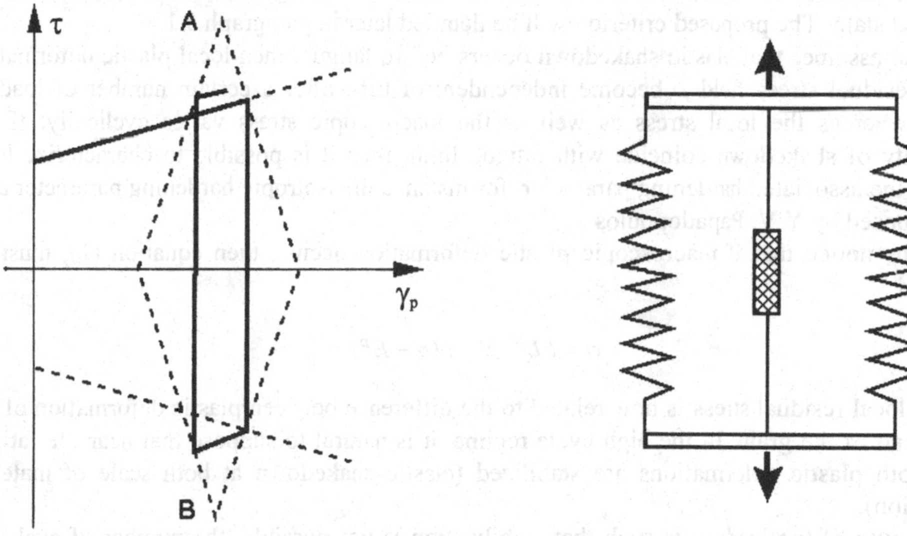

Figure 1.

A very simple generalization of Orowan's model was proposed by Dang Van based on the use of Lin Taylor model.

Let us consider an inclusion submitted to uniform plastic strain p and embedded in an elastic matrix. Let L and l be respectively the elastic compliance of the matrix and the inclusion. Suppose that under external loading, the total strain of the matrix and of the inclusion are the same (Lin-Taylor model):

$$E = \varepsilon^e + p$$

Multiplying both sides of the previous relation by l and taking into account the elastic stress-strain relations:

$$\sigma = l:\varepsilon^e$$
$$E = L^{-1}:\Sigma$$

one obtains:

$$\sigma = l:L^{-1}:\Sigma - l:p \tag{2}$$

This relation is similar to equation (1), where $l:L^{-1}$ is a localization tensor that concentrates the stress (tensor A of equation (1)) and $l:p$ corresponds to the local residual stress induced by the mesoplasticity (plasticity in the grain). Then similarly to Orowan, the fatigue crack initiation criterion in high cycle regime can be formulated using local stress tensor σ in the stabilized state. The proposed criterion will be detailed later in paragraph 5.1.

If one assumes that elastic shakedown occurs before fatigue, then local plastic deformation p and residual stress field ρ become independent of time after a certain number of loading cycles, whereas the local stress as well as the macroscopic stress varies cyclically. If the possibility of shakedown coincide with fatigue limit, then it is possible to characterize limit state by the associated hardening parameter, for instance the isotropic hardening parameter as it was proposed by Y.V. Papadopoulos .

Let us notice that if macroscopic plastic deformation occurs, then equation (2), must be modified:

$$\sigma = l:L^{-1}:\Sigma - l:(p - E^p) \tag{3}$$

The local residual stress is now related to the difference between plastic deformation of the matrix and of the grain. In the high cycle regime, it is natural to suppose that near the fatigue limit, both plastic deformations are stabilized (elastic shakedown at both scale of material description).

However, if the loading is such that stabilization is not possible, the number of cycle for fatigue failure is low. This regime corresponds to low cycle (plastic) fatigue.

4 Elastic Shakedown of an Elastoplastic Structure

Under cyclic loadings, an elastoplastic mechanical structure may have three possible asymptotic responses after a certain numbers of cycles (which could be infinite):

- elastic shakedown which corresponds to stabilization on a pure elastic response;
- elastoplastic shakedown when this response is stabilized on an elastoplastic cycle;
- ratchet when their is no possible stabilization.

The static theorem of Melan gives a sufficient condition for elastic shakedown for a structure made of elastic perfectly plastic material. It can be stated as followed,:
If there exist a time θ and a fixed (i.e. independent of time t) self equilibrated stress field $R(x)$ and a security coefficient m such that $\forall$ point x of the structure and $t > \theta$, $g(m(\Sigma_{el}(x, t) + R(x))) < k^2$, the structure will shakedown elastically.

Σ_{el} is the stress response of the structure under the same external loading, but under the assumption that the constitutive material has a pure elastic behaviour.

(This formulation due to W. Koiter differs slightly from the original formulation of Melan)

Demonstration and discussions of this theorem was given in a famous paper of Professor W. Koiter (Koiter, 1960). In this paper, Koiter draw our attention on the fact that this theorem and its proof do not say anything about the magnitude of plastic deformation which may occur before the structure reaches its shakedown state. It is clear that too large plastic deformation gives a solution, which has no physical meaning. But, Professor Koiter added that if *"the total amount of plastic work performed in the loading is accepted as suitable criterion for assessing the overall deformation, boundedness of the overall deformation may be proved if the structure has a safety factor $m > 1$ with respect to shakedown."*

We do not reproduce more detail of this discussion; *we shall retain the condition that total plastic work must be bounded to ensure acceptable bounds on plastic deformation.*

Melan's theorem was extended by different authors to account for more realistic material behaviour. In particular, generalization to elastoplastic material combining linear kinematic and isotropic hardening by Mandel et al. (1977) (and Q.S. Nguyen but with an other formalism which will be recall hereafter) is particularly interesting. However, these theorems are difficult to apply, because the fixed stress field $R(x)$ must be self equilibrated, a condition which is not easy to fulfil.

It is why Mandel et al. gives an other proposal, which is a *necessary condition* of elastic shakedown. This last condition can be summarized as followed (Mandel et al., 1977):

Shakedown occurs if it exists a fixed stress tensor Σ^* (not necessarily self equilibrated) such that

$$\forall t > \theta,\ g(\Sigma_{el}(x, t) - \Sigma^*) - K^{*2}(P_{eq}) \leq 0 \tag{4}$$

The isotropic hardening parameter K is supposed to be an increasing function of equivalent plastic strain P_{eq} beyond some limit K^*. $K^{*2}(P_{eq})$ is the maximum acceptable value of the yield radius.

Thus, at the shakedown limit, Σ^* is the centre of the of the smallest hypersphere surrounding the local loading path $\Sigma_{el}(x, t)$, the radius of which is $K^*(\varepsilon_p)$.

This theorem is very useful in fatigue.

Let us return to the Melan Koiter sufficient shakedown condition. Koiter's reasoning can be also extended to strain hardening material in the framework of the generalized standard material theory as introduced by Q.S. Nguyen et al. and recalled in Nguyen (2000). (Most of the classical metallic material belong to this class). The boundedness of the dissipation (which corresponds to plastic work plus work induced by generalized strain hardening parameters) ensures that the plastic deformation as well as strain hardening parameters are bounded. (see Nguyen, 2000).Since these results are not familiar to fatigue specialists, it is necessary to recall some important points of these theories:

Let us begin first with the classical case of a perfect elastic plastic material. The elastic domain is defined by

$$g(\Sigma) - k^2 \leq 0 \tag{5}$$

and the associated plastic flow rule can be written :

$$\dot{P} = \mu \frac{\partial g}{\partial \Sigma} = N_c(\Sigma), \quad g \leq 0, \quad \mu \geq 0, \quad \mu\, g = 0$$

μ is the plastic multiplier which is non negative; one can see easily that the plastic flow direction is normal to the plasticity convex at the point Σ, which can be denoted by

$$\dot{P} = N_c(\Sigma). \tag{6}$$

The dissipation rate is

$$D(\dot{p}) = \Sigma : \dot{P}.$$

The proof of the classical Melan Koiter theorem for elastic perfectly plastic material can be done in two steps :

- in the first step, it is proved that under the previous assumptions, the dissipated energy is bounded.
- in the second, one demonstrates that the distance between $\Sigma(x, t)$ and $\Sigma_{el}(x, t)$ tends toward a fixed value.

Only demonstration of the first step is recalled in the following,

Taking account of the property (6), one has for all plastically admissible stress field Σ^*,

$$\int_V (\Sigma - \Sigma^*) : \dot{P}\, \mathrm{d}V \geq 0,$$

choosing $\Sigma^* = m(R + \Sigma_{el}(t))$, which is plastically admissible by assumption, the following inequality is derived :

$$\int_V (\Sigma - R - \Sigma_{el}):\dot{P}\, dV \geq \frac{m-1}{m} \int_V \Sigma:\dot{P}\, dV$$

Since $(\Sigma - R - \Sigma_{el})$ is self equilibrated and the rate of displacement field $\dot{u} - \dot{u}_{el} = 0$, the application of virtual work principle gives:

$$\begin{aligned} 0 &= \int_V (\Sigma - R - \Sigma_{el}):(\dot{E} - \dot{E}_{el})\, dV \\ &= \int_V \left((\Sigma - R - \Sigma_{el}):\dot{P} + (\Sigma - R - \Sigma_{el}):M:(\dot{\Sigma} - \dot{R} - \dot{\Sigma}_{el}) \right) dV \end{aligned}$$

where M is the elasticity matrix ; one concludes that

$$-\int_V (\Sigma - R - \Sigma_{el}):M:(\dot{\Sigma} - \dot{R} - \dot{\Sigma}_{el})\, dV \geq \frac{m-1}{m} \int_V \Sigma:\dot{P}\, dV$$

after time integration, one obtains the following inequality

$$I(t_0) - I(t) \geq \frac{m-1}{m} W_p \qquad (7)$$

W_p is the plastic work and $I(t)$ is defined by

$$I(t) = \int_V \frac{1}{2} (\Sigma - R - \Sigma_{el}):M:(\Sigma - R - \Sigma_{el})\, dV$$

The dissipated energy is thus bounded.

This result can be extended to more general elastoplastic behaviour, thanks to the *general standard material concept* introduced first by Halphen and Nguyen (1975). Beside the usual strain parameters E and P, these authors introduced a set of strain hardening parameters denoted symbolically by β. It is then possible to define a potential energy $W(E, P, \beta)$, from which one derives the family of "associated generalised forces" (Σ, A_p, A_β). More precisely:

$$\Sigma = \frac{\partial W}{\partial E}, \quad A_p = -\frac{\partial W}{\partial P}, \quad A_\beta = -\frac{\partial W}{\partial \beta}$$

The elastic domain is a convex of the "generalised force" space define by the inequation:

$$g(A_p, A_\beta) \leq 0$$

The constitutive equations can be written in a symbolical way like in perfect plasticity:

$$\alpha = N_c(A)$$

where $\alpha = (P, \beta)$. The previous equation corresponds to (generalised) normality rule, like in perfect plasticity (equation (6)). More precisely:

$$\dot{P} = \mu \frac{\partial g}{\partial A_p}, \quad \dot{\beta} = \mu \frac{\partial g}{\partial A_\beta}, \quad g \leq 0, \quad \mu \geq 0, \quad \mu g = 0$$

The maximum dissipation principle which can be written for all $A^* \in C(\alpha)$ (A^* is said to be plastically admissible):

$$A.\dot{\alpha} \geq A^*.\dot{\alpha}$$

results from the normality rule and the convexity of C. This expression is an extension of Hill's maximum principle (or Drucker's postulate) to generalised elastoplastic behaviour.

The corresponding Melan Koiter theorem for generalised standard material is then:

Elastic shakedown occurs, whatever the initial state of the structure, if it exists a field of internal parameters α^* and a security coefficient $m > 1$ such that the associated generalised force field $mA^*(t)$ is plastically admissible $\forall t > T$.

Under these assumptions, Nguyen (2000) demonstrates the following inequality :

$$I(t_0) - I(t) \geq \frac{m-1}{m} \int_0^t \int_V A.\dot{\alpha} \, dV dt \tag{8}$$

where

$$I(t) = \int_V W(E - E^*, \alpha - \alpha^*) \, dV.$$

This expression is similar to inequality (7) corresponding to classical Melan Koiter theorem for elastic perfectly plastic material.

The dissipated energy being bounded, the set of parameters α representing the plastic strain and the internal hardening parameters are also bounded.

Let us notice that the only assumption on the anelastic strain P is that $\alpha = (P, \alpha)$ must fulfil the normality law. Except this requirement P could be purely deviatoric or can include a hydrostatic part.

5 Application of the Shakedown Theory to Fatigue

We propose to apply the shakedown theory to the microstructure in order to derive a unified fatigue model valid for structural applications. The main assumptions of this model is the following:

1. *near the fatigue limit but below it, elastic shakedown takes place at all scales of material description, at the macroscopic scale as well as at the mesoscopic scale. In particular the local plastic dissipation must be bounded.*
2. *If the loading history is such that elastic shakedown is not possible, then the local admissible dissipation is bounded, This bound corresponds to fatigue initiation energy. The number of cycle necessary to dissipated this energy corresponds to the initiation period.*

5.1 Application to Fatigue Limit Criterion

In the high cycle regime, only few misoriented grains (relative to the loading) undergo plastic deformation in localized slip bands. Under the fatigue limit, the dissipation is bounded so that the dissipation per cycle decreases and after a while becomes negligible. In engineering applications, it is therefore not easy to characterize it by evaluating directly the accumulated plastic strain, and consequently it is difficult to calculate dissipated energy cycle by cycle as it was done by Papadopoulos (1999). This calculation is in fact only a way to construct a theoretical model of fatigue then to propose a practical method.

It is the reason why two different ways were explored:

- when the loading characteristics correspond to the fatigue limit, then the asymptotic stabilized stress state is contained in a limit yield surface defined by the limit radius K^*. Papadopoulos in an early work proposes a theory in which K^* depends (linearly) on the maximum hydrostatic tension induced by the loading cycle (Papadopoulos, 1987);
- in 1973, Dang Van, generalizing an idea of Orowan, proposed to consider the mesoscopic current stress state at the apparent stabilized (shakedown) state as the relevant parameters in order to formulate a polycyclic multiaxial fatigue resistance criterion. More precisely, the proposed criterion is a combination of mesoscopic shear $\tau(t)$ and the concomitant hydrostatic pressure $p_H(t)$; more precisely if

$$\tau(t) + a\, p_H(t) - b > 0$$

then fatigue will occur. The two coefficients a and b are material constants that can be determined by two simple types of fatigue experiments; b for instance corresponds to the fatigue limit in simple shear.

General application of this criterion requires:

- first to evaluate the mesoscopic stress tensor knowing the macroscopic stress cycle; this can be done under the assumption of elastic shakedown near the fatigue limit by constructing the smallest hypersphere surrounding the macroscopic loading path as it is suggested the theorem of Mandel et al. Details of this construction is given in Dang Van (1999);
- second one must consider the plane on which the set ($\tau(t)$, $p_H(t)$) is a "maximum" relative to the criterion. This computation can be done as following: the maximum local shear at any time t is given by

$$\tau(t) = \text{Tresca}(\sigma(t)) = \underset{I,J}{\text{Max}} \left| \sigma_I(t) - \sigma_J(t) \right|$$

The stresses $\sigma_I(t)$, $\sigma_J(t)$ are the principal stresses at time t. The quantity that quantifies the danger of fatigue occurrence is defined by:

$$d = \underset{t}{\text{Max}} \frac{\tau(t)}{b - a\, p_H(t)}$$

d is calculated over a period, and the maximum is to be taken over the cycle.

It is frequent in the applications in high cycle fatigue (elastic regime) to use *the concept of local equivalent stress* for a life duration N_i defined by

$$\tau_{0,i} = \tau + a_i\, p_H.$$

For the fatigue limit $\tau_{0,i}$ corresponds to material constant b, but is different from b in general because $\tau_{0,i}$ and a_i depend of N_i if a_i (slope of the fatigue line in τ–p_H diagram) depends weakly on N_i, taking $a_i \approx a$, it is possible to define the local equivalent stress by

$$\tau_0 = \tau + a\, p_H.$$

Very often τ and p_H are maximum at the same time, so that is sufficient in some applications to plot τ_{max} versus p_{Hmax} (cf. Fayard et al. (1996)).

Different applications of this criterion to fatigue analysis of mechanical structures can be found in the litterature. For instance:

- prediction of fatigue resistance of some automotive structures undergoing multiaxial loadings (cf. A. Bignonnet (1999);

- analysis of crack nucleation induced by repeated rolling sliding contacts which necessitates the evaluation of the stabilized mechanical stress and strain states after a great number of loading cycles (cf. H. Maitournam (1999)).

5.2 Application to Low Cycle Fatigue

Plastic fatigue or low cycle fatigue is very much studied since the pioneering work of Manson and Coffin. In order to fit experimental results, these authors proposed to use the amplitude of plastic strain as relevant parameter. These tests were uniaxial and strain controlled, and there are many indications that in that case, the stress is related to the plastic strain amplitude in the stabilised state so that the plastic dissipation can be considered as function of strain amplitude.

However in case of more complex cycles stress state there is no such relation, since it is also well known that in plasticity, the response depends closely of the loading path and of the constitutive equations. Generalisation to 3D formulation of elastoplastic cyclic curve is a convenient way to do, but which is not justified by any theoretical background. In view of plastic fatigue applications, many elastoplastic or elastoviscoplastic constitutive equations were proposed in the eighties (cf. J.L. Chaboche, J. Lemaitre, Mroz...). and summarised for instance in Lemaitre and Chaboche (1990). By numerical computations, it is then possible to evaluate plastic strain amplitude or plastic dissipated energy, but let us notice that for general cyclic loading paths, there is (on the contrary of uniaxial loading mentioned previously) no evident relation between those two quantities. Criteria based on plastic strain amplitude are then not equivalent to those based on plastic dissipation. The following question arises: what is the "good" parameter in plastic fatigue from practical and from theoretical point of view? From previous discussion, we prefer to use a criterion based on a limitation of the local dissipation, since this feature ensures that the corresponding deformation is also bounded which is a natural necessary condition for no rupture. Moreover, dissipation is easy to calculate, without any ambiguity, and presents many advantages particularly in problems involving thermomechnical loadings which are very frequent and important in mechanical industries (engines, power plants...). Such a typical problem is studied by Charkaluk et al. (2001). The mechanical structure is an exhaust manifold which is submitted to gas pressure and temperature varying on a wide range. Since this structure is clamped on the engine body, thermomechanical stresses arise inducing anelastic deformations and low cycle fatigue and even creep fatigue. For such a problem, the approaches deriving from classical L.C.F. are not efficient , since the stress varies with the temperature, for a given plastic strain. The use of a criterion based on a bound on dissipated energy, first identified on laboratory test specimens (isothermal strain controlled LCF tests and thermal fatigue tests on clamped specimens), then applied to the industrial structure for the prediction on the fatigue life (locus of crack and life duration) give very good results. This methodology is successfully applied for the design of structures submitted to thermomechanical loadings (exhaust manifold shown on Figure 2, cylinder head...).

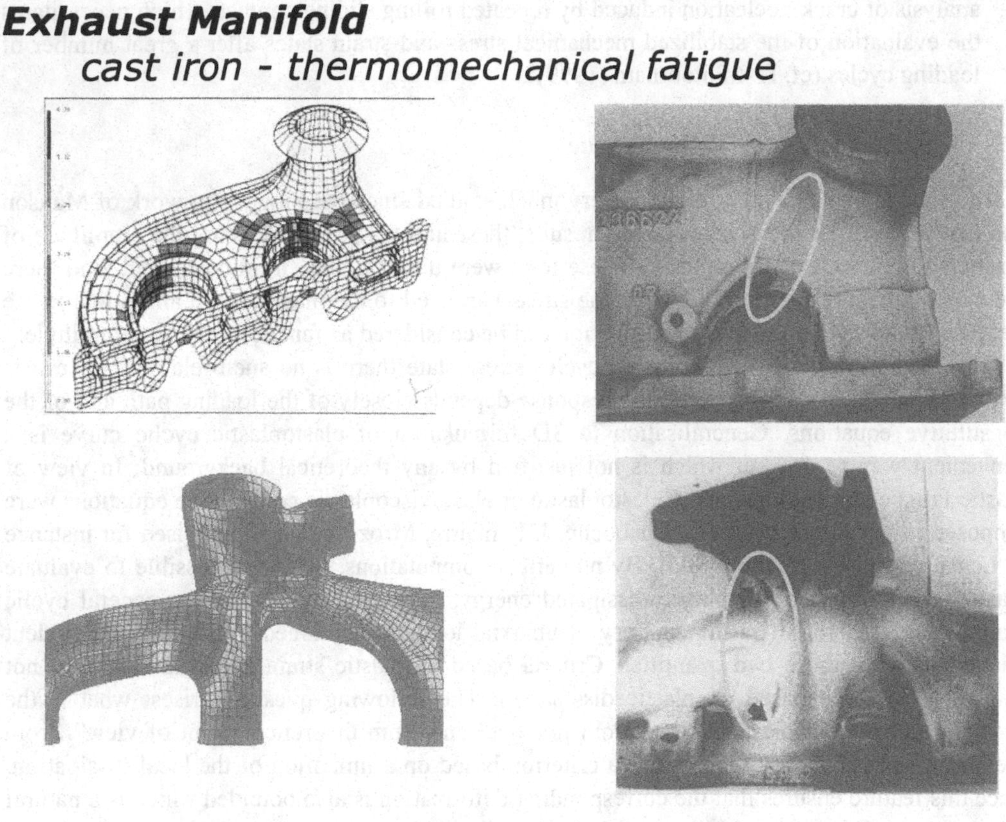

Figure 2. Prediction of crack initiation on exhaust manifolds submitted to thermomechanical cyclic loadings

5.3 Interpretation of the Infrared Thermographic Evaluation of the Fatigue Limit

Evaluation of fatigue limit by conventional testing methods (stair case …) take a lot of time. In order to shorten the experiments, some researchers proposed to use infrared thermography which is a convenient technique for producing heat images from the invisible radiant energy emitted by the test specimen submitted to cyclic loading at an adequate frequency. The temperature rise can be thus captured by the thermographic camera as it is shown on the Figure 3.

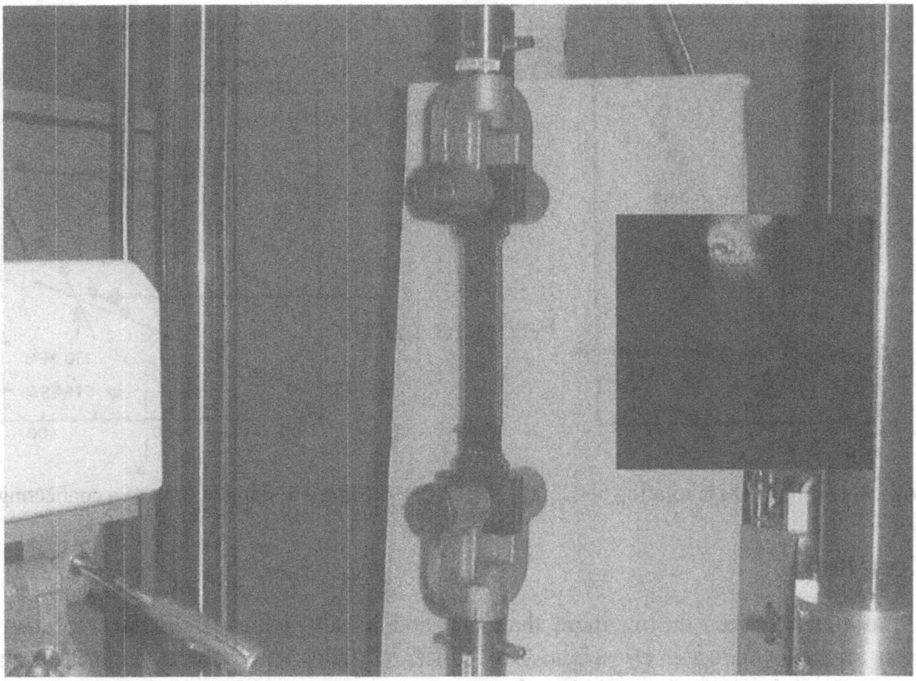

Figure3. Determination of fatigue limit using thermographic camera

When the load is increased, one can observe in the same time an increase of the temperature of the specimen. As it was shown by Luong et al. (1992, 1995), the manifestation of the fatigue damage is revealed by a break of the curve temperature (or also intrinsic dissipation) versus the intensity of the cyclic load as it is shown on Figures 4 and 5. Figure 4 corresponds to determination of the fatigue limit of XC 55 steel by rotating bending test. Figure 5 corresponds to test done in order to evaluate the fatigue limit of an automotive connecting rod. In both cases, the results obtained very quickly by the method using infrared thermography are similar to those obtained by classical fatigue tests. Until now there is no convincing explanation of this coincidence. In the proposed theory of fatigue, the reason of is this coincidence is very clear: the fatigue limit corresponds to a threshold on dissipative energy; below this threshold, dissipation is bounded, and as a consequence plastic strain (and internal strain hardening parameters) are finite as it was shown by W. Koiter for perfectly plastic material and extended by Q.S. Nguyen for generalized standard material (most of metallic strain hardening material). The break on the curves of Figures 4 and 5 corresponds to two different dissipative rate regimes.

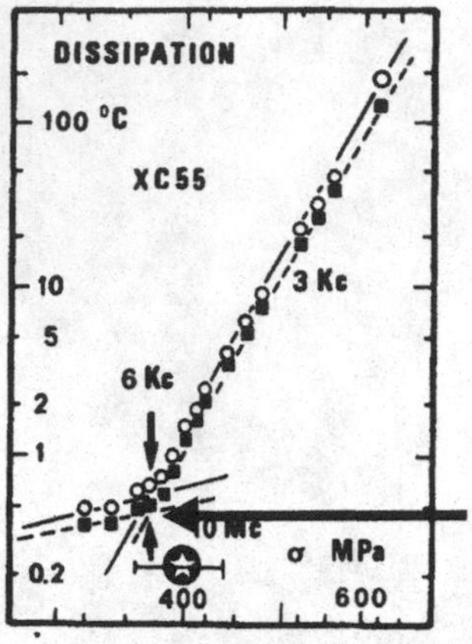

Figure 4. Fatigue limit in rotating bending

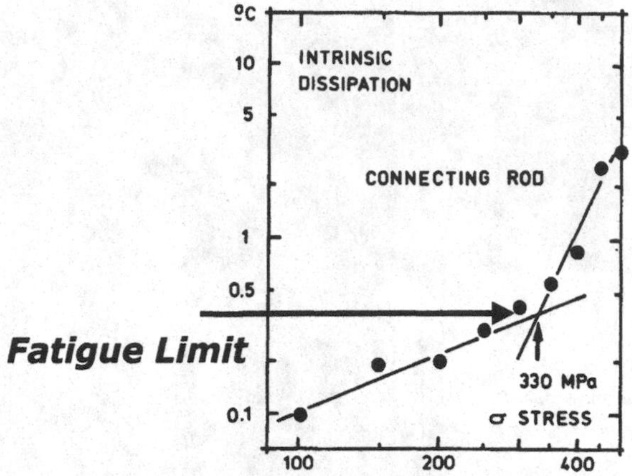

Figure 5. Fatigue limit of a connecting rod

6 Conclusion

Metal fatigue has been during more than one and a half century the subject of numerous research studies conducted by scientists of different fields: mechanical engineers, material scientists, physicists, chemists…Although many progresses were done in the understanding of physical phenomena, many difficulties still exist to achieve an interdisciplinary consensus in the way to modelise fatigue crack initiation. Depending on the discipline, points of view are often very different. A mechanical approach is presented which seems promising: it is based on shakedown hypothesis. Many applications on industrial structures submitted to complex multiaxial loadings are already successfully done; some of these applications, in the domain of high cycle fatigue as well as in plastic or viscoplastic creep fatigue regime are shown this paper.

In final conclusion, lets us recall this sentence written in 1963 by Professor Daniel Drucker which summarizes very elegantly our proposal (Drucker, 1963): *"when applied to the microstructure there is a hope that the concepts of endurance limit and shakedown are related, and that fatigue failure can be related to energy dissipated in idealized material when shakedown does not occur"*. It seems that this sentence was not known by the fatigue scientist community.

References

Bignonnet, A. (1999). Fatigue design in automotive industry. In Dang Van, K. and Papadopoulos, I.V., eds, *High Cycle Metal Fatigue, From Theory to Applications*, CISM Courses and Lectures N° 392, International Center for Mechanical Sciences, Springer: Wien, New-York. 145–167

Charkaluk, E., Bignonnet, A., Constantinescu, A. and Dang Van K. (2001). Fatigue design of structures under thermomechanical loading. *Fat. Fracture of Eng. Mat. Struct* (Accepted for publication).

Dang Van, K. (1999). Fatigue analysis by the multiscale approach. In Dang Van, K. and Papadopoulos, I.V., eds, *High Cycle Metal Fatigue, From Theory to Applications*, CISM Courses and Lectures N° 392, International Center for Mechanical Sciences, Springer: Wien, New-York. 57–88.

Drucker, D.C. (1963). *On the macroscopic theory of inelastic stress-strain-time-temperature behaviour.* Advances in Materials Research in the NATO Nations (AGAR Dograph 62) Pergamon press. 641–651.

Fayard, J.L., Bignonnet, A. and Dang Van, K., (1996). Fatigue design criterion for welded structures. *Fatigue Fract. Engng. Mater. Struct.* 19:723–729.

Halphen, B. and Nguyen, Q.S., (1975). Sur les matériaux standards généralisés. *J. Mécanique*, 14:1–37.

Koiter, W.T. (1960). General theorems for elastic-plastic solids. In Sneddon, J.N. and Hill, R., eds., *Progress in Solid Mechanics*, Amsterdam: Amsterdam: North-Holland. 165–221.

Lemaitre, J. and Chaboche, J.L. (1990). *Mechanics of Solid Materials.* Cambridge: University Press.

Luong, M.P. and Dang Van, K. (1992) Infrared thermographic evaluation of fatigue limit in metal. *Proc. 27th QUIRT Eurotherm Seminar*, Paris.

Luong, M.P. (1995). Infrared thermographic scanning of fatigue in metals. *Nuclear Eng. and Design* 158, 363–376

Maitournam, H. (1999).Finite elements applications numerical tools and specific fatigue problems. In Dang Van, K. and Papadopoulos, I.V., eds, *High Cycle Metal Fatigue, From Theory to Applications*, CISM Courses and Lectures N° 392, International Center for Mechanical Sciences, Springer: Wien, New-York. 169–187

Mandel, J., Halphen, B. and Zarka, J. (1977). Adaptation d'une structure élastoplastique à écrouissage cinématique. *Mech. Res. Comm.*, 4:309–314.

Nguyen, Q.S. (2000). *Stability and Nonlinear Solid Mechanics*. J. Wiley & Sons.

Orowan, E. (1939). Theory of the fatigue of metals. *Proc. Roy. Soc. London, A,* 171: 79–106.

Papadopoulos, I.V. (1987). Fatigue Polycyclique des Métaux: une Nouvelle Approche. Ph.D. Dissertation, Ecole Nationale des Ponts et Chaussées, Paris.

Papadopoulos, I.V. (1999). Multiaxial fatigue limit criterion of metals. In Dang Van, K. and Papadopoulos, I.V., eds, *High Cycle Metal Fatigue, From Theory to Applications*, CISM Courses and Lectures N° 392, International Center for Mechanical Sciences, Springer: Wien, New-York. 89–143.

ADDRESSES OF AUTHORS

Chapter 1

Lecturer:

Géry de Saxcé
Laboratoire de Mécanique de Lille
Université des Sciences et Technologies de Lille
59655 Villeneuve d'Ascq Cedex, France

Co-authors:

Lahbib Bousshine
National High School of Electricity and Mechanics,
University Hassan II, BP 8118 Oasis
Aïn Chock of Casablanca, Morocco

Jean-Bernard Tritsch
Laboratoire de Mécanique de Lille
Université des Sciences et Technologies de Lille
59655 Villeneuve d'Ascq Cedex, France

Chapter 2

Lecturer:

Giulio Maier
Dipartimento di Ingegneria Strutturale
Politecnico di Milano
Piazza Leonardo da Vinci 32
20133 Milano, Italy

Co-authors:

Giuseppe Cocchetti
Dipartimento di Ingegneria Strutturale
Politecnico di Milano
Piazza Leonardo da Vinci 32
20133 Milano, Italy

Valter Carvelli
Dipartimento di Ingegneria Strutturale
Politecnico di Milano
Piazza Leonardo da Vinci 32
20133 Milano, Italy

Chapter 3

Lecturer:

Castrenze Polizzotto
Dipartimento di Ingegneria Strutturale e Geotecnica
Università di Palermo
Viale delle Scienze
90128 Palermo, Italy

Co-authors:
Guido Borino
Dipartimento di Ingegneria Strutturale e Geotecnica
Università di Palermo
Viale delle Scienze
90128 Palermo, Italy

Paolo Fuschi
Dipartimento Arte Scienza Tecnica del Costruire
Università di Reggio Calabria
Via Melissari Feo di Vito
89124 Reggio Calabria, Italy

Chapter 4

Lecturer:
Dieter Weichert
Institut für Allgemeine Mechanik
Rheinisch-Westfälische Technische Hochschule Aachen
Templergraben, 64
52056 Aachen, Germany

Co-author:
Abdelkader Hachemi
Institut für Allgemeine Mechanik
Rheinisch-Westfälische Technische Hochschule Aachen
Templergraben, 64
52056 Aachen, Germany

Chapter 5

Lecturer:
Alan R. S. Ponter
Department of Engineering
University of Leicester
University Road
Leicester LE1 7RH, United Kingdom

Chapter 6

Lecturer:
Andrzej Siemaszko
Institute of Fundamental Technological Research
Swietokrzyska 21
00-049 Warszawa, Poland

Chapter 7

Lecturer:
Ky Dang Van
Laboratoire de Mécanique des Solides
UMR 7649-CNRS
91129 Palaiseau Cedex, France